SCIENCE AND THE
SECRETS OF NATURE

SCIENCE AND THE SECRETS OF NATURE

BOOKS OF SECRETS IN MEDIEVAL AND EARLY MODERN CULTURE

William Eamon

PRINCETON UNIVERSITY PRESS

PRINCETON, NEW JERSEY

COPYRIGHT © 1994 BY PRINCETON UNIVERSITY PRESS

PUBLISHED BY PRINCETON UNIVERSITY PRESS, 41 WILLIAM STREET,

PRINCETON, NEW JERSEY 08540

IN THE UNITED KINGDOM: PRINCETON UNIVERSITY PRESS,

CHICHESTER, WEST SUSSEX

ALL RIGHTS RESERVED

LIBRARY OF CONGRESS CATALOGING-IN-PUBLICATION DATA

EAMON, WILLIAM.

SCIENCE AND THE SECRETS OF NATURE : BOOKS OF SECRETS IN MEDIEVAL

AND EARLY MODERN CULTURE / WILLIAM EAMON.

P. CM.

INCLUDES BIBLIOGRAPHICAL REFERENCES AND INDEX.

ISBN 0-691-03402-8

1. SCIENCE—HISTORY. 2. SCIENCE, MEDIEVAL. 3. SCIENCE—

EXPERIMENTS—HISTORY. 4. SCIENCE—SOCIAL ASPECTS—HISTORY.

5. SCIENCE—PHILOSOPHY—HISTORY. I. TITLE.

Q125.E34 1994

509'.4'0902—DC20 93-31794 CIP

THIS BOOK HAS BEEN COMPOSED IN ADOBE GALLIARD

PRINCETON UNIVERSITY PRESS BOOKS ARE PRINTED

ON ACID-FREE PAPER AND MEET THE GUIDELINES FOR

PERMANENCE AND DURABILITY OF THE COMMITTEE

ON PRODUCTION GUIDELINES FOR BOOK LONGEVITY

OF THE COUNCIL ON LIBRARY RESOURCES

PRINTED IN THE UNITED STATES OF AMERICA

3 5 7 9 10 8 6 4 2

TO THE MEMORY OF

Jerry Stannard
1926–1988

"There is something about a secret
which makes people believe."

(Graham Greene, *Our Man in Havana*)

CONTENTS

ILLUSTRATIONS AND TABLES

Illustrations

Tables

ACKNOWLEDGMENTS

ONE of the pleasures of completing a lengthy project such as this book is recording the debt of gratitude owed to the many individuals who contributed in so many different ways to the work. I especially want to thank two individuals who, over many years, consistently and unflaggingly provided guidance, advice, and encouragement as I worked on the project. Allen Debus gave me much needed support and encouragement at an early stage of my work and has continued to offer his advice ever since. Gundolf Keil's gracious hospitality made my sojourn in Germany a delight, while his vast knowledge of medieval medical writings and his uncanny historical instinct saved me from making many a facile judgment. I am grateful to colleagues who read and criticized sections of the manuscript in various forms, who commented upon portions of the work I presented at seminars and conferences, or who otherwise gave me advice on particular points. I want especially to thank Mario Biagioli, Hal Cook, Brian Copenhaver, Lorraine Daston, Paula Findlen, Bert Hansen, Gundolf Keil, Elizabeth Knoll, Dave Lindberg, Bruce Moran, Bill Newman, Katharine Park, John Riddle, Mike Shank, John Tedeschi, Ed Tenner, and Bob Westman. Mario Biagioli, Brian Copenhaver, and Lorraine Daston kindly gave me manuscript copies of works of theirs that have since appeared in print. Emily Wilkinson and Sara Van Rheenen of Princeton University Press made the production of this book a pleasure. Lauren Lepow, who edited the manuscript, caught numerous errors and made many helpful suggestions to improve my prose. Obviously, I accept sole responsibility for the faults that remain.

The book could not have been completed without financial assistance provided by various grants and fellowships, which periodically freed me from the burden of a heavy teaching load and enabled me to devote my time to research and writing. A Mellon Fellowship at Harvard University gave me a year's freedom in which to begin the project and my study of the Italian language. A postdoctoral fellowship at the Institute for Research in the Humanities at the University of Wisconsin gave me a critical year in which to write. I would like to thank the institute's former director, Robert M. Kingdon, and the other fellows of the institute, for their cordiality. My stay at the institute also enabled me to use the rich holdings of the Duveen collection at Wisconsin's Memorial Library, and to benefit from the assistance of John Neu, whose knowledge of the collection and of the bibliography of the history of science is unsurpassed.

A year as a visiting professor at the Institut für Geschichte der Medizin of the University of Würzburg under the auspices of the Fulbright Foundation provided the opportunity to develop the German material and to undertake with Gundolf Keil a separate project on Nicholas of Poland, parts of which appear in chapter 2. I spent many pleasant hours within and outside the institute with my office mate Ulrich Stoll, while he and Martina graciously hosted me on numerous occasions. Tom and Gabi Minnes introduced me to the wines of Franconia, over which we enjoyed many delightful hours of conversation. My work was also supported by grants from the National Science Foundation, the American Council of Learned Societies, and the Arts and Sciences Research Center of New Mexico State University.

The research on this book was done at several libraries, and could not have been accomplished without the help of competent and courteous librarians at those institutions. I am grateful in particular for the assistance of reference and rare books librarians at Harvard University (Widener and Houghton libraries), the University of Wisconsin, the University of Pennsylvania, the Hagley Museum and Library, the Herzog August Bibliothek, the British Library, the Wellcome Institute for the History of Medicine, the Royal Society of London, the Bibliotheca Universitaria, Bologna, the Newberry Library, the Folger Shakespeare Library, the Huntington Library, the National Library of Medicine, the Spencer Research Library of the University of Kansas, and the Glasgow University Library, who kindly supplied me with copies of imprints from the Ferguson Collection. I especially want to thank the Interlibrary Loan staff at New Mexico State University for their efficiency in obtaining many of the materials used in the study.

Portions of this book were presented at seminars at Harvard University, Brown University, the University of Wisconsin, Heidelberg University, and the University of California at Los Angeles. I am grateful to the audiences of these seminars for constructive criticism. I also want to thank Wolf-Dieter Müller-Jahncke, Bill Crossgrove, and Bob Westman for arranging visits to their institutions.

For permission to incorporate passages and sections of previously published articles into chapters 2, 3, 4, 5, and 10, I am grateful to the editors of *Sudhoffs Archiv*, *History of Science*, and the *Sixteenth Century Journal*, and to the Syndics of Cambridge University Press. For permission to use the epigraph, I thank David Higham Associates, Ltd., literary agents for Graham Greene.

This book would never have come into being had not the late Jerry Stannard suggested to me that the books of secrets might be worth investigating. My deepest regret is that he did not live to see the work published. There are few persons whose reaction to the book I would have enjoyed more. In dedicating it to his memory, I record my gratitude for his joyous friendship, and for having had the privilege to serve my apprenticeship under him.

Finally, I want to thank four women who have enriched my life beyond measure: my mother, my daughter Leslye, Jayne Stuart, and Marcia Bento.

NOTES ON CONVENTIONS AND USAGE

IN citing sources in the notes, to save space I have adopted a short-title system similar to the one employed by Shapin and Shaffer, *Leviathan and the Air-Pump*. Complete titles and publication details are provided in the Bibliography. Quotations in the text and notes are given in the original orthography, usually without modernizing the punctuation. I have generally used contemporary English translations of foreign-language sources when such translations were available to me. Whenever possible, however, I have checked such translations against the original. Unless otherwise indicated, all translations are my own.

SCIENCE AND THE
SECRETS OF NATURE

INTRODUCTION

PRINTING, POPULAR CULTURE,
AND THE SCIENTIFIC REVOLUTION

W HEN the English physician and virtuoso Sir Thomas Browne
sat down to catalog "vulgar errors" in his *Pseudodoxia Epidemica* of 1646, he warned against, among others, authors
who "pretend to write of secrets, to deliver Antipathies, Sympathies, and
the occult abstrusities of things."[1] An avowed Baconian, Browne was
convinced that science would not advance until men's minds were rid of
the errors and prejudices that had blocked the progress of knowledge.
Browne's repudiation of the "books of secrets" signifies an important
break with a tradition of remarkable durability. Although for him the
books of secrets were sources of popular errors, the authors he warned
against had not always written for the people. Indeed, implicit in the
tradition he condemned was the notion that to reveal secrets to the vulgar was to cast pearls before swine. Yet ironically, for a century and a half
prior to Browne's writing, Europe had been inundated with scores of
treatises that professed to reveal the "secrets of nature" to anyone who
could read. Browne singled out two of the most famous of these books
of secrets as being particularly objectionable: Alessio Piemontese's *Secreti*, which had already been published in more than seventy editions
since its appearance in 1555, and Giambattista Della Porta's *Magia
naturalis*, which, though written in Latin primarily for a learned audience, went through almost twenty editions. A campaign against such
suspect authors, "not greedily to be swallowed," seemed to Browne
both necessary and appropriate.

In repudiating the books of secrets, Browne cast into the rubbish
heap of obsolete science a body of literature that for centuries had fascinated, informed, and repelled readers. Alessio and Della Porta were but
the apogee of a literary tradition, going back at least to Hellenistic times,
of writings purporting to reveal secrets jealously guarded by famous
sages and experimenters, or locked up in the bosom of nature itself.
These writings exist in countless medieval manuscripts and in printed
books of almost every European language. Nor did they die out with the
"triumph of modern science" in the seventeenth century. Despite
Browne's warning, books of secrets continued to be written, copied,
published, and read by a sizable portion of the literate public, as well as
by prominent intellectuals. Sir Hans Sloane, who succeeded Newton as

president of the Royal Society, collected scores of manuscript books of secrets. His remarkable library leaves the impression that the printed texts of this genre were but the tip of a huge iceberg.

Why did these writings so capture the medieval imagination, and why did their popularity and authority persist for so long? One reason is that they were linked with a literary tradition of works that promised to reveal the esoteric teachings of revered authorities like Aristotle and Albertus Magnus. Such teachings appealed forcefully to the medieval mind, which was inclined to believe that everything knowable was contained in ancient sources. But they also appealed to Renaissance thinkers, who searched for a *prisca theologica*, an original wisdom rooted in revelation, as an alternative to what they regarded as a bankrupt scholastic tradition. Moreover, the books of secrets promised to give readers access to "secrets of nature" that might be exploited for material gain or used for the betterment of humanity. Underlying these works was the assumption that nature was a repository of occult forces that might be manipulated, not by the magus's cunning, but merely by the use of correct techniques. The utilitarian character of the books of secrets gave concrete substance to this claim. Unlike the recondite treatises on the philosophical foundations of magic, which barely touched base with the real world, the books of secrets were grounded upon a down-to-earth, experimental outlook: they did not affirm underlying principles but taught "how to." Hence they seemed to hold forth a real and accessible promise of power.

Thus the books of secrets were not, perhaps, what the term itself might conjure up in the imagination. Modern readers expecting to encounter some mysterious, arcane wisdom are bound to find these works disappointing. What they revealed were recipes, formulas, and "experiments" associated with one of the crafts or with medicine: for example, instructions for making quenching waters to harden iron and steel, recipes for mixing dyes and pigments, "empirical" remedies, cooking recipes, and practical alchemical formulas such as a jeweler or tinsmith might use. To the modern reader, they more closely resemble how-to books than magic books. Doubtless, many were used as all-purpose household handbooks.

Yet the connotation attached to "secrets" and to books of secrets in the medieval and early modern periods was not as neutral as this characterization might suggest. The books of secrets were not regarded with the same detachment we would have for, say, a cookbook or chemical formulary, the closest modern equivalent of such a work. We do not take very seriously the claim of the cookbook that professes to reveal "all the secrets of the culinary art," or the how-to book that promises to unveil the "secrets of woodworking." Such books may be useful, but few users will imagine they are going to learn more than how to make a tolerable meal or a sturdy piece of furniture. In the medieval and early modern eras, such claims carried much more weight than they do today. It seemed to many readers of the books of secrets that there was much more to be learned from a recipe than merely "how to"—even though in the long run that may have been what they actually learned. Linguistic shifts often signify changes in worldviews: a word's change of meaning may be a clue that the world is also changing. By the eighteenth century, such "secrets" were techniques and nothing more. In the sixteenth century, however, the term was still densely packed with its ancient and medieval connotations: the association with esoteric wisdom, the domain of occult or forbidden knowledge, the artisan's cunning, the moral injunctions to protect secrets from the *vulgus*, and the political power that attended knowledge of secrets. The Scientific Revolution exposed and neutralized nature's "secrets."

The subject of this book is a genre of "scientific" writings: the books of secrets. In a broader sense, however, it is about the process by which European culture divested itself of the tradition of esotericism in natural philosophy. This concern has necessitated a close look at the medieval literature of secrets. In the first two chapters, I trace the development of the tradition of esotericism and the literature it produced from late antiquity until the end of the Middle Ages. I am convinced that the printed books of secrets of the early modern period cannot be understood without our first having some sense of the tradition with which contemporary readers associated them. However, I deal with the medieval texts selectively because I do not think it necessary to guide readers through a tedious rendition of repetitive manuscripts.

The book's focus is the printed books of secrets, principally of the sixteenth century, when the tradition reached its pinnacle. By this time, the literature of secrets was no longer esoteric, but popular. The late John Ferguson, a chemist at the University of Glasgow who was the first to identify the books of secrets as a distinct genre, compiled a two-volume bibliography of hundreds of sixteenth- and seventeenth-century printed editions.[2] Although Ferguson's bibliography is an essential starting point for any investigation of the books of secrets, it is far from complete. Ferguson's goal was to identify as many of these works as he could find; he did not attempt to locate the books of secrets within the context of early modern culture. Nevertheless, in compiling this essential reference work he rendered a valuable service to scholarship.

In taking up where Ferguson left off, I consider the books of secrets as works of "popular science." In addition to describing the contents of these works, I pay close attention to the authors who compiled them, the printers who published and distributed them, and the audiences they reached. I have greatly benefited from recent studies of printing and popular culture in the early modern era, especially Elizabeth Eisenstein's ground-breaking *The Printing Press as an Agent of Change*. Although not without its shortcomings, this work will surely remain, for some time to come, the definitive treatment of printing's impact on early modern culture. Her work provides a valuable strategy for understanding the significance of printing for early modern science. Briefly, Eisenstein's thesis was that the new communications network created by the printing press made possible the recovery of the classical scientific corpus and, by fixing it in print, subjected it to critical scrutiny. Typography thus set the stage for the Scientific Revolution by initiating a process of information retrieval, and by generating a sense of urgency about the need to overhaul inherited schemes. "Old pursuits produced new results after techniques of communications had been transformed."[3]

Although this approach sheds light on traditional interpretations of the Scientific Revolution, it does not fully account for the richness of early modern science. For by adopting the revolution in astronomy and cosmology as her model of the Scientific Revolution, Eisenstein selected disciplines whose development seems to have depended less upon the availability of new printed books than upon new ways of thinking about existing books. She essentially added typography to the explanatory equation advanced by Alexandre Koyré, A. R. Hall, and others, who viewed the Scientific Revolution as a purely intellectual transformation. But the process of reforming classical cosmology began well before the advent of printing. It is not clear from Eisenstein's analysis how the information explosion brought about by typography contributed to the overhauling of the old cosmos. In the classical sciences generally, the printing press appears to have been a relatively passive agent of change in comparison to its demonstrable impact in other areas of thought. Indeed, Eisenstein cautiously states that "early printers play a much less conspicuous and much more problematic role in the Copernican Revolution than in the Reformation."

Although printing may have contributed indirectly to the transformation of prevailing scientific outlooks, it played an instrumental role in generating new scientific interests and disciplines. Recently Thomas S. Kuhn has suggested that not one but two scientific traditions (or "clusters" of research fields) existed side by side in the early modern era. The first, centered in the universities, was concerned primarily with preserving and perpetuating received doctrines. The disciplines making up the university-based sciences—astronomy, optics, mechanics, and medicine—already existed as subjects of scientific investigation in antiquity and were the focus of continuing research throughout the Middle Ages. Alongside but separate from these "official sciences," another group of research fields emerged during the Renaissance. These included subjects such as chemistry, magnetism, electricity, and metallurgy. Although scattered empirical data existed concerning these subjects prior to the sixteenth century, in contrast to the classical sciences they did not inherit a coherent body of theoretical doctrine. To the extent that they existed as fields of research prior to the Renaissance, they were regarded merely as interesting classes of phenomena or as practical arts rather than as coherent scientific disciplines. Kuhn calls this cluster of research fields the "Baconian sciences" after its chief protagonist, Sir Francis Bacon. Bacon insisted that the primary obstacle to the advancement of science in the past had been its over-hasty theorizing. Inherited doctrines, he argued, had to be swept away and replaced by catalogs of facts and natural histories of observations and experiments.[4]

Kuhn pointed out that the aim of experimentation in the Baconian scientific tradition was entirely different from that in the classical sciences. In the latter, experimentation was secondary to logical demonstration. "Experiments" were rarely performed in scholastic science, and when they were carried out, they were designed mainly to provide visual demonstrations of what was already known by other means. The Baconian practitioners of the seventeenth century, on the other hand, performed experiments designed to produce theoretically neutral "matters of fact."[5] They also designed experiments to demonstrate how nature would behave when forced out of its natural state, the better to understand nature in course. According to Bacon, nature was like Proteus, whose true identity lay concealed under a variety of external shapes and forms until he was bound in chains: "So nature exhibits herself more clearly under the trials and vexations of art than when left to herself." Bacon thought the best examples of this kind of experimentation took place in the workshops of craftsmen, who transformed raw materials and gave them new properties. He called for the compilation of detailed histories of the mechanical arts, a project that was given a high priority in the early years of the Royal Society. It is not surprising that these "experimental histories of the trades" usually took the form of compilations of recipes, much like the books of secrets. For recipes are the record of trial-and-error experimentation. They are the accumulated experience of practitioners boiled down to a rule. We trust recipes because we know that behind them stands someone who does not use them.

According to Kuhn, only the classical, university-based sciences were transformed by the Scientific Revolution. However, the availability of new printed editions of older works cannot fully account for the revolution of these sciences. Since the medieval scientific community was relatively small and homogeneous, scientific activity did not depend upon large numbers of copies of particular texts. The relatively few specialists working in these areas of research were usually able to secure manuscript copies of relevant treatises, and when such texts went to press, they were produced mainly for a small audience of specialists and not for general readers.

The "Baconian sciences," on the other hand, were in gestation during this period. Precisely because they were less well-defined in terms of subject matter, less authoritatively governed by established doctrines, these sciences depended for their development on information from a wider range of sources. Nonacademics, amateurs, and craftsmen made important contributions to the development of the Baconian sciences. William Gilbert's experimental studies in magnetism, for example, appear to have owed more to navigators and instrument makers than to medieval natural philosophy. Robert Boyle, in devising his chemical experiments, drew extensively from the empirical information accumulated by metallurgists, dyers, and distillers.[6] The development of these sciences depended directly upon the dissemination of information from such occupational groups, whose activities had been irrelevant, or unknown, to academic scientists. By focusing upon the generation of new scientific interests and disciplines, rather than upon the transformation of the older, traditional sciences, we may discover a more direct role for printing in the Scientific Revolution.

Here Eisenstein's analysis may indeed set the stage for a new understanding of the Scientific Revolution. By showing how the printing press helped to "bridge the gap between town and gown," her work gives important insights into the formation of new scientific disciplines. By pointing out how new kinds of authors were drawn into the business of writing, how printers encouraged the production of certain kinds of scientific and technical writing, and how readers might have responded to these works, her interpretation provides the basis for a more nuanced understanding of the views of Edgar Zilsel and others, who maintained that the modern experimental method was born out of the collaboration between scholars and craftsmen.[7] As Zilsel pointed out more than fifty years ago, academics developed a new appreciation of the mechanical arts in the sixteenth century. Nevertheless, they retained the bookish habits of scholars and probably learned as much about the arts from technical handbooks as from their own workshop observations. In view of this, what is needed is a careful analysis of the technical information that went to press in the sixteenth century, how scholars assimilated it, and the natural-philosophical uses they made of it.

This book is intended as a contribution to fulfilling that need. The books of secrets published scores of artisanal recipes from a broad range of crafts, from metallurgy and "practical alchemy" to dyeing textiles and preparing drugs. One of my purposes is to demonstrate that these works played an instrumental role in disseminating craft information to the virtuosi. However, I also want to make a stronger claim about the significance of the books of secrets for the Scientific Revolution. I hope to show that these works were not merely passive vehicles for the transmission of "raw data" to natural philosophers, but were bearers of attitudes and values that proved instrumental in shaping scientific culture in the early modern era. In particular, I shall argue that the books of secrets articulated a novel concept of experimentation. With their roots in a largely oral and practical tradition, the books of secrets enable us to rummage through the prehistory, so to speak, of the Baconian sciences. In the Middle Ages, the type of experiments Bacon advocated doing (according to a prescribed method involving the collaborative effort of many scientific workers) were still "secrets." That is, they were the "private experiments" of individual practitioners, the sort of events that occurred frequently within the craft tradition but were communicated orally (if communicated at all) and rarely published. Bacon urged that experience be made to "learn her letters"—in other words, that experiments be recorded and communicated to other experimenters. Only when "experience has been taught to read and write," he thought, could scientific progress be hoped for.[8] The books of secrets were examples of such "literate experience," for the recipe distills the arduous trial-and-error experience of practitioners and collapses it into a formula for making. Finally, I want to emphasize and, I hope, to clarify the important distinction between the medieval notion of an "experiment" (or a "secret")—a fortuitous, unexpected, and essentially private experience—and the Baconian concept of experimentation, a scientific program involving the communication of experimental findings to a community of scientists, and the collaborative testing of them. I hope to show that the books of secrets were the "missing link" between medieval "secrets" and Baconian experiments.

Eisenstein's research has also underscored the need for a closer assessment of the impact of printing—and particularly the publication of technical handbooks and treatises—on the process of secularization that took place in early modern Europe. Secularization was a sweeping transformation that involved economic, social, demographic, and political changes, as well as an intellectual revolution.[9] It affected all classes, although villagers remained somewhat immune to its impact relative to city dwellers. By "secularization," I do not mean godlessness, for assuredly most early modern people did not reject religion. It was rather that the boundary between the religious and the secular became more distinct. As Sir Thomas Browne put it, man lives "in divided and distinguished worlds."[10] The sphere of religion was diminished, so that many of the hopes and fears formerly expressed in religious terms came more and more to be expressed in worldly terms. Secularization involved a radical subversion of the traditional view of reality as a single system governed by common principles, in which humanity finds its duly appointed place and its ultimate meaning. William Bouwsma has written, "Secularization rested on a deep conviction that eternal truths are inaccessible to the human intellect, and that only the limited insights afforded by experience in this world are relevant to the earthly career of the human race."[11] The books of secrets, which compressed the lived experience of generations of empirics into simple, time-tested rules, were among the most visible manifestations of such attitudes. The explosion of printed how-to books publicizing the secrets of the trades and the tricks of magic made it clear as never before that a recipe might effectively replace the artisan's cunning, and that most of what had formerly passed as magic could now be seen as mere hocus-pocus and sleight of hand.[12] The sixteenth century was an age of "how to": secrets were spelled out; calculation began to take the place of cunning.

It has often been said that to succeed, a revolution must win the hearts and minds of the people. That statement may also be applied to the Scientific Revolution. To succeed, it had to create a cosmology and an epistemology that were consistent with the secularistic attitudes that emerged, especially among the urban middle class, in early modern culture. The "new philosophy" of the seventeenth century was ideally suited to this purpose. By overruling the action of supernatural forces in the universe, the mechanistic cosmology ratified the boundary between the sacred and the profane. The mechanical philosophy also validated the "maker's knowledge" (*verum factum*) model of scientific explanation, the maxim of reasoning according to which to know something means knowing how to make it.[13] In the new philosophy, the capacity to reproduce nature's effects became a sort of touchstone upon which claims to knowledge would have to be tested. In radical contrast to the scholastic view of knowledge, the new philosophy made how-to knowledge a criterion of truth. From this vantage point, we may begin to see the social and economic history of the Scientific Revolution in a new light. Why did the limited "mechanical know-how" model of natural knowledge win out over the unitary, comprehensive view of knowledge that characterized scholastic natural philosophy? One answer is that "maker's knowledge" mattered more to the emergent middle class than did knowledge based on abstract principles. It was more consistent with bourgeois values. Unpacking the "secrets of nature" may help us make sense of why the Scientific Revolution occurred during the period of early capitalism.

But these perspectives on the Scientific Revolution also raise new and challenging questions: How did the "how-to revolution" gain a foothold among the people, thereby creating a broad, popular base for science? What were the consequences of revealing secrets that were formerly esoteric? What were the sources of these secrets, how were they revealed, and how, once published, were they interpreted? This book does not claim to offer definitive answers to all these questions, but it does, I hope, address them in new ways. I begin my investigation with an analysis of what I shall call the tradition of esotericism in medieval science. As we will see, the metaphor of the "secrets of nature" played upon several different senses of the concept of secrecy.[14] One variety of secrecy was essentially *social*, involving the intentional suppression of information in order to protect knowledge from outsiders who might corrupt or abuse it. I call this form of secrecy the tradition of esotericism. Another kind of secrecy was *epistemological*, implying that secrecy was a given in the order of nature, and that the "secrets of nature" are permanently and fundamentally unknowable. This was the tradition of nature's secrets as *arcana naturae*.

Having described the medieval literature of secrets, I then proceed to a discussion of the process by which the "secrets of nature" were revealed to the general public through printed books. This will lead to a consideration of a third form of secrecy, which might be called *epistemic* secrecy because it included aspects of nature that were taken to be unknown or unknowable for historically contingent reasons.[15] Partaking of both social and natural silences, it involved the recognition that the "secrets of nature" were not a given, but an artifact of a way of looking at the world. Defying traditional restraints against divulging "forbidden knowledge," many popular writers made a full-time occupation of publishing secrets, so eager were readers to know them, and so covetously were they thought to be guarded. I will argue that in the sixteenth-century books of secrets, a new concept of the scientific enterprise emerged: that of science as a *venatio*, or a great hunt after the secrets of nature. This concept of science was essentially a popular image promoted by the "professors of secrets" as a means of selling their books. It was, in other words, a consequence of the transformation of "secrets" and "experiments" into commodities. Nevertheless the conception of science as a *venatio* entailed a new definition of the function of experiments in science. This new conception of experimentation emerged independently of the academic tradition. In due course, it was appropriated by the virtuosi as part of a program for the advancement of science and was implemented in the research program of the Royal Society of London. I do not mean to suggest that the Scientific Revolution was a "revolution from below." But I do believe that any discussion of the "foundations" of the Scientific Revolution must consider a much broader base for it than historians of science have so far attempted.

The path from craft and magical secrets to public science will take us along the Grub Streets of Frankfurt, Venice, and London; through the pharmacies, workshops, and printing houses; and onto the piazzas as well as into the courts and scientific academies of early modern Europe. If our route is the right one, we will discover in the end that it was in the carnivalesque cries of empirics and hack writers, as well as in the polite arguments of the virtuosi, that the Scientific Revolution found its authentic voice.[16]

PART ONE

THE LITERATURE OF SECRETS

ONE

THE LITERATURE OF SECRETS
IN THE MIDDLE AGES

IT has long been unfashionable to use the term "Dark Ages" to describe European culture in the early Middle Ages. To modern historians, that epithet seems too absolute, too suggestive of a prolonged period of hopeless barbarism. As a blanket description of European culture, it is certainly an exaggeration. With reference to scientific culture, however, this unfortunate term describes the early Middle Ages in two important respects. The first is that between the fall of Rome and the revival of learning in the twelfth century, Europeans were by and large ignorant of the major accomplishments of Greek natural philosophy. This does not mean that the early Christians lacked a unified and comprehensive view of the physical world. Even if Greek science had survived intact, the patristic authors would have found Scripture more suitable than philosophy as a model for the interpretation of nature. Nor were the early Christians uniformly hostile to pagan learning, especially as they found it a handmaid to religion.[1] Yet as far as its classical foundations were concerned, the worldview of the early Middle Ages was constructed upon an incomplete and somewhat perverse understanding of ancient science, which by then had been so repeatedly and so thoroughly abridged that it was practically emptied of its original content.

In a second, less obvious or perhaps metaphorical sense, "dark" describes the ideological conditions under which early medieval science developed. In addition to the epitomes of ancient writings, the Middle Ages inherited late antiquity's esotericist attitudes about nature and natural knowledge. To many Hellenistic and Roman authors, nature itself was arcane, just as to early Christians it was full of miracle and packed with symbolic meaning. If nature was a miracle, it was also, according to Hellenistic sources, knowable only by divine revelation. Scientific knowledge was a sacred mystery disclosed only to a chosen few. Thus, much of what survived of ancient science came into the West under an aura of secrecy. Often it came with formulaic injunctions to keep it secret, or with warnings against disclosing it to the vulgar crowd. The tradition of esotericism in science was a cultural inheritance from late antiquity. Even after direct contact with Greek civilization had ceased, esotericism continued to shape the moral economy of medieval science.[2]

The goal of rediscovering the forgotten "secrets of nature," and of re-capturing long-lost "secrets of the arts" provided medieval intellectuals with a powerful motivation for reclaiming the ancient scientific heritage for the West.

The present chapter treats the origin and early development of the tradition of esotericism in science and the literature it produced. The *libri secretorum*, or "books of secrets," were compilations of recipes, for-mulas, and "experiments" of various kinds, including everything from medical prescriptions and technical formulas to magical procedures, cooking recipes, parlor tricks, and practical jokes. The one thing these assorted manuscripts had in common was the promise of providing ac-cess to the "secrets of nature and art." In reality, they were assemblages of traditional lore concerning the occult properties of plants, stones, and animals, along with miscellaneous craft and medicinal recipes, alchemi-cal formulas, and "experiments" to produce marvelous effects through magic. Although many were derived second- or thirdhand from ancient texts, they also included material based upon indigenous folk traditions, the accumulated experience of practitioners, and the discoveries of me-dieval "experimenters." Whatever their sources, and they are often im-possible to ascertain, the largely anonymous and pseudonymous litera-ture of secrets became a prevalent feature of the medieval scientific corpus. Even after the revival of learning in the twelfth and thirteenth centuries, the appeal of this literature did not diminish. If anything, the "twelfth-century renaissance" stimulated rather than retarded the pro-duction of such works, generating hundreds of books of *secreta*, *experi-menta*, and *mirabilia*.

Science and Gnosis: The Secrets of Nature

To understand the widespread popularity and influence of the *libri secre-torum*, we must begin with the formation of the medieval scientific tra-dition in late antiquity, and with the cultural values that shaped it. From its earliest beginnings, Latin science was conditioned more by Hellenis-tic than by Hellenic Greek attitudes and values. For this reason medieval culture inherited certain expectations and assumptions about the nature of science, which were only with great difficulty or reluctance discarded. Among these assumptions was the idea that there existed secret sciences, access to which was privileged, as opposed to the conventional sciences, access to which was relatively open. This conception of science was in turn premised upon the idea that nature itself holds secrets, concealing them from the eyes and the intellect. What is apparent and what is real are two different things, said the Platonists, who schooled Hellenistic culture. But the substance of that message was badly corrupted when it was syncretized with the diverse religious and philosophical influences of the age. Whereas Plato, in a half-serious, half-playful way, had declared that "initiation" into pure philosophical truth was analogous to initiation into the Eleusinian mysteries, in that it purifies the mind of its earthly slough, his Hellenistic interpreters took his ironic metaphor at face value.[3] For they desperately wanted a shortcut through the dense thicket of philosophical doctrines that their classical forebears had bequeathed to them. One shortcut was the encyclopedia, which conveniently summarized the knowledge of the past. But even more alluring to the Hellenistic and early Christian thinkers was the idea that the secrets of nature are made immediately intelligible by divine revelation. It would not be overly simplistic to say that the *libri secretorum* grew out of the conjuction of the encyclopedia tradition and the outlook of the Hellenistic religious mysteries.

When the Romans made their first large-scale contacts with the Greek world in the first century B.C., the speculative philosophical tradition of the Periclean age had already given way to encyclopedism. The sheer quantity of information that had accumulated in the Greek world was by this time so vast that merely digesting it consumed entire academic careers. Thus the polymaths and antiquarians of late antiquity combed the rich collections of the libraries in Athens, Alexandria, and Pergamum, diligently but unimaginatively compiling handbooks of learning for general readers.[4] The Romans, for their part, were impatient with theory. They were content to have the bare rudiments of philosophy and science, enough for polite conversation. Conveniently, when science, along with other forms of Greek culture, became fashionable in the late Republic and early Empire, the Romans found they could learn all they wanted to know by turning to the encyclopedias and epitomes of the Hellenistic polymaths. "At no time in their intellectual history," writes William Stahl, "could the Greeks have been more attractive to the Romans."[5] The handbook tradition surpassed all other forms of literary productivity in volume and popular appeal. "Even the most intellectually curious Romans, like Lucretius, Cicero, Seneca, and Pliny, were satisfied to obtain their knowledge of Greek science from manuals and made no original contributions."[6]

Latin science inherited another trait from Hellenistic culture: its preoccupation with the marvelous. In outward appearance, the spirit of rationalism reached its pinnacle in Hellenistic centers of learning such as the Museum of Alexandria, where some of the most gifted mathematicians, astronomers, and physicians of the ancient world spent their careers. Yet closer inspection reveals a deep and persistent countercurrent of antirationalism. Within certain influential philosophical circles, Greek rationalism gave way almost entirely to a reliance on revelation as the source of truth in matters scientific as well as religious. Increasingly preoccupied with the problem of individual salvation, for which reason held no solution, people turned to other methods, some relying upon sacred books supposedly discovered in Eastern temples, others seeking a personal revelation by an oracle, vision, or dream. Still others looked for security in ritual, whether by joining one of the mysteries or by employing the services of a private magician. A.-J. Festugière described this wholesale "decline of reason" in the following way: "Little by little ancient Greek rationalism, which since the early Ionians had liberated scientific thought from the matrix of myth and the apocalypse, gave way to a very different mentality, where all at once one distrusted reason and relied on means of knowledge foreign to reason. One was not the consequence of the other: rather, both were manifestations of the same spiritual resignation."[7]

Various explanations of this "Greek miracle in reverse" have been offered: the breakup of the *polis*, religious syncretism, the loss of political freedom, spiritual anxiety resulting from continuous warfare and the Roman domination, and plain intellectual exhaustion.[8] Whatever its causes, the symptoms of a general change in the intellectual climate of the Mediterranean world are readily detectable in the scientific literature of the period. To the Hellenistic polymaths, the physical world was a spectacle of the uncanny. Mysterious and impenetrable, nature seemed too full of wonder for philosophy to comprehend. In the last analysis, thought many, it could be known only by divine revelation.

These tendencies are most clearly exhibited in the revelations attributed to the Egyptian god Thoth, called Hermes Trismegistus ("Thrice-Great Hermes") by the Greeks.[9] The Hermetic treatises were composed between the first and third centuries A.D., as the forces of spiritualism and mysticism intensified in Hellenistic Egypt. The writings fall into two classes: the "philosophical *Hermetica*," principally the *Corpus Hermeticum* (the philosophical revelations of Hermes); and the "technical *Hermetica*," tracts on astrology, alchemy, natural history, medicine, and magic.[10] Because of their operative character, the technical *Hermetica* made a deep impression on the Romans. To an age terrorized by angry divinities and the omnipotence of fate, these tracts supposedly gave access to "secrets of nature" that would enable one to gain mastery over nature's occult forces. Such secrets were considered to be opaque and impenetrable by the intellect. Indeed, since it was a question of discovering a vast network of complex and hidden sympathies and antipathies, how could nature possibly be known *except* by direct revelation from God? In the works ascribed to Hermes and his disciples, science was practically indistinct from religion. It was no longer rational under-standing, but *gnosis*, revealed knowledge, an outcome of piety. The sci-entific investigator was no ordinary person having the normal faculties of intelligence and reason, but a man gifted with a special form of knowledge, a magus, someone "in touch with" the occult relations in the universe.[11]

The abandonment of reason in favor of revelation in the search for knowledge is a recurrent topos in the Hermetic literature. Did the liter-ary topos reflect historical reality? That is the question raised by an auto-biographical letter written in the first century A.D. by a medical student called Thessalos of Tralles. The epistle formed the preface to a treatise on astrological medicine attributed to the Egyptian pharaoh Nechepso, one of the fabled recipients of Hermetic revelations. It is addressed, as many such letters were, to the Roman emperor (in this case Claudius). An account of a neophyte's search for truth, Thessalos's letter was plainly influenced by the familiar topoi of revelation and discovery found in the Hermetic literature. But it carries a ring of authenticity almost unique in the *Hermetica*. Not only can its author can be identified with rare probability, the experience it describes can be authenticated by ref-erence to contemporary sources. Even if Thessalos's account is not de-monstrably historical, it is convincing evidence of what was plausible.[12]

Thessalos relates that he began his quest for knowledge with the secu-lar sciences. After studying rhetoric and philosophy in his native Asia Minor, he set out for Alexandria, the cultural capital of the Hellenistic world, to study natural philosophy and medicine. One day, while scour-ing the shelves of a library in Alexandria, he discovered King Nechepso's treatise, which described how to collect, prepare, and administer medic-inal plants according to their appropriate astrological signs. Eager to dis-cover Nechepso's secret of the universal panacea, Thessalos followed the book's instructions to the letter. Alas, his every attempt failed. In de-spair, the young student decided to seek a divine revelation. His search took him to Thebes, where he located a priest skilled in the art of theurgy (invoking visions of deities to obtain oracles from them). Thes-salos persuaded the sorcerer to summon before him Asclepius, the god of medicine, so that he might ask the god, "face to face," the secret of making Nechepso's healing drugs.

On the appointed day, the priest led Thessalos into a darkened cham-ber, recited a magical incantation, and conjured before the student a vision of the god in a bowl of water. Thessalos had come prepared: un-known to the priest, he had brought papyrus and ink to record the god's revelation.

"Oh blessed Thessalos, today a god honors you," proclaimed the apparition. "Soon, when they have learned of your success, men will hold you in reverence as a god! Ask me what you please, and I will gladly answer you in all matters."

"I could scarcely speak," reported Thessalos, "so much was I taken outside myself and so fascinated was I by the god's beauty. Nevertheless, I asked why I had failed in trying out Nechepso's recipes." The god replied,

> King Nechepso, though a very intelligent man and in possession of all magical powers, had not received from any divine voice the secrets you want to know. Endowed with a natural cunning, he understood the affinities of stones and plants with the stars. However, he did not know the proper times and places where the plants must be gathered. Now, the growth and withering of all fruits of the season depends upon the influx of the stars. Furthermore the divine spirit, which in its extreme subtlety can pass through all substances, is poured out in particular abundance in the places successively touched by astral influences in the course of cosmic revolution.[13]

The god then revealed to the eager student the secret of collecting the plants, the places from which they must be gathered, the astrological signs that govern the plants, and the prayers that must be recited while collecting them, without which the drugs have no power. Having revealed the mystery, the god commanded Thessalos not to "reveal [the secret] to any profane person who is a stranger to our art."[14]

Thessalos's letter, although apparently unique as a personal account, illustrates a number of motifs common to the Hermetic corpus: the failure of the quest for scientific truth through rational inquiry, the belief that truth is discovered through divine revelation, the joy of finding out the truth in a face-to-face confrontation with a god, and the admonition by the god not to reveal the secret to the *vulgus*.[15] The conception of knowledge embodied in this account blurs the distinction between religious and scientific truths. Both are on the same plane, both proceed from the same source. Although Hermetism did not constitute a religious brotherhood or cult, it took over from the mystery religions the jargon of revelation and transformed it into a literary motif for communicating esoteric doctrines.[16] The theme of revelation pervades the Hermetic scientific literature. The great majority of its treatises, whether on astrology, alchemy, medicine, or natural history, take the form of revelations from Hermes or some other god to a disciple, or of visions received in a dream.[17] Accordingly truth is always a mystery, a secret concealed within a divinity and revealed only to the god's chosen disciples. It follows that the secrets of nature, which are revealed to the elect by the grace of the gods, should not be profaned by being communicated to the public. Instead, they must be revealed only to the chosen successors of the one to whom the knowledge is revealed. The recipient of the grace of revelation is thus more like a prophet than a scientist, and every prophet of science in turn is commanded to obey the law of silence.

Granted that scientific truth is revealed truth, why did the Hellenistic magi insist so strenuously upon the "law of silence"? Why should the publication of knowledge have been considered blasphemy, instead of, as it would later become, a virtue? Part of the explanation may be that because magic was a capital offense in the Roman Empire, it had to go underground; secrecy was necessary to protect its practitioners from persecution.[18] Yet such fears are almost never alluded to in the Hermetic literature. A more likely explanation is that esotericism served not so much to protect magicians from the authorities as to exalt and legitimize the sciences it served. Thessalos's experience of the epiphany of Asclepius may well have been real; unquestionably others had similar revelation-experiences.[19] Whether genuine or not, however, the similarity between his revelation and those described in the Hermetic literature was certainly no coincidence. The letter employs a series of fictions commonly used in the Hermetic literature to describe revelations and to authorize texts: a letter to a patron (often an emperor), the discovery of a secret book (usually a book hidden away in a temple), an ecstatic revelatory experience, and, of course, the law of silence.[20]

Thessalos, a student of medicine and science, was undoubtedly familiar with these conventions through his reading of the Hermetic literature. Like many of the Hermetic tracts, his preface adopts the ritual language and mythology of the mystery religions. Although the gift of knowledge had no direct soteriological function (rather, it is knowledge of proper times and places), Thessalos underwent a ritualistic death and rebirth (he was "annihilated body and soul" by the god's epiphany). As a result of the revelation, he was "redeemed" from death by his own hands (for he had vowed to commit suicide if his attempts to discover Nechpso's secrets failed). The law of silence preserved the purity of the revelation and hence, just as in the mysteries, of the initiate. But secrecy protects the secular sciences as well: with somewhat convoluted logic Thessalos—or rather the priest speaking through the fabulated god Asclepius—states this explicitly as the reason for invoking the law of silence. Revealing the secret to "strangers" would bring an end to the pursuit of secrets:

> For the very ease with which these recipes can be obtained and used to treat all ailments might lead profane people to despise the learned and admirable acquisitions of the science of medicine. No longer will there be this noble rivalry among different medical schools; indeed, no one will take the

trouble anymore to study the treatises of the ancient physicians. Yes, the
effort that so many marvelous men have taken will be neglected because of
the immediate facility that the present treatise will give, since, as it is said,
all corporal ills will be cured thanks to it.[21]

According to this curious passage, the law of silence not only exalts
the Hermetic sciences, it also preserves the secular sciences. Publishing
the secret threatens scientific knowledge because knowing the secret, a
shortcut to the end sought, makes it unnecessary to know science. Following the familiar topoi of the prophetic literature, Thessalos's revelation came at the end of a long, anxious, and unsuccessful philosophical
quest. The secular search for the secret necessarily fails, but the difficulty
of the intellectual journey proves the worthiness of the initiate to receive
the secret. If the secret were made public, there would be no need to
seek it, and without the search, there would be no one qualified to receive it. Just as in the mysteries the law of silence protects the integrity
of the cult, so in the Hermetic tradition it preserves the authority of the
text. Thus in the Hermetic tradition the language of the mysteries is
deployed to create a new model for communicating esoteric wisdom.
Unlike the cults, with their initiation rites and progressive revelation of
secret doctrine, Hermetism "implies only a certain number of revealed
texts, transmitted and interpreted by a 'master' to a few carefully prepared disciples."[22] This ideology operated even without initiatory cults
to transmit secrets: a text could be lost for centuries, but if it is rediscovered by a competent reader, its message becomes intelligible and
contemporary.[23]

Thessalos's career following his revelation experience attests eloquently to the sociological meaning of being in possession of secret
knowledge. Not long after the foregoing events took place, he went to
Rome to seek his fortune. Pliny reported that Thessalos was one of a
host of ambitious charlatans preying upon the credulous Romans, who
were all too easily duped by the novel medical fads coming in from the
East.[24] Arriving in Rome during the principate of Nero, Thessalos made
a fortune with his newfangled doctrine. Armed with the assurance of
God-given truth, he denounced the theories of Hippocrates and proclaimed that all diseases could be reduced to but three "states" of the
human body.[25] The essence of Methodism, the school Thessalos is said
to have founded, is the principle that all medical doctrines are false;
there are only individual diseases. Such is the sort of radical anti-intellec-
tualism that is preached most convincingly by one, like Thessalos, who
has had an agonizing personal experience of reason's failure. According
to Pliny, Thessalos "swept away all received doctrines, and preached
against the physicians of every age with a sort of rabid frenzy." He attracted many pupils, promising to teach them everything there was to

know about medicine in only six months. Evidently the god's promise of fame came true, for Pliny reports that when Thessalos walked the streets of Rome he was surrounded by a greater crowd than any actor or charioteer. His monument on the Appian Way bore the inscription *iatronices*, "conquerer of the physicians."[26]

These attitudes about the acquisition and dissemination of knowledge represent an extreme departure from the classical view. For Aristotle, natural phenomena were always within the grasp of reason, a faculty common to all people. It took no special magus to interpret experiential data, no special gift of gnosis to make it intelligible. Revelation played no significant part in the acquisition of knowledge. To be sure, the Greeks of the classical period respected oracles, but there was a difference: they went to oracles to find the answer to some specific contingency, not to find out an entire doctrine or science. Once the Greeks had asked Asclepius, "What must I do to be healed of this illness?" Now they asked, "What is the secret to heal all diseases?" The participatory, even confrontational style of classical politics and culture had given rise to a dialectical approach to the acquisition of knowledge, diminishing the importance of revelation as the source of knowledge and diminishing likewise the burden of secrecy it imposed. For the classical philosophers knowledge was not received in the temple by revelation; it was earned in the agora through argument.[27]

The openness of political culture, the habit of public scrutiny and debate in affairs of state, also helps to explain the classical philosophers' relative disinterest in the "marvels" (*mirabilia*) of nature. The aim of classical natural philosophy was to describe and classify ordinary physical events, and to establish laws explaining them. The regularities in nature were Aristotle's chief concern; rarities and marvels were outside the domain of science. The Hellenistic philosophers, on the other hand, were deeply curious about the esoteric and unusual aspects of nature. In fact, they were convinced that the really essential features of nature were hidden from the senses as well as from the intellect. Physical objects, they believed, possessed hidden relations of sympathy and antipathy; stones and plants contained marvelous powers; and the universe as a whole held "secrets" that were accessible only to a gifted few. The causes of these mysterious relations were occult and completely inaccessible to the rational understanding. Yet the magus, who possessed the special gift of being able to "see" nature's secrets, could manipulate them to produce wonders beyond the abilities of ordinary humans.

These symptoms of the transformation of Greek philosophy began to appear as early as the third century B.C. Gradually the Greek mind was seduced by mystical philosophies imbued with astrology, sympathetic magic, and alchemy. Pseudonymous writings supposedly based upon divine revelation became enormously popular; astrology enjoyed a resurgence; and the theory of occult properties immanent in plants, animals, and stones became particularly appealing.[28] Equally significant was the revival of Pythagoreanism, not as a formal philosophy but as a religious cult and way of life. Neo-Pythagoreanism became the principal stimulus to the codification and development of philosophical magic, which grew up in the eastern provinces of the Roman Empire and then spread westward during the early centuries of our era.[29] The history of philosophy was rewritten in light of the neo-Pythagorean revival: now Pythagoras became the foremost Greek magus, the Greek counterpart to the oriental magi and the first in a succession of wizards that included Empedocles, Democritus, and Plato.[30] However, because the practice of magic was illegal, this "learned magic" was always distinguished in the literature from popular sorcery. The most famous neo-Pythagorean, the rigorously ascetic Apollonius of Tyana (first century A.D.), is credited with amazing feats of magic yet according to his biographer "never stooped to the black art."[31]

The Hermetic texts, strongly influenced by this quasi-religious revival, are replete with magic. The *Kyranides*, a treatise of the second century, illustrates the character of Hellenistic magic.[32] Supposedly a compilation of writings by a certain Harpocration of Alexandria and by Kyranos, king of Persia, the work bears two prologues, one each from Harpocration and Kyranos, describing how the treatise was translated from a Persian inscription on an iron stele—another stock fiction of the revelation-literature. The treatise consists of four books divided into chapters arranged according to the letters of the Greek alphabet. Each chapter treats the magical properties of the animals, plants, or stones beginning with that letter of the alphabet. Under the letter alpha, for example, is entered *ampelos* (grapevine), *aquila* (eagle), *aetitis* (eagle stone), and *aquila* (eagle ray). All have marvelous virtues that are cunningly related to one another: from the grape, wine is made; the root of its vine cures epilepsy and drunkenness. The stone found in the head of the eagle ray prevents one from getting drunk no matter how much wine he drinks. If you sketch the form of an eagle upon an aetites stone, then place the stone by your door with the feather of an eagle, it will act as a charm to keep all evil from your house. According to the *Kyranides*, every object and being possesses magical virtues. Even the savage bear has marvelous virtues: its skull cures headache, its eye cures diseases of the human eye, its ear cures earache, and the bear's tooth is prescribed as an amulet to aid children in teething.[33] In such texts, the realm of natural philosophy was scarcely distinguishable from the realm of mysticism, revelation, and the occult.[34]

The conception of nature as esoteric and marvelous penetrated Latin no less than Hellenistic science. No work better illustrates this than the compendious *Natural History* of Pliny the Elder, a Roman administra-

tor who lived from A.D. 23 to 79. In the thirty-seven books that constitute his work, Pliny produced Rome's most impressive example of encyclopedic science. In view of the official Roman suspicion of the Eastern
religious cults, it is hardly surprising to find him scornfully rebuking the
arts of the magi. Whenever he spoke of the oriental magi, he poured out
his invective against their "fraudulent lies." He denounced their art as
"detestable, vain, and idle."[35] His opinions represent the official Roman
view of magic, which was uniformly hostile: harmful acts of magic were
criminal, and so-called white magic, though widely practiced, was always
looked upon with suspicion.[36] Yet Pliny's views of magic were ambivalent. Like many of his contemporaries, he regarded the magi as sources
of profound philosophical wisdom. Democritus, Pythagoras, Empedocles, and Plato, he affirmed, all journeyed to the East to study with the
magi.[37] Nor could Pliny conceal his fascination with nature's marvels.
He reported numerous magical practices without critical comment.[38]
He did not doubt that there were marvelous plants (*herbarum mirabilium*); in fact he listed eighteen of them whose magical virtues he acknowledged were discovered by the magi.[39] He carefully recorded the
rituals that had to be observed in the collection of certain plants, such as
vettonica (betony), which "is gathered without iron with the right hand,
thrust under the tunic through the left arm-hole, as though the gatherer
were thieving."[40] With typical astonishment, he commented on the
menses of women: "Nothing could easily be found that is more remarkable than the monthly flux of women. Contact with it turns new wine
sour, crops touched by it become barren, grafts die, seeds in gardens are
dried up, the fruit of trees falls off, the bright surface of mirrors in which
it is merely reflected is dimmed, the edge of steel and the gleam of ivory
are dulled, hives of bees die; even bronze and iron are at once seized by
rust, and a horrible smell fills the air; to taste it drives dogs mad and
infects their bites with an incurable poison."[41]

And so on. Pliny was not merely credulous; he was inclined by the
temperament of his culture toward the bizarre, the exceptional, and the
marvelous. Reports of fabulous creatures from Africa and Asia, including snake-eating cave dwellers, goat-pans, forty-pound turnips, fantastic
gemstones, miraculous healing plants, and men without heads, their
mouths and eyes attached to their chests—things that "fill us with wonder, and force us to admit that there is still much truth in them"—all
inhabit the pages of Pliny's catalog of wonders. Such a delightfully readable book could not help but cast a spell over ancient readers, who could
leisurely skim its pages without risking uncomfortable confrontations
with difficult philosophical arguments.

Whatever its limitations, Pliny's *Natural History* was at least the
product of assiduous research and honest respect for its sources, a distinction shared by few of his Roman successors. His long and learned

work was cited and epitomized by a host of late Roman and patristic authors, who justifiably saw it as a compendium of ancient scientific knowledge.[42] For some digesters, such as Solinus (third century A.D.), the *Natural History* was plainly and simply a book of wonders.[43] But for most medieval readers, it was a tremendous storehouse of *information*, much of it decidedly practical: instructions on gardening, farming, the care of livestock, the cultivation of herbs and orchards, medical remedies, and technological processes. Because the *Natural History* was arranged topically, readers and copyists could concentrate on those sections of the work that interested them most. The tendency to excerpt portions of the work, a common medieval practice, is already evident in the third century.[44] These considerations, more than any philosophical views Pliny held, explain the *Natural History*'s appeal in the Middle Ages. Medieval readers of Pliny wanted information for specific needs and purposes—facts, instructions, anecdotes—and hence made little effort to expound upon the broader meaning of the text.[45]

The encyclopedia tradition was a mixed blessing. Although the encyclopedias summarized and made accessible to medieval readers much ancient scientific information that might otherwise have been lost, they rarely presented any particular subject in depth. The very comprehensiveness of the enterprise limited the attention compilers could devote to any one field or subject. Moreover, the encyclopedias created a heavy dependence upon written authority and contributed to a willingness on the part of medieval readers to believe that the wisdom of the ancients could be distilled into a few choice phrases. Pliny (of all people) observed that reliance upon written authorities stifles experimentation, contributes to the decline of knowledge, and encourages men to jealously conceal from the public what little they know. Explaining the decline of the knowledge of medicinal herbs among the Romans, he blamed overreliance on books and contrasted ancient values concerning the dissemination of knowledge to those of his contemporaries. "There was nothing left untried or unattempted by [the ancients]," Pliny wrote, "nothing kept secret, nothing which they wished to be of no benefit to posterity. But we moderns desire to hide and suppress the discoveries worked out by these investigators, and to cheat human life even of the good things that have been won by others. Yes indeed, those who have gained a little knowledge keep it in a grudging spirit secret to themselves, and to teach nobody else increases the prestige of their learning."[46]

Pliny's observations might also be applied to the early medieval scientific tradition. The more medieval copyists digested and epitomized encyclopedic knowledge, the more fragmented ancient science became, and the more precious were the pearls of knowledge that survived. As the source of scientific knowledge receded into the distant past, a new kind of literature developed to take its place, a literature consisting of compilations of the supposed "secrets" and "experiences" of the ancients. On the face of it they are little more than miscellaneous assortments of folklore and practical recipes. As a result, modern scholarship has tended to view them as symptoms of a decline of learning. Yet as we will see in the next section, they were in many instances successful attempts to adapt classical knowledge to changing needs. However inadequately they may have replaced the philosophical tradition, these works were serviceable to medieval people in ways that philosophy might not have been. The literature of secrets flourished, above all, because it was useful.

The Secrets of the Arts

Greek philosophy and science, having failed to take root in the arid intellectual soil of Rome, were left to wither in the Latin West during the early Middle Ages. The caretakers of learning, monks and ecclesiastics who had taken refuge in the monasteries and shrinking cities of Europe, were generally indifferent to or suspicious of pagan philosophy, except insofar as it might serve religion.[47] The gap created by the decline of philosophy was partially filled by the literature of an ongoing practical and empirical tradition. Medieval people had to make livings. They had to cope with new and sometimes hostile environments, to combat the ever-present threat of disease, and to make implements and tools for domestic and workshop use. They sought to beautify their homes and churches with articles of wood and metal that visually confirmed their view of the world. For these needs, classical philosophy provided little assistance. However, the ancient world also produced an abundant literature on the applied arts, which, when they survived, were adapted to suit new needs and new materials.

Foremost among the practical arts, from the standpoint of the amount of literature it produced, was medicine. The practical side of ancient medicine was fairly efficiently transmitted to western Europe.[48] Although most of the early medieval medical texts were based upon ancient sources, the manuscripts contain numerous emendations. Traditionally historians have interpreted these textual alterations as symptoms of decline and debasement. However, recent research indicates, to the contrary, that they were often the results of attempts to adapt older techniques and materials to new conditions. The physicians and healers of the early Middle Ages were not slavish copyists but competent practitioners who had to deal with many northern European plants that were unknown to the Greeks and Romans. In their efforts to find substitutes for the medicaments described in the ancient texts, they experimented with ingredients that lay closer to hand, and they altered their texts accordingly.

An early fifth-century pharmacopoeia by Marcellus of Bordeaux, an adviser to the emperor Theodosius, suggests that abbreviated handbooks and guides that eliminated theoretical considerations proved to be more useful to physicians than unabridged ancient texts. Marcellus's *Liber de medicamentis* was essentially a collection of recipes culled from ancient medical texts, to which Marcellus added recipes he had obtained directly by word of mouth.[49] Like Pliny, Marcellus was an avid collector of folk practices. His claim to have recorded remedies "chanced upon by rustics and the people, and simples they have tested by experience," seems genuine and credible.[50] As a result his work contains numerous magical prescriptions, some of which were evidently drawn from indigenous Gallo-Roman popular traditions, while others (including perhaps those based upon astrology and sympathetic magic) seem to be derived from Hellenistic magical books, many of which still survived in Marcellus's time.[51] For Marcellus and his contemporaries, "empirical" meant both "derived from experience" and magical.

Such practical, empirical, and possibly experimental characteristics of early medieval medicine are evident in a multitude of anonymous handbooks of medical prescriptions, called *antidotaria* or *receptaria*, written by and for practitioners. Like the *Liber de medicamentis* of Marcellus, these are mixtures of folk medicine and classical pharmacy. By no means, however, were they mere handwriting exercises for medically illiterate scribes. Not only did medical healers use the *receptaria*, they improved upon them by adding their own observations to them, clarifying the descriptions of plants and their habitats.[52] Medieval writers also made new compilations. They added glosses to existing texts, identifying Latin plant names with their vernacular synonyms, and compiled bilingual glossaries to assist readers in identifying obscure plant names. In an age of rapidly declining facility in the Greek language, the Latin *receptaria* answered an urgent practical need for reliable pharmaceutical information. Writing in the mid-sixth century, the great schoolmaster Cassiodorus advised some monks, "If you have not sufficient facility in reading Greek, then you can turn to the herbal of Dioscorides, which describes and draws the herbs of the field with wonderful faithfulness."[53] Cassiodorus, it turns out, was not referring to the famous first-century herbal by Dioscorides, *De materia medica*, but to a late fifth- or early sixth-century Latin herbal entitled *Ex herbis feminis*. Although attributed to Dioscorides and partially based on his herbal, the treatise is a medieval composition and a typical example of how physicians adapted classical learning to their own needs and circumstances.[54] Innocent of theory, *Ex herbis feminis* was in many ways superior to Dioscorides' herbal: it was shorter and easier to use, and included flora more typical of northern European climates. The popularity of *Ex herbis feminis* was largely due to its simplified format. But its attribution to Dioscorides lent the work authority and prestige, and probably enhanced its appeal to medieval readers.

Such spurious attributions to ancient authorities are entirely characteristic of early Latin scientific and medical treatises. Some of these works, such as the *Medicina* of Pseudo-Pliny, were merely rearrangements of material from earlier authors.[55] Others, such as the *Epistola ad Maecenum* attributed to Hippocrates and Pseudo–Antonius Musa's *De herba vettonica*, were original medieval compositions.[56] Medieval recipes and treatises are attributed to Pythagoras, Galen, Pliny, Dioscorides, Aristotle, and numerous other classical authors. A recipe for a charm in a ninth-century Bamberg codex ends with the claim "This was used by Plato and me."[57] That medieval authors should have so scrupulously avoided claims of innovation, often to the point of concealing their identity under the cloak of some trusted ancient name, may seem puzzling to an age like ours that extolls originality. However, medieval intellectuals were conscious that theirs was a borrowed culture. They had inherited the Roman consciousness of the inferiority of their own intellectual tradition in comparison to that of ancient Greece, while rigid orthodoxy in religion made them distrustful of the new. An intriguing example of such spurious attributions is a popular fourth-century herbal ascribed to "Apuleius Platonicus," or Apuleius of Madaura (second century A.D.), the philosopher and poet who is best known as the author of the picaresque novel *The Golden Ass*. Why the work should have been attributed to Apuleius of Madaura is somewhat mystifying, since his connection with medicine is tenuous at best. On the other hand, he was a passionate devotee of the mystery cults. He was initiated into the cult of Isis and was a priest of the cult of Asclepius, a fact that was well known to the fourth-century author of the herbal attributed to him. Thus it appears likely, as Linda Voigts has suggested, that the Madauran's name was attached to the herbal because of his association with the cult of Asclepius.[58] If so, then we seem to have in the *Herbarum Apulei* a text that is related to, or at least adopts the topoi of, the Hellenistic literary mysteries.[59] In other words, the ascription of the herbal to Apuleius may have been based upon a belief that Apuleius had been initiated into the mysteries of Asclepius, which, as the case of Thessalos demonstrates, could come about only by his receiving a revelation directly from the god. Apuleius is said in the text to have "received" the herbal from Asclepius, and in some manuscripts the god's epiphany is depicted visually. The *Herbarum Apulei* represents not merely a medieval attempt to certify a text by linking it to the earliest classical medical authority but an effort, still common in late antiquity, to authorize a scientific treatise by finding a basis for it in divine revelation. Initiation into the "mysteries" of nature still carried the weight of authority as the Middle Ages opened.

Ancient knowledge of other practical arts came into medieval Europe in a similar fashion, as practical recipes disembodied from their original theoretical matrix. A continuous, unbroken tradition of manuscripts detailing the arts of glassmaking, metallurgy, ceramics, dyeing, and painting extends from late antiquity through the end of the Middle Ages. It is unlikely that these manuscripts were composed by or for working artisans. Although they were preserved in monasteries where these arts were practiced, and in some instances reflect craftsmanship that was still in use, the compilations were the work of scribes, not craftsmen. Until about the twelfth century, when monastic artisans began to record their own techniques, the medieval craft recipe books have to be considered, strange as it may seem, as literary creations within a complex literary tradition.[60]

The literary tradition behind the recipe books dates back to Hellenistic times. Its roots are in the alchemical tradition. The earliest recipes on the chemistry of the arts are contained in two Egyptian papyri dating from the late third or early fourth century A.D. The two papyri (now known as Leyden X and the Stockholm Papyrus) were originally parts of the same text.[61] The purpose of the manuscripts remains something of a mystery. Altogether the two papyri contain some 250 recipes, about a hundred of which describe processes for imitating precious metals, testing their purity, coloring the surfaces of metals, and writing in gold and silver. The remainder are about evenly divided between recipes for dyeing fabrics and those for counterfeiting and cleaning precious gemstones. The overwhelming emphasis upon procedures for creating imitations of precious materials raises the possibility that the two papyri constituted a counterfeiter's manual.[62] Some of the recipes do in fact appear to be fraudulent in intent. A recipe for coloring copper objects gold says, "It is difficult to detect because rubbing gives the mark of a gold object, and the heat consumes the lead but not the gold."[63] Another recipe, for manufacturing silver, cautions that artisans might notice something peculiar about the finished product.[64] Similarly, the dyeing recipes seem intended to produce cheap imitations of "true purple" or other expensive dyes. A recipe from the Stockholm Papyrus, for example, notes that "celandine is costly. . . . You should accordingly use the root of the pomegranate tree and it will act the same."[65]

Although there is ample evidence for the counterfeiting of precious gemstones and costly dyestuffs in ancient Egypt,[66] it seems doubtful the papyri were counterfeiter's manuals. They contain no trace of excessive wear or frequent use, as one might expect of a workshop manual.[67] The elegant courtly hand in which the manuscripts are composed seems to indicate they were intended more for the library than for the atelier. The papyri's connection with the ancient alchemical tradition, on the other hand, has been firmly established. Scholars have shown that the papyri are probably descended from a first-century alchemical treatise entitled *Physica et Mystica* by Bolos of Mendes, who wrote under the pseudonym "Democritus."[68] The two technical papyri originally formed a comprehensive alchemical treatise embracing both practical and theoretical aspects of the art. The Leyden papyrus mentions a "Democritus" as one of its sources, while the *Physica et Mystica* includes recipes—instructions for the manufacture of purple dyes and for making gold and silver alloys—that are identical in character to those of the papyri. In addition, the work contains a lengthy theoretical discussion of the transmutation and "ennobling" of metals. Thus it appears that the compilers of the Leyden and Stockholm papyri were neither artisans nor professional counterfeiters, but alchemists whose chief aim was to transmute base metals into the noblest of metals, gold. Alchemical transmutations were generally linked to color changes, which explains the emphasis in the Alexandrian treatises upon dyeing and alloying. Almost every recipe in the *Physica et Mystica* concludes with the cryptic assertion, "One nature delights in another nature; one nature triumphs over another nature; one nature dominates another nature"; this seems to refer to the law of sympathies and antipathies governing transmutation.[69] Possibly it describes the threefold process whereby one "nature" or quality (indicated by color) is transformed into and "dominated" by another. The succession of operations applied to the dyeing of fabrics described in the Stockholm Papyrus—washing, mordanting, and coloring—may be analogous to this threefold process; in the manuscripts they are applied to stones and metals as well as to fabrics.[70] The language and techniques of dyeing are merely transferred to the "dyeing" of metals and stones. The ancient alchemists, impressed by the observation that certain dyestuffs could completely transform the outward appearance of fabrics, believed they had discovered the secret of the transmutation of all matter.

We are obviously concerned here with a cultural climate in which the distinction between gold making and gold faking was not as obvious as it is to us. As Joseph Needham observed, the distinction between aurifiction (counterfeiting gold) and aurifaction (transmuting "ignoble" metals into gold) is not technological, but cultural: it depends upon what you think you are doing.[71] Indeed, the alchemist's real aim had little to do with either the making or the counterfeiting of gold. His goal was a religious one: to project upon matter the mystical drama of the passion, death, and resurrection of the god of the mystery cult. In the sudden appearance of alchemical texts at the beginning of the Christian era we witness the convergence of two esoteric traditions, one learned and the other popular, one represented by the revealed wisdom of the East and the other by the craft tradition, the guardian of trade secrets.[72] When this literature passed into the Latin West in the early Middle Ages, its theoretical component, disconnected from its original cultural matrix, gradually disappeared. Only faint echoes of the esoteric purpose behind the recipes were heard in the West, as when a tenth-century scribe copied the words "arte et ingenio vinci ingenium"—the hidden alchemical meaning of which almost certainly eluded the copyist.[73]

The occultation of ancient alchemy's theoretical component can be partially explained by religious considerations. Alchemy, a sacred art in the Hellenistic East, employed the metal-tinting recipes as part of a redemptive ritual. Thus it competed with the new redemptive god of the Christians, who suppressed the alchemical texts. Despite persecution, certain heretical Christian sects, particularly the Gnostics, continued to secretly practice the Sacred Art. It has recently been suggested that the alchemical-technological recipes may have passed surreptitiously into the West with Cathar missionaries. The Cathars, in turn, learned the recipes from the Messalians, a medieval Gnostic sect that is known to have practiced the Sacred Art and to have preserved secret books from antiquity.[74] Other recipes may have come with migrating craftsmen who had been in contact with one or another of these heretical sects. Whatever the route of transmission, the purely technological component of the literature entered Europe in the form of recipe books dealing with the arts of manuscript illumination, painting, glassmaking, dyeing, and metalwork.

The earliest of these compilations is contained in a Lucca manuscript, Codex Lucensis 490, which dates from the late eighth or early ninth century. The work, known as the *Compositiones ad tingenda musiva*, contains recipes for making pigments, coloring glass for use in mosaics, dyeing leather and cloth, gilding, soldering, and making alloys. There are numerous parallels between the *Compositiones* and the Alexandrian alchemical papyri. Indeed, one of its recipes appears to be a direct translation of a recipe in the Leyden papyrus. The work also contains traces of Dioscorides, Theophrastus, and other Greek authors.[75] The *Compositiones ad tingenda musiva* was but the first medieval example of what turns out to be a large number of similar works, nearly all of which are compilations drawn from the same great body of recipes. Contemporary with the Lucca manuscript was a work called the *Mappae clavicula* (A little key to painting), whose title was listed in a library catalog of 821–822 of the Benedictine monastery at Reichenau. Although this particular manuscript is lost, an early ninth-century fragment of the work survives, as do manuscripts of the tenth (Sélestat MS 17) and twelfth centuries (Phillips-Corning MS). Recipes from the text appear in more than eighty manuscripts dating from the ninth to fifteenth centuries.[76] No work better illustrates the purely literary character of the early medieval craft books than the *Mappae clavicula*, a compilation based upon written sources that can be traced all the way back to the Alexandrian alchemical treatises. The tenth-century Sélestat manuscript contains almost all of the recipes in the earlier Lucca manuscript, along with a number of new ones of uncertain origin. The Phillips-Corning manuscript of the twelfth century shows further evidence of accretion: it includes most of the recipes from the Lucca and Sélestat manuscripts, which make up about two-thirds of its contents; the remaining third is a miscellaneous assortment of recipes similar in style and content to the older sources. All these works are compilations and not unified treatises. Rarely do they exhibit any conscious attempt to organize the recipes into coherent categories, although the very fact that they were compiled from older manuscripts inevitably led to clusters of recipes on related topics. Most of them bear the mark of having been continuously copied down through the centuries with little technical insight or understanding. Medieval monks were obviously curious about the technical side of the arts, even if they had little grasp of technical details.

What role, if any, did the recipe books play in the development of the early medieval arts and crafts? On the face of it, the compilations look like practical workshop handbooks, and so generations of scholars have taken them, evidently imagining the monastic artisan leaning over his workbench, recipe book in hand, mixing his colors and preparing his materials "by the book."[77] The reality was not so simple. Although almost every monastic library had copies or fragments of one or more of these works, by the time the recipes appeared in the medieval compilations, they were but distorted echoes of an earlier period of technology. The modern translators of the *Mappa*, one a distinguished historian of technology, express disappointment over the work as a source for the history of medieval technology, noting bluntly that "the text was born of drudgery, not inspiration."[78] Some of the recipes are completely unintelligible or badly garbled as a result of having been sequentially recopied by scribes who were ignorant of the technical realities behind the words. Many of them make no technical sense at all, or the sense is submerged under a host of tricks, "experiments," and cryptic expressions. Obviously, these books could not have trained beginning craftsmen and even for experienced artisans would have been quite useless, strictly speaking, as workshop guides.

But only strictly speaking. For the large number of extant copies of the manuscripts suggests that they were regarded as important books and not merely as incidental parts of monastic libraries. Perhaps the best

explanation of their appeal to artisans lies in their connection with a remote classical past rather than with current techniques. It is significant
that the earliest manuscript of the *Mappae clavicula* appears in the Benedictine monastery at Reichenau, an important center of the revival of
artistic activity and classical learning, both stimulated by the Carolingian
court.[79] The library at Reichenau contained more than four hundred
volumes in the early ninth century, when the *Mappa* showed up in its
book list. When placed beside these texts in the monastic library, the
work was perhaps viewed less as a handbook of current techniques
than—as its prologue explains—as a key to unlock the secrets of ancient
arts hidden in the hallowed books of antiquity:

> I call the title of this compilation *Mappae clavicula*, so that everyone who
> lays hands on it and often tries it out will think that a kind of key is con-
> tained in it. For just as access to [the contents of] locked houses is impossi-
> ble without a key, though it is easy for those who are inside, so also, with-
> out this commentary, all that appears in the sacred writings will give the
> reader a feeling of exclusion and darkness. I swear further by the great God
> who has disclosed these things, to hand this book down to no one except
> to my son, when he has first judged his character and decided whether he
> can have a pious and just feeling about these things and can keep them
> secure.[80]

The prologue's references to sacred texts, to a "key" to unlock their
mysteries, to insiders and outsiders, and to the law of silence make no
sense whatsoever as long as the *Mappa* is considered to be a purely technological work. Such references make sense only as the remnants of an
ancient literary tradition. For the prologue leads us directly back to the
literary mysteries of the Hermetic tradition. In the *Mappa*, however, the
"revelation" of sacred knowledge is purely literary. It comes not
through a face-to-face confrontation with a god, but by means of a technical "key," in the form of a recipe book, to unlock the mysteries of
sacred texts. In other words, the prologue suggests that the *Mappa* is a
fragment from, or was once an adjunct to, a larger alchemical corpus. Its
technical recipes were the "little keys" that made it possible to accomplish the ceremonial and initiatory rituals described in the "sacred
books" the work once accompanied.[81] Scattered references throughout
the *Mappa* to sacred arts reinforce this interpretation. Thus a recipe for
making a gold coloring ends, "Keep this as a sacred thing, a secret not
to be transmitted to anyone, and you will not as a prophet have given it
away."[82] Following the principle of dispersing texts as a means of concealing their content, the recipes may have been deliberately separated
from the religious component of the text (or group of texts) in order to
hide the true meaning of the technical and ceremonial aspects of the

Sacred Art. Such techniques were widely practiced in the Hermetic tradition. What medieval scribes and artisans made of the prologue and of other cryptic references in the *Mappa* is impossible to say. They cannot have been aware of the original Hermetic meaning behind them. However, we may well suppose that in the context of the Carolingian literary and artistic revival, the work's repeated references to the "sacred secrets" of the arts (*sanctum laudabile que secretum*), and to the necessity of keeping the recipes secret so as not to corrupt them, struck a responsive chord in the scribes who were responsible for transmitting the work. By the time the recipes came into the Latin West, their original alchemical meaning had disappeared, but the recipes themselves, the treasured literary remains of antique arts, were "sacred things."[83]

A similar outlook is reflected in a work entitled *De coloribus et artibus Romanorum* (The colors and arts of the Romans). Originally a poem in two books, this work is thought to have been written in the tenth century by an otherwise unknown Italian monk named Heraclius. A third book in prose was added in the twelfth century.[84] In his preface to the poem, Heraclius lamented over the loss of the ancient arts of the Romans: "The greatness of intellect, for which the Romans were once so eminent, has faded, and the care of the wise senate has perished. Who can now investigate these arts? Who is now able to show us what these artificers, powerful by their immense intellect, discovered for themselves? He who, by his powerful virtue, holds the keys of the mind, divides the pious hearts of men among various arts."[85]

Now, this work is unmistakably a compilation from earlier literary sources. Despite Heraclius's claim that "I write nothing to you that I have not first tried myself" (*Nil tibi scribo quidem, quod non prius ipse probassem*), it is not the work of an experienced craftsman. Heraclius never speaks of himself as an artisan but instead presents himself as a literary man addressing "you artists" (*vos artifices*). The fact that the work was composed in verse rather than the usual recipe-book format suggests that it was not intended as a workshop guide, nor was it a compilation of craftsmen's recipes; it was a product of classical erudition. Despite Heraclius's repeated claims to have "tried out" the techniques and his assurance that he "took great pains" to discover the secrets of the ancient arts, his discoveries do not inspire much confidence in his experimental skills.[86] For example, he enthusiastically recommends a method for engraving precious stones with the blood of a billy goat, a technique he learned from Pliny and claims to have tried out himself.[87] Generally, when Heraclius speaks of the research that went into the treatise, he means intellectual labor (*gessi cum summa mente laborem*) and not workshop experience, "profound thought" (*sub mente profunda*) and not experiments.[88] Like all of the early medieval compilations on the arts, *De coloribus et artibus Romanorum* drew from a wide range of ancient sources. Its composition seems to reflect more the aims of a humanist revival than practical instruction. In short, the survival of these ancient recipes reflects the interests of scribes rather than artisans, and the belief that what is written is more important than what is done.

Nevertheless, there is evidence in the artistic tradition that the artisanmonks of the early Middle Ages took the manuscripts seriously, and that their influence was felt in the workshop. The handbooks appeared in the West just as the influence of Greek (i.e., Byzantine) styles on European art was beginning to be felt. This is not to say that the Carolingian craftsmen consciously imitated Byzantine styles. The monastic artisans and the craftsmen who worked at the court in Aachen were imitating what they took to be the art of the Christian Roman Empire, a style that fit into the all-pervading idea of the *renovatio imperii Romani*.[89] Of course, their notion of what this "Roman" art actually looked like was rather vague. The northern European artists did not see much difference between Eastern and Western Roman art. What they were after was something "antique" and classical. It was natural that they should look to Italy, and especially to Rome and Ravenna, where Byzantine styles flourished. Possibly it was through this medium that some of the recipes came into northern Europe. The result of these unintentional contacts between East and West was a new, "gorgeous" style in Carolingian art, a style characterized by a lavish use of gold in lettering and painting, rich color schemes in blues, greens, and purples, and the staining of entire pages of parchment in deep purple, making them look like luxurious carpets. The showy, cloisonné effects of the northern European illuminated manuscripts of this period are reminiscent of the goldsmith's and enameler's work, which had been cultivated with consummate skill in antiquity.[90] The desire to embellish and to give the appearance of richness and imperial splendor is everywhere apparent in the *Compositiones* treatises, where the reader is instructed in painting glass, dyeing skins, cutting stones, making brilliant colors, and writing in gold. Surely these precious manuscripts, these "keys" to ancient arts, were not kept in the workshops, where they could be easily soiled or damaged, but were stored in the libraries. The manuscripts show little sign of workshop wear. Yet because the monastic rules defined a close relationship between the scriptorium and the atelier, their voice was certainly heard in the studio. Since monks were assigned to periods of study and reading in the scriptorium and manual labor in the fields and workshops, artisans as well as scribes had access to recipe books like the *Mappae clavicula*. Indeed, for some novices interest in the arts may well have begun in the library. The manuscripts, with their cryptic references to arcane crafts, were like voices from the distant past, taking the place of a long-absent tutor, initiating craftsmen into the mysterious arts of antiquity.

Esotericism was an integral, if perhaps unfortunate, part of the West's scientific inheritance from the ancient world. As we have seen in this chapter, classical Greek science and philosophy were radically transformed during the Hellenistic era. The breakdown of the *polis* and the cynical politics of the Hellenistic world shook the Greek mind's habitual trust in reason. For many intellectuals of that era, divine revelation was a surer hope for certainty than frail human reason. In the Hellenistic East, where the occult sciences were born, "philosophy" no longer meant the exercise of reason aimed at understanding the order and structure of the universe. Instead, it meant a secret doctrine, essentially religious, known only to a select circle of initiates. Since revealed knowledge was sacred, it had to be carefully guarded from pollution by those unworthy of receiving it. Like all forms of gnosticism, Hermetism erected a barrier between the elect "knowers" and the ignorant common people.

By the time the Romans made contact with Greek science, the symptoms of the "decay of reason" had already set in. While Greek philosophy and science fascinated the Romans, it would never become *familiar* to them. Content to dabble in rather than to master Greek theoretical science, the Romans relied heavily upon encyclopedias for their scientific knowledge. The dependence upon literary sources was a trend that would continue into the early Middle Ages. The written tradition did not always drive out original research, as we saw in the case of the medieval herbals. Nevertheless, the overwhelmingly literary character of Latin science shaped the medieval scientific outlook in important ways. It led to the temptation to rely more and more on the written word for an understanding of nature. It made literacy—meaning knowledge of the Latin language—the skill that alone distinguished the learned from the vulgar, the elite from the popular, and the few who were worthy of receiving secret knowledge from the many who threatened to pollute it. For the medieval scholastics, the search for the secrets of nature was a search for lost books.

TWO

KNOWLEDGE AND POWER

IN the twelfth century, the "secrets of nature" took on a new meaning in Western culture. By then, the Europeans had awakened to the realization that a civilization with a scientific tradition vastly richer than theirs flourished in the Mediterranean: the empire of the Muslim caliphs, extending from the Indian Ocean to the Atlantic coast of Spain. Only a few Latin scholars had ventured into Islamic territory before this time, but those who did were profoundly and permanently changed. With the conquest of Toledo in 1085, the Christian world took possession of a civilization next to which the Latin West, thought Daniel of Morley, seemed "infantile" and barbaric.[1]

Arab civilization was itself the product of older cultures, Syrian, Persian, and Greek, which in the course of imperial expansion the indigenous culture assimilated through translation and transformed by the dominating force of its official religion, Islam.[2] Science assumed a prestigious role in Islamic civilization. Affirming the importance of philosophy and natural knowledge in the hierarchy of knowledge, the Arabs gave science a new legitimacy. By broadening its scope to include practical disciplines that had not generally interested the ancients, they reoriented science to serve new purposes. Mathematics aided commerce, alchemy contributed to the development of chemical technology, and medical theory forged a new alliance with pharmacy and public health. Scientific method also changed. By incorporating the work of artisans and instrument-makers, Arab natural philosophers enhanced their powers of observation and measurement, applied mathematics to new problems, and used experimentation as a methodological tool.[3]

From the standpoint of the Latin West, it was in Muslim Spain—al-Andalus, as the Arabs called it—that Islamic science accomplished its most enviable successes. Already in the tenth century the Benedictine monk Gerbert of Aurillac (later Pope Sylvester II) crossed the Pyrenees to study mathematics and astronomy from Arabic sources. So daring did Gerbert's adventure seem to contemporaries that numerous legends developed about the monk who bartered away his soul to the devil in return for heathen learning.[4] Legends notwithstanding, the conquest of al-Andalus was the most significant event in the cultural revival of Western Europe. Its magnificent libraries contained rich holdings of classical writings, and its communities of Christian Mozarabs made Spain an ideal location for establishing contacts between Latin and Arabic scholars. Although the lure of ancient philosophy was not the principal motive behind the West's Crusade in al-Anadalus—crusading hysteria and an appetite for booty were more effective inducements—the acquisition of Arab learning was one of the most important results of the *reconquista*.[5] Among the many treasures brought back from the Muslim world, Europe acquired a recondite and powerfully alluring corpus of writings on Islam's "secret sciences."

The "Secret Sciences" of Islam in the Medieval West

Conscious of the backwardness of Latin culture, European scholars zealously embraced the new learning.[6] The Latin translators who traveled to the Islamic centers of learning went with great enthusiasm, motivated by the expectation of revealing the arcana of philosophy, of bringing back the long-lost secrets of ancient knowledge. A Pisan scholar named Stephen reported that he journeyed to Antioch in the 1120s to learn "all the secrets of philosophy that lie hidden in the Arabic tongue."[7] Hugo of Santalla returned from Aragon in the 1140s telling about the "arcane wisdom" (*tante sapiencie archana*) and the "innermost secrets of philosophy" (*ex intimis philosophie secretis*) he discovered in the Arabic works on the occult sciences.[8] The Europeans understood this "arcane wisdom" to be ancient in origin. They believed that all knowledge stemmed from a single divine revelation that was passed down from the Hebrew prophets to the ancient Chaldeans and Egyptians, then to the Greeks and Romans, to the Arabs, and finally to the Europeans, who were the "sons and successors of the sacred writers and of the wise philosophers."[9] But the Latin scholars had only a vague notion of who the ancients really were, when they lived, and what they wrote. So it was inevitable that when ancient philosophy reentered the West, it came as a heterogeneous mixture of authentic writings, pseudepigrapha, and hybrid compilations of Greek and Middle Eastern texts. Arabs, Greeks, Persians, Chaldeans, Indians, and Egyptians were hopelessly jumbled and conflated.[10] Lacking a critical apparatus to separate genuine from spurious texts, the Latins tended to accept all as authentic, especially when they purported to reveal the esoteric wisdom of the ancients.

In an atmosphere so densely charged with fascination for arcane knowledge, books that professed to reveal the "secrets" and "experiments" of famous (as well as infamous) men of science gained widespread appeal. The Arabs supplied Europeans with an abundance of such works. Drawing upon the voluminous Hermetic literature of the Hellenistic East, they had developed a comprehensive and highly organized system of occult science and an impressive corpus of treatises on astrology, alchemy, magic, and divination.[11] The philosophical foundation of the Arabic "secret sciences" was the doctrine that the world was a network of hidden correspondences and a reservoir of powerful occult forces. This perspective infiltrated Islamic natural philosophy in general.[12] But if the Arabs gave science a prominent place in their educational system, there was no question it took second place to religion. In the Muslim world the supreme goal of philosophy and science, like that of Hermetic wisdom, was to achieve religious understanding, or gnosis.[13] The Muslim scientist (*hakīm*) was essentially a sage to whom knowledge was entrusted. To him it was a sacred duty to guard that knowledge against contamination by the unworthy. Thus the ancient figure of the scientific prophet reemerged in the Islamic scientific tradition, and the Hellenistic literary conventions about revealed knowledge were put to a new use. According to the Arabic treatises on the "secret sciences," God had revealed the secrets of nature to a select few disciples. Highly disciplined and pure of heart, the guardians of arcane knowledge protected their secrets from profanation by writing them down in tortuous symbols and cryptograms. They used mysterious pseudonyms. They enlisted pupils of impeccable moral character and commanded their disciples to obey the law of silence.[14]

These familiar topoi, which were taken over from the Hermetic literary mysteries, took on a new life and meaning once they were appropriated by certain radical Muslim sects. The words *secretum* and *secretum secretorum* occur in numerous Latin translations of Arabic scientific writings. These terms were the usual renderings of the Arabic *sirr* and its superlative *sirr al-asrār*, concepts that occupied a prominent place in the Hermetic literature of the Middle East. The terms appear in the titles of alchemical tracts by Rhazes and Khalid, in books of medical recipes and "experiments," and in numerous works on the occult properties of things. The origins of the term "secret of secrets" in the idiom of Arabic science take us back to the secret fraternities of medieval Islam known as the Ismaili, so called because they supported the imamate of Muhammad ibn Ismāʿīl, the "Hidden Imam," whom the Ismaili alleged had not died but had gone into concealment following his mysterious disappearance in 767.[15] Despite a history of suppression and persecution, this radical Shiite sect exerted a tremendous influence in the Islamic world. The Ismaili movement gave rise to the Fatamid dynasty in Egypt. Elsewhere its esoteric doctrines and its cell-like political organization gave birth to the cult of the Assassins, the Sufi poets, and the religious mystics known as the Brethren of Purity.

Fundamental to Ismaili doctrine was the distinction between exoteric, or apparent, meanings of the scriptures (*zāhir*), which change with each prophet, and the esoteric truths of religion (*bātin*), which are unchangeable, concealed within the Koran, and comprehended only through a cabalistic form of interpretation based on the mystical significance of numbers and letters. Batinism, which makes a radical distinction between exoteric and esoteric knowledge, is common to much of the Islamic intellectual tradition. The religious foundation of Islam is God's revelation to the Prophet. But that revelation, which is contained in the Koran, holds various meanings, which can be interpreted at several levels. Since humans can know only what God has chosen to reveal, the problem of knowledge is one of discovering the hidden meaning of the Koran.[16] This concern with the apparent versus the esoteric meanings of the holy book took on special urgency in the Ismaili context, where it was fused with a revolutionary eschatology. The Ismaili believed that until the end of history, when the *bātin* is proclaimed by the final prophet, the esoteric truths of religion are kept secret and are revealed only to those who are initiated in the true faith. In the final era of history, the seventh Imam following the prophet Muhammad, who is none other than Muhammad ibn Ismāʿīl, will reappear from concealment as the new prophet who reveals the hidden, eternal truths underlying manifest scripture and religious law.[17] This doctrine seriously threatens the traditional basis of Islam, because it proclaims a series of ever-higher truths, each new one abolishing all prior manifestations, including, of course, the revelations of the Prophet himself.

The concept of *bātiniyya*, or esoteric meaning, had a parallel in Ismaili cosmology. The Ismaili were strongly influenced by Neoplatonism. They conceived of the creation as a series of emanations proceeding from the absolutely incomprehensible God. Every facet of the universe mirrored every other, itself, and the hidden God. An intricate and inexhaustible series of hidden affinities and resemblances, nature hid itself within layer upon layer of occult qualities. All aspects of reality possessed both inner and outer meanings. The purely exoteric sciences were merely the vehicles, so to speak, that carry the esoteric meanings hidden within them. The universe was like a cosmic text whose inner meaning can be understood only by symbolic interpretation, and not by a "literal" reading. To the Ismaili, natural phenomena were transparent symbols whose real message was spiritual. Science was a hermeneutics of nature, and the symbolic interpretation of the Koran became the basis for the symbolic study of nature.

The Ismaili, along with other Shiite sects, were the principal purveyors of Hermetic doctrines in medieval Islam. There is an abundant scientific literature connected with the Shiite tradition, most if not all of it Hermetic in origin.[18] Some of these works, such as the famous magical textbook *Ghāyat al-Hakīm* (The Goal of the Sage)—the notorious *Pica-*

trix in the Latin West—were inspired by the teachings of the Brethren of Purity, a secret Ismaili fraternity that flourished in the tenth century. A compendium of knowledge in the form of fifty-two letters addressed to the "faithful brethren," the *Ghāyat al-Hakīm* was translated into Spanish for Alfonso the Wise of Castile in 1256. Soon thereafter the work appeared in Latin.[19] The extensive alchemical corpus attributed to Jābir ibn Hayyān, a collection of around three thousand titles dating from the late ninth and early tenth centuries, are so infused with Ismaili beliefs as to constitute an essentially alchemical exposition of Ismaili religious doctrines.[20] This vast body of literature evidently comprised the collected scriptures of a secret Ismaili brotherhood. According to the Jabirians, the task of alchemy was to make the elixir, a perfectly harmonious substance in which all elements are in balance. Jābir's "science of balance" (*mizan*), the key to alchemy, provided a method by which one might discover the relationship that exists in every body between the manifest (*zāhir*) and the hidden (*batīn*).[21] However, the ultimate goal of alchemy was not to transmute base metals into gold. It was to attain perfect religious knowledge, to discover the most sublime secret of truth: the hidden Imam. Jābir proclaimed the imminent advent of a new Imam who would abolish the law of Islam and replace the revelation of the Koran with Greek philosophy and science. In order to protect themselves against persecution by orthodoxy—for this revolutionary doctrine threatens the very existence of Islam—the Jabirian alchemists used pseudonyms, wrote in obscure cryptograms, and employed the method of "dispersion," breaking up the doctrine and recording it partially in scattered texts that the reader must reunite in order to gain an understanding of the complete doctrine.[22] The "secret sciences" were intimately related to the esoteric teachings of Islam, and the writings connected with them were immersed in the mysticism and piety of these movements.

Since the religious component of Islamic hermeticism entered the West as an integral part of its science, the Latin translators were unavoidably influenced by Islamic batinism, and by the esoteric doctrines of the Ismaili. To the Latins, Arabic science seemed to hold out the promise of unlocking secrets of nature that were long hidden and inaccessible to ordinary people. When Hugo of Santalla described his discovery of an astronomical treatise in the secret, innermost part of a library (*inter secretiora bibliotece penetralia*), intentionally or not he identified himself as a descendant of the ancient line of scientific prophets going back to the discoverers of Hermetic books in temples, libraries, and stelae.[23] Hugo knew the topos well, having encountered it repeatedly in the Arabic scientific treatises. The genre presented seemingly endless variations on themes involving the discovery of sacred books and the revelation of scientific knowledge. A "Book of the Moon" describing the engraving of magical images was supposedly discovered in a golden ark by an "investigator of wisdom and truth."[24] A ninth-century Arabic compilation of medical recipes, supposedly the "experiments" of Galen, is said to have been saved from a great fire "that descended from the sky upon the altar [and] burned the King's books."[25] A work containing Solomon's vast knowledge of magic was found in his son's casket; its discoverer was unable to comprehend the book's secrets until an angel of God revealed them to him on the condition that he conceal them from the *vulgus*.[26]

One of Hugo's more interesting contributions to Latin science was a translation of Balīnās's (Pseudo-Apollonius of Tyana) "Secret of Creation" (*Kitāb sirr al-halīqa*). The work contained a lengthy allegory on the search for the esoteric truths underlying public knowledge.[27] The fable, which combined stock Hermetic themes with Plato's allegory of the cave, told of Balīnās's discovery of an underground tunnel beneath a public statue of Hermes. Guided through the passageway by a magical lamp, Balīnās arrived at a chamber, where he found an old man seated on a golden throne. In one hand the old man held the book on the "Secret of Creation," in the other an emerald table inscribed with obscure symbols and the caption "The art of reproducing nature."[28] In this fable the two texts represent two kinds of scientific knowledge, one public and the other esoteric. The "Book of the Secret of Creation" (also called the "Book of Causes"), explaining the causes of natural phenomena, was open to all students of philosophy. The emerald table, containing "the art of reproducing nature," held the knowledge that enabled its possessor to influence the course of natural events, giving him power over nature. This knowledge was written in cryptograms and obscure symbols, and was reserved for initiates in the Hermetic school.[29] In the Latin West the distinction between the two kinds of knowledge was essentially that between *scientia* and *magia*: science, the knowledge of the causes of natural phenomena, and magic, consisting of the techniques by which nature is controlled, manipulated, and made to serve human ends. The esoteric doctrine contained in the emerald table was necessary to illuminate the public doctrine contained in the "Book of the Secret of Creation." Balīnās received both the book and the emerald table, and with the knowledge gained he became a wonder-working magus. The Arab author ended the treatise with the familiar admonition to "anyone into whose hands this book might fall . . . to guard it as he would his own soul, and not to give it to any stranger."[30]

Ironically, these supposedly "secret books" were far from secret in the Latin West. Highly fashionable among Europe's emerging class of university graduates, the *libri secretorum* appeared in dozens of Latin manuscripts and were widely circulated. They represent the popularization, so to speak, among intellectuals, of the tradition of esotericism dating back to the Hellenistic era. Few university students of the twelfth and thirteenth centuries, however orthodox they may have become later in life, were unswayed by the seductive lure of the Arabic "secret sciences." William of Auvergne (1180–1249), who was bishop of Paris from 1228 until 1249, was an avid reader of Hermetic books during his student days at Paris, although later he repudiated everything in them. William recalled numerous magical and astrological works he had seen and handled as a student, and expressed grave concern that the same books and others like them were still openly available to students and teachers at the university.[31] Many other students, including Robert Grosseteste, John of Salisbury, and Michael Scot, flirted with these intoxicating and dangerous books: the legendary monk Gerbert is a type representing the overly curious clerical magician.[32] The promise the *libri secretorum* held out of privileged access to secret knowledge was a powerful psychological incentive for Western intellectuals to accept the treatises as genuine. But the real appeal of these works was not their secrecy; it was their promise of virtually unlimited power: over nature, over one's enemies, over the uncertainties of chance. *Picatrix*, the most famous of all the Arabic treatises on learned magic, declared that when a student "apprehends all the intelligences and compositions of the things of the world," which the book promised to reveal, "all things will serve him and he will serve none of them."[33] The medieval books of secrets reinforced the self-image of the scholastic intellectual. For as the *Picatrix* asserted, knowledge of the "properties of things," which is the foundation of magic, cannot be gained without philosophy: "No one can reach an understanding of the manner and method by which the heavens produce effects in the terrestrial world without first mastering natural philosophy and mathematics. Whoever lacks this training will not understand the heavenly motions, nor will he ever reach the goal he strives for, because the components and the foundations of the science he wants to know are extracted from these disciplines."[34]

The ancient image of the learned magus who holds the secrets of the universe (updated with a respectable university degree), was admirably suited to advertise the skills of liberal arts graduates, who were eager to find positions in courts and civil administrations. The prospect of gaining privileged access to arcana not only heightened curiosity about Arabic science among Western intellectuals, it also helped to promote a sense of the exclusiveness of scholastic culture vis-à-vis the rest of society. Esotericism, which had served to protect the Islamic practitioners of the secret sciences from persecution by orthodoxy, was deployed in the Latin West as an instrument to promote the interests of the university masters. To the growing number of university graduates in the late Middle Ages, knowledge of the Arabic secret sciences offered significant social and political advantages: for the more esoteric knowledge is, the easier it is to maintain the distinction between those who know it, hence can use it, and those who do not, who are powerless by comparison.

Roger Bacon and the *Secretum secretorum*

The extent to which the pseudonymous books of "secrets" captured the medieval imagination is best exemplified by the extraordinary popularity and influence of the pseudo-Aristotelian *Kitāb Sirr al-Asrār* (Book of the secret of secrets), known to Europeans as the *Secretum secretorum*. From the time it came into the West in the middle of the twelfth century, the *Sirr* aroused passionate interest among intellectuals. More than six hundred Latin and vernacular manuscripts of the work, in full or fragmentary form, have been identified, making it far more popular and more widely known than any of the genuine works of Aristotle. It may well have been, as Thorndike characterized it, "the most popular book in the Middle Ages."[35]

The origins of the *Secretum secretorum* are obscure. It is doubtful there was ever a Greek original, although the work does contain elements of Greek philosophy, including a certain amount that derives from genuine Aristotelian doctrine. All known versions go back to an Arabic original composed in the tenth century, of which there are two recensions.[36] The older, an eight-book version known as the Short Form, originated as a "mirror for princes" based upon the supposed letters of Aristotle to Alexander the Great, which the philosopher is said to have sent to the young king, then on his Persian campaign.[37] During the eleventh and twelfth centuries, this version took on a proem and additional layers of scientific and occult material, emerging as the so-called Long Form in ten books. The bulk of this accreted material was derived from the *Rasā'il*, or philosophical epistles, of the Brethren of Purity (*Ikhwān al-Safā'*), a secret religious and political fraternity devoted to the Ismaili cause.[38] Thus the *Secretum secretorum* gradually developed into an encyclopedic reference work. In addition to its original moral and political component, the work contained sections on medicine, health regimen, astrology, physiognomy, alchemy, numerology, and magic. Both versions of the *Secretum* were known in the Latin West, although the Long Form was the more popular of the two. In the mid-twelfth century, John of Spain, working in Toledo, translated the Short Form into Latin. Some 150 manuscripts of this version are known. Philip of Tripoli, a scribe working for Guido de Vere, the bishop of

Tripoli, translated the Long Form in the first half of the thirteenth century. It exists in more than 350 manuscripts and in numerous printed editions.[39]

What accounts for the extraordinary popularity of this work? Certainly the vogue of Aristotle was one factor. The *Secretum* came into the Latin West with the tidal flow of Aristotelian translations. Despite the Church's attempts to restrict the teaching of the Peripatetic philosophy, enthusiasm for the new Aristotle grew feverishly, peaking about the time (ca. 1250) that the long version of the *Secretum* reached scholars at the University of Paris, one of the centers of Aristotelian studies. The work's esotericism and its relation to the genre of secret writings enhanced its appeal. To students already familiar with Aristotle's philosophy, the *Secretum* was like a revelation. In contrast to Aristotle's public doctrine contained in the works on logic, metaphysics, and natural science, it professed to reveal the philosopher's esoteric teachings, which he had reserved for a few intimate disciples.[40] To this topos, which links the *Secretum* to the Hermetic "literary mysteries," is added the esotericism of the Ismaili doctrine of the two sciences, one apparent and the other secret (*intrinseca et extrinseca*).[41] As the Pseudo-Aristotle explains in one of his letters, statecraft depends on two kinds of knowledge, public and secret:

> Harmony and cooperation . . . between the ruler and the ruled, are gained by two means; one of them is evident and apparent, and the other is secret and mysterious. With the former I have already acquainted you. . . . The secret means is one peculiar to the saints and sages whom God has chosen from amongst His creatures and endowed with His own knowledge. And I shall impart to you this secret as well as others in certain chapters of this book, which is outwardly a treasure of wisdom and golden rules, and inwardly the cherished object itself. So when you have studied its contents and understood its secrets you will thereby achieve your highest desires and fulfill your loftiest expectations.[42]

According to the *Secretum*, Aristotle's esoteric doctrine was the knowledge of how to put philosophical understanding (*scientia*) to practical use. Thus it promulgated the view that knowledge of the secrets of nature enabled the knower to accomplish limitless things in the material world. The work's language reinforced the conviction that the *Secretum* really did hold Aristotle's deepest secrets: its elusive, enigmatic terminology made the *Secretum* even more alluring to medieval intellectuals, who were already convinced that the secrets of nature were esoteric, and that works revealing secrets hid them in parables and riddles in order to conceal them from the unworthy. In one of his letters, Pseudo-Aristotle explained:

I am revealing my secrets to you figuratively, speaking with enigmatic examples and signs, because I greatly fear that the present book might fall into the hands of infidels and arrogant powers, whereby they, whom God on high has deemed undeserving and unworthy, might arrive at that ultimate good and divine mystery. I would then surely be a transgressor of divine grace and a violator of the heavenly secret and occult revelation. Because of this, I expose this sacrament to you in the manner in which it was revealed to me, under the seal of divine justice. Know therefore that whoever betrays these secrets and reveals these mysteries to the unworthy shall not be safe from the misfortune that shall soon befall him.[43]

Despite the heavy overtones of Islamic religious doctrines, such claims were eminently believable to medieval intellectuals. The distinction between Aristotle's public teachings and the esoteric doctrine he reserved for his intimate disciples was one that dated from his own lifetime and was universally known in the Middle Ages. The existence of a sizable Latin literature of apocryphal letters from Aristotle to Alexander lent credibility to the work and its spurious attribution.[44] No medieval student could be blamed for receiving the *Secretum* as genuine, however different it may have been from the books he met with in the schools.

It is difficult to convey a sense of the profound impact the *Secretum secretorum* had on the medieval West. One measure of its influence is the deep impression it made upon the Franciscan friar and philosopher Roger Bacon (ca. 1220–ca. 1292). One of Bacon's biographers, Stewart Easton, has suggested that upon reading the work, Bacon "awoke to a new world." The *Secretum* "awakened his dormant sense of wonder" about the mysteries of nature, and caused him to shift his interests from philosophy to what he would call "experimental science" (*scientia experimentalis*). "He had a vision of what was to be the activity of his life, and he found it 'beautiful and good.'"[45] The vision, according to Easton, was of a universal science, a body of knowledge that God revealed to the patriarchs and prophets, whence it was dispersed among the pagans, who, because they lacked the gift of God's saving grace, were unable to comprehend its truths.[46] Bacon made his own edition of the *Secretum*, added to it a long introductory preface, and wrote extensive glosses in which he attempted to clarify the work's more obscure passages. The *Secretum* fascinated Bacon. It seemed to him that the treatise was a key to the "true sciences" of astrology, alchemy, and physiognomy, which work "by art assisting nature" (*per artem juvantem naturae*) and not by magic or old women's charms (*set non carminibus magicis nec vetularum*). As we have already seen, the topos of a book as a "key" to the knowledge contained in other texts went all the way back to the ancient Hermetic sciences. Profoundly influenced by it, Bacon was convinced that whoever read and understood the *Secretum secretorum* would find in it "the greatest natural secrets to which man or human invention can attain in this life."[47]

The *Secretum secretorum* was above all a book of revealed knowledge, of arcana hidden from ordinary people and reserved for the select few. The original discoverers of the secrets of nature, wrote the Pseudo-Aristotle, "observed extreme caution and miserliness in communicating them to others," even while knowing they are of universal benefit. "They did so from the fear that they may come to share this knowledge with those who did not possess sufficient understanding for it, and because God's wisdom has decreed that His gifts should not be equally divided among His creatures. But, thanks to God, you are not one of those who are debarred from knowing these mysteries, but are fully worthy of it."[48] According to the *Secretum*, knowledge is a sacrament that must be guarded against corruption by the *vulgus*. Bacon returns to this theme repeatedly in his scientific writings. In the *Opus Majus*, he wrote, echoing the *Secretum*:

> The wise have always been divided from the multitude, and they have veiled the secrets of wisdom not only from the world at large but also from the rank and file of those devoting themselves to philosophy. . . . Aristotle also says in his book of Secrets that he would break the celestial seal if he made public the secrets of nature. For this reason the wise although giving in their writings the roots of the mysteries of science have not given the branches, flowers, and fruits to the rank and file of philosophers. For they have either omitted these topics from their writings, or have veiled them in figurative language or in other ways, of which I need not speak at present. Hence according to the view of Aristotle in his book of Secrets, and of his master, Socrates, the secrets of the sciences are not written on the skins of goats and sheep so that they may be discovered by the multitude.[49]

Bacon found in the *Secretum* potent secrets that he thought might be used for the advancement of Christendom. The methodology spelled out in the work, he informed his patron, Pope Clement IV, might be used to bring about a reformation of learning and a strengthening of ecclesiastical authority. In the writings he dedicated to Clement, Bacon portrayed himself as the new Aristotle, a sage giving intimate counsel to a powerful prince.[50] He employed the secretive tone and language of the *Secretum*, noting that "the man is crazy who writes a secret unless he does it in a way that conceals it from the crowd, so that it can be understood only by effort of the most studious and wise."[51] He worried that some kinds of scientific knowledge, such as the knowledge of the manufacture of gunpowder, might pose grave dangers to the peace and welfare of Christendom if allowed to become public.[52] Responsible philosophers, he thought, were morally bound to conceal such secrets from the multitude. Confronted with the heretical sections on astrology and magic in the *Secretum*, Bacon was confident that they contained something of God's original revelation. He strove to bring the work under the cover of orthodoxy, arguing that faulty translation had distorted its true meaning. Astrology does not undermine the doctrine of free will, he explained, because true astrologers do not presume to foretell future events with certainty. Instead they foretell possibilities, which are contingent upon God's will and sufficient causes. Astrology is an essential science for princes because it enables them to prepare for events that are likely to occur. Although these events might be inevitable, foresight might mitigate their impact.[53] Similarly Bacon distinguished between legitimate experimental science and magic. Denouncing the fraudulent claims of the magi and the hoaxes of jugglers and cunning women, Bacon insisted that "art using nature for an instrument" (*ars utens natura pro instrumento*) is legitimate and more powerful than magic.[54]

It has been suggested that one reason for the popularity of the *Secretum secretorum* was that it was " pocket edition of all the subjects that most interested the Middle Ages." Another scholar has characterized the work as "the great middle-brow classic for the layman."[55] The *Secretum* does contain material that would have interested a broad segment of medieval readers, including lengthy, if entirely conventional, discussions of the different branches of science and medicine, capsule treatments of health regimen, astrology, physiognomy, and the calendar. In addition, its advice on the qualities of justice and its shrewd observations on politics, written in a homiletic yet courtly style, held an obvious fascination for medieval readers. Philip of Tripoli's translation, if not always accurate, was clear and readable, and of the right flavor to make it semi-popular reading. But for all its attractiveness as "light reading," the *Secretum* carried a serious message to the educated classes of medieval Europe. The message was that knowledge is power. "Understanding is the head of government," said the Pseudo-Aristotle in a maxim that summarized the kernel of his doctrine. "It is the health of the soul, the preserver of virtue, and the mirror of vice. By it are hateful things cast out and worthy things chosen. It is the fountainhead of virtue and the root of all good, praiseworthy, and honorable things."[56] Describing the *Secretum* as a mirror for princes misses the work's true significance. It might be better characterized as a handbook for medieval courtiers. The treatise urged rulers to appoint ministers with special intellectual qualities: the perfect councillor should have "a good understanding, and a quick apprehension of what is said to him"; he should have a good memory and be able to speak eloquently; he should be "intelligent and quick-witted." On top of this, he should be "skilled in all sciences, especially in arithmetic, which is the most certain demonstrative art by which nature is comprehended."[57] These were intellectual, not military qualities, and they were gained through university study. The *Secretum*'s description of the ideal councillor was tailor-made to match the skills of the liberal arts graduates and would-be philosopher-advisers of the Middle Ages. No one could have expressed their ideology more authoritatively than (as they believed) Aristotle. The *Secretum*'s pragmatic, almost Machiavellian political advice and its requirement that political advisers know philosophy was a message to which ambitious university graduates could eagerly respond.

Equally important was the *Secretum*'s constant refrain that there is both an exoteric and apparent means by which the prince gains his subjects' confidence, and one that is esoteric and secret. The distinction between public knowledge and esoteric council was, in reality, the secret of power and advancement for late-medieval intellectuals. As ecclesiastical and secular administration grew increasingly centralized, developing more complex bureaucratic structures, the need for trained manpower grew accordingly. But as the demand for a skilled work force increased, so too did the number of university graduates, thus intensifying the competition for positions. The *Secretum* entered the Latin West in the wake of political and economic changes that created a magnetic field for scholars, who, though generally of nonnoble status, were taking their places in a power structure that was traditionally reserved for the nobility. We know medieval intellectuals read the *Secretum* for the pertinence of its teachings because they updated the work to make it fit current political circumstances. Thus in one version Aristotle becomes "Philip of Paris," and in certain versions of French and English provenance the destruction of a kingdom through overspending has been altered from "destructio regni Chaldaeorum" to "destructio regni Anglorum." Christendom's intellectuals were invading high politics, in the Church as well as in secular governments. Knowing the secrets of effective government was for them an essential tool for success.[58]

The influence of the *Secretum secretorum* upon scholastic philosophy and science is somewhat more difficult to ascertain. The work's imprint upon medieval Latin Aristotelianism is visible, although its mark can be detected more easily in attitudes and methodologies than in specific doctrines. There is no question, for example, that the *Secretum* influenced Bacon's formulation of an "activist" or utilitarian interpretation of Peripatetic philosophy. Bacon's idea of experimental science (*scientia experimentalis*) bears the unmistakable mark of the Pseudo-Aristotle, whom the friar cites repeatedly and reverentially when discussing the subject.[59] According to Bacon experimental science had three "prerogatives" that placed it above the conventional scholastic sciences. First, it completes the speculative sciences by using experience to certify the conclusions arrived at by deductive reasoning. Second, experimental science takes up where the existing sciences leave off, adding new knowledge that cannot be found out by deductive reasoning alone. Finally, and most important, experimental science enables investigators to discover secrets of nature that are completely outside the existing sciences. Only by experiment, for example, do we discover the marvelous virtues of plants and animals, or ingenious inventions such as ever-burning lamps, machines for hoisting great loads, war engines, and fireworks.[60]

It was especially when discussing the second and third "prerogatives" of experimental science that Bacon invoked the authority of the *Secretum secretorum*. The "Aristotelianism" of the *Secretum* enabled him to make far more sweeping claims for "experimentation" than the authentic writings of Aristotle would ever have allowed him to make. Above all, the *Secretum* shaped Bacon's lofty vision of the ethical and political dimensions of science. Experimental science, he thought, is superior to the purely speculative sciences by virtue of the power and utility it lends them. Bacon made much of this utilitarian doctrine, urging Pope Clement IV to support research aimed at producing weapons and devices for the defense of Christendom against Antichrist. The Pseudo-Aristotle, who advised Alexander on how to use science to win battles and how to rule conquered countries by changing the air to make the peoples' complexions more pliable, inspired Bacon's conception of the utility of *scientia experimentalis*. As he reminded the pope, "by the paths of knowledge Aristotle was able to hand over the world to Alexander."[61] Bacon's notion of experimental science was something close to what would become known in the Renaissance as "natural magic": the manipulation of nature by the application of art, the medieval conception of what today we would call scientific technology. The success of *scientia experimentalis* would depend upon a knowledge of the speculative sciences. But its true value, Bacon maintained, lay in its utility: "For this science teaches how wonderful instruments may be made, and uses them when made, and also considers all secret things owing to the advantages they may possess for the state and for individuals; and it directs other sciences as its handmaids, and therefore the whole power of speculative science is attributed especially to this science."[62] Bacon speculated wildly about some of these "wonderful instruments" in a letter on "the secret works of art and nature," where he imagined the possibility of constructing airplanes, submarines, portable bridges, mighty cranes, and powerful optical devices. Although his famous list of technological wonders may seem quaint today, Bacon grasped a fundamental principle, indeed the secret, of modern scientific technology: if you know how nature works, there is no limit to what you can do with it. "Even if nature is powerful and marvelous," he wrote, "art using nature as an instrument is more powerful."[63]

In addition to confirming his view of the utility of *scientia experimentalis*, the *Secretum secretorum* was instrumental in shaping Bacon's view of the relation between reason and revelation.[64] Bacon believed that all knowledge, including science, was originally revealed by God to his prophets. Philosophy "is merely the unfolding of divine wisdom by learning and art."[65] However, because of humanity's sinful nature, the pristine truth of divine revelation became corrupted by unbelievers and distorted by faulty translation as it passed, successively, from the Hebrews to the Chaldeans, the Greeks, the Arabs, and finally to the Latin Christians. Bacon thought the chief purpose of experimental science was to confirm revelation, not in the sense of testing it, but in the sense of uncovering its true meaning, its *batīn*, as the Arabs would say. As evidence for his theory of the origin of knowledge, Bacon noted that God gave all creatures the knowledge of how to preserve the body against the effects of old age. Although humans lost this knowledge with Adam's fall, the animals, being uncorrupted by sin, retained it by instinct. Hence experimenters, guided by the revealed Word of God and carefully observing the habits of animals, can recover these lost secrets for humankind.[66] Bacon even thought it would be possible experimentally to approximate the elemental qualities of the fruit borne by the Tree of Life.[67] The *Secretum* confirmed his theory of the origin of knowledge. He accepted without reservation the work's claim to hold secrets that God had revealed to his prophets, and through them had transmitted to the ancient philosophers.[68] In the *Opus Majus*, he elaborated on the special place the *Secretum* occupied in the history of revelation:

> And especially was this wisdom granted to the world through the first men, namely, through Adam and his sons, who received from God himself special knowledge on this subject, in order that they might prolong their life. We can learn the same through Aristotle in the book of Secrets, where he says that God most high and glorious had prepared a means and a remedy for tempering the humors and preserving health, and for acquiring many things with which to mitigate such evils; and has revealed these things to his saints and prophets and to certain others, as the patriarch, whom he chose and enlightened with the spirit of divine wisdom, etc. . . . But these matters and the most hidden secrets of this kind have always been hidden from the rank and file of philosophers, and particularly so after men began to abuse science, turning to evil what God granted in full measure for the safety and advantage of men.[69]

Plainly, the *Secretum secretorum* was for Bacon a key text. A book of revealed wisdom, it stood midway between the Hebrew prophets and the Latin Christians, and supplied an essential link in the unbroken descent of God's revelation to his people. Because it contained the esoteric teachings of the ancient world's wisest philosopher, it confirmed Bacon's view of the essential unity of philosophy and revelation. It presented a compelling image of the philosopher as a man of practical affairs. Most important, it led Bacon to think of experimental science as a powerful means for advancing the cause of Christendom.

In a famous passage in the *Opus Tertium*, Bacon wrote that over a period of twenty years of research he had spent more than two thousand pounds for secret books, experiments, instruments, and other materials relating to his scientific work.[70] We do not know what "secret books," besides the *Secretum secretorum*, Bacon may have owned. Given his often-stated fascination with contemporary "experimenters," most assuredly he was familiar with the Latin compilations of *secreta*, recipes, and *experimenta* that by his time were widely available to the academic community of Western Europe. William of Auvergne, who was bishop of Paris when Bacon was a professor at the university, spoke repeatedly of the "experimenters" (*experimentatores*). He left the impression that their books (*libri experimentorum*) were generally accessible to university students. The massive amount of evidence about these works assembled by Lynn Thorndike leaves little doubt that they made up a significant portion of the scientific literature of the Middle Ages. It is now time to examine the relationship between the *libri secretorum* and scholastic science.

Secrets and *Scientia*

Despite its nearly ubiquitous presence in the West after the twelfth century, the literature of secrets did not find a place among the official sciences of the universities. To understand why this was so, we must consider what constituted "scientific knowledge" in the scholastic tradition. Aristotle, the dominant medieval authority on scientific methodology, had established fairly rigorous criteria for what kind of knowledge qualifies as true science.[71] *Scientia*, or "unqualified scientific knowledge," he defined as "a state of capacity to demonstrate" (*Ethics* 1139b31). It is not merely empirical knowledge of a fact (*demonstratio quia*) but a demonstration of the reason why (*demonstratio propter quid*). Aristotle defined demonstrative knowledge as knowing "the cause on which the fact depends, as the cause of that fact and no other, and, further, that the fact could not be other than it is" (*Posterior Analytics* 71b10ff.). Furthermore, Aristotle maintained a firm distinction between what can be the object of unqualified scientific knowledge and what cannot: "Scientific knowledge is judgement about things that are universal and necessary, and the conclusions of demonstration, and all scientific knowledge, follow from first principles" (*Ethics* 1140b31ff.). Thus for Aristotle *scientia* meant demonstrable knowledge of the universal and necessary causes of normal, quotidian natural phenomena. It required *propter quid* or demonstrative explanations, ordinarily in terms of the manifest qualities of the four terrestrial elements, earth, air, fire, and water.

Secreta, on the other hand, referred to phenomena of an altogether different sort. They included, in the first place, manifestations of occult qualities, or events that occur unexpectedly or idiosyncratically as a result of insensible causes. Magnetic power, for example, was an occult quality. Unlike the manifest qualities of taste and color, which can be immediately apprehended by the senses, the attractive virtue of the magnet is hidden from the senses. The effect of magnetism upon iron is evident, but its cause is occult.[72] *Secreta* also referred to certain events that took place as a result of artificial instead of natural causes, "tricks of the trades," as it were. Technical operations do not produce scientific knowledge because "art is concerned neither with things that are, or come into being, by necessity, nor with things that do so in accordance with nature" (*Ethics* 1140a13ff.). *Secreta* could be experienced, but because they were not demonstrable, they could not be the objects of scientific knowledge. The power of the lodestone to attract iron can be experienced, but it cannot be deduced from the nature of the element earth; the craftsman's ability to harden iron to various degrees by submerging it in quenching baths composed of certain natural ingredients, such as the juice of radishes, is something that can be experienced, but the technique is the craftsman's "secret": the hardening cannot be deduced from the known qualities of the radish plant.[73] From the standpoint of philosophy "secrets" were idiosyncratic, in that they were peculiar to a relatively narrow range of phenomena, or they could be effected only by some special insight, skill, or cunning, whether it be of the artisan or of the magus. For these reasons secrets could not, properly speaking, be the objects of scientific knowledge. Indeed, knowledge of secrets was *stricto sensu* impossible: they could be experienced, and could be found out "experimentally," but they could not be understood or explained according to the canons of logic and natural philosophy. Under the ground rules of scholastic discourse, artisans and experimenters could at best produce demonstrations *quia*, but they could not produce demonstrations *propter quid*. Hence the "secrets" connected with these arts lay outside the boundaries of official science.

Nevertheless, the existence of a sizable literature on *secreta* in the form of craft, medicinal, and magical recipes tended to raise uncomfort-

KNOWLEDGE AND POWER 55

able questions about the limits of the knowable and the scope of *scientia*. Medieval Scholastics could not help noticing, for example, that many secrets attributed to known and respected scientific authorities bore a striking resemblance to ordinary folk beliefs. Normally authority itself would provide the stamp of approval, distinguishing truths based upon accepted texts from false opinions circulating among the people through oral tradition.[74] The medieval scholar was prepared to accept as fact a wealth of data found in written texts that to us appears to be nothing but foolish superstition. At the same time, he could just as readily dismiss, and relentlessly attack, "superstitions" identified with rustics, farmers, and craftsmen.[75] The criteria by which the Scholastics distinguished between fact and superstition were quite specific and decidedly unempirical: "facts" were the data reported or confirmed by the *litterati* (those who wrote in Latin for a scholarly audience), while "superstitions" were the data that circulated among the *illiterati* (identified with oral and vernacular traditions).[76] Under this criterion the credibility of a given "fact" was often a question of whether or not the datum showed up in an authoritative text.[77] The goal of scholastic science in general was not to uncover new data by empirical research, but to assign causes to data already accepted as factual: that is, to make "facts" intelligible by explaining them in terms of causes. Thus Albertus Magnus (1206–1279) in *De mineralibus*, after presenting a fairly routine "empirical" account of the varying hardnesses of stones, concluded, "It is [the task] of natural science to assign causes for these accidental properties, based on the material and efficient causes, in the manner described elsewhere."[78]

Successful as this attempt may have been in some instances, physicians and experimenters frequently turned up events that could not be confirmed by reason or certified by authority. Often experience alone had to be one's guide, as suggested by the familar phrases *probatum est* and *expertus est* that often accompany the recipes. These conventional expressions should not, of course, be taken to connote deliberate or controlled experimentation in the modern sense. Medieval Scholastics used the words *experimentum* and *experientia* interchangeably to refer to knowledge acquired or confirmed by direct observation, as opposed to truths demonstrated by rational argument.[79] To say that something had been proven *per experimentum* (or *per experienciam*) meant only that it had been witnessed; the locution had nothing to do with a deliberate methodology designed to test hypotheses. Nevertheless, although "experiment" and "experience" were nearly synonymous (both referring to general sense experience), it is possible to identify at least two additional meanings of an "experiment" in scholastic science: an empirical test and a contingent event. As empirical tests, experiments included anything from a simple empirical verification of a proposition to a trying-out of a medical prescription. Bernard de Gordon (fl. ca. 1283–1308), a professor of medicine at Montpellier, maintained that "any consideration, no matter how reasonable, has no value unless it is proven by experiment (*nisi comprobetur experimento*)," and insisted that all medical prescriptions, "unless tested by experiment (*experimento probetur*), are speeches of rhetoricians who offer fancy propositions that have nothing to do with science."[80] Despite such modern-sounding claims, Bernard was inconsistent in his attitude toward experimentation, in general preferring rational prescriptions to experimental remedies, which he associated with untrained empirical healers.[81]

More frequently, the term "experiment" referred to events that were indeterminate or purely contingent, and hence could be known only by experiencing. As Michael McVaugh observes in relation to a fourteenth-century collection of medical *experimenta*, the word "referred not to an event planned to illustrate the rational order of nature, but to an event lying outside that rational order."[82] Sharply distinguished from *scientia*, or demonstrable causal knowledge, empirical or "experimental" data were consistently associated in academic discourse with popular practices and the occult tradition. "Empirical" remedies (*empirica*), for example, generally referred to remedies used by popular healers, whose experience was unregulated by reason. Although such remedies were generally regarded as untrustworthy, the Scholastics acknowledged that rustics, like the brute animals, had a certain "natural instinct" (*ex quodam instinctu nature*) that enabled them to acquire empirical remedies to heal themselves.[83] Finally, "experiments" included a large and heterogeneous body of magical recipes and methods for manipulating occult forces and qualities. Ever problematic in scholastic science, magical *experimenta* underscored the limitations of *scientia* understood as purely demonstrative knowledge. To summarize, the relation between reason and experiment in the scholastic tradition was a matter of two different roads to truth, and the way of reason (*via rationis*) was thought to be manifestly superior to the way of experiment (*via experimentalis*).

The fact that compilations of *secreta* and *experimenta* make up such an immense body of literature, distinct from the conventional scientific corpus, illustrates two important characteristics of late-medieval science. On the one hand, there was an ongoing "experimental" tradition in the Middle Ages. Although the Scholastics may have been men of the book, they were not unmindful of the need to bring theory in line with experience. Yet, on the other hand, medieval experimentalism remained by and large distinct from the speculative, theoretical sciences taught in the universities. Roger Bacon described *scientia experimentalis* as a separate discipline whose purpose was to confirm and extend knowledge gained in the speculative sciences. However, Bacon's positive evaluation of experimental science was untypical. Most Scholastics agreed that the truths confirmed by reason were more certain than those discovered by experience. In a revealing passage, Bernard de Gordon implied that "experiments" represent immature, imperfect science: after summarizing the appropriate dietary treatment for tertian fever, he indulged a portion of his readership: "Because the young greatly enjoy *experimenta*, let me give some here." Immediately Bernard added that "safer than all these, however, is to proceed by the method described above."[84] This attitude prevailed in academic circles until at least the sixteenth century. Writing in 1529, Joachim Fortius Ringelbergius described the *experimenta* he collected from common people between trips to various universities as "very agreeable playthings" (*nugas iucundissime*), which scholars might amuse themselves with in their spare time. Ringelbergius, a public lecturer who gave cram courses on natural philosophy, noted that the experiments, while perhaps not entirely useless, were not to be taken very seriously by university students.[85]

The radical distinction between the *via rationis* and the *via experimentalis* in medieval science was a manifestation of the isolation of theory from practice in Scholasticism generally. This characteristic of scholastic science is perhaps most clearly visible in medicine. According to Nancy Siraisi, academic medicine "was a rigorous mental discipline requiring of its devotees a high degree of skill in handling abstract concepts and theoretical systems; simultaneously, it was a fairly simple technology or craft."[86] The university medical professors developed sophisticated theories to explain the origin of diseases and the action of drugs, but their theories often had little influence upon medical practice. Even as they debated over the rationale for the action of compound drugs, the academic physicians recorded the recipes they actually prescribed to patients in books of *experimenta*. These "empirical" remedies bore little resemblance to the treatments recommended for similar conditions in the more formal medical treatises. The *experimenta* recorded by Arnald of Villanova (ca. 1240–1311), for example, were pharmacologically conservative and scientifically less sophisticated than the therapeutic measures he introduced into the treatises he addressed to an academic audience. Arnald, a medical professor at Montpellier, constructed an elaborate scientific system based upon humoral doctrine, which theoretically enabled him to determine "rational" drugs for diseases. In practice, he prescribed medicaments that were simple, practical, and empirical. So completely absent was Arnald's theoretical medicine from his medical practice that, as Michael McVaugh concluded, Arnald "actually practiced two medicines: one learned and formal, carefully prepared and polished for circulation among his professional colleagues or for presentation to royalty, and one more empirical or, in Arnald's own sense, 'experimental,' practiced as a matter of daily routine when scholarly learning was not required."[87] Arnald theorized about one medical system and practiced another. His *experimenta*, it might be added, were of the type Scholastics commonly referred to as "secrets," in the sense that they lay outside the discourse of academic science. They were, so to speak, "private experiments," or records of unique, isolated, and sometimes unexpected occurrences, not experiments consciously designed to test general propositions. It was readily conceded that such recipes and experiments worked on grounds that could not be demonstrated by reason, especially in cases of those calling for somewhat bizarre ingredients, or for substances whose virtues were considered to be "marvelous."[88] Arnald made no attempt whatsoever to explain the action of his *experimenta*. Lacking *demonstrations*, he recorded *instances* of successful treatments: "I cured the lord Cardinal Jacob of this infirmity," "I tried this on a certain Cardinal of Ostia," and "I tried this one many times on myself." Similarly, Scholastics sometimes had difficulty giving rational accounts for techniques they observed or heard about in the manual arts. Thus Albertus Magnus, a keen observer and one of the thirteenth century's leading scientific authorities, was at a loss to explain the action of certain special quenching waters used by smiths, such as the juice of radishes mixed with crushed earthworms.[89] Such "secrets" lay outside the rational ordering of nature; they were purely contingent, and could not be predicted or explained by theoretical science.

Secrets, Magic, and the Polemic against Curiosity

Outlawed from the dominion of science, *secreta* fell more naturally into the domain of magic. Magical recipes and experiments show up in scores of medieval manuscripts, and are especially abundant from the fourteenth and fifteenth centuries.[90] Frequently they appear alongside technical and medicinal recipes, because like the latter, magical *experimenta* expressed attempts to bring about concrete changes in the physical world. Certainly many of these "experiments" were completely fanciful. However, what concerns us here is not the efficacy of magic but scholastic culture's evaluation of it. In the twelfth century, following the introduction of the Islamic "secret sciences" into the Latin West, scholars began hesitatingly to incorporate learned magic into the formal schemes they developed for classifying knowledge. When they attempted to find a place for the magical arts within these classificatory schemes, the Scholastics usually grouped them with the mechanical arts rather than with science, considering them to be part of *techne* rather than *scientia*. Like technology, magic was "artificial" in the sense that it involved the manipulation of occult qualities in order to produce "marvels."[91] Writing about 1150, Domingo Gundisalvo listed "necromancy according to physics" as one of the eight subdivisions of the mechanical arts, together with alchemy, medicine, agriculture, navigation, optics, and the science of images and judgments. Although he did not condone magic, neither did he condemn it; he merely classed it, along with *honores seculares*, as one of the worldly vanities.[92] In another scheme, apparently derived from William of Conches's *Philosophia mundi*, magic is placed parallel to the mechanical arts as one of the studies that overcomes the human evil of physical infirmity.[93]

All such attempts to domesticate magic met with firm resistance. To the growing number of scholars whose curiosity was aroused by the newly discovered treatises on learned magic, the religious and academic establishment issued a stern warning. Hugh of St. Victor, writing in the 1120s, categorically denounced all forms of magic, cautioning that "magic is not accepted as part of philosophy, but stands with a false claim outside it: the mistress of every form of iniquity and malice, lying about the truth and truly infecting men's minds, it seduces them from divine religion, prompts them to the cult of demons, fosters corruption of morals, and impels the minds of its devotees to every wicked and criminal indulgence."[94] Hugh's denunciation of magic, like virtually all official medieval pronouncements on the subject, was essentially a restatement of the position of St. Augustine (354–430), for whom magic (theurgy) was a form of idolatry, since it involved cooperating with demons.[95] However, underlying the medieval hostility toward magic lurked a deep suspicion of intellectual curiosity in general. In contrast to legitimate intellectual inquiry, magic was considered to be a form of aimless erudition, or *curiositas*, the "passion for knowing unnecessary things" (*libido sciendi non necessaria*).[96] In the Middle Ages the word "curiosity" (*curiositas*) had a far more pejorative meaning than it has today. To be "curious" about something was neither innocent nor virtuous. Instead, it implied being a meddlesome intellectual busybody who pries into things that are none of his business. Nor, according to patristic opinion, could scientific curiosity be considered fully legitimate, for God *intended* nature to be a mystery and had so fashioned the world as to make many of its secrets occult and unintelligible. In order to protect the secrets of nature from man's prying eyes, Lactantius (ca. 250–ca. 325) pointed out, God made Adam the last of his creations so that he should not acquire any knowledge of the process of creation.[97] In confirmation of this, the popular image of the goddess Natura implied that nature covers herself with a veil in order to hide her secrets from mortals.[98] To pry into mysteries of nature that God chose not to reveal, as the brazen magi attempted to do, was to trespass the boundary of legitimate intellectual inquiry, to challenge God's majesty, and to enter into the territory of forbidden knowledge.

Since patristic times theologians had condemned curiosity as a vice. Numerous scriptural references supported this interpretation, including Wisdom's ominous warning, "What the Lord keeps secret is no concern of yours; do not busy yourself with matters that are beyond you."[99] Most of the medieval pronouncements against curiosity concerned searching out the mysteries of God's providence, as when Adam and Eve sought to obtain knowledge of good and evil. Exhortations to practical piety in contrast to speculation about theological mysteries, they were not necessarily intended as condemnations of scientific curiosity. But Augustine, Christendom's chief authority on *curiositas*, made inquisitiveness in general the subject of a particularly vicious polemic, thereby setting the tone for the medieval debate over intellectual curiosity. In the *Confessions* Augustine included *curiositas* in his catalog of vices, identifying it as one of the three forms of concupiscence that are the beginning of all sin (lust of the flesh, lust of the eyes, and ambition of the world).[100] Curiosity, he wrote, is a kind of "lust of the eyes" (*concupiscentia oculorum*) because the eyes are our chief source of knowledge, and because "seeing" is used by analogy to describe finding out any kind of knowledge. The "empty longing and curiosity for acquiring experience through the flesh," and the "lust to find out and know" things not for any practical purpose but only for the sake of knowing (experiencing) them: these are the marks of the overly curious mind. It is because of the "disease of curiosity" that people go to watch monsters and freaks in theaters and circuses. Yet Augustine saw no essential difference between such perverse entertainments and the "empty longing and curiosity [that is] dignified by the names of learning and science" or, for that matter, magic. All exhibit the same lust to know merely for the sake of knowing. In this respect, according to Augustine, there is no difference between gawking at a mutilated corpse or going to a freak show and making investigations in natural philosophy and magic: "From the same motive men proceed to investigate the workings of nature, which is beyond our ken—things which it does no good to know and which men only want to know for the sake of knowing. So too, and with this same end of perverted curiosity for knowledge, people make inquiries by means of magic."[101] Inevitably, thought Augustine, curiosity—learning for learning's sake, not aimed at salvation—degenerates into pride, one of the seven deadly sins and the cause of Adam's fall. Augustine's severe judgment of intellectual curiosity, linking it with the sin of pride and the

Fall, and ascribing to it the origins of heresy and the black arts, became conventional in medieval thought.[102]

Although *curiositas* referred to any form of intellectual inquiry carried to excess, for several reasons magic was the medieval world's paradigmatic example of forbidden knowledge. In the first place, the boundary between "natural" and demonic magic was ambiguous, and hence magic of any kind might tempt practitioners into making pacts with demons in order to learn the secrets of creation.[103] So in the Renaissance, Faust, out of vain curiosity, would sell his soul to Satan in order to know the secrets of nature. However, complicity with the Evil One did not merely imperil the individual soul; it affronted the foundations of the cosmic order itself. To the medieval and Renaissance mind, which was predisposed to seeing things in terms of binary opposition, nothing could be more perverse than to seek to know "high things" by means of the low.[104] But that, according to medieval writers, is precisely what the practitioners of magic attempted to do. Augustine, whose view of the black art was authoritative in the Middle Ages, stressed the perversity of theurgic magic, the pagan attempt to use demons to "circuitously win the favor of deities."[105] What sort of beings are these, he asked, whom the magi claim can mediate between man and God? Although demons have both a soul and a body, and thus have elements of the divine and the human, perversely it is their bodies that are eternal, while their souls are human: they are suspended "as it were upside down, so that their lower part, the body, unites them with beings above them, and their higher part, the soul, binds them with men below. . . . They are exalted by their lower part and humbled by their higher."[106] Demonic magic involved a reversal of the natural order, an inversion of the divinely ordained relationship between high and low. Any attempt to attain knowledge of divine things through demons, or to win God's favor through demons, was a delusion and a threat to cosmic order.

Of all intellectual activities, magic more than any other exhibited curiosity's gravest danger, pride. Not only did the magus pry into nature's hidden recesses and steal its secrets, he used his illicitly won knowledge to glorify himself and to impress the world with his "marvels." There was a persistent tendency in the Middle Ages to associate intellectual curiosity with pride. Carlo Ginzburg pointed out that St. Paul's condemnation of moral pride in Rom. 11:20, rendered in the Vulgate as "noli altum sapere," was quoted century after century as the standard biblical authority against intellectual curiosity. As a result of what Ginzburg calls a "collective slip" or misunderstanding of the Vulgate passage, Paul's "be not high-minded" was consistently rendered as "do not seek to know high things."[107] It is not difficult to understand how the "slip" occurred, for the link between curiosity and pride was firmly established in the patristic tradition. Augustine's discourse on *curiositas* in the *Confessions*, book 5, was specifically a commentary on the sin of pride. "The proud cannot find you [Lord]," wrote Augustine, "however deep and curious their knowledge, not even if they could count the stars and the grains of sand, or measure the constellations in the sky and track down the paths of the stars."[108] This interpretation of *curiositas* occurs repeatedly in the medieval Christian tradition. Bernard of Clairvaux (1090–1153) made curiosity "the first step of pride [and] the beginning of all sin." It was pride that led Adam to seek "forbidden knowledge by forbidden means."[109] Pride, according to medieval accounts, caused Gerbert of Aurillac to leave his monastery and journey to Spain in order to study astrology and magic under Saracen teachers—at the price of his soul. Gerbert, whose insatiable thirst for knowledge was legendary, was but the most famous medieval example of the overly curious cleric who crossed the boundary of legitimate intellectual inquiry to dabble in the forbidden art. Similar stories implicated Roger Bacon, Albertus Magnus, Robert Grosseteste, and Michael Scot. Indeed, any medieval scholar who had a reputation for his knowledge of natural science was a potential antihero in this rich legendary tradition.[110] Like the Renaissance Faust, they sought to know secrets that are "beyond us" (*quae praeter nos*) or, as in the case of astronomers, things "above us" (*quae supra nos*), matters that have nothing to do with our destiny and that God has chosen not to reveal to us.

Finally, Augustine condemned magic as vain *curiositas* because it was a purposeless "trying out" or "tempting" (*tentatio*) of nature. Magic was a perversion of legitimate science in the sense that it "tempts" nature and causes nature to do what it would not, under ordinary circumstances, do on its own. Augustine compared this "tempting" of nature to the "appetite for experience" in religion on the part of those who "experiment" with God in seeking miracles and portents: "Even in religion itself," he wrote, "this prompting drives us to make trial of God when signs and wonders are demanded, not for any saving end, but simply for the purpose of seeing them."[111]

Augustine's polemic against intellectual curiosity gained new relevance during the scholastic period, when reason reared its prideful head to challenge faith. Although theologians, following Augustine, could readily dismiss ancient magic as a pagan superstition, the learned magic that developed after the twelfth century proved to be a more formidable adversary. It was supported by impressive philosophical authority and was grounded upon metaphysical principles that the Schoolmen accepted with few reservations. Nevertheless magic, even when ostensibly it did not involve demon worship, appeared dangerous because it glorified human power and raised it too near to the divine. Philosophy's attempt to "naturalize" marvels was tantamount to claiming that miraculous effects could be produced without any supernatural agency. Ominously, astrology and magic threatened to take the power of prophecy and miracle making away from God and to place them into human hands.

Many theologians believed that magic's Promethean effort to wrest divine power from God was the inevitable result of the pervasive and growing influence of secular philosophy. God's sovereignty over the creation was at stake. Already in the eleventh century Peter Damian sounded the warning: "Conclusions drawn from the arguments of dialecticians and rhetoricians should not be thoughtlessly addressed to the mystery of divine power; dialecticians and rhetoricians should refrain from persistently applying to Sacred Scripture the rules . . . of the syllogism . . . , and from setting their inevitable conclusions against the power of God."[112] The recovery of ancient philosophy stirred up waves of rationalism and skepticism. Bold new doctrines proclaimed human reason's competence to understand the mysteries of the creation without the assistance of revelation.[113] Equally threatening was Scholasticism's attempt to restrict the scope of the miraculous. Whereas for Augustine all events proceed from the will of God and hence are miraculous, the Scholastics made a clear distinction between nature and the miraculous, arguing that miracles proceed directly from God and occur only rarely, while everyday natural events have purely physical causes.[114] Indeed, the claim of scholastic philosophy—that the universe is rational and that events can be predicted (e.g., by the science of astrology)—implicitly limited God's absolute power over nature. It implied that God was circumscribed by the rules of logic, that even God cannot do the logically impossible.[115] Bernard of Clairvaux lashed out against this dangerous principle and the philosopher whom he considered to be its most intemperate perpetrator, Peter Abelard (ca. 1079–ca. 1142). Denouncing the popular Paris teacher as a "scrutinizer of majesty and fabricator of heresies," Bernard warned Pope Innocent II that "Peter Abelard is trying to make void the merit of Christian faith, when he deems himself able by human reason to comprehend God altogether."[116]

Between Saint Bernard the mystic and Peter Abelard the rationalist there could be no common ground. The scholar's most grievous sin, thought the preacher, the sickness of the age, was pride of intellect: "The man is great in his own eyes." Its terms thus set, the debate over *curiositas* climaxed in the thirteenth century, when the Scholastics, armed with a virtually complete Aristotelian corpus, countered the theologians' dour Augustinianism with arguments based on the Philosopher's *Ethics*. Albertus Magnus, contrasting curiosity to prudence, argued that "curiosity is the investigation of matters which have nothing to do with the thing being investigated or which have no significance for us; prudence, on the other hand, relates only to those investigations that pertain to the thing or to us."[117] In other words, it is not the pursuit of knowledge itself that constitutes curiosity, but the pursuit of irrelevant knowledge. Similarly, Albert's pupil Thomas Aquinas (1224–1274) distinguished between *studiositas*, or devotion to learning, and *curiositas*, an "appetite or hankering to find out." *Studiositas*, he argued, is not an appetite but a virtue whose purpose is to moderate the natural human desire to know, whereas *curiositas*, an appetite, is by its very nature immoderate. For Thomas, as for Albert, the pursuit of intellectual knowledge was fundamentally unlike curiosity. Knowledge of the truth is inherently good (*cognitio veritatis bona est*), although it may incidentally lead to evil because of its possible consequences, including pride. *Curiositas*, Thomas maintained, has to do only with the "hankering" to know; it has nothing to do with devotion to intellectual knowledge. "That which makes man similar to God and which he receives from God cannot be wrong," he wrote. "Abundance of knowledge is from God. . . . However much it abounds, knowledge of the truth is not bad, but good. The desire for a good is not wicked. Therefore no wrongful curiosity can attend intellectual knowledge."[118]

Against such dialectical hairsplitting, against all the forms, essences, and quiddities propounded by the Scholastics to define God's nature, late-medieval theologians proclaimed the absolute contingency of the world. God's essence is his freedom. Since God can will anything, argued the proponents of the fourteenth century's "New Way" (*via moderna*), nothing, except God, exists by necessity; all else exists solely because God wills it to be. Something is good because God wills it; he does not will it because it is good. This outlook, pronounced in late-medieval theology and characteristic of nominalism, makes God *deus absconditus*, a hidden God, and nature intrinsically occult. For, because God wills everything into being, knowledge of all other existence than God (which is unknowable) cannot be necessary and demonstrative knowledge. Insisting on the inherent contingency of all creation, the theologians of the *via moderna* sought to release nature from the bondage of pagan philosophy. In doing so, they drastically restricted the scope of what reason could know. The redrawing of the boundary between what can be known and what must be believed implied that any effort by the intellect to penetrate God's existence, or even that of nature, constituted *vana curiositas*.[119] John Gerson (1363–1429), chancellor of the University of Paris, summarized this late-medieval perspective in a sermon preached in 1402: *Against Vain Curiosity in Matters of Faith*.[120] The root of all philosophical error, he remonstrated, is pride, which leads to the desire to attain demonstrations of matters that arise solely from God's will—which is to say, all matters. At the center of Gerson's critique stood inquiry concerning the origin and end of the world: "How and when the world began, or if it is about to come to an end, cannot be known from any experiences whatsoever that philosophy leads to, because that [knowledge] is located in the freest will of the creator. Why then do the philosophers fail when they try to penetrate this secret of the divine will? Because just as the divine will is its own reason, so by that will alone is granted the knowledge of what it might wish to reveal of itself."[121] Those philosophers err grievously who claim that God acts out of natural necessity or through some divine "property" whose nature can be deduced by logical reasoning. They "drink too freely from the golden cup of Babylon," pagan philosophy. God is not a nature but a free will, Gerson asserted. As a result we cannot know anything about him or his creation except what he chooses to reveal to us through his Word. Natural knowledge has its limits; we must not seek to go beyond them.

The medieval polemic *contra vanam curiositatem* was premised on the assertion that secrecy was an inherent aspect of the creation, an idea that followed from, and in turn reinforced, the doctrine of the omnipotence of the divine will. Classically formulated by Augustine and reaffirmed in the late Middle Ages by the nominalists, the doctrine of the omnipotent divine will implied that God, as the author of the universe, had the sole and sovereign right to the secret of his creation. God intended his creation to be a mystery, its secrets impenetrable. These ideas had two important consequences for medieval philosophy and science. First, they tended to affirm the conception of nature as a repository of occult powers, and to that extent substantiated the fundamental claims of the occult sciences. At the same time, by radicalizing the conception of nature as a miracle, the polemic against curiosity placed strict limitations on the investigation of nature. Ironically, magic was both an enemy and a partner to theology: an enemy because it magnified human and demonic power, a partner because it supplied a striking demonstration of one of Christianity's basic tenets, the fall of man through his lust for forbidden knowledge. It is no coincidence at all that when seculars and liberal arts students discovered that knowledge is power, theologians reconstructed the Augustinian doctrine of *curiositas*, according to which the secrets of nature are divine secrets and curiosity is a form of pride, indeed the pride that went before Adam's fall. No one guilty of delving too deeply into the mysteries of nature could enter Heaven.

The ideological meaning of the polemic against intellectual curiosity is not difficult to discover. Bernard of Clairvaux stated it unambiguously in his advice to novice monks, *The Steps of Humility and Pride*. "Seek not what is too high for you, peer not into what is too mighty," he warned. "Stay in your own place lest you fall if you walk in great and wonderful things above you." Bernard wrote his admonition against "stepping out of one's station" with reference to Lucifer, who "fell from truth by curiosity when he turned his attention to something he coveted unlawfully and had the presumption to believe he could gain." But his real aim was to warn novices against prying into secrets that did not concern them, lest they upset the peace and stability of the monastic community. Developing a political analogy, he expanded upon the danger to the commonweal of the prideful curiosity of even one of its members: "All else in heaven's courts are standing: you [Lucifer] alone presume to sit. You disturb the concord of your brethren, the peace of the heavenly country, and if it were possible, even the peace of the Blessed Trinity. Wretch! to this your curiosity has led you. With reckless insolence you shock your fellow-citizens and insult your King. . . . You pry with insatiable curiosity, push yourself forward without respect, and would place your throne in heaven and make yourself the equal of the Most High."[122] The polemic against curiosity had weighty political overtones. For Bernard, humility was next to godliness and was a necessary condition of social order.

The Uses of Magic

When Hugh of St. Victor wrote his famous and much-quoted condemnation of magic, about 1120, the impact of Graeco-Arabic philosophy was just beginning to be felt in the Latin West. During the next century and a half, as the trickle of Arabic scientific treatises turned into a flood, his warning became increasingly more pertinent. In the 1230s, William of Auvergne, bishop of Paris and himself a former devotee of magic, felt compelled to caution students against the pernicious influence of these works. Evidently William's admonishments had limited effect, because a few decades later Roger Bacon reported that there were still many magical books in circulation. Calling attention to the pseudepigraphical character of the *libri secretorum*, Bacon noted that despite their ascription to famous authors, most were "new inventions," the work of "seducers" who "fabricate their lies under the cover of a text."[123] Bacon, who thought these works ought to be banned, asserted that one of the main purposes of *scientia experimentalis* was to test the claims of magic and to provide a method by which to separate false magical books from those containing true wisdom.[124] However, Bacon also worried that the mounting attack on magic, part of a general reaction against the secularization of knowledge, threatened legitimate experimental science because of the tendency to associate "experimentation" with magic. In a letter to an unnamed colleague *On the Secret Works of Art and Nature* (ca. 1260), he attempted to neutralize this danger by making a careful distinction between magic and experimental science. Magic, he argued, is always illicit and sinful because it is either fraudulent, as in the deceits perpetrated by jugglers and ventriloquists, or else it is accomplished with the aid of demons. Fraudulent magic is worthless and without power; it is simply sleight of hand. Demonic magic, while powerful, cannot be controlled by human agency; instead, through it demons exercise their power over human souls. Against magic Bacon upheld the power of nature and of "art using nature as an instrument": "Whatever is beyond the operation of nature or of art either is not human or is a fiction and accomplished by fraud."[125] In a spirited defense of experimental science against the false claims of the magi, Bacon firmly distinguished between the supposed marvels of magic and the real marvels accomplished by experiment, concluding that "it is unnecessary for us to aspire to magic since nature and art suffice."[126]

Despite the condemnations of the black arts that issued like so much conventional wisdom from the schools, interest in magic continued. Judging by the number of treatises devoted to the occult sciences, it increased rather than diminished.[127] Nor was interest in learned magic restricted to academic circles. About 1270 an Italian soldier wrote to Thomas Aquinas at the University of Paris asking the master's opinion concerning "the occult workings of nature." Thomas responded in a letter that assumed a surprisingly sophisticated philosophical understanding of the issue outside the universities.[128] About the same time we begin to hear reports of clerical diviners and magicians conjuring demons for paying clients.[129] However, learned magic may have gained its most devoted following in the courts. As early as 1159 John of Salisbury warned that magicians were particularly active in the courts, where ambitious servants used whatever devious means available to them to curry favor with princes. To John's dismay, courtiers were actively consulting soothsayers, diviners, and magicians.[130] By the thirteenth century, it appears, their numbers and influence had increased alarmingly.

Why was the medieval court such a fertile breeding ground for magicians? Some scholars have argued that the clash between two rival systems of power at the court opened up a space for magic. One power system comprised officials who were formerly invested with political office and whose power rested upon the traditional criteria of rank, status, and title. Running counter to these officeholders were the throngs of courtiers who had no formal claim to power in the traditional sense, but who for various reasons wielded informal, but nonetheless real, power. Such holders of informal power may have earned princely favor by wholly arbitrary and incomprehensible means, by virtue of their friendship with the lord, for example, or because of the lord's whim of the moment; or they may have gained power because they possessed certain skills: engineers, poets, entertainers, and scribes, for example. In the demimonde of the late-medieval court, where servants, gossips, pimps, panderers, and rising favorites competed for places, professional magicians found a clientele eager for their services. For magic promised some measure of control over a situation governed by a climate of envy, deceit, and intrigue, where attendants, often lowborn and without family connections, struggled for the precarious favor of patrons. Under circumstances like these, where one's fate hung on a ruler's whim or impulse, the professional magician was a powerfully attractive figure: he offered shortcut techniques for predicting the future, being victorious against adversaries, reading another person's mind, or winning honor and favor.[131] Moreover, his magic was not the superstitious sorcery of the village cunning-woman; it was the learned magic of the scholar. Though dangerous and suspect, it carried enough weight of philosophical authority to make it acceptable to someone in "sophisticated" court circles. Some evidence suggests that during the thirteenth and fourteenth centuries, times of chaotic social and economic change, ambitious courtiers increasingly resorted to sorcery to advance themselves to positions of power and influence. The period witnessed a series of politically motivated sorcery trials involving royal servants, papal prelates, and public officials who were accused of using magic to gain influence at court. The rise of such humble men disrupted established political relationships and challenged the customary role of the nobility as courtly officeholders. The traditional holders of power, the nobility, fought back against their lowborn competitors by bringing charges against them of using sorcery—that is, of hiring professional magicians—to gain political influence. The conventional motif of "the king's wicked adviser" took on a new dimension by being fused with that of the learned magician.[132]

The case of Conrad Kyeser (1366–1405) is instructive. A native of Eichstatt, Kyeser was trained as a physician and was connected with several courts, including those of Duke Stephan III of Bavaria-Ingolstadt and King Wenceslas of Bohemia. Either as a physician or as an engineer, Kyeser accompanied King Sigismund of Hungary on the last general European Crusade against the Turks, which ended in bloody disaster at Nicopolis in 1396. In 1402, political circumstances drove Kyeser from the imperial court and forced him into exile in Bohemia, where, shortly before his death, he composed a great work on warfare, the *Bellifortis*, which he dedicated to the emperor Rupert. We do not know the exact circumstances behind Kyeser's sudden reversal of political fortune, but the strange content of the *Bellifortis* invites a speculation. This beautifully illustrated work contains, in addition to some of the medieval West's most inventive (if sometimes fanciful) examples of military technology, dozens of magical recipes and formulas: magic lanterns that cause enemies to see strange visions, amulets to ward off evil spirits, and a torch made of hair from the tail of a rabid dog mixed with the dog's fat, which is used to drive the enemy away. Kyeser deliberately projected an image of himself as a powerful sorcerer. In one picture, he appears summoning spirits from a castle tower while two naked goblins, one riding a broom and the other carrying a taper made from the fat of a hanged man, advance toward the castle.[133] Kyeser fits perfectly into the mold of those who held "informal" power in the medieval courts. Born of burgher parents and university-educated, he acquired skills as a military engineer that put him in high demand, propelling him into several princely courts. For those traditional power-holders who saw their political influence fading, it must have seemed as if the meteoric rise of men like Kyeser was possible only through some kind of special power, perhaps sorcery. Was Kyeser one of those who were implicated in charges of using sorcery to gain influence at court? Was he branded as one of the "king's wicked advisers"? Certainly he did nothing to dispel such suspicions. Indeed, if Kyeser gained a reputation as an emissary from the devil, as Bertrand Gille suggested, the *Bellifortis* suggests that he himself was largely responsible for it.[134]

Plainly, the new knowledge introduced into Europe from Arabic sources could not be kept within the boundaries set by the Scholastics. The alarming proliferation of magical treatises in the thirteenth and fourteenth centuries is perhaps related to the emergence of what R. R. Bolgar has called the "intellectual proletariat," a group composed of university-educated laymen who had failed to find useful or permanent employment and who were, consequently, cast upon the treacherous seas of the patronage system.[135] Members of this "underworld of learning," angered by civil and ecclesiastical structures that denied them preferment, and receptive to new ideas, may have been particularly attracted to the achievements of Arab science—its occult as well as its experimental component. Moreover, literacy was increasingly becoming a requirement for craftsmen and engineers, who were apt to be influenced more by magic than by conventional scholastic philosophy. Medieval engineers enthusiastically appropriated magic as a theoretical framework for technology. Indeed they regarded magic as technology's sister art. Not only did learned magic give technology a theoretical matrix, it served an important ideological function by promoting the image of the professional engineer as a magus who, with his inventions, manipulates nature's occult forces and gains mastery over the physical world. Kyeser declared that the magical arts, or *artes theurgices*, were a branch

1. Magical activities at a medieval castle, from Conrad Kyeser, *Bellifortis* (ca. 1405). This page of Kyeser's manuscript depicts the author summoning from a castle tower two goblins, one riding on a broomstick and the other carrying a taper made from the fat of a hanged man.

of the mechanical arts, ranking them just below the military arts. He saw no inconsistency in dedicating entire sections of his work on military technology, the *Bellifortis*, to magical formulas, amulets, experiments to produce "marvels," and incantations to summon demons. Evidently for Kyeser, and doubtless for others as well, the usefulness of the occult sci-

ences in this world overcame any consternation about the dangers it may
have held for the soul in the next. For this engineer, whose infatuation
with the legends of Alexander conjured images of a new Hellenic conquest of the East, magic and technology were complementary means for
regaining the Holy Land for Christendom.[136]

Kyeser's main authority on magic, it turns out, was not some obscure
Arabic treatise on the occult sciences but a popular Latin book of secrets
spuriously attributed to Albertus Magnus. No work better illustrates the
assimilation of the occult sciences in the Latin West than the *Secreta
Alberti* (The secrets of Albert), also known as the *Experimenta Alberti*
(The experiments of Albert).[137] The work was doubtless one of the *libri
magicorum* that Roger Bacon censured, for its composition reflects the
methodology of the Latin authors of occult books who "put famous titles upon their works and impudently ascribe them to great authors in
order to more powerfully allure men to them."[138] By far the most famous medieval book of "experimental" magic, the work was, as its title
implies, a compilation of "secrets" and "experiments" drawn from various sources, the most respectable of which was Albert's own *De mineralibus*, from which the editor constructed the lapidary section of the
work. Other standard authorities, including Pliny, *Physiologus*, and the
Pseudo-Aristotle of the *Secretum secretorum*, were also close at hand.[139]
Nothing is known about the compiler of the *Secreta Alberti*. Thorndike
pointed out that it "pretends to be a product of [Albert's] experimental
school among the Dominicans at Cologne."[140] It is indeed possible that
the author was a pupil or follower of Albert. The earliest surviving manuscripts date from the late thirteenth century and were therefore composed either during Albert's lifetime or soon after his death. The compiler was obviously an avid student of the Arabic occult sciences. He
mentions Kyranides, the "Book of Alcorath" (evidently another Arabic
"experimental" book attributed to Hermes), and "the books of necromancy, images, and magic." Possibly he was a university student like the
one William of Auvergne described when speaking of his own youthful
fascination with magic. Whoever the author was, the work represents a
fairly successful attempt to use Albert's ideas to develop a rational theory
of magic.

The *Secreta Alberti* was essentially a treatise on employing the "secret" or marvelous virtues of plants, stones, and animals. The first herb
mentioned, the heliotrope, if gathered when the sun is in the sign of Leo
and wrapped in a laurel leaf with a hound's tooth, insures that the bearer
will be addressed with only friendly words. The stone opthalmus (opal)
renders a man invisible because its virtue "blinds the sight of men that
gaze upon it." Eating the heart of a weasel allows a man to foretell
things to come. It is possible, using the marvelous virtues of natural ob-

jects, to cause terrifying dreams, to create illusions of a room full of serpents and headless men, and to make enemies do whatever you will. Despite the obvious potential for certain of these "experiments" to be used
with malicious intent, the author insists that magic is not inherently evil,
but is good or evil depending on the operator's intentions:

> As the Philosopher says in several places, there is something good in all
> sciences. However, sometimes good is accomplished and sometimes evil,
> according to whether the science is directed toward good or evil ends.
> From this two things are concluded: The first is that the science of magic
> is not evil, for by the knowledge of it evil may be avoided and good fol-
> lowed. The second conclusion is that an effect is to be praised according to
> its end; hence the end of a science is censured when it is not appointed to
> good or to virtue. From this it follows that every science or operation is
> sometimes good, sometimes evil.[141]

The *Secreta Alberti* was obviously indebted to the occult tradition
leading all the way back to the Hellenistic era. However, what made the
work so different from the ancient Hermetic books was the author's unwillingness to accept that marvels are *merely* marvelous. Instead, he attempted to explain them according to the principles of scholastic science. Many of the manuscripts of the *Secreta Alberti*, and practically all
the printed editions of the work, are accompanied by two other pseudo-
Albertine tracts: a work on the "secrets of women" (*De secretis mulierum*) and a brief theoretical treatise entitled *De mirabilibus mundi*
(The marvels of the world).[142] The author of the *Women's Secrets*, a
cleric, wrote the work for priests who had inquired about "certain hidden, secret things about the nature of women." A commentary accompanying the text notes that the work was composed "so that in confess-
ing [women] we might know how to give suitable penances for their
sins."[143] The "secrets of women" were essentially the causes of such
"mysterious" phenomena as menstruation, conception, and childbirth,
which the author rationalized according to the principles of Aristotelian
natural philosophy. The *Secreta mulierum* also contains various "experiments" (*experimenta*), including techniques to determine whether a
woman is pregnant—and, if so, whether she will bear a male or a female
child—or whether her childlessness is the result of impotency or infertil-
ity. Some of the "experiments" were items of folk medicine, perhaps
picked up from midwives. To help a woman become pregnant and bear
a male child, the author recommends giving her a powder made from
the dried intestines of a hare; or, alternatively, "let her place a goat's hair
in the milk of a female donkey and let her tie this around her at the navel
while she has sex with her husband, and she will conceive." A commen-

tator explained the former experiment on the grounds that the potion "alters the temperament of the woman's seed," making it hotter and more prone to conception; as for the latter, "The reason for this is unknown to us, however it is known to nature."[144]

The theoretical principles underlying nature's "marvels" are spelled out more fully in the *De mirabilibus mundi*. According to the unknown author of this text, marvels are natural and caused by the "rational virtues" in things, even though these causes may be hidden from the intellect. In making this argument, the author was following an approach to occult qualities that by the thirteenth century had become fairly standard in scholastic circles.[145] As M.-D. Chenu pointed out, the "desacralizing of nature," which placed severe limitations on the province of preternatural phenomena, was well under way in the twelfth century: "Criticism of the preternatural, whether in nature or in everyday life, continued to grow from this point on despite the permanent attraction that the marvelous held for men."[146] Although certain qualities in nature may be insensible or idiosyncratic, the Scholastics argued, it is nevertheless possible to find rational, physical explanations for them—unless, of course, they are caused by demons. Thomas Aquinas, in his letter on the "occult works of nature," restricted the scope of preternatural powers by referring them all to the demonic, thus attempting to preserve the rationality of nature as well as the proper sphere of the supernatural.[147] In the fourteenth century, Nicole Oresme (ca. 1320–1382) devoted an entire quodlibetal treatise, *De causis mirabilium*, to the subject of marvels and their causes, arguing that "marvelous" phenomena do not require supernatural causes to explain them. Oresme advanced detailed arguments to prove that events people generally regard as marvelous proceed instead from natural causes that are overlooked, or they result from perceptual errors. Once the causes are known, such phenomena no longer appear marvelous.[148]

"The philosopher's work is to make marvels cease," wrote the author of the *De mirabilibus mundi*—that is, to search for an understanding of their causes.[149] Certain causes of presumed marvels can be known, he continued, such as the characteristics of the four elements, the influence of the heavens, or the force of sympathy and antipathy. As for the properties that appear idiosyncratic, not reflecting the universal characteristics of nature, we can at least be assured that some are innately present in the entire species (*secundum totam species*), while others occur only in certain individuals (*secundum individuum*) at certain times. However, like most writers on marvels and the occult, this author conceded that for some marvelous events no rational account is possible. In such instances, only experience can confirm their existence:

Certain things are to be believed only by experience, without reason, for they are concealed from men; others are to be believed only by reason, because we lack sensations of them. For although we do not understand why the lodestone attracts iron, nevertheless experience shows it, so that no one should deny it. And just as this is marvelous and made certain only by experience, so likewise should one suppose in other things. One should not deny any marvelous thing because he lacks a reason for it, but should try it out (*experiri*); for the causes of marvelous things are hidden, and follow from such diverse causes preceding them that man's understanding, as Plato says, cannot apprehend them. . . . Thus the philosophers declare that marvelous things are [known to be] in things by experience (*per experientiam*), which no one ought to deny until it is tried out (*experiti*), according to the fashion of the philosophers who discovered it.[150]

Although this academic strained himself to explain marvels, unwilling to relinquish his conviction that nature was rational, his commentary underscored the ambiguity of the scholastic resolution of the problem of occult qualities.

Books like the *Secreta Alberti* raise perplexing questions about the credibility and the survival of magic. For on the face of things, it is difficult to imagine any practical uses these works might have served. Surely, it might be argued, the futility of magic should have been, and would have been, obvious to any intelligent observer. Yet the widespread distribution of the books of magical "experiments" is ample testimony to the fact that they were widely known and read in medieval Europe.[151] William of Auvergne spoke repeatedly of the "experimenters" (*experimentatores*) and their books, and his numerous references to these works leave little doubt that he had studied them carefully.[152] How then do we account for the survival and influence of the medieval magical books when, manifestly, magic does not work?

Part of the explanation is that, for a variety of reasons, the futility of magic may not have been obvious to medieval observers. For magic is not usually employed as a *substitute* for empirical techniques, but as an *aid* to them. Quite commonly in manuscripts magical recipes and empirical techniques are juxtaposed and undifferentiated.[153] Moreover, occasionally magic may have been successful. It seldom promises to produce a result by itself, but only to enhance the effectiveness of empirical actions that do produce results. For example, no one would be foolish enough to rely on magic alone to win a battle or a debate without actually fighting the battle or making every effort to argue convincingly. In this sense, magic may be effective in the way that Bronislaw Malinowski suggested, not as causally changing nature, but as affecting the motivations and expectations of the actor, enhancing his confidence that his actions will succeed.[154] In addition, magic was part of a coherent system that had the weight of tradition behind it. One of the characteristic features of magical science is its self-confirming character. A doctrine about the occult, it is compatible with virtually all empirical evidence. It cannot be directly refuted. Hence failure of one instance of magic does not prove that all forms of magic are useless, only that experience is peculiar. Finally, medieval people did not consider experience as a *test* of theory. They did not pool their unsuccessful experiences with magic in a way that caused them to challenge the entire magical art. Since there were always stories circulating confirming the efficacy of magic, belief in magic was supported both by intellectual tradition and by other peoples' "experience."[155]

Above all, magic responded to a new intellectual climate developing in Western Europe in the late Middle Ages: the growing emphasis on the idea that knowledge could be exploited for practical gain.[156] The flowering of science after the twelfth century was due not only to the introduction of new sources; it was also a consequence of the revival of commerce and industry, the spread of technology, and the strengthening of the European economy. These developments contributed to the emergence of a new attitude toward scientific knowledge, one that stressed the possibility of exploiting nature through an understanding of it, of mastering nature rather than submitting passively to it. Roger Bacon, in arguing for *scientia experimentalis*, emphasized the utilitarian function of knowledge, stressing the idea that limitless benefits would accrue from discoveries in medicine and the mechanical arts. "The Church should consider the employment of these inventions against unbelievers and rebels," Bacon urged Pope Clement IV, "in order that it may spare Christian blood, and especially should it do so because of future perils in the times of Antichrist, which with the grace of God it would be easy to meet, if prelates and princes promoted study and investigated the secrets of nature and of art."[157]

Yet, as Bacon's words seem to suggest, the widespread optimism about the potential of experimental science barely concealed deep anxieties over the future of Christendom and also, perhaps, a growing sense of uneasiness about the scholastic program. For many intellectuals of the late Middle Ages, the limits of reason had been reached. In philosophy, the reaction against the scholastic program expressed itself in the form of nominalism, whose proponents refuted the claim that the general terms and classifications used in scientific explanations are real. The nominalists asserted that all mental conceptions are only convenient ways of ordering sense experience. Ultimately, they maintained, our knowledge about the world is limited to our perceptions of it. The intention of the nominalist critique was to refute what many intellectuals thought were dangerous implications of scholastic philosophy: in rationalizing the world, purging it of miracles, philosophy had inevitably reduced the scope of God's power in the world. By challenging the reality of universals, the nominalists attempted to put God back into the world; and as a consequence of their radical critique of reason, they asserted the primacy of experience over theory, of faith and intuition over rational understanding.[158]

Another reason, then, for the proliferation of books of secrets in the late Middle Ages may have been a growing dissatisfaction, on essentially religious grounds, with the aggressive movement by scholastic philosophy to limit God's almighty power in the universe. For the books of secrets asserted, in example after example, that the world was full of marvels that human reason could never hope to explain, any more than man's feeble intellect could explain the ultimate mystery of God's creation. Marvels are known only by direct experience, just as the heart alone, by intuition, knows God's presence. The excessive stress on the authority of experience, which is so pronounced in the medieval books of secrets, may thus have been symptomatic of the anti-intellectual and antiauthoritarian tendencies that surfaced in late-medieval culture. It suggests growing frustration, perhaps especially among that "underworld of learning" whose presence Bolgar detected, with scholastic philosophy, and in the case of books of medical secrets, with official therapeutics. Occasionally these critiques erupted into violent polemics against scholastic learning. Toward the end of the thirteenth century, a Dominican friar and medical practitioner called Nicholas of Poland (fl. ca. 1270) wrote a fierce invective against scholastic medicine in a work whose title, *Antipocras* (Against Hippocrates) reveals the principal target of his attack.[159] In this versified polemic, Nicholas took the role of an advocate defending his empirical methods, which he claimed brought miraculous cures all over the Latin world, against the bankrupt methods of the physicians. Attacking the physicians' tendency to rely on reason and authority over experience, Nicholas wrote, "Here the advocate rejects the authority of Galen, who says, 'Physician, how can you cure, when you are ignorant of the causes?' Galen, I show that a cure can well be effected without knowledge of the cause."[160]

Nicholas of Poland was no charlatan. An alumnus of the University of Montpellier, he was a resident of the city for twenty years (probably as a liberal arts teacher at the Dominican *studium*) before returning to his native Silesia to practice medicine in a convent near Cracow. Neither was he a conventional practitioner; he was the founder of an "alternative medicine" movement that flourished in Upper Silesia in the late thirteenth century.[161] In addition to the *Antipocras*, Nicholas left a collection of medical *experimenta*, wherein it is reported that the friar "was a man of such experience that before him there is not believed to have been his like, nor is it hoped for the future, as is plain in his marvelous works in making great and sudden cures in various provinces and regions."[162] Nicholas's practice was certainly unorthodox. Urging a return to "natural" methods of healing, he attributed extraordinary virtues to toads, scorpions, lizards, and snakes. He dedicated an entire section of his *Experimenta* to various preparations and uses of serpent's flesh. To break a bladder stone, he recommended that the patient drink his "snake powder" (*pulverem serpentis*) in wine twice daily. To make the drug even more effective, he added powdered toads; best of all was a concoction made of powdered serpents, toads, and scorpions. He made pills of dried frogs, which he prescribed as a remedy for weak hearts and sore eyes, and promised would make the troubled sleep. His favorite remedy (*curat universa*) was serpent's flesh, carefully prepared according to detailed instructions spelled out in the *Experimenta*. Nicholas recommended that kings, dukes, and other noblemen eat it at every meal. But, he insisted, snake meat was good for everyone: "Briefly, according to the doctrine of friar Nicholas, it is advantageous for all people, of whatever station, to eat serpents whenever it is possible to get them."[163]

Nicholas of Poland's strange and revolting arsenal of drugs was based upon the principle that, while God had conferred marvelous virtues on all of nature, "the more filthy, abominable, and common things are, the more they participate" in these marvelous virtues.[164] Hence, remedies made of the commonest, most contemptible creatures contained far greater medicinal virtues than the "precious and famous" drugs recommended by the physicians. The methods of the "Hippocratic" physicians were nothing more than lies masked by polished words, serving only to line the pockets of the physicians and multiply the deaths of patients.[165] This radical critique of official medicine was grounded upon an essentially religious doctrine, reflecting Nicholas's deep skepticism of the scholastic effort to explain the mystery of the creation and his distrust of the physicians' attempts to heal by naturalistic means alone, without God's intervention. The effort by scholastic physicians to exclude the miraculous from medicine was, according to Nicholas, an impossibility, because all true healing agents, since they emanate from God, are miraculous. That God had conferred the most marvelous virtues on the least-esteemed creatures was further proof of the fallacy of reason. Contemptuous of philosophy, Nicholas asserted that knowledge of causes was unattainable. Ignorance, except of that which God has revealed, is the human condition.[166] His insistence on the primacy of revealed, intuitional truth as opposed to the inventions of philosophy reflected a desire to return to a religion of humble piety. His conviction that the secrets of nature are discovered by experience instead of reason went hand in hand with his preference for faith over theology.

Nicholas's assertion of the poverty of philosophy and the superiority of experience over reason carried a pronounced ideological bias. Why, he asked, is "Hippocratic" medicine not instructed to "pluck the fruits" of the marvelous properties God implanted in things? "Perhaps because in this lot he [Hippocrates] was a pauper," Nicholas answered, "or rather, perhaps because the prophet prayed that there should be none more like Hippocrates."[167] God loves the humble, Nicholas proclaimed. He chose to reveal his deepest secrets to ordinary people, just as he had conferred the most marvelous medical virtues on the meanest beings in nature. Hence the common people of the villages had deeper insights into the secrets of nature than did the learned physicians: "The people love empirical things," Nicholas declared, "because none of them are harmful; but the physicians are ashamed because great works prefer the villages, where the marketplaces resound in their praises of empirical remedies."[168] Nicholas had indeed discovered profound secrets, not only of medicine, but also of God, nature, and politics: God reveals his truths to the common people. The secrets of nature are known empirically, not by philosophy. Scholastic science, which constructs elaborate fictions as explanations, is nothing more than an effort to conceal the truth and keep the elites on top.

Such a radical critique of the academic establishment could only have come from one who was a product of it. The twenty years that Nicholas spent at Montpellier were critical years for the development of the university's medical school. Between about 1250 and 1270, when Nicholas resided at Montpellier, the intellectual foundations of the school's imposing system of theoretical medicine were being laid.[169] The philosophical discussions that Nicholas heard at Montpellier—exemplified by dense theoretical treatments of Galenic complexions and degrees and by elaborate efforts to explain the "marvelous" virtues of the drug theriac—resulted in a theory so complex that it was, practically speaking, unworkable. In many ways, the occult virtues of theriac symbolized the limits of the scholastic program. Like the magnet, its unique virtues could not be predicted from the nature of the ingredients composing it, the most outstanding of which was serpent's flesh prepared according to elaborate and carefully controlled procedures. Nor could reason understand or explain the cause of theriac's powers. It was a *secretum*, and its virtues were known only by experience. To the scholastic physicians at Montpellier, theriac was an anomaly, a troubling problem that would require years of concentrated philosophical effort to resolve. But to Nicholas of Poland, it was just snake meat. All the other ingredients that went into making it were as worthless as the arguments constructed to explain away the miracle that God had revealed, through him, to the people.

Nicholas insisted he wrote nothing that he had not first proved by experience (*non sit aliud quicquam insertum, nisi quod ex usu expertum est*). Yet significantly, he invoked the authority of "master Albert" to confirm his doctrine.[170] Pseudo-Albertus's stress on experience over reason in dealing with occult qualities is repeated throughout Nicholas's medical writings; and like Pseudo-Albertus, Nicholas believed in the curative powers of magic rings and amulets. Moreover, Nicholas, following the general discussion of the *Secreta Alberti*, used magnetism as a model to illustrate the influence of heavenly virtues on terrestrial objects.[171] The *Antipocras*, written around 1270, was one of the earliest medieval texts to bear the imprint of Pseudo-Albertus's ideas. The work also illustrates the extremes to which, in the hands of a radically anti-scholastic thinker, speculations about "secrets," "marvels," and "empiricism" could be taken. Unlike the Pseudo-Albertus, Nicholas attempted not to rationalize marvels or to bring them under the cloak of *scientia*, but to preserve them as a separate domain, distinct from *scientia*. Thus according to Nicholas an *empiricum* (i.e., an "empirical" remedy or phenomenon) was like Saint Agnes, who, despite being mutilated, did not lose her marvelous, saintly virtue. It is something "whose innate virtue is such that, by pouring itself out from afar, is itself not subjected to diminishment."[172] According to this Neoplatonic conception, an "empirical" virtue is not merely one that is known by experience (or intuition), but is something akin to the divine. By its very nature, it is marvelous.

No work better illustrates the extremes to which medieval intellectuals—and anti-intellectuals like Nicholas of Poland as well—went to find practical uses for secrets than the *Secreta Alberti*. The work's emphasis was overwhelmingly upon techniques by which a person might exploit the occult forces in nature in order to gain practical advantages in the world: "If you want to forejudge or conjecture of things to come. . . . If you want to overcome your enemies. . . . If you want to be acceptable and pleasant. . . ." Books of secrets and experiments, precisely because they related the most "marvelous" phenomena conceivable, confirmed the medieval conviction that knowledge was power. They purported to show how the occult qualities in things could be put to concrete, practical use, whether it be to foretell the future, win a debate, or heal the incurable, forsaken by the physicians. The message implicit in the literature of secrets was that nature was power-laden, and that this power could be exploited by those who knew, by experience, its secrets.

Science, Technology, and the Ideology of Secrecy

Medieval science was a corporate system of knowledge. Access to it was restricted to a relatively small population of university-educated adult males, who, as academics, enjoyed privileges other social groups did not enjoy. From the end of the twelfth century, natural philosophy found its primary institutional home in the universities, whose development as autonomous corporate bodies coincided with the formation of craft guilds in commerce and industry, and of self-governing communes in the civic sphere. Like other occupational groups in the Middle Ages, the academic community governed itself by its own rules and strictly regulated entry into its ranks. Academics held high status in medieval society. They were accorded special rights and privileges, such as immunity from the jurisdiction of secular authorities and exemption from taxes, levies, and tolls.[173] The academic elite acquired the status of a separate order or estate: the *Studium*, coequal with the two familiar orders of medieval polity, the *Sacerdotium* and the *Regnum*. The guardians of a formerly esoteric body of knowledge, academics became virtually a "new nobility" in medieval society.[174]

It was no secret to academics that significant social and economic advantages attended their status as the self-appointed guardians of the new learning. The educated elite became a kind of intellectual aristocracy, which sought to consolidate its gains by cultivating a moral code distinguishing it from other orders in the social hierarchy. Academics set themselves apart from the *illiteratus* below by proclaiming the greater honor of intellectual labor over manual labor. They distinguished themselves from the nobility above by equating "true" nobility with virtue, and virtue with learning. Scholastic depreciation of "rustics" and "the crowd" became particularly virulent in the thirteenth century, as the educated elite attempted to reinforce its status and to set itself above the herd of ordinary men.[175] "The common crowd, because of its multitude and paucity of intellect, as well as because of other evil dispositions, lives almost like brute animals," wrote William of Auvergne.[176] With this well-worn formulation, academics described the uneducated masses as creatures of passion in contrast to the wise, who dominated their passions. "It is an old proverb," wrote an academic around 1160, "that as far as men are removed from beasts, so far are the educated removed from the illiterate."[177] More cautiously, but with similar intent, intellectuals attacked "unlettered" priests and noblemen. As intellectuals fought against persons below them on the social hierarchy in order to rise above them, they also fought against those above them in order to replace them.

Besides putting a premium on virtue acquired through learning, besides stressing the dignity of intellectual labor over manual labor, the scholastic ethos fostered an ideology of esotericism, which conceived of intellectual knowledge as the special preserve of the "philosopher," not to be revealed to the *vulgus*. If, as Roger Bacon maintained, experimental science could be used to further the cause of Christendom, that did not imply sharing knowledge with the public at large, who could not be expected to understand it, much less contribute to its advancement. "It is foolish to feed an ass lettuces when thistles suffice him," Bacon wrote. For the multitude, "the rude, cheap, imperfect food of science is sufficient." Because of the powerful and potentially dangerous ways in which *scientia experimentalis* could be used, it is all the more necessary to safeguard it, indeed to keep it secret from all but those who are morally and intellectually equipped to employ it beneficially. "For the wise have always been divided from the multitude, and they have veiled the secrets of wisdom not only from the world at large but also from the rank and file of those devoting themselves to philosophy. . . . Nor ought we to cast pearls before swine; for he lessens the majesty of nature who publishes broadcast her mysteries."[178] The literature of secrets, with its repeated injunctions to keep secrets from the *vulgus*, was nourished by, and in turn reinforced, the ideology of esotericism. Bacon's expression of the code bears the unmistakable mark of the *Secretum secretorum*. The polemic against *curiositas* also promoted esotericism, as did the literary tradition of the goddess Natura. The goddess was always portrayed as being modest. She covered herself with a veil, was ashamed to be seen naked, and resented any attempt to pry into her secrets.[179] The philosopher had a moral duty to approach nature discreetly. He had to speak of her in veiled images and fables (*fabulosa*), so as not to expose her nakedness to public view. The moral obligation to be circumspect when dealing with the secrets of nature was a conviction woven into the fabric of medieval intellectual life.

The corporate structure of the medieval craft economy also influenced attitudes toward public disclosure of technical secrets. Technical knowledge was the craftsman's most valuable property, even more valuable than his materials or his labor. Economic realities compelled craftsmen to keep the secrets of the arts guarded from public view. In specialized and highly skilled crafts like dyeing and glassmaking, success depended upon precise and detailed knowledge of the kind and quality of materials for a process, the manner and proportions of combining them, and the often subtle effects of temperature on the materials. Such "trade secrets" were valuable intellectual property. Disclosing them to the public threatened to undermine guild monopolies over specialized crafts.[180] Guild ordinances often enforced restrictions against openness—for instance, by forbidding artisans to teach the craft to anyone but sworn apprentices.[181] In the early seventeenth century, the London Pewterers' Company passed an ordinance against "the abuse of dyvers of the company who worketh openly in the shopes." The company forced one of its members to put up a partition in his workshop after a complaint was filed against him for "suffering of a Goldsmith to worke openly in his shopp."[182] The Venetian glassworkers' guild prohibited its members from plying the trade outside of Venice.[183] Keeping the technical secrets of glassmaking hidden provided the Venetian glassworkers with significant competitive advantages. Yet if craft secrets were valuable as intellectual property, they might also be valuable commodities. Medieval craftsmen may occasionally have sold technical secrets for a quick profit, but in the long run such a practice would have endangered the artisan's livelihood.[184]

Scholastic philosophers were by and large uninterested in the empirical data generated by unlettered artisans. During the Middle Ages, technical information was transmitted primarily through an oral tradition and preserved within an exclusive, guild context. "Science," on the other hand, was the preserve of the universities, where the authority of the written word held sway. The two traditions rarely merged, and there was little fruitful exchange between them. Moreover, the Scholastics inherited the ancient distinction between the *artes liberales* (identified with the subjects of the trivium and quadrivium) and the *artes serviles* (identified with the mechanical arts), a distinction that reflected the deep social divisions between freemen and slaves. The mechanical arts, according to this ideology, were incompatible with the education of a "free man"; they were practiced exclusively by manual workers. Since *techne* merely imitates nature, it remained outside the proper sphere of the sciences.[185] Consequently, medieval scholars exhibited little interest in the technical knowledge of the crafts. A famous passage in Roger Bacon's *Opus majus* has been mistakenly regarded as an important exception to this attitude. "More secrets of knowledge have always been discovered by plain and neglected men than by men of popular fame," Bacon wrote, "and I have learned more useful and excellent things without comparison from very plain people unknown to fame in letters, than from all my famous teachers."[186] However, the context of Bacon's statement makes it clear that his intention was not to praise the mechanical arts or to promote serious study of the crafts. Rather, he was issuing a warning to academics not to glory in their own learning, for "God's conversation is with simple folk according to the Scriptures" and not with science. Far from being a paean to the mechanical arts, this passage was a warning against intellectual pride.

Nor should we be misled by the tendency of scholastic authors to begin including the mechanical arts under the umbrella of the liberal arts. Much has been made, for example, of Hugh of St. Victor's restructuring of the traditional classification of the liberal arts. Lynn White, jr. thought Hugh's willingness to include the mechanical arts among the liberal arts indicated a revolutionary change in the culture's attitudes toward labor and technology.[187] Yet Hugh could barely conceal his disdain for the mechanical arts. He retained the classical view that the crafts are "adulterate, because their concern is with the artificer's product, which borrows its form from nature."[188] Hugh's rationale for including the mechanical arts in his classification scheme was to show that they, as much as the liberal arts, helped overcome the human evils of ignorance, vice, and need, thereby helping prepare the student for blessedness. It was decidedly not to argue for a closer relationship between theory and practice.[189]

Academic attitudes toward the arts are no mystery. But what of medieval craftsmen? How did they understand science? Unfortunately, we know very little about the attitudes of early medieval craftsmen, either toward their own work or toward that of scholars. In the late Middle Ages, however, as literacy spread, craftsmen began more frequently to record their technical secrets in writing. They composed handbooks to train other artisans and to stake claims to their inventions. Some of the manuals were written by monastic craftsmen, who naturally tended to be more literate than secular artisans. But urban artisans also saw advantages in writing. From literally dozens of examples of the writings craftsmen produced, in the following paragraphs I sample three texts, the first by a monastic artisan, the second by an artisan who worked in both secular and clerical contexts, the third by a Nuremberg alchemist and metallurgist.

In the 1120s, a German Benedictine monk writing under the name of Theophilus composed a detailed treatise on the monastic arts of painting, glassmaking, and metalwork. Theophilus's treatise, entitled *De diversis artibus* (On diverse arts), was entirely new.[190] It was completely different from the early medieval arts-handbooks, such as the *Compositiones variae* and the *Mappae clavicula*, which continued to be copied and recopied down through the Middle Ages. Whereas the earlier works were anonymous, random compilations based on ancient technology, *De diversis artibus* was an original and systematic instructional manual written by an accomplished artisan from his own workshop experience. Theophilus's work is equally significant for its positive evaluation of the mechanical arts and its novel attitude toward revealing artistic secrets. Theophilus gave craftsmanship the highest sanction medieval culture could give it: holiness. Whoever beholds the marvelous works of art in churches, he wrote, is made to "praise God the Creator in this creation and to proclaim Him marvelous in his works." Art makes God palpable. Seeing works of art, the eye "grasps" God's truths in their beauty, inspiring the heart with hope. "God delights in embellishment," wrote Theophilus, "the Spirit of God has filled your heart when you have embellished His house with such great beauty and variety of workmanship." Because of this, the secrets of the arts should not be hidden but should be taught openly: "Let [the artisan] not hide his gifts in the purse of envy nor conceal them in the storeroom of a selfish heart but . . . let him simply and with a cheerful mind dispense to those who seek." Craftsmanship, a gift of God, should be given freely to all who wish to know. "Do not hide away the talent given you by God, but, working and teaching openly and with humility, . . . faithfully reveal it to those who desire to learn."[191] There is no trace in *De diversis artibus* of the aura of secrecy or of the forbidden that surrounded the artistic literature of the early Middle Ages: no tightly guarded "keys" to unlock secrets, no expressions of anxiety over revealing secrets to the vulgar. A huge divide separates the Benedictine ethic from the ancient mysteries.[192] Like Thessalos of Tralles, the novice whom Theophilus addressed in his book was "blessed" with the artistic knowledge that the master had attained only by "intolerable effort." However, unlike Thessalos, who was bound by the law of silence to keep the secret hidden, Theophilus's pupil was bound by the covenant of openness to teach "openly and with humility" all he has learned.

Such a view of craftsmanship might seem unexceptional in a monastic context, where charity was valued above material gain. However, secular artisans also wrote treatises on the arts, albeit none as eloquent as that of Theophilus. By the fourteenth century, there were growing signs that writing was becoming an important means of communicating technical knowledge in nonmonastic as well as in monastic contexts.[193] One example of such a craftsman was Gottfried of Franconia, a Bavarian vintner and orchard-master who lived in the vicinity of Würzburg. His *Pelzbuch* (Book of grafting, ca. 1350), a treatise on the care of vineyards and orchards, illustrates some of the ways in which oral and written traditions intersected.[194] Very little is known about Gottfried. In addition to having been an experienced vintner and orchard-master, he may also have been a cleric. Most of the orchards and vineyards around Würzburg were owned by the prince-bishop and were administered by ecclesiastical ordinaries. Such a background would explain Gottfried's ability to compose the treatise in a somewhat rough-and-ready Latin generously interspersed with German technical words. Like all medieval craftsmen, he learned his trade primarily through an apprenticeship under an experienced gardener, imitating and following instructions delivered by word of mouth and example. However, Gottfried's mentor, a certain "Master Richard" (*magister meus*), was somewhat unusual in that he had written a manual for apprentice gardeners. Frequently Gottfried referred to techniques recorded by "Master Richard in his book." He opened his own treatise with the precepts of Master Richard, then shifted to his own experience. In writing the *Pelzbuch*, he was following his master's example.[195]

Gottfried's extensive travels, while perhaps untypical for ordinary artisans, were not unusual for monastic craftsmen. After completing his apprenticeship in Franconia, Gottfried struck out on his own to learn techniques from gardeners and vintners throughout Germany and Europe. His journeys took him eastward as far as Brabant and as far south as Calabria. He learned techniques from fruitkeepers in Greece, from German knights, and from craftsmen, gardeners, monks, and housewives. In Flanders he met Nicholas Bollard, an English Benedictine monk and orchard-master. Nicholas also wrote a treatise on planting and grafting fruit trees. The parallels between his and Gottfried's work suggest that the two freely exchanged information and techniques.[196] Arguably, Nicholas and Gottfried were beginning to consider themselves more as authors than as artisans.

In addition to making professional contacts throughout Europe, Gottfried was familiar with ancient and medieval horticultural writings, particularly the literature that filtered through northern Italy. He read Martial, Pseudo-Aristotle (*De plantis*), Isaac Judaeus, and a Latin translation of the *Geoponika*, a tenth-century Greek compilation of agricultural excerpts. He also mentioned "the men of Salerno," possibly referring to the Salernitan medical writers. He made frequent references to the Roman author Palladius, whose *De re rustica* was his main literary authority on the care of fruit trees. He cited Socrates as an authority on celestial influences on wine.[197] (Craftsmen as well as scholars appropriated the names of classical authorities to enhance their prestige.) But Gottfried was no mere compiler. A literate craftsman, possibly a cleric, he was a cultural broker, an intermediary between lay and learned cultures and a mutual interpreter of each to the other. Gottfried traveled easily between the two cultures. He knew Latin well enough to familiarize himself with the ancient agricultural writers and was fluent in the specialized technical *Fachsprache* of German vintners and fruit-growers. He was familiar with the academic discourse on occult qualities. Writing about the influence of thunder and lightning on wine must, he noted, "Many such thynges ther ben of which al a mannes mende may not suffyce to assigne the reson."[198] Instead of merely copying from ancient books, he checked traditional techniques by his own experience and that of contemporary gardeners, thus giving new life to ancient practical knowledge.[199] The information exchange also went in the opposite direction. Gottfried reported advice he picked up from contemporary gardeners and funneled it into the literary tradition. More than eighty Latin manuscripts of his work have been identified, including translations into German, Middle English, and Czech. Gottfried's influence continued into the age of printing, turning up in German, English, Czech, and Polish books on gardening and wine making.[200]

Medieval craftsmanship was guided primarily by experience and the unwritten rules of oral tradition. However, as Gottfried's *Pelzbuch* suggests, it was also influenced (in ways that are difficult to measure) by the written word. The abundance of technical recipe books from the late Middle Ages strongly suggests that the relationship between the practical and written traditions within the crafts was closer than is generally supposed. Metallurgical manuals, for example, frequently contained influences of alchemical theory. A manuscript written in 1389 by a Nuremberg blacksmith and experimenter begins "Nv spricht meister alkaym. . . ." The work continues with recipes for hardening iron and steel that bear the unmistakable influence of alchemical terminology and methodology.[201] Cold water, wrote the Nuremberg smith, is the "common way" (*dy ist gemeyne*) to harden iron and steel. However, specialized waters containing mixtures of various herbs and animal substances work better for certain tools. Medieval smiths knew that iron and steel could be hardened by being quenched "the common way," in cold water. By carefully observing color changes in the heated metal, they could obtain satisfactory results.[202] But such a quench often resulted in brittle steel because of the rapid cooling of the metal. One way of solving this problem was by briefly interrupting the quench, removing the steel from the quenching bath in order to allow the metal to cool down slowly. Another method was to mix water with various animal and plant materials to make an oily or pulpy quenching bath, which would allow the heat in the metal to dissipate slowly. The Nuremberg smith listed more than a dozen recipes for such compound quenching baths. Scythes, for example, were best hardened in suet, while files should be quenched in a mixture of linseed oil and goat's blood. To quench knives and other cutting tools he recommended crushed bugloss leaves, and for quarry hammers the juice of caterpillars. The variety of these and other recipes suggests that smiths continually experimented with new ways to control the tempering process.

By what criteria were such ingredients selected? Practical experience was certainly the most common guide, and probably the most reliable. But it appears that simple trial and error was supplemented by folklore and by analogical reasoning about the properties of materials.[203] It has been suggested that the idea of adding plant and animal materials to quenching baths was based upon an analogy with medicine, the idea being to "treat" deficiencies in metals with various "drugs." The Nuremberg metallurgist evidently believed that the characteristics of various materials could be transferred to the heated iron during the quench. Plants such as radishes and horseradish may have been selected for quenching baths to harden knives and swords because of their inherent "sharpness," giving the metal a sharp cutting edge. The iris (*Schwertlilie*, or "sword lily"), which is prescribed in some recipes for hardening steel blades, may have been used because of its leaf's resemblance to a sword.[204] The Nuremberg manuscript also reflects the presence of learned tradition. The reference to "meister alkaym" in the manuscript suggests at least nominal acquaintance with alchemical doctrine. The fact that some of the recipes in the manuscript are written in Latin is a fairly certain sign of the presence nearby of a text. Even more interesting, however, are the ways in which the Nuremberg metallurgist used his materials. A recipe for hardening the edges of steel cutting tools prescribes one part each of radishes, horseradish, earthworms or larvae, and billy goat's blood "when the goat is in rut."[205] This recipe stems from sources going all the way back to Pliny, who believed that by virtue of the billy goat's inherent "hardness" and libidinous aggressiveness, its blood could break a diamond.[206] The Nuremberg metallurgist went even further by trying to give this piece of folklore a scientific explanation. After listing the four ingredients to be used, he noted that "this hardening has all four elements" (*dy herte hat dy vier elementen gar*). Other evidence of the influence of alchemical methods is found in recipes calling for various distilled ingredients.[207] Late-medieval metallurgy was not a simple empirical process, but a craft involving practical experience combined with elements of alchemy, magic, and analogical thinking.

By the fifteenth century, writing had become an important method for conveying technical information in the crafts. Economic expansion, particularly in the metallurgical industries, increased the demand for skilled labor, while changing technologies rendered older methods obsolete. The use of writing was especially prominent in the new crafts, such as those created by changes in military technology. The introduction of gunpowder and cannon created the office of the *Büchsenmeister*, or munitions master, one of the earliest "scientific" technicians. Engineers discovered that sketching could be an effective means of creating new designs and of solving engineering problems. They also found they could use their design books to advertise their services to prospective patrons. In the new technologies, books not only communicated important technical information, they also defined professional identities. The anonymous author of an early treatise on the *Büchsenmeister* discussed, besides technical information, the new profession's importance to society, the diverse knowledge and skills required by the art, and the history of the discovery of gunpowder. So important had literacy become for the munitions master that this author declared, "The master should also be able to read and write, for otherwise he cannot comprehend all the things that pertain to this art."[208]

Despite the advantages of the written word as a means of communicating technical information, publication was not universally regarded as advantageous. As we have seen, the artisans' economic well-being depended upon maintaining craft secrecy. Engineers and inventors were also reluctant to publish their discoveries because they knew that others might steal their secrets and claim themselves as the inventor. In most cases, a "secret" was more valuable than public knowledge. Although we are no longer inclined to explain Leonardo da Vinci's use of mirror writing in his notebooks as a method to avoid prosecution—it came naturally to him as a left-handed autodidact—the practice protected his inventions by making casual copying difficult. Other Renaissance engineers employed similar tactics. Giovanni da Fontana (ca. 1395–ca. 1455), an Italian engineer, composed all his technological treatises in cipher to prevent their being read.[209] The Florentine architect Filippo Brunelleschi (1377–1446) warned Mariano Taccola (1382–ca. 1453):

> Do not share your inventions with many, share them only with the few who understand and love the sciences. To disclose too much of one's inventions and achievements is one and the same thing as to give up the fruit of one's ingenuity. Many are ready, when listening to the inventor, to belittle and deny his achievements, so that he will no longer be heard in honorable places, but after some months or a year they use the inventor's words, in speech or writing or design. They boldly call themselves the inventors of the things that they first condemned, and attribute the glory of another to themselves.[210]

Taccola, ignoring Brunelleschi's advice, was repeatedly plagiarized. So was the architect Francesco di Giorgio Martini (1439–1501), who complained bitterly:

> Such knowledge [of my inventions] as I have has been acquired with great toil and at the sacrifice of my means of livelihood, so I am reluctant to show them forth to all, for once an invention is made known not much of a secret is left. But even this would be a lesser evil if a greater did not follow. The worst is that ignoramuses adorn themselves with the labors of others and usurp the glory of an invention that is not theirs. For this reason the efforts of one who has true knowledge is oft retarded. If in all epochs this vice hath abounded, in our own it is more widespread than in any other.[211]

In the absence of effective copyright provisions, Renaissance engineers were justifiably reluctant to publish their discoveries.

The idea that craft knowledge constituted a form of intangible property developed in the late-medieval guilds.[212] Craft secrecy was the instrument by which the guilds maintained the integrity of such "intellectual property." In the fifteenth century, city governments also began to realize that technical knowledge was valuable intellectual property. In order to ensure that the local economy would incur the benefits of inventions, city governments took measures to protect inventors' rights. Patents emerged in response to a growing awareness that knowledge could be put to practical use, and that as long as new discoveries were kept secret, the advancement of knowledge, and hence of profit, would be retarded.[213] The earliest known patent was one issued in 1421 by the Council of Florence to the architect Brunelleschi for a design for a cargo ship. The council's order expressly forbade any person, "wherever born and of whatever status, dignity, quality, and grade," to use Brunelleschi's design for a period of three years. The order was intended as an encouragement to Brunelleschi to "open up what he is hiding and . . . disclose it to all." It also aimed to stimulate the inventor, "so that he may be animated more fervently to even higher pursuits and stimulated to more subtle investigations."[214] The precedent established by the Council of Florence was emulated by other Italian cities, frequently by the grant of a monopoly on the use of a device to the inventor. The first patent law was enacted in 1474 in Venice. The law justified the protection of intellectual property on strictly economic grounds:

> We have among us men of great genius, apt to invent and discover ingenious devices; and in view of the grandeur and virtue of our City, more such men come to us every day from divers parts. Now if provision were made for the works and devices discovered by such persons, so that others who may see them could not build them and take the inventor's honor away, more men would then apply their genius, would discover, and would build devices of great utility and benefit to our commonwealth.[215]

The statute enjoined anyone from imitating patented inventions for a period of ten years. More important, it recognized the social utility of patents and viewed the protection of intellectual property as necessary for advancing technological knowledge.

The number of patents granted increased dramatically in the sixteenth century. Throughout Europe the system insured that inventors, by taking out a patent, need not fear losing their claim to priority of discovery. The concept of intellectual property rights was also extended to the realm of pure ideas, as printers were accorded the same privileges as inventors. In 1469 the Venetian Senate granted John of Speyer, a German printer, a patent on publishing books in the Republic of Venice, giving him a monopoly on the printing and sale of books for a period of five years.[216] By the following century copyrights had come into being, according authors as well as printers intellectual property rights.[217] Scribal culture offered little incentive for men of science to make their discoveries known. As long as secrets were kept secret, they were valuable; once they became public property, they could be exploited by anyone and hence were worthless. However, with the advent of printing, an author, even without a formal copyright, could at least be assured that his discoveries would be publicly acknowledged, since his name would be prominently displayed on the title pages of his books. As we shall see in the following chapters, the growing market for printed books soon made it apparent that one could reap not only fame but also profit by publishing "secrets" for a new, vastly expanded reading public.

PART TWO

THE SECRETS OF NATURE
IN THE AGE OF PRINTING

THREE

ARCANA DISCLOSED

MEDIEVAL science was defined and shaped by university culture and its official language, Latin. Science was the body of knowledge created in the universities by professors, passed down to students through lecture, commentary, and disputation, and embodied in a textual tradition whose language was Latin, the lingua franca of the learned elite. A language spoken only in the classroom and completely controlled by writing, Latin symbolized the barriers that divided learned from popular culture in the Middle Ages.[1]

The textualization of science in Latin served among other things to legitimize it, setting science apart from local, popular, and oral traditions.[2] Since in the Middle Ages "literacy" almost always meant knowing how to read and write Latin, knowledge of the Latin language became the norm that separated the scientific elite from the rest of society. The terms Scholastics used to describe the illiterate reveal much about the cultural meaning of latinity. In addition to calling them *illiterati*, academics referred to them as the *laici* and the *indocti*, signifying their secular status and their ignorance of science and doctrine. Perhaps most instructively, they called the uneducated *rustici* or *idiotae*, or referred to them as being *simplices* or *pauperes*.[3] As Brian Stock has pointed out, one of the consequences of the emergence of written culture in the Middle Ages was the notion that "literacy is identical with rationality." Illiteracy, by contrast, connoted credulity, superstition, and, at best, rustic simplicity. With the development of formal languages in science, theology, literature, and philosophy, textual analysis emerged as a general and preferred methodology, against which Scholastics invariably contrasted the "hearsay" of popular tradition, the dubiousness of local custom, and the naive empiricism of the unlettered *idiotae*. They dismissed as "popular," and hence unreliable, anything not supported by textual authority. In scholastic science the "idea of nature," authenticated by texts, supplanted the concreteness, physicality, and tangibility of nature that oral tradition accepted without question. The terms *litteratus* and *illiteratus* thus carried cultural meanings that transcended measuring the ability to read and write Latin. They signified the divide between two cultures. The Latin language, which during the Middle Ages was "increasingly a foreign tongue employed by a minority of *clerici*," drove a wedge between the rational, scientific culture of the educated elite and the concrete, "phenomenal" culture of the laymen, rustics, commoners, and "simple folk" at the bottom of the social scale.[4] Even in the presence of large numbers of vernacular manuscripts on the arts and crafts, and on alchemy, surgery, and other quasi-scientific subjects, medieval intellectuals maintained the conception of *scientia* as demonstrative knowledge, and of science as the exclusive preserve of learned culture.

Print Culture and the Divulging of Secrets

The advent of printing did not, by any means, erase the boundary between learned and popular cultures. To some extent it merely formalized that boundary. Yet printing permanently altered the distribution of cultural materials in society and facilitated exchanges of information between groups formerly kept apart by social and cultural barriers. The culture that grew up around the printing press—"print culture," as Elizabeth Eisenstein calls it, distinguishing it from "scribal culture"— brought together scholars, craftsmen, merchants, and humanists engaged in common pursuits.[5] The printer's workshop was the prototypical locus for such exchanges, but the cultural impact of printing resonated far beyond the publishing houses. The spread of printing resulted in the creation of a host of new occupations, some directly connected with the production of books, others related to the distribution of books or to the regulation of the printed word. Thousands of men and women became involved in an activity that had formerly employed only a few hundreds. Publishers, printers, typefounders, engravers, compositors, woodblock cutters, proofreaders, booksellers, and even peddlers, whose traditional stock was enhanced by pamphlets and broadsheets, all worked at trades that were either new or significantly altered by printing. Printing also transformed the oldest activity connected with the production of books: writing. For it is anachronistic to speak of the professional writer before the advent of printing: the profession of the author was "bound to the press and born because of it."[6] When apothecaries, potters, sailors, distillers, and midwives got into print along with scholars, humanists, and clerics, the Republic of Letters was permanently changed. No longer did authors write books only for a small audience of academic readers. A broader and more diverse readership had arisen, and those involved in the production of books could not afford to ignore the varied interests of Europe's newly literate.[7]

If on one level printing opened up new avenues of communication among scholars, craftsmen, and the general public, plainly it did not fully "bridge the gap between town and gown."[8] In the traditional university-based disciplines such as medicine, printing tended instead to formalize the distinction between learned culture and lay culture. Sixteenth-century physicians, harassed by a flood of popular medical tracts streaming from the presses, responded with vicious attacks against quacks, charlatans, and popularizers. The physicians discovered that they, just as effectively as their competition, could use the printing press as a propaganda weapon and an instrument to demarcate the boundary between themselves and unlettered popular healers.[9] But the literacy barrier no longer coincided with the Latin barrier. With the advent of professional translators, the growth of printing houses specializing in vernacular literature, and the explosion of vernacular publication, the old distinction, Latin/lettered versus vernacular/unlettered, began to break down.

Printing gave a voice to high and low cultures alike. It also mediated between the two cultures. To better understand this process and its implications for early modern science, we must take a closer look at print culture. What kinds of cultural materials did authors and printers appropriate from the oral and the written, and from the Latin and the vernacular traditions? How did they modify and adapt these materials for their imagined audience of popular readers? Who read the "popular" scientific books and why did they read them? What resulted from the encounter between the rationalism of the educated elite and the naive empiricism of popular culture? The present chapter addresses these questions. My aim is not to offer a comprehensive treatment of "popular scientific culture" in early modern Europe. Instead, I want to examine how printing mediated between two traditions: one official, Latin, and guided by a methodology premised upon understanding nature by abstraction; the other unofficial, vernacular (and in some cases oral), and more "physical" and empirical in its approach to nature. For a variety of reasons, Germany offers a convenient starting point for such an exploration. Early modern Germany inherited a large and diverse vernacular scientific and technical literature, consisting of manuscript works in medicine, surgery, craft and industrial technology, architecture, and engineering.[10] This literature, the work of professionals writing for other professionals, was largely devoid of theoretical content. It was, on the other hand, a literature that spoke to immediate practical needs. It possessed a hands-on, empirical character rarely met with in the scientific books written for a scholarly audience. Germany also had a rapidly growing lay readership, a flourishing printing industry, a small but influential community of humanist scholars, and a language not well suited for expressing classical scientific concepts. Hence it was a culture in which printers, professional authors, and translators would play key roles in disclosing to laymen the arcana of medieval science.

The German printing industry experienced phenomenal growth in the sixteenth century. By the end of the fifteenth century, presses had been established in about sixty German towns. A century later, there were over three hundred printers at work in more than a hundred towns.[11] Cologne, Frankfurt, and Nuremberg were the leading centers of printing, having in the 1590s a combined total of seventy-four printers. Augsburg and Strasbourg, with eleven printers each, and Wittenberg with ten also boasted flourishing printing industries. Even small towns and villages such as Schaffhausen and Torgau had printers. Several factors contributed to this expansion. Economic prosperity and urban growth brought about sharp increases in literacy rates, especially in the cities.[12] Changing technologies pressured craftsmen to acquire new skills, many of which they could gain or improve by reading books. In the wealthy German towns, a rich and cultivated bourgeoisie acquired books for moral and religious edification, for practical instruction, and for entertainment. The winds of intellectual change were also propitious. The spread of humanism furthered the growth of schools and furnished a growing supply of scholarly texts for the nascent industry. The Reformation, which kindled massive amounts of spiritual and propagandistic literature, fueled the growth of printing and shifted its center of gravity from the south to central and northern Germany.[13] Books were merchandise to be sold at a profit. So the competition of the marketplace, rather than purely intellectual considerations, determined which titles went to press. As the demand for books grew—especially for new titles—so too did need for new copy, and the industry could meet only a small portion of it by translating classical works or, as frequently happened, by pirating the editions of other printers. Hence there emerged onto the literary scene the professional author who produced copy on demand. Whether working for wages as in-house authors or free-lancing and dependent on patronage, these obscure but prolific scribblers produced a sizable share of the books that streamed from the sixteenth-century presses.

Scientific Authors and Their Publics

For most scientific authors, writing was an adjunct to some other occupation.[14] Yet the advent of printing also made it possible for some individuals to make a living by the pen alone. One such author was Walther Hermann Ryff (ca. 1500–1548), who was by far Germany's most prolific and best-known scientific writer. Although his career as a scientific and medical author was a mere decade long, from 1538 to 1548, Ryff published forty-three books, at a dizzying pace of over four per year. In all, his works went through more than two hundred sixteenth- and seventeenth-century editions.[15] Like many popular writers, Ryff has a poor reputation among modern historians: "archplagiarist" is one of the kinder names reserved for him.[16] However, such characterizations have tended to blind scholars to his positive contributions to German culture. He plundered existing works freely, often without acknowledging his sources, and unabashedly claimed for himself authorship of works he only edited or translated. Yet in Ryff's day compilation, translation, and the retrieval of past knowledge were considered legitimate literary pursuits. Almost every author "borrowed" material from others, with or without acknowledgment, and thus, almost inevitably, risked being called a plagiarist. Moreover, during the first century and a half of printing, literary property rights were not clearly defined and were rarely protected by law. Even imperial privileges, Vesalius lamented, were "not worth the paper they cover."[17] Pirated editions abounded because most books were unprotected by copyright, and because it was cheaper to reprint a previously published work than to compose a new one.[18] That Ryff put his name down as the author of works he only edited or translated was hardly unreasonable for a professional writer working at an entirely new trade in an age having ambiguous standards of literary ownership. More important, Ryff and others who wrote for the popular press created a new kind of scientific literature. Drawing upon standard classical and medieval authorities and a rich store of German technical literature, they edited, annotated, and adapted these specialized works to the needs of untutored readers, and translated them into a language that had a limited scientific vocabulary. In contrast to the elevated philosophical language of academic discourse or the tightly abbreviated prose of the vernacular technical treatises, Ryff wrote in an informal, conversational style that was laced with moralisms and proverbs, repetitious in its use of synonyms, and tied to the concrete.[19] Ryff knew his audience, he knew its language and its interests, and he made his voice heard by hundreds of thousands of readers. Although his "vulgarizations" drew bitter denunciations from the academic establishment, from the perspective of the *vulgi* his books revealed scientific knowledge that had hitherto remained the exclusive property of learned culture.

A Strasbourg native, Ryff was apprenticed as an apothecary. He went on to the Basel medical school but returned to Strasbourg after only a year to begin a new career as a professional author.[20] Writing under the pseudonym "Quintus Apollinarius," he quickly established a reputation as one of Germany's leading translators of Latin scientific and medical works. His best-selling translations of Pseudo-Albertus's *Liber aggregationis* and the *Secreta mulierum* went through more than thirty editions, making it one of the most popular German scientific books of the sixteenth century.[21] His magnificent translation of Vitruvius's *De architectura*, which he executed for the Nuremberg printing firm of Johannes Petreius in 1548, made available to German architects and craftsmen the most valuable classical work on the rules of proportion.[22] Ryff also edited and translated contemporary scientific works. His books on surgery drew on recent works by the Strasbourg surgeons Hieronymus Brunschwig (ca. 1450–1512) and Hans von Gersdorff (ca. 1455–1529).[23] To make the Vitruvius translation more accessible to craftsmen, he included a manual on perspective, ballistics, and "geometrical measurement" (e.g., using geometry to calculate the height of a building). In writing the manual, he drew from a variety of contemporary works, including ballistics treatises by the Italian mathematician Nicolo Tartaglia.[24] Ryff's popular pharmacopoeia, the *Kleine deutsche Apotheke* (1542), was an extensive revision of a pharmaceutical treatise by Otto Brunfels (1488–1534), a Strasbourg physician and reformer who had been one of Ryff's professors at Basel.[25] Brunfels's *Reformation der Apotecken* (1536) was part of an effort to reform pharmaceutical practice along the lines of the contemporary religious reform. Just as the Church had drifted away from its true scriptural principles, Brunfels asserted, so had medicine and pharmacy "wandered away from the correct ancient fountains, Hippocrates and Galen, and fallen into the stinking puddles of the Arabs, Avicenna, Serapion and their like."[26] Brunfels was not alone in fulminating against the "scandalous" condition of the pharmacies. Medical reformers throughout Germany polemicized against the sale of adulterated drugs and urged using fresh native plants instead of exotic imported ingredients.[27] Such polemics, like those touched off by the religious reform, appealed to a nascent sense of cultural nationalism, and to a growing sentiment in favor of basing pharmaceutical practice upon a medical "scripture" purified of medieval (especially Arabic) contamination.[28] Ryff's *Kleine deutsche Apotheke* played an instrumental role in disseminating the "pharmaceutical reformation" in sixteenth-century Germany. Indeed, it sold far better and was more widely distributed than the work upon which it was based: Brunfels's *Reformation der Apotecken* appeared in only one Strasbourg edition, whereas Ryff's various versions went through ten Strasbourg editions and another thirteen at Frankfurt. Later Ryff published a version of the book expressly for ordinary householders, a work that went through more than a dozen editions.[29] While Brunfels initiated the pharmaceutical reformation at Strasbourg, Ryff disseminated its principles throughout Germany.

However we are inclined to judge Ryff's editorial practices, it is clear his works played a key role in disseminating scientific information to the German people. Indeed, his most notorious "plagiarisms" were in some instances the works that made the deepest and most lasting impression on early modern German culture. In his popular book on human anatomy (1541), he reproduced (in reduced form) the drawings from Vesalius's recently published *Tabulae anatomicae* (1538), a preliminary version of the famous *De humanis corporis fabrica* (1543).[30] The *Tabu-*

lae anatomicae, comprising six large woodcut illustrations on anatomy and physiology, was a genuine novelty. Printers immediately recognized its value for medical students. At least five unauthorized versions of the work appeared, drawing an angry response from Vesalius, who chastised "the miserable slaves of sordid printers" for plagiarizing his work.[31] Yet Ryff's work was not a mere reissue of the *Tabulae sex*; it was an attempt to produce a comprehensive anatomical work based on the most advanced research available. Combining Vesalius's drawings with Johannes Dryander's drawings of the brain and skull and Eucharius Roesslin's illustrations of the fetus in utero, Ryff created an essentially original work that doubtless did more to acquaint German readers with human anatomy than did any of the originals. Ryff made numerous improvements on Vesalius's work. His most important innovation was in the way he handled Vesalius's figures of the veins and arteries. Whereas Vesalius had represented the venous and arterial systems schematically, without attempting to illustrate their anatomical topography, Ryff superimposed the two systems on seated male and female figures, so that the relation of the vessels to the trunk and limbs of the human body was immediately apparent (see figure 2). Ryff was obviously trying to make an abstract representation concrete and understandable to an audience with a limited knowledge of human anatomy. He used Dryander's more accurate depiction of the sternum as a single plate, rather than Vesalius's seven-plated sternum, a relic of Galenic anatomy. Ryff informed readers that the pictures were "of yourself" (*dein sebst*) and were meant to reveal the marvelous handiwork of God. "Come along and look at the Lord's work," he beckoned, "for the Lord is wonderful and his work unfathomable." As a result of Ryff's popularizations, the ancient Socratic dictum "Know thyself" took on a new meaning.

Ryff stated repeatedly that he wrote his books for the *gemeine Mann*, the "common man." But he did not mean "everyman." In the preface to one of his works, he characterized his intended audience in the following way:

> As I don't want to carry water to the Rhein, I do not write this little book for the educated people, for they already know this art. Nor do I write for those ignorant blockheads whose brains you could make into pig's troughs. I write only for the simple, respectable, and devout little people who have until now, through God, asked for my advice and help. Some of them have not reached me only because they are too distant, or because poverty makes the way too hard by which they might give themselves help or at least comfort, until God sends them another kind of help. Because of these various requests and cheerful supplications I have . . . very briefly compiled [this book] from many old and esteemed writers.[32]

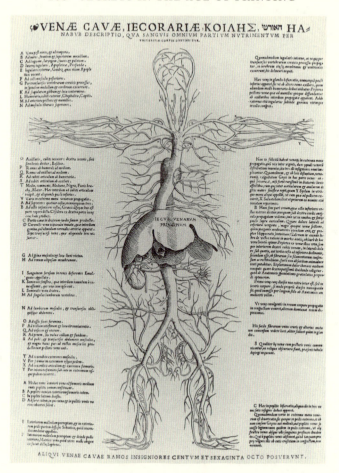

2. Venous system as depicted by (above) Vesalius, *Tabulae anatomicae sex* (1538), and (opposite) Walther Hermann Ryff, *Omnium humani corporis partium descriptio* (1541).

Elsewhere Ryff identified the common man as a *Hausvater* or as an "industrious householder."[33] For Ryff, the "common man" was an urban citizen or a member of a village community, hence a person with a secure, if modest, income and an established position in the middle ranks of society. Generally in the sixteenth century, the term *gemeine Mann* referred to the urban *Burgertum* (citizens, excluding patricians) and to nonnoble landowners and householders (*Allgemeindenutzer*) in the countryside.[34] The essential condition of being *gemein* was having legal rights in a municipality or village corporation (*Gemeinde*). The term "common man" thus excluded those at the top as well as those at the

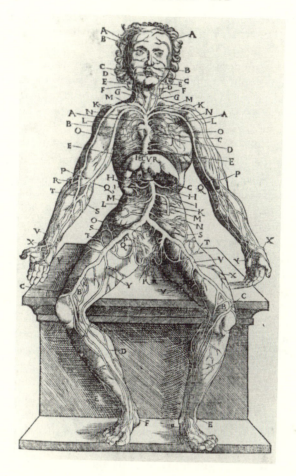

bottom of the social and economic scale: urban patricians and rural nobility on the one hand; on the other hand, the urban poor, landless peasants, Jews, and Gypsies. But the "common people" did not constitute a homogeneous class. Disparities of income and property differentiated individuals of "common" status. We may assume, however, that in addition to rural *Hausvätern*, the "common people" included journeymen and masters in the trades, merchants, shopkeepers, and growing numbers of women. Ryff assumed the common man was literate, but not learned. He could read German, but not Latin, and he had to rely on his own resources because he could not always avail himself of the help of others. Although Ryff offered plenty of advice for treating the poor (*armen Leute*), he did not imagine they were the main readers of his books, which they could rarely afford to buy. He was more concerned with providing medical advice to the apothecaries, surgeons, empirical healers, and *Hausvätern* under whose care the poor might find themselves.[35]

In publishing medical advice for the "common man," it was certainly not Ryff's intention to challenge the physicians' status or to question the efficacy of official medicine. If anything, he wanted to bring established medical practice and reformed pharmaceutical principles into "common use" (*gemeynem nutz*). Nevertheless, by publishing his works in German, he inevitably entered into an ongoing dispute with the physicians, who charged the vernacular writers with "prostituting" medicine. Responding to learned criticisms of his works, Ryff justified his "vulgarizations" on the grounds of improving the health and general welfare of the people. In reality, he implied, it was the physicians, not the popularizers, who threatened the growth of medical knowledge: "I know very well that this and my other works, which I have put into print in the German language for the simple man, have brought upon me the anger and scorn of many learned people. . . . I answer them in this way, that if good things are going to be thrown out for the sake of avoiding their misuse, many magnificent creations will have to be eradicated."[36] Not once, however, did Ryff challenge the legitimacy or the efficacy of academic medicine. His objectives were quite different from those of Paracelsus and his disciples, who deployed the German language as part of a wholesale political attack upon the "tyranny of Latin" exercised by the medical establishment.[37] Unlike Paracelsus, Ryff had no alternative school of medicine to offer, no theoretical ax to grind. Like most popular medical authors, he wanted to "reform" medical practice only in the sense of bringing it in line with traditional medical theory, not replacing the old theory with a new one.

Nevertheless, the proliferation of vernacular medical treatises drew an angry response from academic physicians. Even Lorenz Fries, a staunch defender of academic medicine whose academic credentials none could question, felt compelled to apologize to his fellow physicians for revealing the secrets of "Apollo's science" in the language of the common people. Fries insisted that he had written his popular *Spiegel der Arzney* (Strasbourg, 1532) to extirpate the errors of empirics and charlatans, and to inculcate respect for learned medicine among the people. Besides, he thought, the glory of medicine would surely not suffer for being communicated in German:

> It occurs to me that the German tongue is no less worthy, that everything might be written in it, than Greek, Hebrew, Latin, Italian, Spanish, or French, into which all sorts of things are translated. Is our own language supposed to be inferior? . . . The greater part of this art of medicine was written in the mother tongue, and only lately did it come into Latin: in

Greek by Hippocrates and Galen, in Hebrew by Isaac and Rabbi Talmud and Moses, in Arabic by Avicenna and Rhazes. . . . Should it be unfair, then, that the German nation . . . also have instruction in the holy art of medicine?[38]

Obviously, many academics thought otherwise. In the preface to a later edition of his book, Fries lamented that he was "very much hated and persecuted by many learned physicians, because I published the contents of this art in the German language."[39] Of course, the issue was not really the worthiness of the German language to convey medical knowledge. Underlying that hotly debated sophism was a more urgent concern: the proliferation of vernacular medical books enhanced the repertoire of empirics, the physicians' chief rivals in an increasingly competitive medical marketplace. In order to combat the "errors" circulating in the popular medical tracts, physicians had little choice but to write in the vernacular themselves. Johannes Dryander, professor of medicine at Marburg, made a special point of noting that his *Artzenei Spiegel* was no imitation of the other vernacular medical works but was firmly grounded upon established academic principles. Warning readers to stay away from "inexperienced mountebanks, uneducated monks, Jews, and foolish old women," he advised them instead to seek the help of legitimate physicians:

> For such bums and cheats have a drug, potion, salve, plaster, or some other absurd thing for every ailment, so that many a life is squandered away by them. Saying this, I do not want to dismiss the *experimentatores*, that is, the experienced physicians, insofar as their experience corresponds to the ancient physicians, or whenever they use science. . . . But wherever you can find a pious, God-fearing, experienced, learned physician, as one who owes his position to some authority certified by the community, him you may fully trust to treat any illness, and thus you may have, [and not disdain] God's means in a physician and in medicine.[40]

If the popular medical writers avoided overtly criticizing the medical establishment, their works nevertheless opened up an important debate over the "tyranny of Latin" in the professions and implicitly challenged the hegemony of the physicians over health care.[41] Defending the use of the vernacular on the grounds that the ancients had written in their own *Muttersprache*, the vernacular writers claimed that knowledge should no longer be the private reserve of the learned elite but should be made accessible to everyone. Taking their cue from the religious reformers, they drew a parallel between the restoration of the true Scriptures and the need for such a reformation in science and medicine. Jobst de Necker, one of Vesalius's plagiarists, wrote that while "Almighty God in His great wisdom and mercy has in our times loudly proclaimed His holy word to us, cleansed from all human obscuration, . . . [medicine] has been overlain and obscured especially because of the great handicap of ignorance, as the majority of physicians did not know anything except what they had sucked superficially from official pandects."[42]

In taking up this debate, the popular writers promulgated new values about medicine and health. While acknowledging that sickness is ordained by God, they taught that individuals could take measures to maintain their own health, and provided tangible advice on how to do it. When all around them friends and relatives fell victim to sickness and death, people were obsessed with health. They exchanged recipes, interrogated their physicians about remedies, and, increasingly in the sixteenth century, turned to medical self-help books to expand their repertoire of cures. Although it is difficult to determine what impact these tracts had upon the state of health in sixteenth-century Germany, at least they enabled people to take control over their lives and to do something tangible to improve their well-being.[43] The vernacular authors also valorized practical experience over theoretical schooling. The challenge to the prevailing negative evaluation of the manual arts was especially forceful in the treatises on surgery. Hieronymus Brunschwig noted that "surgery is a work of the hand which belongs to the *Wundarzt* and not to the physician," implying that if the surgeon is not competent to practice "physic," neither is the physician qualified to do the surgeon's work.[44] Even though surgery is a handicraft, Ryff maintained, like medicine it has two parts, the theoretical and the practical. He left little doubt about where in the equation his own sympathies lay: "Theory is the art unto itself, through which we find out something afterwards, and diligently reconstruct the cause of every effect. But practice is that by which something is grasped by the hand, and is actually accomplished."[45] While printing may have reinforced the distinction between theory and practice, it also helped to make "invisible" (i.e., nonliterate) arts visible, lending them textual legitimacy. The printing press was deployed to assert the dignity of the arts and the value of experience gained by practice.

In the preceding discussion I have characterized Ryff as a popularizer, a writer who attempted to make current medicine accessible to a popular audience. It would be a mistake, however, to conclude that he merely parroted the university professors, or that the effect of his books was to "acculturate" the people with the values of the learned elite. As we have seen, his knowledge of theoretical science was shaped largely by the classical tradition, but he also used vernacular sources for subjects about which academic writings were silent. He consulted medical empirics and craftsmen—surgeons, distillers, oculists, and reckonmeisters, for example—to obtain clarification of specialized terms and concepts. A medical school dropout, he had only a smattering of formal training in medicine. His culture was not that of the academy but that of the printing house, a shifting space in which academic, "popular," craft, and perhaps even folk influences and traditions converged. These considerations have implications for our current understanding of how popular culture changed during the age of printing. They substantiate Peter Burke's critique of the conventional "acculturationist" model of cultural change, which implies a dominant culture acting as the donor and a subordinate culture as the recipient of its values. Such a model does not adequately describe the complex interaction between elite and popular cultures. A better model, Burke suggests, might be "negotiation," in the sense of the adaptation or interpretation of the dominant culture's values and ideas to suit the needs of ordinary people.[46] Such a model certainly comes closer to accommodating the activities of popular authors like Ryff, who are all too often, and erroneously, viewed simply as "vulgarizers" of the knowledge of the learned classes. Creating scientific books for popular readers was no simple, mechanical process of translating scientific words and concepts into ordinary speech. Translators had literally to rewrite existing scientific works, and to create a vernacular scientific language.[47]

Clearly, it makes little sense to speak of "elite culture" and "popular culture" as monolithic entities. Quite apart from the difficulty of identifying such-and-such cultural forms (texts, beliefs, or practices) as "popular" or as "elite," cultural consumption of supposedly "popular" forms was rarely uniform. Ryff's medical books were read not only by barbers, bone-menders, and *Hausvätern*, but also by merchants and doctors. The vernacular scientific literature of the sixteenth century resulted from a discourse that took place along a continuous spectrum, not from a dialogue between two monolithic groups, "lay" and "learned." Even if much of that literature was watered-down academic science, it was filtered through the lens of cultural brokers such as Ryff. The result was a mediation between natural philosophy and the people's immediate, concrete, and empirical experience of nature.[48] At one end, the "idea of nature" penetrated vernacular literature, an extension into the popular domain of the psychological power over the concrete that abstraction gives. At the other end, the empirical and phenomenal reality of the craftsmen's experience was brought to bear upon theory.

Scientific Publishing

Although professional authors were essential for the creation of popular scientific literature, they would have had no role at all without printers, who published and marketed their books. The printer was the key figure around whom all the relations and arrangements within print culture revolved. He was responsible for raising capital, obtaining supplies and labor, organizing the workshop, and mobilizing the talent behind the production of books. The printer decided which books went to press and determined the format—whether cheap octavo or expensive folio—in which they were published. He commissioned translators and editors to compose new copy from existing texts, and sought out authors of original works. Printers made literary celebrities out of obscure authors like Girolamo Cardano, who wrote that being discovered by the Nuremberg printer Johannes Petreius was "the beginning of my fame."[49] Printers also publicized and marketed books. They had to estimate the audience for their publications and to arrange for their distribution. When they made wrong decisions about any of these matters, the results could be disastrous. It was in large measure the printer's willingness to accept the risks associated with the competitive and uncertain vernacular book market that made sixteenth-century "popular science" a reality.

If becoming a printer was easier than becoming a baker, as Erasmus said, staying a printer was another matter altogether.[50] Since the trade was new and almost entirely without the regulations that controlled other crafts, many were encouraged to try their luck. The early recruits into the industry included former craftsmen, scribes, bookbinders, merchants, painters, students, and scholars, a mosaic of the Renaissance petite bourgeoisie.[51] But because the market for books was highly competitive and unpredictable, many failed. The secret to success in publishing, as in authoring, was to strike a balance between specialization and diversification in the types of books one produced.[52] Specializing in selected subjects, such as law, music, or medicine, enabled a printer to gain expertise in those subjects and to establish a reputation as a reliable publisher of certain kinds of books. It also gave him a modicum of control over a selected literary market. Yet there were limits to how far such a strategy could be pursued. In the uncertain market conditions of the nascent printing industry, specialization was risky. It restricted a printer's audience of readers and increased the possibility of saturating the marketplace. For many printers, specialization was a practical impossibility, because it presupposed knowledge of esoteric languages and subjects, required special type-fonts, and usually represented large investments in material and labor. Given the relatively restricted market for specialized books, most smaller publishers found it difficult to justify such major capital outlays. The general rule in sixteenth-century printing was to produce works aimed at several markets. By diversifying his output, publishing several different kinds of books at once, a printer broadened the potential audience for his books and decreased the risk of saturating the marketplace. Most smaller printers diversified out of practical necessity, printing whatever would sell on the local market or job-printing for universities, secular governments, or the Church. But even the larger and more prestigious printers followed such a strategy. Christopher Plantin, one of northern Europe's leading printers, supplemented his extensive output of humanistic works with liturgical manuals and popular literature, and did job-work for the Spanish government and the Antwerp City Council.[53]

Whatever his publishing profile, no sixteenth-century printer could afford to ignore the fact that the readership of Europe was rapidly changing. The traditional reading public—the clergy, professionals, and university professors and students—was being amplified by a swelling population of literate laymen. Reformation polemics had uncovered the enormous potential of the popular book market, which printers were eager to exploit. Even as interest in religious subjects began to flag in the late 1520s, the demand for vernacular books remained high. Only it shifted to new subjects, including science, medicine, and technology. Miriam Chrisman notes that in 1530 publication of scientific books in Strasbourg outstripped religious publication, and in subsequent years scientific publications consistently outnumbered religious ones.[54] Understanding the material world was becoming increasingly important to lay people.

Initially a response to the realities of a changing market, scientific publishing proved to be such an effective strategy that many printers adopted science as their principal subject. Ryff regularly sent his books to such publishers. By focusing on three of Ryff's printers, Jacob Cammerlander of Strasbourg, Christian Egenolff of Frankfurt, and Johannes Petreius of Nuremberg, we can develop a profile of scientific publishing in sixteenth-century Germany. Cammerlander, Ryff's first publisher, exemplifies the medium-sized printer of popular literature for whom scientific publication was a practical business strategy but not a dedicated literary program. Egenolff was Germany's largest and most famous publisher of vernacular scientific books. Petreius, who was also Copernicus's publisher, was a fairly typical Renaissance scholar-printer whose leading projects were Latin works for academic readers.

By the standards of the Strasbourg printing industry, Jacob Cammerlander was not a large publisher.[55] Operating a single press at near-peak capacity, he published about 140 editions between 1531 and 1548. Ninety-two percent were in German, and more than half were on scientific, medical, and technological subjects. The remainder was an eclectic blend of vernacular novels, travel books, histories, almanacs, calendars, and, when the Reformation came to Strasbourg, violently anti-Catholic pamphlets.[56] Drawing upon Strasbourg's rich tradition of vernacular scientific writing, he published works by local authors such as Ryff, Lorenz

Fries, and the alchemist Peter Kertzenmacher. His success was due to his remarkable ability to repackage works that had already appeared in print, almost invariably advertising them as "new": *Ein newer Albertus Magnus, Eyn newes Complexion Büchlein, Ein news Kunstbüchlein*. The German printing industry supported numerous small to medium-sized scientific printers like Cammerlander. Some were physicians or men of science, such as Heinrich Seybold, a professor of medicine who published medical books as an adjunct to his practice. But most had little or no scientific training. They published scientific books because they made a profit from them. Most German cities had one or more printers with publishing profiles resembling that of Cammerlander. Hermann Gülfferich, who operated a press at Frankfurt am Main from 1542 until 1556, followed a strategy almost identical to Cammerlander's, publishing predominantly vernacular works and specializing in reprinting popular scientific books.[57]

Certainly few printers exploited the market for popular scientific and technical books more effectively than Ryff's Frankfurt printer, Christian Egenolff (1502–1555). Unlike some of Europe's more famous printers, Egenolff did not attempt to make his mark as a publisher of scholarly books. He published only a few, carefully selected Latin titles, including works by Erasmus and Melanchthon. His target audience was made up of the new readers who had benefited from the educational advances accompanying urbanization, the growth of industry, the spread of humanism, and the Reformation. Over a career spanning some thirty years, Egenolff published more than five hundred titles, slightly more than half of which were on scientific, medical, or technological subjects; the remainder were about equally divided among devotional tracts, humanistic works, and didactical manuals on subjects ranging from letter writing to law.[58]

Egenolff's career exhibits the hybrid character of the early modern printer.[59] Born in 1502 at Hadamar, near Mainz, he attended the University of Mainz from 1516 until 1519 but did not obtain a degree. In 1524, he moved to Strasbourg and worked as a typefounder in the shop of Wolfgang Köpfel, a leading Protestant printer. Four years later, he set up his own press and began publishing assorted medical and scientific books, popular novels and chronicles, and a smattering of humanistic works. Egenolff may have realized that the business environment in Strasbourg, with its nineteen printers, was not favorable for small publishers. In December 1530 he sold his shop to Jacob Cammerlander and moved to Frankfurt am Main, where he set up shop as the city's only printer.[60] Eager to establish the new trade, the Frankfurt City Council immediately granted Egenolff's application for citizenship and assisted him in the purchase of a house in which to set up his press. The move to

Frankfurt was auspicious. It gave Egenolff a virtually competition-free environment in which to establish his business and provided him with an ideal center for marketing his books.[61] A flourishing commercial city, Frankfurt was a hub of transalpine and east-west European trade. Its two annual book fairs drew publishers and bookdealers from throughout Germany and Europe, making Frankfurt the most important center for the international sale and distribution of books.[62] Attendance at the Frankfurt book fair was imperative for Europe's leading printers. Not only was Frankfurt a major distribution point for their own books, the fair also alerted them to new foreign titles they might reprint and market at home. For Egenolff, who reprinted or had translated numerous books already published by other printers, the fair was an indispensable source of new titles. Thanks to his ideal location, his acute business sense, and timely patronage from the Frankfurt City Council, Egenolff prospered. Soon he expanded his operations by opening an office in Marburg and became the university's official printer. By 1550, he owned, besides his two printing establishments, seven houses in Frankfurt and a paper mill in the Black Forest.[63] When he died in 1555, he bequeathed to his heirs one of Germany's most successful publishing firms, which continued to operate under the Egenolff name until 1667.

Egenolff's background, combining craft training with a university education, ideally prepared him for the role of a cultural broker. As the proprietor of Frankfurt's most prestigious printing firm, he was a patron and member of the city's humanist community. The Egenolff circle included the humanist Jacob Moltzer (Micyllus); Johannes Fichard, the principal figure behind the Reformation in Frankfurt; Eucharius Roesslin the Younger, the city physician and a prominent medical author; and Philip Melanchthon, who visited Frankfurt in 1539.[64] Egenolff also maintained contacts with the artisanal community, which supplied him with several of his most popular titles and in turn provided a steady market for his books. Egenolff's prominence as a scientific and medical publisher drew numerous authors from Marburg and Frankfurt, including Adam Lonicer, Eucharius Roesslin, and Johannes Dryander, whose careers were propelled by their association with the Egenolff press. In 1544, Egenolff hired Ryff as a house author. During the three years of their collaboration, Ryff produced ten best-selling works, including his popular pharmaceutical manual, the *Confectbuch und Hausapotheke*, treatises on distillation and surgery, a midwife's manual, and an all-purpose health-care manual. The Ryff-Egenolff collaboration was mutually beneficial, for the Egenolff firm continued to publish and profit from Ryff's works down into the seventeenth century.

Johannes Petreius of Nuremberg (1497–1550) was one of sixteenth-century Europe's most important scientific publishers.[65] Petreius, who is best known as the publisher of Copernicus's *De revolutionibus* (1543), was an altogether different kind of printer from either Cammerlander or Egenolff. Unlike these, both of whom published cheap editions for popular readers, Petreius, working in tandem with leading Nuremberg humanists, dedicated his press principally to the production of scholarly editions of major scientific and medical books.[66] The majority of his publications were products of the humanist scientific revival, and it was as a humanist scientific publisher that he made his reputation. But like all printers, he also had to respond to the broader demands of the local book market. In addition to his humanistic projects, he printed a number of vernacular works, including travel books, Reformation tracts, Bible translations, almanacs, and medical handbooks. The project for which he hired Ryff, the *Vitruvius Deutsch*, was aimed at a mixed audience including humanists, craftsmen, and general readers. Even the most "scholarly" printers had to cater to the popular book market, if for no other reason than to finance the books they printed for academics.

The sixteenth-century printing industry flourished in an atmosphere of almost complete laissez-faire. For the vast majority of printers, the market alone determined what to publish and how many copies to print. Aggressive printers plagiarized one another freely and as a result often became involved in violent literary disputes. Since the principle of literary property was but nascent, there were practically no effective legal measures that could be taken to prevent a printer from reprinting a work already published. Nor, evidently, did moral qualms about plagiarism dissuade printers from pirating already-published works. Once a work was in print, it was considered part of the public domain. Not surprisingly, fierce disputes erupted over literary property. In 1533, the Strasbourg printer Johann Schott filed suit against Egenolff, accusing the latter of plagiarizing the woodcuts from one of Schott's printed herbals.[67] Egenolff replied that the suit was entirely unfounded. The two herbals were completely different, he contended; the one he printed was not based on one of Schott's publications but was a reprinting of an earlier work by Johannes von Cuba. Egenolff pleaded innocent to the charges on the grounds that his version of the herbal was printed for the common good: "No one is prohibited from reprinting old books, especially not such books as are of benefit to mankind, inasmuch as unprecedented and serious diseases have arisen in the present age."[68]

Charges and countercharges of plagiarism flew back and forth over the literary landscape like a hail of cannonballs. Meanwhile printers continued the profitable practice of reprinting the work of other printers at will. The literary free-for-all of emergent print culture taught authors a hard lesson: in order to get published they had to sell their manuscripts outright to a printer, who had practically no way of preventing other printers from pirating the book. Scientific and technical books were especially vulnerable to literary piracy, first of all because the growing demand for practical literature gave printers a powerful incentive to raid the bookstalls for copy. Moreover, plagiarism was especially difficult to prove in a work that ostensibly described nature or technical processes, in contrast to works expressing the author's own ideas. In his response to Schott's lawsuit, Egenolff compared an herbal with a painting or other work of art imitating nature. Egenolff argued, "Although Albrecht Dürer, Jacob Maller of Wittenberg [Lucas Cranach] and others have privileges that prohibit others from copying their paintings, it does not follow from this that if they paint a figure based on the story of Adam and Eve, Actaeon, or Achilles, no other painter should be allowed to make a painting after the same fable."[69] Evidently Egenolff's plea prevailed, since we know he printed several more editions of the book in question. While the gray market in ideas may have robbed authors of their rightful property, from the readers' standpoint it facilitated access to new ideas and techniques.

With the advent of printing, another scientific tradition emerged alongside academic science. The realities of early modern publishing compelled humanistic and vernacular printers alike to compete in the market for popular literature, and to print scientific books for the "common man." The bulk of this literature was made up of summaries and translations of classical works, books of medical remedies, prognostications, almanacs, and compilations of technical and trade "secrets." Does this mean that the popular scientific literature of the period was part of a concerted effort by elite culture to "acculturate" the masses? So, at least, have some historians read the popular literature of the period. Robert Muchembled argued that the literature of colportage "filled an ideological gap between learned culture in its pure form and what remained of popular culture. In popularizing the ideals of the elites for the masses, . . . it worked insidiously to bring to perfection the system of submission that was already in operation." Popular literature "presented a discourse on the validity of the dominant system," claims Muchembled, and thus reinforced the values of the elite even while pretending to speak to the people in their own language.[70] Certainly popular culture changed under the impact of printing. But the acculturationist thesis fails to take into account the relative autonomy of print culture in the early decades of typography, before effective copyright laws and the machinery of state- and Church-sponsored censorship went into effect. By focusing primarily upon popular literature and not upon how it came into being or how people read it, the acculturationist thesis does not give sufficient weight to the influence of popular writers, who adapted classical works to popular tastes and reading abilities. Nor does it take into account the business strategies of printers, who responded to the shifting popular book market by selecting and marketing the books that went to press. Nor, finally, does the acculturationist account recognize the aggressively original way in which the people read books and rearranged their content to create new meanings.[71] Popular culture was not a wax tablet upon which elite culture stamped its values.

A better model for the interaction of learned and popular cultures in the early modern era might be that of negotiation and coaptation. Such a model emphasizes the active participation of printers, vernacular writers, and readers in the production of "popular culture." On the basis of an intimate knowledge of the vernacular book market, printers and popular writers selected works from the classical repertoire and adapted them to their imagined public. At the same time, readers appropriated material selectively, giving it their own meaning independent of the meanings intended by the original authors. Moreover, the audience for vernacular books was considerably more heterogeneous than historians have generally acknowledged. As Roger Chartier points out, "the popular cannot be found readymade in a set of texts that merely require to be identified and listed; above all, the popular qualifies a kind of relation, a way of using cultural products such as legitimate ideas and attitudes."[72] The interaction of learned and lay cultures was not a stamping of one culture upon the other; it was a process of negotiation between the two. In the early modern era, when orality still held sway, reading was not strictly a relation between the private self and a text, but, more commonly, a public affair mediated by the spoken word.[73] Unintentionally and almost invisibly, the popular and learned traditions coapted to one another, producing a novel cultural synthesis. Some key elements of early modern science developed out of the copenetration of elite and popular cultures. The meeting of scholarship and craftsmanship, for example, resulted in a new definition of the relationship between nature and art. Empirical, "manipulative" knowledge confronted theoretical, analytical knowledge. The theoretical did not domesticate the empirical, nor did craft empiricism conquer theory. Instead, the meeting of the two gave rise to a new conception of scientific knowledge: the idea of knowledge as making.[74] By the seventeenth century, the capacity to reproduce nature's effects by artificial means had become a goal of science and an epistemological guarantee of its validity.

Printing and the Secrets of the Arts:
The *Kunstbüchlein*

Traditionally the artisan's world—the world of things and techniques—existed independently of the written word. Although there were technical manuals in some of the trades before the advent of printing, craftsmen learned primarily by apprenticeship rather than by following instructions from a book. Literary works on practical activities such as farming tended to be reflective and aphoristic rather than descriptive and operational.[75] One of typography's most important contributions to sixteenth-century literature was to produce a barrage of how-to-do-it manuals and of technological treatises detailing the manual side of the arts. These handbooks have received little attention from historians. Yet they are deserving of study for several reasons. First, the proliferation of the craft manuals made technical literacy a practical possibility. Indeed, for a growing number of people, including artisans as well as nonartisans, literacy, not membership in a guild or a formal apprenticeship, became the precondition for knowing the secrets of the arts. Moreover, the technical handbooks gave artisans information about novel, "scientific" procedures and about techniques developed by innovative craftsmen. At the same time, they contributed to standardization of procedures in the crafts. They also brought the craftsman's "laboratory"—his workshop—closer to the intellectual, enabling scientists to compare theoretical claims with technological results.[76] Finally, the publication of large numbers of technical handbooks made a permanent impact upon the mentality of the European middle class. In the following sections, I want to show how the sixteenth-century technical recipe books removed the veil of mystery surrounding the crafts and, in doing so, demonstrated that technical proficiency was not a matter of cunning, but merely of "how to." The "disenchantment" of technology in turn created new expectations about science: any true theory, according to the proto-Baconians of the sixteenth century, had to produce concrete results.

The most widely circulated of these printed how-to-do-it books was a group of craft manuals known collectively as the *Kunstbüchlein* (Skills-booklets), which appeared in various German towns in the early 1530s. Originally issued as four separate pamphlets, the booklets became immediate best-sellers, appearing in more than a dozen editions between 1531 and 1533. They were reprinted at least fifty times in the sixteenth and seventeenth centuries.[77] Although originally written for craftsmen, their influence did not end at the workshop. The recipes making up the booklets appeared in numerous works that appealed to a broad middle-class readership, including the famous *Secrets* of "Alexis of Piedmont," Johann Jacob Wecker's work of the same title, Wolfgang Hildebrand's often-reprinted *Magia naturalis* (1610), and the *Hausväterbucher* of the seventeenth century, those earnest encyclopedias of domestic economy that German householders consulted for all sorts of information, from caring for pigs to caring for infants.[78] The *Kunstbüchlein*, all anonymous publications, were printers' compilations. Popular printers like

Christian Egenolff compiled them from workshop notes, from word-of-mouth sources, and from various "experimental" treatises. They offer a glimpse of the printer as a disseminator of technical information, and as a divulger of craft secrets.

The *Kunstbüchlein* first appeared between 1531 and 1532, in response to the demand for technical information fueled by Germany's expanding industrial economy. Christian Egenolff, an innovator in technical publication, printed the first, a fifty-page booklet entitled *Rechter Gebrauch d'Alchimei* (The proper use of alchemy). Egenolff based the work upon an alchemical treatise by Petrus Kertzenmacher, "a famous alchemist" from Mainz. No manuscript of Kertzenmacher's treatise has ever been found. However, Jacob Cammerlander published the complete treatise in 1534, apparently in its original form.[79] Egenolff may have owned a unique copy of the Kertzenmacher manuscript and brought it with him when he emigrated from Mainz to Strasbourg in 1530. Cammerlander, who purchased Egenolff's Strasbourg printing shop when Egenolff moved to Frankfurt, probably acquired the manuscript as part of the purchase. A comparison of the two works shows that Egenolff made numerous revisions on the Kertzenmacher manuscript, eliminating its esotericism and metaphysical content, thereby completely changing the work's intent. For Egenolff, the "proper use of alchemy" was not gold making, nor the search for the philosopher's stone, nor the fixing of alchemical spirits, but practical alchemy, what today we would call industrial chemistry. The *Rechter Gebrauch*, he explained, was "not just for clever alchemists, but for all skilled workmen."

Kertzenmacher's treatise began with a preface expounding upon the nobility of alchemy and its esoteric character:

> All art proceeds from God, and is in Him without end. He imparts it, according to his pleasure, to all who desire it from Him. Now men desire only what is useful to them; therefore they seek out strange crafts, but only for their own advantage, not taking into account the glory of God. Therefore in some arts people rarely succeed and lose a lot of money in them. Among the arts, alchemy is the best, as Marogines, Hermes, Rhazes, and Albertus Magnus show. It is the highest of all, for whoever has it and is willing to find out its secrets will overcome all things. But it is very obscure, for the old masters who found out this art would not teach it even to their children and friends. Happy therefore is he who finds it, for it is not easily found. However, hard work conquers all things. Whoever seeks the proper arts with steadiness and diligence will find them.[80]

Kertzenmacher described alchemy as the art of separating out the "spirits" (*spiritus*) of metals, substances such as sulphur, mercury, and ammonia, which normally flee from the fire, and fixing them alongside me-

3. Title page of *Rechter Gebrauch d'Alchimei*,
showing vignette of a jeweler's workshop
(Frankfurt am Main, 1531).

tallic "bodies" (*corpora*). "Whoever wants to know this art correctly," he
wrote somewhat enigmatically, "must be able to make *Spiritus* become
Corpora, so as to remain permanently together in the fire, without losing
everything." Kertzenmacher's intent was to describe procedures that
would allow alchemists to accomplish the traditional goal of transmutation—as he put it, to "overcome all things." But Egenolff omitted
Kertzenmacher's preface entirely and eliminated the recipes that were
inconsistent with the more utilitarian purposes he had in mind for the
Rechter Gebrauch. He added recipes for making artificial gemstones and
for additional metalworking procedures. By editing the tract and supple-
menting it with new recipes, Egenolff converted a treatise on extracting
and fixing "spirits" into a workshop manual on practical chemistry. He
designed the work specifically for an audience he thought would benefit
from a knowledge of alchemical procedures, but one that had little inter-

est in speculative alchemy. Evidently he thought the booklet would be especially useful for goldsmiths and jewelers: a vignette of a jeweler's workshop adorns the title page, and the work contains numerous recipes applicable to these arts, including directions for making artificial amber and pearls, instructions on gilding, tinning, and silvering metals, on separating gold from alloys by liquation and by aquafortis, on assaying ores, on cementation and cupellation, and on softening gold so it could be easily coldworked for making gold leaf. The *Rechter Gebrauch* contained no speculative or theoretical discussion of any kind. It made no reference to transmutation, nor did it have any of the metaphorical language or cryptic symbolism so typical of the late-medieval alchemical tracts. Indeed, Egenolff made every effort to demystify alchemy, including, for example, a glossary of alchemical symbols and terms. Driving home the distinction between the "proper use of alchemy" and abuse of the art, Egenolff concluded the work with a doggerel verse, warning:

> Eight things follow alchemy:
>> Smoke, ash, many words, and infidelity,
> Deep sighing and toilsome work,
>> Undue poverty and indigence.
> If from all this you want to be free,
>> Watch out for Alchemy.[81]

By "purifying" alchemy, purging it of its metaphysical elements, Egenolff hoped to instruct artisans in its legitimate uses.

The *Rechter Gebrauch d'Alchimei* opened up at the popular level a debate that had long preoccupied the academic community. Late-medieval Scholastics had passionately debated the "question of alchemy," that is, whether or not alchemy, despite its dubious scientific merits, might still yield practical benefits in the mechanical arts.[82] The proponents of alchemy argued that art (*techne*) has the potential not merely to equal but to outdo nature, a revolutionary idea for a culture conditioned by the classical idea that art is inferior to nature. However, such radical views never attracted a wide following within the academic community. Despite widespread interest in alchemical procedures, the alchemical enthusiasts remained a minority, particularly after the fourteenth century, as religious authorities took an increasingly hostile stance against the art. To many churchmen, the claim that humans could transmute species created by God seemed arrogant in the extreme. William Newman has suggested that mounting criticism of alchemy, culminating in Pope John XXII's papal bull of 1317 condemning the art, forced the alchemists to go "underground," to write under pseudonyms, and to otherwise conceal their identities. Only in the sixteenth century, when an intellectual climate more receptive to the art had developed, did alchemy resurface as an intellectually respectable discipline.[83]

While academics acutely felt the constraints of theological condemnations of alchemy, and hence regarded the subject with caution, artisans used alchemical techniques freely, unconcerned with philosophical justifications or apologies. Conrad Gesner (1516–1565) reported that empirical doctors were more receptive to the use of distillation for preparing drugs than the learned physicians.[84] Although apothecaries were probably the first professionals to distill alcoholic drinks on a large scale, lay people also took up the trade. In Germany, the aquavit distilleries (*Wasserbrennereien*) began as home industries run chiefly by women. So many "women confectioners and housewives" were brewing *Branntwasser* in sixteenth-century Nuremberg that the city council passed an ordinance against the practice. Despite attempts to curb the trade, the *Wasserbrennerinnin*, or aquavit women, continued to practice their trade, and to make products in simple kitchen stills.[85] Traditional crafts also adopted alchemical techniques. In the rapidly expanding metallurgical industries, alchemical apparatuses and procedures were being used in a wide variety of functions. Georgius Agricola noted, for example, that the alchemists had discovered a method for parting gold from silver by nitric acid.[86] So similar were the arts of alchemy and metalworking that Vannoccio Biringuccio wrote, almost apologetically, that the goldsmith must possess certain secrets "which in truth are parts of alchemy, such as softening gold when it is brittle and raw, coloring it when it has little color, soldering, enameling, working in niello, blanching, gilding. . . . Except for the manual work, the art of the goldsmith has a close connection with that of the alchemist because it often makes a thing appear what it is not, as is seen in setting gems, in heightening the color of gold, in whitening silver, and also in gilding things that really are of silver, brass, or copper but appear to be of gold."[87]

Biringuccio distinguished between two different "pathways" toward alchemy. One was "the just, holy, and good way" that artisans took. The other was the road taken by fraudulent alchemists, who deceived and robbed people. Good alchemists were "but imitators and assistants of Nature." Their art "gives birth every day to new and splendid effects such as the extraction of medicinal substances, colors, and perfumes, and an infinite number of compositions of things." "Sophistic" alchemy, on the other hand, though born from practical alchemy, was entirely different. "Usually only criminals and practicers of fraud exercise it," Biringuccio maintained. "It is an art founded only upon appearance and show. . . . It contains only vice, fraud, loss, fear, and shameful infamy. Since its result is mean and poor, this art is followed by persons of a like nature."[88] The *Rechter Gebrauch* made a similar distinction between "proper" and "improper" alchemy: not a scientific, but a moral distinction. "Proper" (*recht*) alchemy, the alchemy of the artisan's workshop, had honest intentions and produced useful results. Improper alchemy, because of its false promises and its fraudulent intent, placed it-self outside the moral economy.[89]

The second tract making up the *Kunstbüchlein*, entitled *Artliche Kunst* (Pretty skills), was a twenty-two-page booklet of recipes for inks and paints used principally to illuminate books and manuscripts. In 1531, three different printers—Simon Dunckel of Nuremberg, Peter Jordan of Mainz, and Melchior Sachs of Erfurt—published imprints of the work. The *Artliche Kunst* was probably compiled from an illumina-tor's workshop manuscript. Many such manuscripts existed and would have been readily available to most printers.[90] Far from putting book decorators out of business, the printing industry kept them busier than ever by producing hundreds of copies to illuminate. Many early modern readers treasured printed books as if they were manuscripts, and often had them embellished and decorated.[91] The intensification of book pro-duction brought about by printing led to changes in workshop organi-zation, since the number of experienced illuminators would not have been sufficient to meet the increased labor requirements. Hence printers had to hire less experienced craftsmen. To train the new work force, printers and master illuminators created workshop manuals and model books. The so-called *Göttingen Model Book*, recently discovered by Hellmut Lehmann-Haupt, is an example of such a manual.[92] It set forth step-by-step instructions for drawing and coloring acanthus leaves and checkered backgrounds, two of the basic elements of manuscript deco-ration. Its simple, methodical arrangement ideally suited it to the "as-sembly line" mode of manufacture that was being introduced into this stage of book production. Possibly the *Artliche Kunst* was created to meet similar needs. However, such a booklet would also have found a market outside the printing establishments. Even after woodcuts re-placed pen-and-ink drawings, owners frequently decorated their own books, and printers occasionally provided instructions on the colors to use.[93] The demand for such handbooks increased as more books became available. The *Artliche Kunst* was reprinted continuously down into the 1540s, until it was superseded by Valentin Boltz's *Illuminierbuch* (1549). The latter work went through six sixteenth-century editions and another ten in the seventeenth.[94]

The last two *Kunstbüchlein* appeared in 1532. *Allerley Mackel und Flecken aus . . . zu bringen* (How to remove various stains and spots from clothing), a household manual on dyeing and cleaning fabrics, was printed by Sachs, Jordan, Meierpeck, and Kunigunde Hergot of Nurem-berg.[95] Most of the recipes in the *Artliche Kunst* are for spot removal; the dyeing recipes are fewer in number and technically not very sophisti-cated. In the sixteenth century, dyeing was a highly skilled art whose secrets were carefully guarded by specialists. The *Allerley Mackel* would have been useless for professional dyers, but quite handy for general readers.[96] Stains were a constant problem for ordinary people, who owned few garments. The stain-removing agents in the handbook employed mild acids, starch pastes, and lye, especially effective on grease spots. The dyeing recipes specified ingredients that were easy to obtain locally, such as safflowers, elderberries, gallnuts, and brazilwood, which could be purchased from the apothecary. A typical recipe, one for dyeing linen blue, made a good, serviceable dye: "Take elderberries, dry them in the sun, and then soften them in vinegar for twelve hours. Rinse them by hand. Strain them through a cloth, and mix with verdigris and ground alum. But if the color is to be bright blue, put in more verdigris. Put the thread or cloth in the mixture."[97] With the exception of the formulas that employed brazilwood, none of these recipes would have made a very fast dye, although they were probably satisfactory for renewing the color to older fabrics or for concealing spots left by removing stains.

The final tract, *Von Stahel und Eysen* (On steel and iron), was a booklet on metallurgy, explaining techniques for hardening iron and steel, and for soldering, etching, and coloring metals. Sachs, Jordan, and Hergot all published the work in 1532. Since *Von Stahel und Eysen* appears to reflect a Nuremberg metallurgical tradition, it is probable that Kunigunde Hergot was the work's original publisher. The booklet's title page announced that it was printed "for all armorers, goldsmiths, girdle-makers, engravers of seals and dies, and all other skilled artisans who use steel and iron." The work was not intended for alchemists, wrote the printer; while the alchemists discovered these arts, "they are only the beginnings and are child's play compared with the practices of the alchemist's art." The printer's aims were more modest, as were the sources of the recipes: "Since these recipes were brought to me and taught me by reliable folk, I did not wish to keep them for myself but wished to make them available to everyone to whose work they pertain and who may profit from them, in particular armorers, locksmiths, engravers of seals, etc., together with all others who work iron and steel."[98] Among the work's most interesting recipes are the ones for etching on steel, a relatively new technology in the sixteenth century. Etched ornament first appeared on late-fifteenth-century armor. The art reached its peak in Nuremberg, where artists such as Peter Flotner and Albrecht Dürer created designs.[99] However, recipes describing the process were rare, appearing in print for the first time in *Von Stahel und Eysen*. The work thus described a process that was almost completely unknown outside of professional workshops.[100] Extreme specialization in the metalworking trades tended to confine knowledge of the "secrets" of smithing to small groups. As Vannoccio Biringuccio noted, "there are as many kinds of special masters [of ironsmithing] as there are things that can be made of iron." Some masters worked iron well and steel poorly, while others worked steel well and iron poorly. Consequently, he marveled, the art "has more secrets and perhaps more ingenious secrets than the art of any metal."[101] The first printed book on the technology of iron and steel, *Von Stahel und Eysen* revealed to the general public the secrets of an esoteric and highly specialized art.

About half the recipes in *Von Stahel und Eysen* prescribed quenching waters for hardening and softening steel. The recipes are similar to those in the Nuremberg metallurgical manuscript examined in the previous chapter.[102] As we have seen, such "secrets" were quite specialized, specifying different combinations of organic and inorganic materials for each type of tool or desired hardness, not unlike a medical recipe book that prescribes specific compound drugs for various illnesses. The recipes, improvements over the "common" methods of working steel, suggest a melding of alchemy with the craft tradition. According to *Von Stahel und Eysen*, the "common way" to harden steel was to quench it in cold water. However, special baths are more effective. Files, for example, should be hardened in linseed oil or the blood of a billy goat. Cutting tools are best quenched in a bath made of the juice of radishes, horseradish, earthworms, cockchafer grubs, and billy goat's blood. Augers and drill bits should be hardened in a mixture of urine, verbena, and cockchafer grubs, while a water made from distilled snails makes "a hardening that cuts through everything." As I suggested earlier, it is possible that some of these concoctions resulted from attempts, derived empirically, to vary the cooling rates of the hot metal during the quenching process. Pure water might have cooled the metal too rapidly, causing it to become brittle, whereas the addition of organic materials to the quenching bath may have created surface deposits on the metal that tempered the cooling process. The only explanations possible for this in the sixteenth century were analogies based upon sympathetic magic or the doctrine of signatures. According to the former, the juice of horseradish hardens knives and swords because of its inherent "sharp" qualities, making the quenching water more "biting," while, according to the latter, water of iris leaves (*Schwertlilie*, "sword lily") does so because of the leaf's resemblance to a sword. None of these explanations appears in *Von Stahel und Eysen*, and we have no way of knowing whether sixteenth-century readers extrapolated such meanings from the text. We can be sure, however, that the work represents the cumulative practical and theoretical experience of generations of medieval craftsmen, whose knowledge and skills were, for the first time, being revealed to the general public.

Science and Popular Culture in
Early Modern Germany

Who was this public? Who read the *Kunstbüchlein*? Who could read them? At present we can give only tentative answers to these questions because our evidence concerning literacy rates, book ownership, and reading practices in the sixteenth century is fragmentary. Important clues to a book's audience can be found in printers' and authors' dedications, for certainly they knew the book market better than we do. When Egenolff said he intended *Rechter Gebrauch* for "all skilled workmen," and when Kunigunde Hergot stated she published *Von Stahel und Eysen* for metalworkers, we ought to take seriously the possibility that the *Kunstbüchlein* were in fact read and used above all by artisans and by those who aspired to enter one of the crafts. It may seem unnecessary to stress this point, but the relatively unsophisticated nature of the recipes in these tracts has led to the view, widely shared among historians of technology, that craftsmen would have found little that was new or useful in the *Kunstbüchlein*. This conclusion, based upon a positivistic technical history of the crafts, assumes that the craftsman's sole criterion for selecting technical information was how up-to-date it was. It ignores important social and economic changes that transformed the status of artisans in early modern Germany.

One of the most important developments in sixteenth-century German urban history was the emergence of a lower middle class (*Kleinburgertum*), consisting primarily of craftsmen who had lost control of the means of production. The principal cause of this transformation, according to a recent study by Christopher Friedrichs, was the emergence of a new manufacturing system, the *Verlagsystem*, or putting-out system. Under this system, a *Verleger*, a merchant who "put out" work, provided artisans with raw materials and a part-wage, the remainder being paid on delivery of the finished product. Even though some craftsmen continued to work independently in their own shops, the *Verleger* organized and controlled the production process, which often involved several different crafts. As his control over the means of production increased, the *Verleger* was able to concentrate production into protofactories, or manufactories, hiring relatively unskilled or semiskilled workers as wage laborers.[103] Although the *Verlagsystem* is usually associated with the textile industry, it arose in connection with many other types of production as well. By the early sixteenth century in Nuremberg, it had been introduced into the metal industries, printing, and the manufacture of purses, gloves, brushes, paper, nails, bottles, knives, and other articles.[104] The *Verlagsystem* caused sweeping changes in guild structure and workshop organization. Guild restrictions tightened, making it increasingly difficult for apprentices and journeymen to advance to the rank of master craftsman: "the more economic pressures the craft masters were subjected to from above, the more they tried to defend themselves from further competition by blocking off the admission of new masters from below."[105] Eventually these changes resulted in the disappearance of the artisan as an independent producer. As wealth concentrated in the hands of merchant entrepreneurs, the political and economic power of the guilds declined. Master craftsmen protected trade secrets more jealously than ever, journeymen lost their sense of solidarity with the guilds, and the industrial work force gradually took on the mentality and character of an unskilled proletariat, which it was becoming.

Printers were not unaware of these economic and social changes. Many of them had spent years working for wages as pressmen, typefounders, and proofreaders in a trade that was relatively new and without traditions or guild restraints. They had entered a career that was wide open to all talents: some had previously been goldsmiths or woodblock-cutters, while others had been painters, bookbinders, barbers, or even tavern-keepers. As "self-made men" eager to prove the power of the printed word, printers optimistically (and perhaps naively) responded to the changing economic currents by producing self-help manuals to enable aspiring craftsmen to learn trade secrets on their own. Although the *Kunstbüchlein* would not have been adequate substitutes for a formal apprenticeship, they might have supplemented on-the-job training and served as workshop reference books to novices. Moreover, the new mode of industrial organization brought about by the expansion of the *Verlagsystem* and by the rise of manufactories required fewer, simpler, but specialized skills—the type that might be acquired by reading recipes. Indeed, printers may have designed some of these craft booklets for their own workers, for the printer's workshop was in many respects an early model of the manufactory that was soon to take shape in almost every industry.

This hypothesis, of course, assumes that artisans and aspiring artisans could read. Are we justified in making this assumption? Although evidence of literacy in early modern Europe is notoriously difficult to interpret, modern scholarship makes three key points: (1) literacy was much higher in most places in sixteenth-century Europe than has formerly been supposed; (2) the rate of literacy increased steadily throughout the century; and (3) the incentives for learning to read and write were especially felt in urban areas, where economic advancement often depended upon the acquisition of complex new skills.[106] In Germany as elsewhere in Europe, literacy was being recognized as an indispensable skill for almost every urban occupation tied to commerce and exchange. In almost every German city in the sixteenth century, private schoolmasters set up shop to teach the rudiments of reading, writing, and arithmetic to ordinary citizens.[107] Hundreds of artisans from all over Germany could sign their names in the *Bruderschaftsbuch* of the city of Frankfurt am Main between the years 1417 and 1524. In conservative Bavaria, the authorities were alarmed to discover that a large portion of the "common burghers" (*gemeine bürger*), artisans, "ordinary men" (*der gemein man*), and even peasants possessed Bibles, prayer books, and seditious religious tracts.[108] While some authorities considered reading to be dangerous, city officials in many communities fervently resisted attempts to close schools. Much was at stake: as the town fathers of Heidenheim expressed it, knowledge of reading and writing "do honor to God and promote trade and an honorable walk of life."[109] Literary evidence from all over Germany suggests that literacy was taken for granted in the higher ranks of society and regarded as normal among artisans. Reading was so commonplace by the end of the sixteenth century that the Augsburg minnesinger Daniel Holzman could proclaim it was a proverb everywhere in Germany that "he is but half a man who cannot read or write."[110] Indeed, it would have been pointless for printers and Protestant propagandists to generate thousands of religious tracts, how-to-do-it books, and self-improvement manuals for a population that was generally illiterate.[111] The fact that so many printers made successful careers publishing exclusively vernacular literature is in itself a strong indication of widespread literacy.

Such evidence points to another, larger and more diverse audience for the *Kunstbüchlein*: nonspecialists and the general public, that mosaic of new readers that printing itself helped to create. The majority were urban people, since in the cities literacy was relatively common and books were readily available. The ability to read and write was less widespread in the countryside, although still not unusual. In any case, as Roger Chartier has pointed out, "in the sixteenth and seventeenth centuries, . . . a relationship with the written word did not necessarily imply individual reading, reading did not necessarily imply possession of books, and familiarity with the printed word did not necessarily imply familiarity with books."[112] The people's utilization of printed material could be collective when read aloud in religious assemblies, at public festivals, in workshops, and at evening winter gatherings such as the *Spinnstuben* and the *veillée*. Nor did one have to own books to read them. Borrowing books was as common then as it is today. In numerous and varied ways the common people encountered the printed word, even if they did not own many books.

An indication of how popular literature circulated can be gained from the account book kept by the Frankfurt bookdealer Michael Harder for the Frankfurt book fair of 1569.[113] Harder sold more than 5,900 books to bookdealers all over Germany. He had customers in cities as far north as Braunschweig and as far south as Zurich, and in more than twenty cities and towns in between, including Nuremberg, Münster, Magdeberg, and Schwäbisch-Hall. Georg Willer of Augsburg purchased 1,092 books from Harder through his agent, while a shipment of 527 books returned to Wittenberg with two printers. Harder's clients bought predominantly vernacular books, and their purchases reflected the literary tastes of the readers whom they in turn served. The inventory makes it clear that "ordinary" German readers bought books for entertainment, information, and moral instruction. The three most popular books on the inventory were a German version of the *Gesta Romanorum* (233 copies), Walther Ryff's medical handbook, *Das Handbüchlein Appollinarius* (227 copies), and Johann Pauli's *Schimpf und Ernst* (202 copies), a joke book. Harder sold 141 cookbooks, 343 arithmetic and accounting books, and numerous agricultural treatises, calendars, herbals, marriage manuals, and medical formularies. He sold 27 copies of the *Artliche Kunst* at 3 shillings each, about the price of a cheap herbal. By comparison, a reckonbook (arithmetic book) cost a bookdealer between 8½ and 12½ shillings, while a cookbook could cost as much as 18 shillings. The most expensive item on the list was Wilhelm Kirchhof's *Wendenmuth*, a collection of historical tales; Harder sold 118 copies of the work at 65 shillings each. The technical and scientific books in Harder's inventory were considerably less costly than some popular romances and religious tracts.

Private estate inventories provide additional information about popular book ownership. Recently, Michael Hackenberg made a detailed study of several hundred postmortem estate inventories of sixteenth-and seventeenth-century German book owners.[114] Not surprisingly, he found that the largest libraries were owned by traditional readers such as medical doctors, academics, and clergymen. However, new types of readers appeared with greater frequency as the century progressed. Hackenberg found only two postmortem inventories of artisans or day laborers containing books before 1521, while at the end of the century, between 1571 and 1600, more than sixty artisans' inventories contained books. Two-thirds owned libraries of fewer than ten books.[115] The number of women owning books also increased, although their libraries were small. Scientific and technical books turn up most frequently in the libraries of people for whom they served immediate professional needs. Anton Weidenteich, a goldsmith whose estate was inventoried in 1595, owned thirty books, including an unspecified "Kunstbuchlein in gelb pergament gebunden," an "Experiment buchlein" (possibly an alchemical book), and Adam Riese's arithmetic book.[116] The Braunschweig mason Balzer Kircher owned a "Historien boeck" and two "Kunstbücher." Medical practitioners tended to have larger libraries than other artisans, and they usually contained a high proportion of medical manuals. Joachim Wittenheder, a barber-surgeon from Braunschweig who died in 1567, owned thirty-one books. In addition to a Bible, his library contained twenty-seven medical books, including eight by Paracelsus and three by Walther Ryff. In addition, Wittenheder owned a copy of Agricola's *De re metallica* and six manuscript *Kunstbücher*. Another surgeon, who died in 1583, owned twenty-four books, including a work by Paracelsus, a distillation manual, and three manuscript *Kunstbücher*.[117] Certainly many more artisans owned or used books than appear in the estate inventories. Tens of thousands of printed copies of manuals such as the *Kunstbüchlein* circulated in the sixteenth century; most perished through constant use, leaving no trace of their owners. Yet Hackenberg's evidence shows conclusively that book ownership among all occupational groups extended beyond immediate professional interests. The Braunschweig cobbler Peter Crüger left a library of forty-one books that included humanistic literature, a Latin-German dictionary, an accounting book, local legal codes, a book on chivalry, the German edition of Albertus Magnus, and Johannes Coler's handbook on domestic economy.[118] A joiner who died in 1583 left twenty-eight books, all but four of which were Lutheran theological works.[119] Craftsmen owned romances, histories, and devotional literature, while merchants and notaries as well as artisans bought Ryff's *Hausapothek* and the *Kunstbüchlein*.[120]

It should be stressed that printers, not craftsmen, created the *Kunstbüchlein* and were largely responsible for disseminating technological information in the early sixteenth century. Although these works had their origins in the workshops, the information was gathered, assembled, and distributed by printers who recognized the needs of a new body of readers. To the random assortments of recipes that had formerly circulated among craftsmen, printers added title pages, tables of contents, glossaries of technical terms, and prefaces bringing them to the attention of a new public. They rearranged existing texts to make them respond to different needs, deleted obsolete recipes, and updated the books with new recipes. Frequently the recipes had to be translated into standard vernacular languages, since the technological manuscripts of the late Middle Ages contained many terms and expressions familiar only in certain trades, dialects, and localities.[121] It was the printer who first interpreted these terms for readers, either by adding synonyms or by having them translated into a language accessible to a broad cross section of readers.

Although the technical information contained in these how-to books was rarely innovative, the new mode of disseminating it was. The publication of techniques for making steel, formulas for dyeing textiles, and the secrets of distillers and alchemists made the mysteries of the trades a little less mysterious. Technical information of this kind was now within the reach of average readers who were not members of a guild. The *Kunstbüchlein* were not simply examples of the transmission through typography of an older manuscript tradition; they represented an entirely new genre of technical literature. The distinct impression one gains from reading these works is of a group of highly competitive publishers constantly on the lookout for new practical information to speed into print. With one eye on the oral tradition of the workshop and the other on a newly literate public, printers became the mediators between folk, popular, and learned cultures. Recipes that were traditionally confined to the workshops became part of the public domain, and philosophical traditions such as alchemy were given a new relevance by being placed within the reach of general readers. On behalf of the middle class, printers penetrated secrets from both directions.

The Age of "How To"

The publication of vernacular technical handbooks was certainly not limited to Germany, nor even to areas where the Reformation had its deepest impact. Although Protestantism was a powerful stimulus promoting ideals of self-improvement, it was by no means a precondition for them. Far more effective was the rapidly changing European economy, which opened up new employment opportunities for literate men and women. The Italian cities provided exceptionally favorable markets for the sale of vernacular literature. By the middle of the trecento, nearly every city in northern and north-central Italy had schools—whether private or communal—to teach the rudiments of reading and writing to children of practically every social station. Paul Grendler estimated that in Venice, approximately 4,600 boys plus a handful of girls attended 245 schools in 1587. A little more than half of these children attended vernacular schools, usually run by single teachers who taught "lezer, scriver, abbaco et quaderno" (reading, writing, computing, and keeping accounts), subjects designed to prepare pupils for careers in a growing commercial economy. Grendler reckoned that about 33–34 percent of Venetian boys were literate in 1587.[122] He reported similar figures for Florence, where about 30–33 percent of the male population was literate.[123] Although schooling was much more common in the cities, where even working-class children and girls learned to read and write, elementary schools also existed in smaller towns and villages. Carlo Ginzburg found it "astonishing that so much reading went on" even in the small village of Montereale, in the Friuli region of northern Italy in the 1570s.[124] In addition to schools, there were various handbooks that, as one 1524 publication promised, could instruct "anyone who can read to teach his son, daughter, or friend who cannot read, in such a way that anyone can learn."[125] The evidence permits the conclusion that a substantial, and growing, portion of the urban population in Italy could read, or were learning to read, in the sixteenth century.

The growth of literacy drew numerous entrepreneurs into the publishing business. A conservative estimate has it that about 500 publishers operated in Venice alone in the sixteenth century, producing an astonishing 15,000 to 17,000 editions.[126] These ranged from giants like Gabriel Giolito, who published around 900 editions from 1541 to 1578, to small printers who published but a single title and then disappeared without a trace. Large and small publishers alike entered the expanding market for technical and household recipe books. The earliest Italian printed tract of this type was a work printed in 1525 entitled *Opera nova intitolata Dificio di ricette* (A new work entitled the house of recipes).[127] The *Dificio di ricette* went through at least twelve editions by midcentury and was picked up by other printers, including Marcio Sessa and Francesco Bindoni, operators of two of the largest Venetian printing establishments.[128] Like the German *Kunstbüchlein*, this anonymous pamphlet was undoubtedly a printer's compilation. Its preface, "To My Good Reader," explains that the secrets had been collected over many years' time, and that at the urging of his friends, the printer had decided to publish them "per beneficio universale." The work, however, is more like a general household recipe book than a specialized craft handbook. It contains "many very useful recipes, which will also be pleasing," including a method to determine whether a wife or husband is impotent, a recipe to make wine taste like muscatel, instructions for making artificial windows out of paper, gardening hints, and a quenching water for hardening iron. Also included are various "parlor tricks" and illusions, such as making a candle burn under water, making a ring dance, and making a room appear to be full of grapevines. Another portion of the work is devoted to recipes for soaps, perfumes, and cosmetics (including a water to remove stains from an artisan's hands), along with more household and gardening hints. The last chapter contains medical recipes, "each one of which is truly worth a treasure"—as indeed they are, one being a remedy "for a sick man abandoned by the physicians."[129]

The *Dificio di ricette* was the best known and most popular title in what eventually grew to become a large family of Italian all-purpose manuals for household use.[130] Mere pamphlets at the outset, they gradually accreted new recipes, often duplicating one another, and eventually provided urban popular culture with a fairly uniform body of scientific and technical lore. Like the German *Kunstbüchlein*, they were apparently published for local book markets. Two nearly identical chapbooks of this kind came out in the 1540s. The first was entitled *Colleta de molte nobilissime ricette* (Collection of many very noble recipes), a medical recipe book published "for those who want to avoid many expenses, and for the comfort of the poor." The other was entitled *Opera nova nella quale ritroverai molti bellissimi secreti* (A new work in which you will find many very pretty secrets). Both were cheaply produced octavo pamphlets printed on one signature (eight pages), destined for the commonest sort of readers and for the peddler's pack. These works were imitated by a host of medical empirics and hack writers in sixteenth-century Italy.

Similar handbooks appeared all over Europe. In 1513, the Brussels printer Thomas van der Noot published a miscellaneous compilation of technical recipes under the title *Tbouck van Wondre* (The book of wonders).[131] This anonymous work was compiled from several manuscripts, including, evidently, a workshop manual for apprentice dyers. The dyeing recipes in *Tbouck van Wondre*, which make up the bulk of the work, are considerably more sophisticated than those found in its German counterpart, the *Allerley Mackel*. Like the *Kunstbüchlein*, the "Book of Wonders" accreted new recipes and became the basis for more comprehensive handbooks. When Simon Cock of Antwerp reprinted the work in 1544, he added recipes on wine making, grafting fruit trees, etching metals, and hardening and softening iron and steel, thus producing a comprehensive craft and household recipe book. Meanwhile, the *Kunstbüchlein* became part of the common inheritance of European popular culture. Almost immediately after their appearance, they were combined to form an all-purpose general recipe book. In 1532, the Leipzig printer Michael Blum assembled the three booklets on metallurgy, dyeing, and pigment making, and published them under the title *Drei schoner kunstreicher Büchlein* (Three pretty booklets of ingenious skills). Three years later, all four tracts were published together in a work entitled *Kunstbüchlein, gerechten grundlichen gebrauche aller kunstbaren Werckleut* (The little book of skills: Proper, basic practices for all skilled workmen). Two nearly identical editions of this tract, by two different printers, appeared in 1535. Egenolff printed one (evidently the first of the two); the other (dated 16 June 1535) was printed at Augsburg by Heinrich Steiner.[132]

Other combinations of the booklets also appeared. In 1539, Jacob Cammerlander brought out *Von Stahel und Eysen* and *Allerley Mackel* under a new title, *Mangmeistery*. The following year, Cammerlander

4. Title page of *Kunstbüchlein*
(Frankfurt am Main, 1535).

published the ink-making recipes from *Artliche Kunst* in his edition of
Fabian Frank's *Orthographia*, a handbook on writing in German. The
Kunstbüchlein recipes also appeared in various "housefather books" of
the sixteenth and seventeenth centuries, and were thus marketed as skills
for everyday use. Eventually the recipes migrated beyond Germany to
became the common property of European culture. In 1583, the English author and translator Leonard Mascall combined what he thought
to be the best of the German and Dutch recipes, thereby creating an
English counterpart to the series. His *Profitable booke . . . very necessarie
for all men*, a selection of recipes from the *Kunstbüchlein* and the *Tbouck
van Wondre*, was intended to serve both as an instructional manual for
workmen and as a help to ordinary householders. In 1549, the *Kunstbüchlein* recipes were translated into Dutch (and slightly rearranged) by
Simon Andriesson.[133] This translation went through more than a dozen
editions in the late sixteenth century and several more in the seventeenth

century. In 1559, the Antwerp printer Christopher Plantin had the recipes translated from Dutch into French and added them to his French edition of the *Secrets* of Alessio Piemontese.[134] It was through this work that the *Kunstbüchlein* recipes gained their widest exposure in Europe: Alessio's *Secrets* went through more than ninety editions in almost every European language by the end of the seventeenth century. Not all of these editions carried the *Kunstbüchlein* recipes, but their association with the legendary name of "Alexis of Piedmont" undoubtedly enhanced their credibility and respectability among middle-class and even scholarly readers.

The little Italian pamphlet *Dificio di ricette* was also widely circulated. Translated into French around 1541, the work was reprinted many times under the title, *Bastiment des receptes*. A Dutch edition, translated from the French, appeared in 1549. It too was frequently reprinted.[135] In the seventeenth century, the *Bastiment des receptes* was printed by Jacques Oudot of Troyes and became a standard offering in the famous *bibliothèque bleue*. Cheaply printed on coarse paper and wrapped in rough covers of blue paper, these chapbooks were peddled by the thousands in the French countryside by colporteurs in the seventeenth and eighteenth centuries.[136] It is not difficult to see why the booklets were so popular. As William Coleman pointed out, the popular medical literature of early modern France was characterized by an "activist" approach to health, rather than by the "expectant" approach promulgated within academic medicine. Recipe books promised cheap and fast cures for specific ailments, an approach that was much more compatible with the medical needs of popular readers than the approach of traditional academic medicine, which stressed lengthy, drawn-out, and often expensive dietary regimens. Besides, a household recipe book enabled people to doctor themselves and their families, bypassing altogether the need for a physician.[137]

It is difficult to measure the cultural impact of the flood of technical manuals and how-to-do-it books that streamed from the sixteenth-century presses. On the one hand, the handbooks made the crafts less arcane: although trade secrets did not become public knowledge as soon as they were printed, at least they were accessible to any literate person. For a significant portion of society—including intellectuals and the middle class—technical literacy became possible without workshop experience. On the other hand, written accounts of technical processes were necessarily abstractions of what actually went on in the workshops. No description of assaying metals, however detailed, can substitute for the experience of standing over the smoky fire of an assayer's furnace. Yet remoteness from the workshop was not necessarily an obstacle to the construction of technological knowledge. Indeed, it was a precondition for science. Walter J. Ong pointed out that writing separates and distances the knower from the known and promotes "objectivity": "Between knower and known writing interposes a visible and tangible object, the text."[138] Whereas workshop instruction took place within a dense, buzzing continuum of speech, gesture, and performance, handbook learning objectified these fleeting experiences and fixed them in a text.

The conventional format for recording technical processes in the early modern how-to books was the recipe, formally a list of ingredients along with a set of instructions describing how they were to be employed. There are important linguistic and semiotic differences between the recipe and other means of conveying technical information, such as the descriptive-historical method employed by Pliny and by Renaissance technological authors like Biringuccio and Agricola. The description (or "history") of a technical process, like a historical account, is a narrative that is self-contained and exists independently of the process itself. Since the actual event is anchored in past time, the description serves as a proxy for the experience of it. The recipe, on the other hand, does not have the same integrity.[139] It does not describe an event or process, nor can it stand as a substitute for experiencing a technological act. Unlike a description, a recipe implies a contract between the reader and the text. It is a prescription for taking action: *recipe* is the Latin imperative "take." Because it prescribes an action, a means for accomplishing some specific end, the recipe's "completion" is the trial itself. A recipe is a prescription for an experiment, a "trying out."

If recipes differ in important ways from descriptive-historical accounts, no less significant are the differences between recipes and oral instructions, such as those an apprentice artisan learns. Unlike oral instructions, recipes exist independently of the teacher. Once they are recorded in a book, they become depersonalized and acquire a more general, universal quality.[140] Collecting recipes and compiling them in reference books makes it possible to compare, test, and discern what is common in them. Such comparing and testing opens the door to the discovery of general formulas underlying a myriad of ad hoc rules. The idea of "standard" practices—as opposed to those taught by a particular master—came into being: thus the *Kunstbüchlein* was a comprehensive book "of the correct, basic practices of all skillful workmen."[141] Most important, whereas in the craft tradition instructions were communicated orally from master to apprentice and were the property of exclusive corporate groups, printed recipes were accessible to anyone who could read. The general availability of craft, medical, and alchemical recipes tended to demystify trade secrets, reducing them to simple formulas and procedures. Did it also give readers a sense of empowerment, of having some tangible means of controlling the forces of nature? Did technical literacy contribute in some measure to the "disenchantment of the world" that Weber noted as one of the hallmarks of modern culture? It is difficult to know. Certainly the avalanche of self-help manuals showing readers how to draw and paint pictures, assay metals, dye fabrics, make gunpowder, and handle all manner of tools and instruments contributed to a sense of independence on the part of many who became self-taught. Even if these books did not measurably affect practices, "one can scarcely avoid noting that self-reliance was being encouraged in new ways."[142] And if recipes may be regarded as "explanations" of what has already been made or constructed, then it seems reasonable to conclude that the dissemination of technical manuals promoted a greater understanding of the artificially created world. Recipe books translated craft "secrets" into simple rules and procedures, and replaced the artificer's cunning with the technologist's know-how.

The essential thing about a recipe, however, is that it is written down—or, in the case of books of secrets, printed and broadly distributed. Of course, anything that can be done using a recipe can also be done, and has been done, without one. Artisans traditionally practiced their trades according to unwritten rules they had learned through an oral tradition and by imitation. Practice according to such "implicit recipes" is, however, quite different from following a written formula. For one thing, it allows greater latitude and flexibility with regard to ingredients and procedures. Indeed, it is this very freedom that characterizes peasant cooking, which (in contrast to cookbook cooking) is less tied to specified ingredients, quantities, and techniques. Yet this very flexibility has its own contraints, since it is tied to the ingredients of the field and cupboard, and its range of techniques is essentially limited to those learned by imitation or rote. Moreover, without some means of preserving them, new discoveries tend to get lost. The recipe, on the other hand, imposes constraints of a different kind: it prescribes specific ingredients and an ordered sequence of events to be followed in preparing them, often specified in exact quantities and measured lengths of time. Following recipes both standardizes and widens the repertoire of the practitioner. By requiring precision in measurement and orderly, methodical procedures to be followed, it disciplines the practitioner and provides a basis for comparative testing of results. Since they can be collected in one place, recipes can be classified and tried out by others, who might evaluate them in different ways or according to new criteria. Such testing and comparison not only separates grain from chaff, it also opens the door to new discoveries and improvements. As Jack Goody characterizes the distinction between proceeding according to unwritten rules and following a recipe, "it is the constraints and freedom of the *bricoleur* as opposed to those of the scientist, where exact measurement may be a prelude to discovery and invention."[143]

For most sixteenth-century readers of books of secrets, however, the availability of technological recipe books simply had immediate practical significance: it meant they could learn new craft skills without entering a guild, or that they could master techniques applicable to everyday needs without having to rely on the specialized (and relatively more expensive) skills of professional artisans. The effectiveness and sophistication of the techniques described in these early manuals varied wildly, and it is difficult to gauge the extent to which they may have actually influenced technical know-how at the popular level. Some readers of the *Dificio di ricette* may have been disappointed by the results they experienced from trying out the prognostication instructions and magical tricks they encountered in the book. On the other hand, lack of success may have encouraged more skeptical attitudes toward magic, especially when compared to the results of the more strictly technological recipes, such as those found in the *Thouck van Wondre* or the *Kunstbüchlein*. Readers may not have known why certain materials hardened iron or made fast dyes, but they knew how to do these things.

Modern scholars have stressed that forces of secularization were at work in the early modern era to transform European society from one dominated by clerical and aristocratic elites to one in which laymen and burghers held a significant share of wealth and political power.[144] Certainly the principal causes of these far-reaching changes were the emergence of a market-oriented economy, the rise of the middle class, and the growing dominance of cities in the political and economic life of Europe. No less important, however, was the inculcation of new attitudes toward materials and material life. The distinction between the sacred and the profane was increasingly more sharply drawn. Moreover, the idea of a "desacralized" nature was shared by the lay public as well as by men of letters. Formerly sealed off from the republic of letters by the barrier of illiteracy, lay folk now became part of it, not only sharing in the new attitudes that came to predominate, but also influencing them. The concern with the material needs of everyday life, the emphasis upon hands-on experience, the confirmation of the greater efficacy of technology over the sacred, and the availability of self-education through reading—all these forces contributed to a growing awareness that humanity's lot could be bettered, not by magic, cunning, or the grace of God, but by knowing "how to."

FOUR

THE PROFESSORS OF SECRETS

AND THEIR BOOKS

IN 1555, the last printed edition of the Latin *Secretum secretorum* was published at Naples and Venice. Its editor, Francesco Storella, a professor of rhetoric at the University of Naples, gave the work far more serious attention than had any scholar since Roger Bacon—far more, given the state of scholarship at the time, than the work merited.[1] For by the time Storella's edition appeared, the *Secretum secretorum* was so generally regarded as spurious that sixteenth-century editors of the Aristotelian corpus rarely bothered to mention the work. Although the *Secretum* continued to be popular in vernacular editions, for most humanists it was simply another example, among many so determined during the Renaissance, of a medieval work falsely attributed to a famous ancient author.

As one tradition of secrets-literature came to a close in 1555, another opened. In the same year, the bookstalls of Venice displayed a book of secrets whose form, content, and intended audience were entirely different from that of the *Secretum*. The work bore the name of one Alessio Piemontese—"Alexis of Piedmont"—a devout and learned man whose anguish over the condition of his soul led him, albeit reluctantly, to publish his precious secrets, the product of his life's labor, for the world's benefit. The *Secreti del reverendo donno Alessio piemontese* was an instant best-seller. The first edition sold out within a year and was reprinted by three different publishers in 1557. Alessio became internationally famous through the dozens of editions of his book that were published in almost every European language. In Italy, the work unleashed a torrent of "books of secrets." Within four years of the work's publication, two more volumes of Alexian secrets appeared, and these were followed by a multitude of books of secrets modeled upon the original *Secreti*.

The "Professors of Secrets"

The appearance of the printed books of secrets caught the attention of Tommaso Garzoni (1549–1589), one of the most astute, if eccentric, observers of Italian social and cultural life. In 1585, the Augustinian monk and social commentator published a huge survey of all the professions and trades he reckoned made up the social world of sixteenth-century Italy. Among the more than five hundred different "professions" included in the kaleidoscopic *Piazza universale di tutte le professioni del mondo*, Garzoni acknowledged the arrival on the scene of a new group of authors whom he called *i professori de' secreti*, the "professors of secrets."[2] According to Garzoni, the professors of secrets were relentless seekers of obscure, veiled, and occult things. Secrets, he wrote, are things "whose reasons are not so clear that they might be known by everyone, but by their very nature are manifested only to a very few; nevertheless they contain certain seeds of discovery, which facilitate finding out the way toward discovering whatever the intellect may desire to know."[3] Garzoni, who thought the world was filled with every possible kind of fool (and wrote the book on it), was amazed there were so many who would dedicate their lives to seeking such elusive and chimerical things. "There are some who attend to this profession of secrets so zealously," he observed sardonically, "that they yearn for it more than for the necessities of life itself." Professing secrets could be an honorable and even noble pursuit, he acknowledged, as long as one dedicated himself to the pursuit of "good secrets" (*buoni secreti*). Too often, however, the professors of secrets chased after the "ridiculous and vain secrets" of magic, and from such a pursuit proceeded "more smoke than victuals."[4]

Who were the "professors of secrets"? Garzoni identified several of them, including Alessio Piemontese, Girolamo Ruscelli, Leonardo Fioravanti, Gabriele Falloppio, Giambattista Della Porta, and Isabella Cortese, all of whom were familiar to readers by the time of Garzoni's writing. But there were more. Garzoni could have added to his list the names of Pietro Bairo, Giovanni Battista Zapata, Timotheo Rossello, Giovanni Ventura Rosetti, and a swarm of lesser-known peddlers of secrets (see table 1). Little is known about most of these authors. With the sole exception of Bairo, they were not, by any means, representatives of the Italian academic establishment. Gabriele Falloppio, the renowned professor of anatomy at Padua, has to be eliminated from the list because the book of secrets attributed to him is spurious. (It was probably a printer's attempt to capitalize on the famous anatomist's reputation.) Pietro Bairo (1468–1558) was the only other author mentioned by Garzoni who is known to have had a university degree or to have held a university position. Bairo was a professor of medicine at the University of Turin and principal physician to the dukes of Savoy. His compilation of medical secrets was based upon standard authorities such as Galen and Avicenna, and thus reflects the viewpoint of the medical establishment.[5] The other medical men—Giovanni Battista Zapata (ca. 1520–ca. 1586) and Leonardo Fioravanti (1518–1588)—were empirics and not

TABLE 1
The "Professors of Secrets" and Their Books

	Occupation and Dates	Title and Date of First Edition	No. of Editions 1555–1599
"Alessio Piemontese"		*Secreti del reverendo donno Alessio piemontese* Venice, 1555	17
Giovanni Ventura Rosetti	*provisionato* at Venetian Arsenal	*Notandissimi secreti de l'arte profumatoria* Venice, 1555	2
Giambattista Della Porta	nobleman (1535?–1615)	*Magia naturalis* (1558; 1589); first Ital. ed., Venice, 1560	3 Italian 15 Latin
Isabella Cortese	noblewoman?	*I Secreti* Venice, 1561	7
Pietro Bairo	physician (1468–1558)	*Secreti medicinali* Venice, 1561	7
Timotheo Rossello	unknown	*Della summa de' secreti universali* Venice, 1561	5
ps.-Gabriele Falloppio		*Secreti diversi et miracolosi* Venice, 1563	9
Leonardo Fioravanti	surgeon (1518–1588)	*Secreti medicinali* Venice, 1561	1
		Capricci medicinali Venice, 1561	7
		Del compendio de i secreti rationali Venice, 1564	6
Girolamo Ruscelli	professional writer (1500–ca. 1566)	*Secreti nuovi* Venice, 1566	1
Giovanni Battista Zapata	physician (ca. 1520–ca. 1586)	*Li Maravigliosi secreti di medicina e cirurgia* Venice, 1577	9
		Total Italian editions	74

physicians. Zapata was a successful Roman practitioner who was well known and widely admired for his devotion to the poor. Two of his disciples honored his work among Rome's destitute by publishing his medical secrets and dedicating the collection to the master.[6] Fioravanti was a famous Bolognese surgeon and popular medical author who generated tremendous controversy, especially among physicians, with his supposed marvelous cures and his unconventional methods. He was called everything from a charlatan to a maker of miracles.

Girolamo Ruscelli (1500–ca. 1566) was a professional writer employed by one of Venice's prominent publishing firms, the Valgrisi. Giovanni Ventura Rosetti, a native of Venice, worked at the Venetian Arsenal, or state shipyard, as a *provisionato*.[7] Rosetti's commercial activities took him to other Italian cities, where he learned about the materials and methods of dyeing and perfumery from local artisans. In 1548 he published an important treatise on dyeing, the *Plictho de l'arte de tintori*, and in 1555 a treatise on perfumes and cosmetics, *Notandissimi secreti de l'arte profumatoria*. Not content to allow these "plebeian arts" to remain "locked up under that regrettable custody of secrecy used by the ancients until now," Rosetti published the secrets of dyeing and perfumery in order to promote the development of the arts in Venice.[8] About Isabella Cortese, the only woman among the professors of secrets, nothing is known outside of what she reveals in her *Secreti*, her only known work. Cortese dedicated the book to her "beloved brother," the archdeacon of Ragusa (Dubrovnik), and mentioned having traveled extensively in eastern Europe, where she learned the arts of alchemy and perfumery, the principal subjects of her book. Although Cortese's social status is uncertain, her book bears the mark of a woman of the Venetian nobility. Timotheo Rossello is equally obscure. His book, the *Secreti universali* (also dedicated to the archdeacon of Ragusa), is a typical collection of recipes relating to the arts of distillation, metallurgy, alchemy, and dyeing. Giambattista Della Porta (1535–1615), a Neapolitan aristocrat, was the only one of the professors of secrets to earn an important reputation among contemporaries for his scientific work. He did not attend a university or obtain a degree, but with an insatiable curiosity dedicated himself to a lifelong pursuit after the "secrets of nature." The precocious Della Porta published the first edition of his famous *Magia naturalis* in 1558, when he was only twenty-three, and devoted the remainder of his life to expanding the scope and findings of that early work.[9] His efforts drew praise from Kepler, Peiresc, and other members of the European scientific community. Along with Galileo, Della Porta was an early member of the famous Accademia dei Lincei, which was founded at Rome in 1603.[10]

The professors of secrets were thus on the fringes of academic culture. Often self-taught, they were more accustomed to the clamor of the piazza and workshop than to the solitude of the study. The predominance of medical recipes in the books of secrets, and of medical empirics among the professors of secrets, is an indication of important changes taking place in the medical economy and in medical practice.[11] By the sixteenth century, it had become essential for practitioners to have a large arsenal of drugs at their disposal. The medical economy, no less than the economy at large, had been transformed by the intense competition among the providers of medical services. As competition grew more intense, the demand for certain kinds of services accelerated, particularly the demand for drugs to treat specific ailments as opposed to complex regimens for living. Changing medical fashions also created a demand for new drugs. "Paracelsian" remedies, distilled products, and various medical "secrets" became increasingly popular.[12] The changing medical marketplace attracted a multitude of medical empirics and charlatans, who sold nostrums and powders in the piazzas and taverns of Italy's cities. Venice, in particular, was renowned for its swarms of *ciarlatani* and "mountebanks," who peddled their miracle drugs in St. Mark's square and delighted passersby with their music, dance, and entertainment.[13] The printing press accelerated these trends, since it was but a small step from purchasing a remedy from a pharmacist to buying a book of remedies to cure oneself. Physicians naturally grew nervous about the trend toward self-cure, because it threatened their livelihood. Printers, on the other hand, delighted in it, and willingly published any book of medical secrets they thought would sell. Medical empirics discovered that broadsides and printed books could be effective ways to advertise their medical wares and to reach new markets. Leonardo Fioravanti, Venice's most famous empiric, set the standard for using the book of secrets as an advertising medium. He gave his nostrums exotic trade names like *elettuario angelico* (angelic electuary), *oleo benedetto* (blessed oil), and *siroppo maestrale Leonardi* (Leonardo's magistral syrup), wrote detailed relations of the marvelous cures he had accomplished with them, and invited customers to his shop and to the pharmacies where his remedies were sold. Fioravanti even offered to dispense his medications free of charge in the hope of luring new clients. "If there is anyone in the world who would like to have himself cured of [the gout]," he wrote in his *Capricci medicinali*, "I offer myself with the help of God to heal him very quickly and happily without making him spend anything, but only so that the incredulous might understand this matter, and that after me the world might understand such remedies and the people might be able to free themselves from such evil infirmities. Anyone who would like to avail himself of my remedies will find me in Venice at San Luca, where I will always be ready to the service of everyone."[14]

If the motives of the vendors of medical secrets seem all too transparent, we should not too hastily conclude that the professors of secrets were masters of hype and nothing more. In addition to being a medium for marketing empirical remedies, the book of secrets was also a practical "how-to" book and, not least, a compilation of experiments. Rosetti stated that his compilation of dyeing recipes was "a work of public Charity that I bequeath for the public benefit, and which has been imprisoned for a great number of years in the tyrannical hands of those who kept it hidden."[15] By publicizing craft secrets, Rosetti hoped to encourage competition among craftsmen and to promote the growth of the Venetian dyeing industry. It was this perception of themselves as "divulgers" of secrets that perhaps best characterizes the professors of secrets. They proclaimed themselves to be the revealers of secrets jealously and selfishly guarded by practitioners—or, as Della Porta put it, of secrets locked up in the bosom of nature. With the sole exception of Della Porta, who wrote in Latin and not for the people, the professors of secrets did not maintain that secrets are the rightful property of the privileged few, as Roger Bacon did. They insisted that everyone ought to have access to the secrets of nature, and they used the printed book to make this goal a reality.

The prototype of all these works was the famous *Secreti* of Alessio Piemontese, the most popular book of secrets in early modern Europe. By the end of the sixteenth century, more than seventy editions of the work had been published, including translations into Italian, Latin, French, English, German, Dutch, Spanish, and Polish (see table 2). By then, the name of Alessio was so legendary that printers all over Europe were eager to publish editions of his secrets—and even to publish fabricated compilations attributed to him, as did the Venetian printer Giorgio de' Canalli when he brought out the "fourth and last part" of Alessio's secrets in 1568. Christopher Plantin, the famous Antwerp printer, published French and Dutch translations of the *Secreti*, and protected his valuable copy by securing dual privileges from the Council of Brabant and the Privy Council of Brussels, a practice he normally followed only for popular religious tracts and works whose high sales might arouse the envy of competitors.[16] For middle-class readers in early modern Europe, Alessio Piemontese was a familiar and friendly name.

The Secrets of Alessio Piemontese

"Truely, I would not sette my selfe (beeyng in the age, and disposition, bothe of bodie and mynde, that I am nowe in) to write fables and lies, that should continue alwaies. . . . There is nothing in this boke but is true and experimented."[17] So "Alessio Piemontese" assured his readers in a preface that was perhaps as famous as the compilation it preceded. Despite Alessio's protestations, the work that bears his name is consid-

TABLE 2
Alessio Piemontese, *Secreti*: Editions and
Translations, 1555–1699, by Decade

	Number of Editions	Cumulative Editions
1555–1559	17	17
1560–1569	28	45
1570–1579	13	58
1580–1589	7	65
1590–1599	5	70
1600–1609	4	74
1610–1619	10	84
1620–1629	5	89
1630–1639	3	92
1640–1649	3	95
1650–1659	3	98
1660–1669	1	99
1670–1679	2	101
1680–1689	1	102
1690–1699	2	104
Total	104	

ered by modern bibliographers to be a complete fabrication, the creation of the Venetian humanist Girolamo Ruscelli. There is no question that Ruscelli had a hand in the composition of the work, but, as I shall explain below, in all probability as its editor and not as the original author. His role in the publication of Alessio's *Secrets* is itself fascinating, and we shall explore it in more detail in the next section. For the moment, let us take Alessio at his word and see where his preface leads us. For his self-portrait is the most detailed description we have of the sixteenth-century professor of secrets. More than any other work, it was responsible for creating the image of a characteristic sixteenth-century topos.[18]

In the preface, "Don Alessio" reported that he was a member of the Piemontese nobility. As he modestly put it, "I have alwaies had my pleasures, and great plentie of richesse, yea farre passing the smalness of my desertes." His education included instruction in Latin, Greek, Hebrew, Chaldaic, and Arabic. But his passion was for "philosophy and the secrets of nature." Giving up humanistic studies, he spent his entire life traveling from place to place, solely with the purpose of collecting secrets, which he gathered "not alonely of men of great knowledge and profound learning, and noble men, but also from poor women, artifi-

5.   Title page of Alessio Piemontese, *Secreti*
(Venice, 1555).

cers, peysantes, and all sortes of men." Alessio spent more than fifty-
seven years traveling throughout Italy and the Middle East, "and sondry
times travailed almost all other partes of the world, without resting or
sojourning at any time in one place above five moneths." One of his trips
commenced about 1514, when he went to the Levant. In 1518 he was
aboard a ship from Jerusalem that was attacked by pirates. In the ensu-
ing battle, he treated a sailor's gunshot wound with his "oil of a red
dog," a balsam made by boiling a red-haired dog until it falls to pieces,
combining it with scorpions, worms, marrow of an ass and a hog, and a
number of plants in a definite order. The ointment was Alessio's favorite
secret, which he always carried with him. With it he treated a monk's
withered arm, "dried up like a stick," and a Portuguese man who suffered from a painful case of gout. In 1521 Alessio went to Syria, where
he observed the medical and cosmetic practices of the Moorish nobility
and healed a lord's daughter of some undesignated illness. In 1523 he

was living in Aleppo, where he learned about a remedy to cure the plague.[19] Returning to Italy, he was in Bologna in 1547, in Venice the following year, and a few years later in Milan. The portrait Alessio sketches of himself perfectly matches the one Garzoni drew of the "professors of secrets": the lonely, wandering scholar who turns away from traditional learning in order to pursue his passionate quest for rare secrets, one whose pursuit takes him to every corner of the world, and to men and women of all ranks and stations. Alessio was the prototypical professor of secrets. His self-portrait profoundly influenced Garzoni's characterization.

Alessio cherished his precious secrets and was loath to reveal them to anyone. Widely known as he was for his knowledge of rare secrets, his fame rested on the fact that he alone knew them: "I [have] always been noseled up [nurtured] by a true ambition and vainglorie, to knowe that, whiche another should bee ignorante of; whiche thyng hath grafted in me a continuall niggardnesse or sparing, to distribute or communicate any of my secretes, yea unto my moste singuler friends that I had: saiying, that if the secretes were knowen of every man, thei should no more be called secretes, but publike and common."[20] Alessio was compelled to change his mind by an incident that occurred in Milan, when he was eighty-two years old. A surgeon, who knew Alessio's reputation as a possessor of rare secrets, went to him for a remedy to cure an artisan tormented with a painful bladder stone. The old professor of secrets refused, perceiving that the surgeon "would use other mennes thinges, for his own profite and honour." The surgeon refused Alessio's offer to administer the cure gratis, fearing it would ruin his reputation if word got out he had consulted someone else. The poor artisan paid for the vanity of both men, for by the time Alessio arrived at the patient's bedside, "I founde [him] so nigh his ende, that after he had a little lifted up his iyes, castyng them pitiously towarde me, he passed from this into a better life, not having any neede, neither of my secrete nor any other receipt to recover his health." Gripped with remorse over his part in this unfortunate incident, Alessio forsook all his wealth and retired to a secluded villa, there to live a monkish life with a single servant. Renouncing his devotion to secrecy, he resolved to publish his secrets to the world:

> But yet not havyng the power to put it out of my fantasie, but that I was a verie homicide and murtherer, for refusing to give the Phisician the receipt and remedie, for the healyng of this poore manne, I have determined to publishe and communicate to the worlde, all that I have, beyng assured that fewe other menne have so many as I. And mindyng to set forth none, but soche as bee moste true and proved, I have these dais past (taken partly out of my Bookes, and partly out of my memorie, all those came to hand)

made a collection of soche as I am certaine bee veritable, true, and experimented, not caryng if some of them be written or printed in any other
Bookes than this.[21]

This famous preface, which created the topos of the Renaissance professor of secrets, raises some intriguing questions. First of all, the similarities (and the differences) between it and Thessalos of Tralles's letter,
written fifteen centuries earlier, are striking. Both prefaced a book that
revealed "secrets." Both were written by authors who were passionately
devoted to the search for the secrets of nature. Both writers began their
search with traditional science and philosophy, but found those subjects
wanting. Abandoning these studies, they continued their search beyond
the boundaries of the academy. But there was an essential difference between Alessio's search for knowledge and that of Thessalos. Whereas the
Hellenistic student sought truth in a divine revelation, the Renaissance
empiric discovered secrets through travel abroad, and through contacts
with empirics and practical people. Alessio and Thessalos also differed in
their attitudes toward publicizing knowledge. While Thessalos wrote for
the emperor's eyes alone (doubtless making a bid for patronage), and
accepted the law of silence with dead earnestness, Alessio renounced
esotericism and converted to the ethic of publicity. In retrospect, Alessio's most important contribution to the development of early modern
scientific culture was his discovery that publishing secrets was ethically
superior to concealing them from the "unworthy." In the age of printing, that reversal would be decisive.

But if, as is generally thought, "Alessio Piemontese" was a pseudonym invented by the popular writer Girolamo Ruscelli, does this not
also raise the possibility that his preface was nothing more than a grandi-
ose fiction created to promote the book's sales? The resemblance of
Alessio's self-portrait to the characteristics of Paracelsus—the lonely,
wandering scholar seeking secrets of nature from empirics everywhere—
certainly raises the eyebrows of the skeptical reader. Or was the story of
Alessio's dramatic conversion to the ethic of publicity a tale whose moral
was about the vanity of attempting to keep secrets in the age of printing?
If so, Garzoni, among others, got the message.

Yet the preface and the text accompanying it, so full of details about
Alessio's adventures, carry a ring of authenticity. Moreover, its authenticity is corroborated by an unexpected source. Ambrose Paré (1510–
1590), the famous French surgeon, was in Piedmont during the 1530s
while serving as a surgeon in the French army. In his *Apology and Treatise* (1585), Paré described an ointment to treat gunshot wounds that he
claimed was the secret of a famous Piedmontese surgeon. This glorious
and grotesque balsam, called "oil of whelps," was made by boiling live

puppies with earthworms, oil of lilies, and turpentine—in other words, essentially the recipe for Alessio's "oil of a red dog." It took Paré two years to convince the surgeon to give him the secret. Afterwards, Paré reported, "he sent me away as rewarded with a most pretious gift, requesting me to keepe it as a great secret, and not to reveale it to any."[22] This is not to suggest that the famous Piedmont surgeon was Alessio Piemontese. Nothing in Alessio's autobiography indicates he was a professional surgeon. But surgeons were among those from whom Alessio collected secrets. Possibly Paré and Alessio obtained the secret from the same source. In any case, Paré's account makes Alessio's Piedmont origins more credible. It also supports the general tone of Alessio's preface, suggesting that the ethic of secrecy was shared by other empirics.

What were Alessio's secrets? Of the approximately 350 recipes contained in the first edition of the work, about a third (108) were medicinal. The recipes are entirely different in character from those authorized by the official pharmacopoeias and sold in the pharmacies. The latter were compounded according to the principles of classical pharmacology. Their composition was strictly regulated by local governing bodies of physicians and apothecaries, which insured that official drugs were composed of the ingredients authorized by Theophrastus, Galen, and Dioscorides.[23] Alessio's medical "secrets," on the other hand, represent an alternative pharmaceutical tradition that bore almost no relation whatsoever to classical pharmacology. His recipes called for ingredients that probably would not have appeared in a respectable pharmacy—for instance, wild boar's teeth, skin of a dog, and "dung of a blacke Asse, if you can get it; if not, let it be of a white Asse."[24] According to Alessio, the recipes were the tried-and-true secrets of surgeons, empirics, gentlemen, housewives, monks, and ordinary peasants—in other words, almost anyone but representatives of the medical establishment. They were not rationalized by classical rules but instead were proven by "experience."[25] Although Alessio did not overtly express any disenchantment with traditional practices, he was fully aware of the unorthodox character of his sources. He warned readers that physicians, "moved with a certaine rustick and evyll grounded envie, with a passion of jalousy, are wont to blame and contempne thinges that come not of themselves."[26]

Alessio's medical secrets were for the most part treatments of common ailments that afflicted the general populace: burns and bites, discomfort of the eyes, toothache, sores, abscesses, intestinal wounds, stomachache, hemorrhoids, fevers, warts, rabies, and superficial wounds. Several entries addressed disorders connected with menstruation and pregnancy, including recipes to bring on menstruation, which may have been abortifacients.[27] In addition to addressing such everyday kinds of medical problems, Alessio gave twenty-six recipes to treat plague, still among the most dreaded diseases. But if the ailments to be treated were common, the ingredients in Alessio's recipes definitely were not. The recipes were detailed and complex, some running to several pages of instructions. Many called for exotic ingredients and expensive spices from the East. For example, Alessio recommended a mouthwash made of rosemary, myrrh, cinnamon, and bengewine (benzoin). A "heavenly water" that has fifteen medical applications is made from a mixture of more than forty different fruits, oils, gums, peppers, and spices. With only a few exceptions, it is difficult to imagine "ordinary people" benefiting much from Alessio's remedies.

Approximately one-third of the *Secrets* is devoted to the subject of domestic economy. An entire chapter contains recipes for making perfumes, scented soaps, skin lotions, body powders, and suffumigations to scent rooms and clothing.[28] Another chapter brings the reader into the kitchen, giving recipes for making preserves and confitures of oranges, quinces, and other fruits by cooking them in honey or sugar and storing the mixture in vessels through the year. Finally, there is a section devoted to cosmetics, hair tonics and dyes, depilatories, skin treatments, and cleansers for the teeth. This last section makes it clear that the sixteenth-century middle class was as concerned with the body's appearance as we are today, and sometimes went to extreme lengths to improve it.

The last third of Alessio's *Secrets* bears a close resemblance to the *Kunstbüchlein*, although Alessio's recipes are considerably more detailed than the ones that appeared in the German handbooks. Some of the recipes were entirely new, having never before appeared in print. For example, Alessio gave a detailed description of the process for making ultramarine azure, an expensive blue pigment. Originally imported from the Levant, ultramarine was a costly luxury material, and its lavish use in conjunction with gold gave Renaissance paintings an intrinsic value that was an essential component of contemporary aesthetic tastes. Cennino d'Andrea Cennini described ultramarine as "a color illustrious, beautiful, and most perfect beyond all colors."[29] While to modern observers ultramarine looks no different from ordinary dark blues, the Renaissance public knew its exotic qualities. Its presence was a hallmark of the Renaissance Italian style. But if ultramarine was a luxury item in Italy, it was even rarer in northern Europe. As late as 1549, Valentin Boltz remarked that "ultramarine is prized as the choicest of all, but in German lands is seen seldom, and in small quantity."[30] Doubtless one reason why Alessio's *Secrets* became so popular outside of Italy was that for northern

Europeans it presented the broadest range of Italian artistic techniques available in any single recipe book. For German artists, the *Secrets* was an important source of *welsch* techniques.

Alessio's chapter on alchemy and metallurgy is in some respects the work's most interesting section, for it contains detailed descriptions of technical operations that were typical of sixteenth-century workshops. The alchemical recipes, like those in the *Kunstbüchlein*, describe operations undertaken by painters, jewelers, goldsmiths, cutlers, and blazoners of seals. Although most of these procedures do not seem particularly novel, Alessio introduced a few secrets of unusual rarity. One was a method to make artificial vermilion in large quantities, "which hitherto has not been known in Italy." Another was the process for refining borax, a secret not generally known in Europe until the eighteenth century. Borax, an exotic chemical, was an important material in metallurgical industries for its use as a flux. It was imported into Europe through Venice, whose refineries held a virtual monopoly over its purification and distribution. Indeed, borax was commonly thought to be an artificial material, a compound of ingredients known only to the Venetians. Alessio appears to have been the first to record the process for refining borax, thus violating an important trade secret.[31]

It should be apparent from the nature of Alessio's recipes that he wrote not for the "common person" but for upper-class readers. His recipes for sweet-smelling soaps, cosmetics and "precious waters" for the face, pomanders, and perfumed lamps for the bedchamber do not, by any means, seem destined for the masses. His readers were curious about esoteric arts like alchemy. They wanted to learn distillation techniques in order to make perfume, scented oils, and essences. Alessio also collected recipes relating to the arts of painting, etching, jewelry, the making of armor, and the casting of bells. However, not a trace of the "base" mechanical arts such as carpentry or blacksmithing is to be found in his book. On the contrary, his attitude toward the manual arts was conditioned by the concept of "virtuosity," an ideal that tended increasingly to fashion upper-class tastes in the early modern period. As long as the arts were pursued out of pleasure and curiosity, or used to fill idle moments, they could be considered the marks of a gentleman. Only when they were practiced to make a living did they become sordid. For the virtuosi, the arts were not so much useful as an avocation that set one apart from the crowd.

Among the numerous incidents Alessio claims occurred during his travels, one in particular stands out. It concerns a water to cure pleurisy, which, Alessio recalled, "was given me at Bologna the yeare 1543, of a gentleman called Girolamo Ruscelli, with the which, the same yeare, he was healed of the same disease in a short space without letting blood or using any other medicine, but onlie this water."[32] What makes this statement so intriguing is that twelve years after the publication of the *Secreti*, Ruscelli claimed that he himself was Alessio Piemontese.

Girolamo Ruscelli and the Academy of Secrets

Although Ruscelli is to modern scholars an obscure figure, he was a popular and respected writer in sixteenth-century Italy. Born about 1500 of a poor family in Viterbo, Tuscany, he was taken into the court of Cardinal Marino Grimani at Aquileia, where he received his early education in the classics. After further studies at the University of Padua, he followed his patron to Rome, where in 1541 he formed a literary academy called the Accademia dei Sdegnati (the scornful). Not long after, Ruscelli moved to Naples and entered the service of Alfonso d'Avalos, the powerful marquis of Vasto. By all accounts Ruscelli was a model courtier. He served his patron with such distinction, as both an ambassador and a court poet, that he was immortalized as a perfect courtier by Torquato Tasso in his dialogue *Minturno*.[33] But when Avalos died in 1546, leaving him without a patron, Ruscelli was forced to find other means of employment. So he moved to Venice and found a job as a writer and proofreader in the publishing house of Vincenzo Valgrisi. Like many humanists, he was lured to Venice by the city's political independence, by its prosperity, and above all by the opportunities that were opening up for writers in that publishing capital of Italy. He died in Venice in 1565 or 1566.

When Ruscelli arrived in Venice, he found a city alive with opportunities for literary adventurism. It was the "golden age" of the *poligrafo*, or professional writer.[34] Venetian printers made vernacular literature profitable, and their success provided Italian writers with the opportunity to shake themselves loose from the courts and to live and write independently. Authors, including many drawn from the lower classes, found more independence and freedom of expression in Venice than anywhere in Italy—at least until the 1560s, when the Inquisition clamped down with its rigid censorship.[35] Within broad limits, the Venetian *poligrafi* could write what they pleased and get paid for it without having to continually curry the favor of patrons. Following the example of Pietro Aretino, the first and most famous *poligrafo*, who had come to Venice in 1527, aspiring writers from all over Italy flocked to Venice to breathe the free air of that "glorious republic." Some of them entered Aretino's household in the Palazzo Bolani, where Aretino gathered about him a community of hungry bohemians. The *poligrafi* wrote whatever their printers demanded: plays, travel books, popular histories, social and literary criticism, burlesques, and editions of the works of writers more renowned than themselves. Their books, inexpensively printed in octavos or duodecimos, went through hundreds of editions, and while a few *poligrafi* became famous writers, they were unabashedly honest about their role as popular writers. They had no pretensions of making serious contributions to scholarship but saw their books simply as profitable and pleasing to ordinary readers.

Like most *poligrafi*, Ruscelli was a prolific author. His literary efforts were devoted mainly to the sort of light reading that appealed to upper-middle-class tastes, including editions and commentaries on Italian poetry, collections of the letters of Italian authors, and social commentaries. Ruscelli translated Ptolemy's *Geography* into Italian and authored treatises on warfare, orthography, and the dignity of women. He wrote a sumptuously illustrated work on coats of arms, for which he designed the engravings himself. His contributions to the world of letters, though now long forgotten, drew lavish praise from his contemporaries.[36] Yet Ruscelli's activities were not restricted to philosophy and belles lettres. Francesco Sansovino reported that he was also a devoted student of alchemy and the secrets of minerals and medicine. Ruscelli was an avid collector of alchemical, medical, and technological secrets. One of the manuscripts he left at his death was a compilation of over a thousand recipes, which his nephew published in 1567 under the title *Secreti nuovi*. In the proem to the work, Ruscelli claimed that he was the real author of the *Secrets* of Alessio Piemontese, having published the work under that "feigned name." The two books, Ruscelli explained, contained recipes that were all collected and tried out by an experimental academy he founded in "a famous city" in the kingdom of Naples some years before. Ruscelli's lengthy, detailed description of the organization, finances, and operation of this academy is an account that bears close scrutiny, for it is the only contemporary description we have of a sixteenth-century Italian scientific society.[37]

The academy, according to Ruscelli, had twenty-four active members. Its patron was a local nobleman, who along with a member of his family and one of his ministers, brought the society's total membership to twenty-seven, "a number deemed perfect and highly mystical among the best pagan philosophers and wisest theologians."[38] Each of the members contributed to the finances of the society according to his ability to pay, dues ranging from one hundred to seven hundred scudi a year. The society also received a generous endowment and regular financial support from an unnamed "magnanimous Prince" of the city. Although the prince's participation in the society seems to have been primarily financial, Ruscelli reports that he "committed himself to coming to a general meeting the first Sunday of every month."

Supported by the patronage of this unidentified nobleman, Ruscelli and his group built a meetinghouse, a three-story structure called the Filosofia, on a parcel of land donated by the prince. The house, which Ruscelli describes room by room, included an herb garden, a courtyard with fountains and bird cages, and a laboratory (*lavoratorio*) for conducting the experiments. The society hired several artisans to assist them in doing their experiments, including two apothecaries, two goldsmiths, two perfumers, and four herbalists and gardeners. There were also domestic servants and a team of "choremen" assigned to the laboratory itself, where they attended the furnace, cleaned vessels, ground herbs and chemicals, and made lute for the distillation apparatus. Each of these workers was assigned to specific tasks according to his special skills, while the academicians acted as overseers, giving directions for conducting experiments: "All those who stood above these attendants gave orders to the apothecaries, goldsmiths, and perfumers, according to the tasks they were performing; that is, if they were things of chemistry, they ordered the chemists, of perfumery, the perfumers, of colors, the painters, etc." Even so, the academicians were not averse to dirtying their own hands, for as Ruscelli explicitly states, "these higher people never failed to lend a hand willingly or to busy themselves where necessary."[39]

Ruscelli called his society the Accademia Segreta, a name meant to signify two of its predominant characteristics. First, it was quite literally a secret society, each of its members having taken an oath not to breathe a word of its existence to anyone without first obtaining permission from the entire group. Only the town physicians were let in on the secret, possibly because they assisted the academy in its experiments with new medicaments. Ruscelli insisted, however, that the society did not intend to keep its work permanently secret:

> On the contrary, his Excellency's intention and ours was that within a few years it would be manifested and publicized to everyone as a thing most honored, most virtuous, and most worthy to elicit the noblest rivalry from every true Lord and Prince in his state, and from every beautiful and sublime mind. But we wanted it secret . . . because while reducing it to perfection, we would be able to do it more quietly, without being disturbed, impaired, or troubled all the time by this one and that one running to come and see it. And above all it seemed to us worthy of scholars to want the world to see and hear the fruit of our labor rather than mere rumors or extravagant promises such as many make.[40]

The name also signified the nature of the society's activities, which was "to make the most diligent inquiries and, as it were, a true anatomy of the things and operations of Nature itself. . . . In addition to our own pleasure and utility, we devoted ourselves equally to the benefit of the world in general and in particular, by reducing to certainty and true knowledge so many most useful and important secrets of all kinds for all sorts of people, be they rich or poor, learned or ignorant, male or female, young or old."[41]

The contents of the *Secreti nuovi* allow us to make a plausible reconstruction of the academy's experimental activities. The work contains more than a thousand "secrets," including medical recipes for ailments ranging from eye sores, burns, and abscesses to cures for leprosy, epilepsy, and contagious diseases. The work also contains recipes for cosmetics such as hair treatments, facial powders, and compounds for cleaning teeth. One section is devoted to technical recipes for making paints, pigments, dyes, articles of metal, perfumes, soap, and artificial beads and gems. Distillation and other alchemical techniques dominate the work. Yet despite the practical orientation of the *Secreti nuovi*, the academy's main goal, as Ruscelli put it, was to "reduce secrets to certainty." The members of the Accademia Segreta considered themselves to be natural philosophers who had adopted a novel experimental method. Their aim, Ruscelli wrote, was not merely to collect recipes from books, as others had done, but to try them out in an organized fashion, and to publish only the ones that had passed the experimental test:

> During all those years, we continually experimented on all the secrets that we could recover from books, whether printed or written, be they ancient or modern. And in doing such experiments, we adopted an order and method, one better than which cannot be found or imagined, as will be recounted next. Of all those secrets we found to be true by doing three experiments on each . . . we, by the command of our Prince and Lord, chose the ones that are easiest for everyone to do, of minor expense, and most useful for all kinds of people.[42]

The "method" articulated in this passage is obviously rather primitive, judged by modern standards. But historically it is quite significant, for it reveals a stage in the development of the concept of experiment that stands midway between the medieval connotation of *experimentum* as ordinary experience and Galileo's method of using experiments to test hypotheses. For Ruscelli, experiments were not merely random experiences. Nor, on the other hand, were they deliberately designed to prove or disprove general propositions. They were tests, or trials, of recipes and techniques that were recorded in books or had circulated by word of mouth. The academy's requirement that three trials of every secret be made before the stamp of approval was given was a control mechanism, a recognition, perhaps, of the variability of experimental situations. Although the purpose of the society's experimentation differed from Galileo's and our own, Ruscelli realized that establishing standard procedures for testing secrets was a way of guaranteeing their effectiveness.

Philosophical Ideals and Political Realities

Did this extraordinary society really exist, or was it merely the product of a fertile Renaissance imagination? Unfortunately, we have only Ruscelli's testimony to go by, although his account is substantiated by his contemporaries Francesco Sansovino and Girolamo Muzio, and by the historian Girolamo Ghilini, all of whom affirm that Ruscelli was the author of Alessio's *Secreti*. Nevertheless, Ruscelli's proem, no less than Alessio's preface, strains credibility. The repeated references to the numbers three, seven, and twenty-seven (a "most highly mystical number," according to Ruscelli) certainly arouse suspicion. It seems equally suspicious that an academy enjoying such lavish patronage as Ruscelli claims his did should have gone entirely unmentioned by contemporaries. Ruscelli's explanation of this—that the academy kept its deliberations secret—is also difficult to accept at face value, since the group is supposed to have built its meetinghouse near the central piazza. On the other hand, it is well known that the Neapolitan nobility organized, participated in, and patronized academies whose activities went beyond the merely literary interests of most Renaissance Italian academies.[43] Nor does the substantial financial support that Ruscelli claims was extended to his academy seem implausible in light of what is known about other academies, such as the Accademia Venetiana founded by the Venetian patrician Federigo Badoar.[44] Moreover, as we will see later, the scientific aims and methods of Ruscelli's group were entirely consistent with those of another secret Neapolitan academy founded by Giambattista Della Porta in the 1560s (and called, coincidentally, the Accademia dei Secreti).[45] The similarity of the aims, motives, and even the names of these two groups leads us to wonder: Why all this preoccupation with secrecy in the Neapolitan academies? What was it about the pursuit after "secrets" that caused those engaged in it to be so cautious?

Ruscelli's reasons for concealing the existence of the academy, it appears, had little to do with the group's strictly scientific activities. The Neapolitan academies were more than just centers of literary and philosophical discussion. In the mind of the Spanish viceroy, Don Pedro of Toledo, they were also seedbeds of political and religious subversion. Toledo had legitimate reasons for his suspicions, for the academies were patronized by the powerful Neapolitan nobility. Proud, fiercely independent, and jealous of the traditional autonomy they enjoyed as subjects of the king of Naples, the barons chafed under the rule of the Spanish viceroys, who attempted to force them to submit to the absolute rule of the Spanish monarchy. Toledo's difficulties with the Neapolitan nobility had begun almost immediately upon his appointment to the viceroyalty in 1532, when he tried forcibly to reduce the barons' power and to curb their rowdy behavior.[46] In doing so, he made himself a bitter enemy of the aristocrats and came to believe, with some justification, that they secretly plotted against him.

Ruscelli does not tell us the name of the "famous city" where his academy met, nor does he identify the "magnanimous Prince" who financed it. We know only that his patron was a powerful Neapolitan baron, a "Lord of the land," and that for some reason Ruscelli was reluctant to identify him. Two likely candidates for the role emerge: Ruscelli's patron, Alfonso d'Avalos, the marquis of Vasto (1502–1546), and Ferrante Sanseverino, the prince of Salerno (1507–1568). Both fit Ruscelli's rather vague description of the academy's benefactor as a powerful local baron and a generous patron of artists and letters.[47] Avalos, however, died in 1546, too early to have financed the academy for the ten years Ruscelli claims it existed, and in any case the principality he ruled was distant from the city of Naples and unlikely to have been characterized by anyone as famous. Sanseverino, on the other hand, was certainly the prince of a justly famous city, Salerno. He was the most powerful baron of the kingdom of Naples and a Maecenas of learning and the arts. As we will see, Ruscelli had ample reasons for concealing his identity. Avalos and Sanseverino were both sworn enemies of the Spanish viceroy. In 1535, they had been coconspirators in an attempt to persuade the emperor, Charles V, to have Toledo removed from office.[48] The plot failed, but the Neapolitan barons continued to regard the emperor as a better ally of their political interests than either the viceroy or the pope. Ruscelli's sympathy with the imperial party was well known to contemporaries, and it may be surmised that he reflected the views of his patron.[49]

The local interests of the Neapolitan nobility also intertwined with the conflicting religious policies of the great European powers. In the 1530s, the barons tended to support the emperor's attempts to achieve a reconciliation between Protestants and Catholics through the convocation of a general church council. That position was sternly opposed by Pope Clement VII, grudgingly acceded to as a matter of necessity by his successor Paul III, and continually sabotaged by the king of France.[50] Underlying the dream of unification, which the Neapolitan barons supported largely for political reasons, lay strong religious sentiments. In Naples, the movement known as Evangelism, a combination of Erasmian reformist ideas and Spanish mysticism, produced high expectations among the nobility for the success of a general council.[51] In 1534, the Spanish humanist Juan de Valdes, whose teachings inspired Italian Evangelism, settled in Naples and attracted a celebrated group of aristocratic admirers that included the prince of Salerno; Alfonso d'Avalos; Avalos's wife, Maria of Aragon, and her sister Giovanna; Avalos's aunt Vittoria Colonna; and Giulia Gonzaga, the young widow of Vespasiano Colonna, count of Fondi.[52] Although the Evangelists (or Valdesians, as the followers of Valdes are also called) were not religious revolutionaries, some of their ideas—in particular their belief in justification by faith—bordered dangerously on Protestantism.[53] The Valdes circle also included Fra Bernardino Ochino, a Capuchin monk whose fiery sermons ignited religious zeal all over Naples. Ochino later fled Italy, converted to Calvinism, and ended his days as an anti-Trinitarian.

The effects of Fra Ochino's electrifying sermons were soon manifested. The Neapolitan historian Pietro Giannone relates that Ochino was so eloquent that when he preached the Lenten sermons in 1536 he emptied all the other churches. Charles V, visiting Naples at the time, commented that Ochino "preached with such Spirit, and so much Devotion, that he made the very Stones weep." According to Giannone, by the time Ochino departed the city in 1541, "Every Cobler in his Stall took the Liberty to discourse of St. Paul's Epistles, and of the most difficult Passages in them; and, which was worse, at his Departure, he left some faithful Disciples behind in Naples, and many infected with his pernicious Doctrine, as he had done through all Italy where he had preached."[54] The apostasy of Fra Ochino cast suspicion on his followers in Naples and convinced Toledo that something had to be done to check the growth of heretical doctrines. In 1544 he issued edicts requiring the confiscation of heretical books and the establishment of press censorship.[55] But he warned the emperor that even sterner measures were needed, and in 1546 received permission to establish the Inquisition in the kingdom of Naples.[56] Whereas earlier attempts to establish the Spanish Inquisition in Naples had failed because of the citizens' fear and hatred of the institution, the viceroy hoped to mollify these fears by bringing in the papal Holy Office. The general belief was that Pope Paul III, anxious over the growing ascendancy of the emperor, willingly complied with the request because he was eager to create civil strife in Naples: "Knowing how odious the Inquisition was to that People," wrote Giannone, "and bearing the Emperor a secret Grudge, he believed that the Attempt to introduce [the Inquisition] into that Kingdom would occasion Revolutions, Tumults, and Seditions in that City."[57]

The pope's expectations and the viceroy's worst fears were almost immediately realized. The very mention of the Inquisition was enough to inflame the people, and on 11 May 1547, when the edict announcing its establishment was posted, the city exploded in revolt. Amid cries of "Armi, armi! Serra, serra!" the edict was torn down, houses and shops were closed, and the people, urged on by several prominent noblemen, stormed the Spanish garrison. Hundreds were killed or arrested during the rebellion, which erupted periodically throughout the summer. The prince of Salerno, Toledo's bitterest enemy, was delegated by the anti-viceroyalist "Association" to plead the city's case to the emperor. Although Sanseverino failed in his main mission, which was to play off the viceroy against the emperor and let the viceroy destroy himself, the attempt to introduce the Inquisition was, for the moment, abandoned.[58]

Although the Inquisition was forestalled by the bloody "tumult" of 1547, Toledo took strong measures to stem the tide of popular revolt. Among the victims of the repression that followed were the various literary and philosophical academies that had arisen in the years just prior to the insurrection. Regarding the academies as centers of religious and political subversion, Toledo closed them all down. According to Giannone, "The Reason why these Academies were so suddenly suppressed, was a Rule laid down, That every Member should read a Lecture, in Disputing, upon which afterwards (although the Subjects were Philosophy or Rhetorick) they often dropt the Subject, and fell into Questions of Scripture and Divinity, and therefore all those Schools were forbidden and abolished."[59] The historian and notary Antonio Castaldo, who was at the time secretary of one of these academies (the Sereni) observed, perhaps more to the point, that the academies were closed because "it did not look good to have under the pretext of literary exercise so many gatherings and such continuous meeting of the wisest and loftiest minds of the city, since the pursuit of letters makes men more shrewd, bolder, and more headstrong in their actions."[60] Although there is no positive evidence of political intrigue in the academies, Toledo's suspicions are understandable in light of the deliberate cloak of secrecy under which many of the groups operated. Not only were the academies under suspicion of promoting unorthodox religious doctrines, they were patronized and sometimes organized by those whom the viceroy most feared, the powerful barons who resented Toledo's attempts to break the nobility's quasi-independent status. These groups certainly constituted, in Toledo's mind at least, potential centers of political conspiracy.

Against the background of these tumultous political and religious disturbances, we can now attempt a fuller reconstruction of the Accademia Segreta described by Ruscelli. Supposing the academy met in Salerno under the patronage of Ferrante Sanseverino, since Ruscelli says it met continuously for ten years, we can establish the period during which the academy flourished as between 1542 and 1552. After 1552, it cannot have existed under the prince of Salerno's patronage, because in that year the ambitious Sanseverino, frustrated by his inability to convince Charles V to dismiss Toledo, abandoned the imperial cause and left Naples to enter the service of King Henry II of France. Hoping to secure for himself the crown of Naples, Sanseverino convinced Henry to put him in command of an expedition to conquer the kingdom for France. The rebellion was a dismal failure. The commander of the Turkish fleet, whom Sanseverino counted on as an ally in the enterprise, was bought off for 200,000 ducats, which Toledo, it is said, deemed a good price for such a fine principality as that of Salerno. The viceroy declared Sanseverino a rebel, sentenced him to death, and confiscated his estate. The ruined prince of Salerno wintered over in Constantinople (living in debauchery, according to Giannone) and then returned to France. Humiliated by his defeat and condemned as a traitor, Sanseverino converted to the Huguenot cause and died in Avignon, disgraced and deprived of his fortune, in 1568.[61]

Meanwhile, in 1546, Alfonso d'Avalos died, leaving Ruscelli without a patron. By then, the political situation in Naples had become so charged that—his reputation perhaps tainted by his association with the Valdesians—his most prudent course was to leave the city. He probably did so soon after Avalos's death, for in 1548 we find him working in the Valgrisi printing house in Venice. Although the evidence for this reconstruction is admittedly circumstantial, it would explain not only Ruscelli's move to Venice, but also his publication of the *Secreti* of Alessio under a pseudonym and his reluctance to publish his account of the Accademia Segreta. Even though Ruscelli published the *Secreti* in Venice, he would have been concerned that revealing his true identity so soon after the events of 1547 and Sanseverino's rebellion might reopen old wounds and cause trouble for friends still living in Naples. In any case, the association of Ruscelli's beloved academy with the name of Sanseverino and his notorious treachery would only have tainted the reputation of that group. These concerns evidently remained with Ruscelli through the rest of his life. Despite the popular success of Alessio's secrets, Ruscelli withheld publication of the *Secreti nuovi*, with its description of the academy, leaving the book to his heirs to send to the printer after his death. Ironically, none of the works printed under Ruscelli's own name were remotely as popular as the one he published under the pseudonym Alexis of Piedmont.

The foregoing reconstruction of Ruscelli's activities in Naples sheds additional light on the research program of the Accademia Segreta. According to Ruscelli, the academy's aim was not simply to produce a collection of useful and "proven" recipes, but also to acquire *conoscenza di se stessi*, knowledge of themselves. Since the human being is a microcosm reflecting all the attributes of the macrocosm (the universe), the route to "self-knowledge" is through the study of nature. As Ruscelli explained, "our intention was first to study and learn about ourselves, there being no other study or discipline that is truer in natural philosophy than this, of making the most diligent inquiries and, as it were, a true anatomy of the things and operations of Nature according to itself."[62] The goal of studying nature "in and of itself" (*in se stessa*) and the identification of self-knowledge with knowledge of nature were dominant motifs in the intellectual milieu of late-Renaissance Naples.[63] The roots of the twin ideal of *conoscenza di sè* were in the tradition of philosophical naturalism, a worldview shared by thinkers as diverse as Girolamo Cardano, Giordano Bruno, and Tommaso Campanella. For Neapolitan thinkers, the most influential source of the naturalistic metaphysics was the philosophy of Bernardino Telesio of Cosenza (1509–1588), whom Bacon dubbed "the first of the moderns."[64] Telesio's *De rerum natura* was a frontal assault on Aristotelian metaphysics. Rejecting the forms, essences, and categories of Peripatetic philosophy, Telesio insisted that nature must be understood and investigated "according to its own principles" (*iuxta propria principia*). Observation alone, he asserted, not reason or authority, was the path to true knowledge. When we let things speak for themselves, thought Telesio, nature presents itself as a single, unitary physical substratum activated by the forces of heat and cold. Forms are produced by a Heraclitean "struggle" between the active forces of heat and cold to impress themselves in matter. Telesio thus repudiated the Aristotelian view of prime matter as pure potency, a mere logical subject of change induced by forms. Nature is concrete and physical. However, it is not inert; nor do the forms of things arise by blind chance. On the contrary, all of nature is endowed with sensation by virtue of a subtle, corporeal "spirit" that penetrates all things. Sensation enables nature to perceive which active principles are appropriate to it, to accept whatever tends to its preservation and to reject whatever militates against it. This principle of self-preservation (*conservatio sui ipsius*), which is universal, determines all individuation in nature. Telesio's radical panpsychism implies that subject and object, knower and known, are of the same nature. Thus to "know" a thing means to become one with it.[65]

The Accademia Segreta was, I believe, inspired by Telesian metaphysics and was dedicated to experimental research aimed at fulfilling the promise that, in the minds of Ruscelli's Neapolitan contemporaries, Telesio's philosophy held out. For if, as Telesio claimed, nature is unitary and capable of taking on many forms, if self-knowledge and knowledge of nature are one, and if subject and object are one, it follows that the knower, in knowing nature, acts upon and alters nature. For Telesio, the supreme sensation was touch, not sight. Knowing nature meant literally to "grasp" nature in all its being. *Conoscenza di sè* thus led to the possibility of using knowledge to improve the moral and material condition of humanity. As Ruscelli explained, self-knowledge is accomplished by making "true anatomies" of nature through experiment. From this principle proceeded the second goal of the academy's investigations, which was to apply their research to making improvements beneficial to man: "For it is seen that the origin and development of medicine as well as all the other disciplines important to human life and the ornament of the world were helped by art. And so, in addition to our own pleasure and utility, we devoted ourselves equally to the benefit of the world in general and in particular, by reducing to certainty and true knowledge so many most useful and important secrets of all kinds for all sorts of people, be they rich or poor, learned or ignorant, male or female, young or old."[66] To accomplish this aim, the entire ancient and medieval scientific encyclopedia had to be retried and subjected to the authority of experimental tests. In place of the "mere rumors" and "extravagant promises" of scholastic philosophy, the Segreti would substitute "the fruit of our labor."[67]

The meaning of this last statement becomes clearer in light of Telesian epistemology and metaphysics. The focus of Telesio's thought was upon nature's "operations" rather than its "essence." Observation reveals that the "operations" of nature consist of active forces influencing passive but sensitive matter. Telesio maintained that while observation yields the only knowledge we have of the world, it reveals only nature "operating" in various ways; it does not reveal the operator or the ultimate cause of things. Such knowledge must remain a mystery to us.[68] Ruscelli defined the object of the Segreti's research as making "anatomies" of the "things and operations of Nature" (*delle cose & dell'operationi della Natura*). How Ruscelli and his fellow academicians thought such a "true anatomy" (*vera anatomia*) of nature ought to be undertaken is revealed by the recipes in Alessio's *Secreti* and Ruscelli's *Secreti nuovi*. The vast majority of the operations described in these works were alchemical in nature. Distillation occupies a prominent place, and considerable space is given to metallurgical processes involving alchemical operations. The analogy between alchemy and anatomy was a common theme in the nature philosophies of the sixteenth century. For Paracelsus, anatomy was above all *chemical* anatomy: "the examination of the composition of various parts of the body in order to discover the affinity of individual parts with individual substances in the universe."[69] "True anatomy" had nothing to do with the dissection of corpses. Instead, it studied the parallelism between the microcosm and the macrocosm. Similarly, distillation was an "anatomy" of nature because it captured the hidden virtues or "essences" in things and revealed analogies to natural and celestial events, such as weather phenomena. The Italian naturalists tended to interpret the capturing of "quintessences" by distillation in materialistic and celestial terms. The products of distillation, they thought, were identical to the celestial spirits radiating from the heavens.[70] Thus alchemy, and in particular distillation, was a way of investigating, directly and empirically, the celestial virtues at work in the terrestrial world.

In the Neapolitan intellectual tradition, these elevated aims of alchemical and experimental research were frequently associated with efforts to implement moral and religious reforms. Such associations were made as early as the 1520s, when astrologers predicted that a general deluge would occur as a result of a conjunction of the planets in Pisces anticipated for the year 1524.[71] As the date of the conjunction drew nearer, controversy over the question intensified in Italy, fed by the debate over Lutheranism, the French invasion, and the election of Charles V as Holy Roman Emperor. Among the many prognostications for the year 1524 was one published in Naples in 1523 by Giovanni Abioso da Bagnolo.[72] Although Abioso denied there would be a universal flood, he argued, more alarmingly, that natural signs indicated the coming of Antichrist. Abioso thought Antichrist's advent might be delayed, however, by a general reform of human manners, and by the restoration of Christian unity. In order to accomplish such reforms, he proposed an experimental research program aimed at the discovery of "new secrets of nature" (*venari nova naturae secreta*) that might assist humankind in finding ways to become more "in touch with" the celestial and divine forces in nature. As an example, Abioso wrote enthusiastically about a marvelous drug he discovered by separating the elements and extracting its quintessence through distillation. Abioso explicitly linked experimental science with an Erasmian program of religious and social reform. Similar tendencies are evident in the writings of the *poligrafo* Nicolò Franco, who was also active in the southern Italian academies. Franco's naturalism, Erasmianism, extreme anticlericalism, and subversive political views eventually landed him before the Inquisition, which condemned him to death in 1570.[73] Other Neapolitan academies evidently adopted research programs similar to that of the Accademia Segreta. The short-lived Incogniti (Unknown ones), which was founded by a group of Neapolitan patricians in 1546 and closed during the "tumult" of 1547, expressly stated that its aim was *conoscenza di sè*. Among the writings the society sponsored was a work contributed by a member (known only by his academic name, "Il Segreto") on the "mysteries of nature." Another work discussed by the academy was a treatise by "Il

Silenzio" on "all the constitutions, rites, and privileges observed and enjoyed in the world."[74]

Toledo, "a great admirer of the scholastic divines," was deeply troubled by the spread of heresy in his realm. There is also evidence that he was concerned about the growing interest of intellectuals in natural philosophy, particularly their fascination with alchemy and the occult sciences. In 1544, when the subject of alchemy came up in discussions at his court, the viceroy asked Benedetto Varchi, a prominent Aristotelian and one of the participants in the discussion, to communicate to him in writing his opinion whether, "according to the principles of Aristotle, alchemy can be proved or demonstratively confirmed to be either possible or impossible." Varchi's response, contained in his *Questiones sull'Alchimia*, was that some forms of alchemy, such as the alchemical methods used by artisans, were useful and valid. However, Varchi warned his patron, "sophistical" alchemy, the kind that claims to transmute gold or make marvelous drugs that restore youth and cure all diseases, promotes all kinds of wickedness. Such forms of alchemy are "more than deservedly banned by all good princes and by well-ordered republics, and persecuted with fire."[75] We do not know how seriously Toledo took Varchi's advice, or whether it gave him additional reasons for closing the academies. It must certainly have added to his worries about those secretive groups, whose natural-philosophical research may have masked subversive political and religious activities.

If the prince of Salerno was in fact the patron who financed the Accademia Segreta, as seems probable, the convergence of two traditions—socioreligious reform and self-knowledge through knowledge of nature—helps to explain what may have motivated his patronage of that group. The character of Sanseverino presents a picture of contrasts. Ambitious, vain, and prone to fits of violence, he was the quintessential Machiavellian prince of the Renaissance. Princely too, however, was his devotion to the ideal of courtesy: he was truly, as Ruscelli characterized him, a "magnanimous Prince" and patron of letters. At his court in Salerno lived some of the brightest lights of Neapolitan letters. His patronage of private theatrical troupes contributed to the flourishing of dramatic literature that nourished Giambattista Della Porta's literary career. Sanseverino's active support of the School of Salerno contributed to the revival of the university's reputation in the early sixteenth century. He attracted to its faculty such notable scholars as the astrologer Luca Gaurico, to whom Sanseverino gave an ecclesiastical benefice, and the philosopher Agostino Nifo, who received from the prince a pension for life.[76] After the viceroy of Naples closed down the Academia Pontaniana in 1543 under suspicion of heretical activities and dismissed its leader, Scipione Capece, from the University of Naples, Sanseverino brought

Capece to Salerno and appointed him professor of law at the school. Thus it appears that Sanseverino's support of the Accademia Segreta was part of an ambitious plan to make the principality of Salerno the center of Neapolitan cultural and intellectual life.

The prince of Salerno's interest in the academy may have been motivated by even greater ambitions, however. Ruscelli and his fellow academicians linked alchemy with the exalted spiritual ideal of *conoscenza di sè*, which, as we have seen, had implications for both religious and social reform. The ideology of "self-knowledge" brought together and unified the diverse intellectual traditions of southern Italy: Telesian metaphysics, with its conception of a single, unitary substance underlying all forms; the alchemical ideal of the possibility of harnessing and exploiting these cosmic energies; evangelism; and the idea of a "return to nature," which would result in a restoration of manners, customs, and the Church *in capite et membris*. The academy's claim that experiment is the ultimate test of tradition's value established a new standard of authority that challenged the scholastic order, upon which the power of Church and state rested. Although it may be too extreme to argue, as Nicola Badaloni does, that the Neapolitan nobility actually conspired to restructure society through violent revolution, there is little doubt that their hopes for reform focused upon Toledo's most prominent enemy, the prince of Salerno.[77] Sanseverino's ambition—to become the king of Naples—surfaced only in 1552, but his hostility toward the viceroy was long-standing. Like most Renaissance princes, he also solicited views in support of his political ambitions from the intellectuals in his court.[78] Although the Segreti, according to Ruscelli, were sworn to secrecy "on the love and grace of our Prince," they did not plan to keep the society permanantly secret but intended to publicize it "within a few years." Then, writes Ruscelli, the society "would be manifested and publicized to everyone as a thing most honored, most virtuous, and most worthy to elicit the noblest rivalry from every true Lord and Prince in his state, and from every beautiful and sublime mind." The aims of the Segreti toward a social and religious renewal of southern Italy thus coincided with Sanseverino's political goal of ruling the kingdom of Naples himself.

But the prince of Salerno's ambitions were never realized. His abortive rebellion failed, and with the confiscation of his principality by the viceroy, the Accademia Segreta's hopes for implementing social and religious reform under the banner of *conoscenza di sè* came to an abrupt and disappointing end. Yet other Neapolitan intellectuals carried on the program inspired by the Segreti, among them the precocious Giambattista Della Porta, who may have been one of the privileged observers of the experimental activities of Ruscelli's group. Ruscelli moved on to Venice, a republic relatively immune to the Inquisition's oppressive censorial arm. There, among the *poligrafi*, Ruscelli found others who shared his ideal of publishing secrets tested by experience and "useful for all kinds of persons." And he found that printers made more reliable accomplices in this enterprise than princes.

The Heirs of Alexis

Alessio Piemontese's *Secrets* spawned scores of imitations. The books of secrets were among sixteenth-century Italy's most popular scientific writings. It is not difficult to understand why this was so. First of all, the books of secrets were handbooks of practical information, containing recipes for arts ranging from dyeing and ink making to making perfumes and cosmetics. Although they were by no means "everyday" household books, they were handy reference guides for a growing middle class whose interests focused more upon the material than the spiritual or the purely intellectual. Moreover, these works revealed the secrets of exotic arts such as perfumery, of arcane arts such as alchemy, and of specialized crafts such as dyeing and metallurgy, all subjects that were supremely important to the burgeoning sixteenth-century urban economy. Finally, the professors of secrets brought into sharp relief the contrasting authorities of experience versus book learning, and came down firmly on the side of the former. Their secrets, they insisted, came not from conventional academic authorities, but from careful observation, long experience, and extensive travel. They did not disdain going to ordinary people, empirics, and housewives for information. Alessio postured as a man who had turned away from traditional academic pursuits in order to gain experience of the world, while Ruscelli's academicians put the claims of books to the test of experiment. Leonardo Fioravanti, one of the most outspoken critics of the medical establishment, advocated taking medicine back to the "first physicians," who learned medicine purely empirically, by observing how the animals healed themselves.[79]

The appeal to nature and experience was one of the most characteristic expressions of Renaissance scientific culture. "Whoever wishes to explore nature must tread her books with his feet," wrote Paracelsus. "Writing is learned from letters; Nature, however, (by travelling) from land to land: one land, one page. Thus is the Codex Naturae, thus must its leaves be turned."[80] Cornelius Agrippa von Nettesheim wrote that whereas the physicians acquire their remedies from books, the "old wife" searches nature for plants, learning their colors, tastes, shapes, and odors, and according to her experience of their virtues administers sure remedies free of charge to anyone.[81] Paracelsus's Danish pupil, Peter Severinus, exhorted naturalists to "sell your lands, burn up your books, buy yourself stout shoes, travel to the mountains, search the valleys, the deserts, the shores of the sea, and the deepest depressions of the earth. . . . Be not ashamed to study the astronomy and terrestrial philosophy of the peasantry. Purchase coal, build furnaces, watch and operate the fire. In this way and no other you will arrive at a knowledge of things and their properties."[82]

The idea of studying "the astronomy and terrestrial philosophy of the peasantry" must have sent up howls of laughter from the academic establishment. Nevertheless, the trashing of academic learning was a serious matter, not easily tolerated by those whose social identity and economic interests were vested in monopolies sanctioned by the academy. Laurent Joubert, chancellor of the Faculty of Medicine at Montpellier, observed that "no arts are more subject to calumny than the military and the medical."[83] According to Joubert, the most erroneous of all popular beliefs was that physicians regularly prolonged illnesses. Even if the physician were evil, he protested, there were so many others competing with him that it would be pure folly for any physician to prolong an illness. "If [the physician] heals in less time than the others," he argued, "he will be in greater demand and will have such a wealth of patients that he will not be able to treat them all. . . . For who would not rather pay double, even triple or quadruple, and be cured sooner?"[84] Joubert shrewdly identified the physician's dilemma in the changing medical marketplace: if "everybody makes medicine his business," how can the ordinary person tell the true physician from the charlatan? As it was, so many "surgeons, barbers, apothecaries, attendants, midwives, charlatans, and other quacks" peddled nostrums and quick cures on the streetcorners that some of the physicians had begun to serve up dubious panaceas of their own.[85] There was a brisk market for medical "secrets," which might be any sort of nostrum, electuary, or quick cure for what ails you. Printing added a new variable to the bullish market for secrets. While it cheapened individual secrets—for now you could buy a whole *book* of secrets for a mere tenpence—it enabled practically any medical empiric to exploit the vast potential of the printed book market to sell his "secrets."[86] Physicians continued to hold onto the traditional view that publication should be restricted to a professional audience. But that did not prevent others from publishing physicians' secrets for popular readers, or from publishing their own secrets spuriously, under some famous physician's name.

Pietro Bairo's *Secreti medicinali* was a popularization of the famous physician's personal *veni mecum*, or book of recipes he took with him when he visited patients. A professor of medicine at the Turin medical college, Bairo was also a personal physician to Dukes Charles II and III of Savoy.[87] Bairo wrote the work for a professional audience rather than for general readers. The Latin version of the handbook, *De medendis humanis corporis malis Enchiridion* (Turin, 1512), was widely used by physicians as a clinical guide: it was printed in Basel, Lyon, Leyden, and Frankfurt, as well as in Turin and Venice. The book is arranged, in traditional academic fashion, by ailments from head to foot for easy reference. Under each entry Bairo gave an analysis of the ailment's causes according to the theory of humoral imbalances, a detailed regimen to counteract the humoral imbalance, and recipes for drugs to help restore the body's constitution to its normal state. Bairo's *Medical Secrets* was thus a rather conservative work reflecting a traditional outlook toward medical practice.[88] The *practicum* attributed to Giovanni Battista Zapata (ca. 1520–ca. 1586), by contrast, devoted little attention to theory. More typical of the printed books of secrets, the work merely gave remedies for common ailments ranging from dropsy to ringworm. Zapata was a celebrated Roman surgeon and empirical doctor. It is said he was a student of Hippolyto Salviani, a papal physician and a professor of practical medicine at the Roman Sapienza.[89] Although Zapata did not obtain a medical degree, he was a famous practitioner who gathered around him numerous disciples. Indeed, he seems to have been the focus of a cultlike following because of his dedication to Rome's poor and suffering. The *Maravigliosi secreti* (Marvelous secrets) was not actually written by Zapata, but by a group of his students (they reverentially refer to Zapata as *il nostro Precettore*), who collected the master's medical secrets and published them under Zapata's name. One of the students, the surgeon Giuseppe Scientia, wrote that the secrets were the ones Zapata had proved and prescribed daily in his practice. Scientia begged physicians to follow the master's example by devoting a share of their practice to the treatment of the poor, widows, and orphans. The very first recipe in the work was one for "potable gold" for poor people: a solution of sugar in aqua vitae. Though an empiric, Zapata was well versed in school medicine. He invoked the authority of Galen, Hippocrates, and Avicenna. His pupils insisted they were "methodical doctors, not empirics" (*Medici Methodici e non Empirici*), thus distinguishing themselves from the wandering empirics and street healers that were such a familiar feature of the Italian cities.[90] Even though they did not possess medical degrees, Zapata and his disciples were outspoken critics of the *ciarlatani* and their quick cures. Zapata believed that nature had its own medicines. The natural healing process could only be assisted, not overcome or improved upon, by the physician's remedies. Compared to nature's healing powers, he maintained, the *ciarlatani*'s bogus panaceas were completely useless.

While generally conservative in his approach to therapy, Zapata also used alchemical drugs and was a strong advocate of distillation. Alchemy, thought the professors of secrets, was the science par excellence for discovering the secrets of nature. Giambattista Della Porta wrote that the art of alchemy, though much abused and slandered, should be "embraced and much sought after, especially by such as apply their minds to Philosophy, and to the searching out of the secrecies of Nature."[91] Isabella Cortese's all-purpose *Secreti* (1561) was dedicated entirely to various applications of the art of alchemy.[92] Cortese was an avid alchemical practitioner. She traveled widely in search of the secrets of alchemy and learned techniques firsthand from alchemists throughout Italy and central Europe.[93] Her recipe book gives us a precious glimpse at the alchemical underworld of early modern Europe. What kind of discourse was carried out among the alchemical "operators"? It did not seem to have much to do with a "chemical philosophy" of nature. Although Cortese's work is an extraordinarily rich source of information about the practical applications of alchemy, she was openly contemptuous of alchemical theory. Writing to her brother-in-law, Mario Chaboga, who was, like herself, a disappointed alchemical aficionado, she advised Chaboga not to bother reading the books of the famous alchemists:

> I tell you, dearest brother, if you want to follow the art of alchemy and to operate in it, it is not necessary to follow the works of Geber, nor of Ramon [Lull], nor Arnaldo [Villanova], nor any of the other philosophers, because they have not recorded anything truthful in their books, but only fictions and riddles. . . . I have studied these books for more than thirty years and have never found anything good in them. I wasted time and almost lost my life and all my possessions. . . . And because, dearest brother, I know that you have wasted a lot of time and spent much, out of compassion for you I beg you not to lose any more time on these books of the philosophers. Just follow the rules I write down for you. Do not increase or diminish anything, but do what I say and write, and follow my commandments written below.[94]

Cortese's "ten alchemical commandments" warned operators never to work with any supposed "great master" of alchemy. Use sturdy, well-crafted equipment, carefully regulate the fire, have a trusty servant, and take precautions to insure the secrecy of your work. If someone asks you about something relating to alchemy, pretend not to know. Don't let anyone into your workshop. Don't ever leave your servant in the workshop alone. Above all, don't teach the art to anyone, "because revealing the secret of something causes it to lose its efficacy" ("perche il revelare de secreti fa perdere l'efficacia"). Secretive attitudes still governed alchemical practice. Indeed, it is not clear whether Cortese ever intended to publish her work, or whether the letter was intended for her brother-in-law's eyes alone and fell accidentally into a printer's waiting hands.

Nor did alchemical practice have much to do with resurrecting the soul or making the philosopher's stone. Cortese was more interested in making perfumes, cosmetics, artificial pearls and other bijouterie, in distilling oils and essences, making jewelry, and working with metals. There was nothing metaphorical or allegorical about the operations themselves. They were aimed strictly at producing objects for immediate practical use. Not that alchemical products were intended solely or mainly for the common sort. In Italy, in contrast perhaps to Germany, the alchemical operators catered to an aristocratic audience. Cortese's publisher thought her secrets belonged "to all the great ladies," since they included recipes for making perfumes, cosmetics, sweet-smelling soap, pomades, body powder, and hair dyes, as well as for the more mundane operations with metals. Timotheo Rossello, another alchemist, arranged his *Secreti universali* "for men and women of high genius, as well as for physicians, for all kinds of artisans, and for all the virtuosi."[95] Rossello dedicated the work to two noble patrons. He presented the first edition to Cortese's brother-in-law, Mario Chaboga, the archdeacon of Ragusa. Chaboga, Rossello hoped, though frustrated in his own alchemical research, might still be forthcoming in his patronage of more successful operators. In 1575, Rossello enlarged the work, dedicating the added part to Giovanbattista Romano, a Paduan nobleman, who he thought would take delight in the work's "copious variety."[96]

One of the main exceptions to this aristocratic orientation among the alchemical operators was the Jesuati religious order.[97] The "aquavit-brothers," as the Jesuati were sometimes derisively called, dedicated themselves to providing health care to the poor. The brothers specialized in making distilled "elixirs" and cordials, which they believed preserved the body from corruption and putrefaction. This medical doctrine was inspired by the writings of John of Rupescissa, a fourteenth-century Franciscan friar who taught that spirit of wine (alcohol) was the incorruptible "fifth essence" of substances: it is related to the four qualities as heaven is related to the four elements.[98] The Jesuati set up distilleries throughout Italy and manufactured remedies, which they distributed gratis to the poor. The only work that has survived relating to Jesuati alchemy is a sixteenth-century *Libro de i secretti e ricette* composed by one of the order's brothers, Giovanni Andrea di Farre de Brescia.[99] This remarkable manuscript contains numerous illustrations of alchemical apparatus and detailed descriptions of distillation procedures. It enumerates hundreds of remedies for ailments affecting all parts of the body. Much space is devoted to the dreaded *mal Francese* (syphilis), and there are repeated references to guaiac wood (lignum vitae), one of the most famous new drugs from the New World.[100]

If the alchemical operators were skeptical of alchemical theory, they were enthusiastic about the practical benefits of the art. The image of *homo faber* pervades the books of secrets. Cortese wrote of humanity's drive to overcome nature, and to accomplish things that nature unassisted could not. Alchemy made man godlike:

> Man being of all creatures the most perfect, . . . and retaining in his essential form the greatest similitude with his creator, to whom idleness has no place, it follows that nothing can remain idle in the human intellect. Hence arise the speculations from which the sciences are made. Hence also arises the investigation of the occult secrets of nature. But why do I say investigation, when man is not content with investigation? For he strives, in putting everything into works, to make himself the Ape of Nature, indeed to supersede nature, as he tries to do that which to nature is impossible. And the truth is, it is possible to extract Secrets that every day we hear about and see put into execution.[101]

The professors of secrets fervently endorsed the idea of humanity's "natural desire" to discover new inventions and new "secrets of nature." Rossello believed that history showed "it is a natural thing to search after the secrets of nature." Indeed, the wise men of antiquity busied themselves so much with the search for nature's secrets that "they were counted among the many gods who discovered things necessary for human life."[102] Nothing could be more characteristic of the views of the professors of secrets, for whom the search for the "secrets of nature" was a scientific and moral crusade.

In 1563, the Venetian printer Marco di Maria published a work entitled the *Secreti diversi et miracolosi* (Diverse and miraculous secrets), ascribing the work to Gabriele Falloppio, the famous Padua anatomist. Di Maria explained that the compilation had "fallen into his hands" after Falloppio's death in 1562. In his dedication of the work to Girolamo Prioli, Di Maria wrote glowingly of Falloppio's lifelong devotion to experimentation. The anatomist collected a "sea of very rare experiments" from various literati and empirics, and put them all to the test in his own practice. He found many to be completely useless and cast them aside. Of the ones that passed the test, he made the present collection.[103] Di Maria may have believed these statements, but in reality the work was entirely spurious, evidently an attempt to capitalize on Falloppio's reputation.[104] A few tantalizing clues may lead us to the identity of the work's author. The *Secreti diversi* makes numerous references to other empirics, including Benedetto Faentino, Matteo Ungaro, the surgeon Zorzi, and, most frequently, Leonardo Fioravanti. Indeed, the work praises the famous Bolognese surgeon so effusively that it reads like an extended advertisement for Fioravanti's books. The work lauds Fioravanti as a *huomo raro* for his marvelous ability to cure syphilis. It recommends Fioravanti's water for leprosy above all others, extols the virtues of his *oleo benedetto*, and fairly rhapsodizes about his *balsamo artificiale* for wounds, exclaiming, "Of all the balsams that have been made, there have not been any that have come as near the true one as this balsam made by the excellent master Leonardo Fioravanti of Bologna."[105] The author of the *Secreti diversi* records several recipes from Fioravanti's *Capricci medicinali*, adding that "many other lovely experiments" can be found in that book. Fioravanti, the compiler notes, is also the author of an excellent work on surgery. But his latest publication, the *Specchio di scientia universale*, contains many new and wonderful secrets, including the specifications for an unsinkable ship—one of Fioravanti's prized conceits. Countless other wonders could be found in Fioravanti's books. The *Secreti diversi*'s ample use of personal testimonies, its theatrical descriptions of healing events, and its repeated references to trade names of drugs invented by Fioravanti all point toward Fioravanti as the work's probable author.[106] It would certainly not have been out of character for Fioravanti, an apprentice writer in 1563, to use Falloppio's reputation to advertise his books in order to boost their sales. But as we shall see in the next chapter, his rising success as a popular medical writer soon made it unnecessary for Fioravanti to use any other name than his own.

FIVE

LEONARDO FIORAVANTI,

VENDOR OF SECRETS

L IKE many Italian popular writers, Leonardo Fioravanti was driven by restless curiosity, intellectual discontent, and brazen opportunism. His reputation was, and remains, as checkered as his career. An outspoken critic of the physicians, he drew down the wrath of the medical establishment, while modern historians have repudiated his bombastic claims as the bluster of a charlatan. Yet to contemporaries outside the medical establishment—and to some within it—Fioravanti was one of the wonders of the age. His skill as a surgeon and his unorthodox medical practices made him the focus of a cultlike following. He had an insatiable thirst for fame. In a world dominated by courts, fame alone bridged the gulf between obscurity and respectability. He never secured for himself a permanent place at court, although he tried more than once. He never earned the respect of the medical establishment because he never concealed his contempt for it. As Aretino had declared himself the scourge of princes, Fioravanti was the scourge of the physicians. Passionately and righteously he lashed out against the physicians' abuses, charging them with having cheated the people and having extinguished the light of "true medicine." He was no less vehement in his critique of the evils and vices of the courts. His polemics mirrored the anger of the Venetian *poligrafi*, whose moral indictment of highbrow culture gave voice to the discontent of the people. Fioravanti won fame because his writings focused on a question that was at the center of the age's consciousness: the relation between the elites and the people, between the rulers and the ruled, between privilege and powerlessness.

Fioravanti *Cirurgico*

Little is known about Fioravanti's early life. Beyond the date of his birth in Bologna, 1517, his first thirty years are almost a complete mystery.[1] For the period after that, however, a good deal can be gleaned from the detailed autobiography Fioravanti left in his *Tesoro della vita humana* (Treasury of human life, 1570). It was not an autobiography in the conventional sense, but an account of some of the more notable cures he accomplished with his newfangled remedies. Since it was intended to document his successes and to advertise his medical secrets, the work has to be taken with a healthy dose of skepticism. Nevertheless, the *Tesoro* is a rich source of information about the life of a famous empirical doctor, and if used in conjunction with other documents, it gives us an incomparable insight into medical practice in sixteenth-century Italy.

The autobiography begins in October 1548, when Fioravanti was thirty years old. Following the example of Hippocrates, Galen, and Pliny, he wrote, "I left my home sweet home, Bologna, solely with the intention of going traveling throughout the world in order to gain knowledge of natural philosophy."[2] Since it is apparent from his writings that he had learned some formal medicine, it is possible that prior to his departure from Bologna Fioravanti attended lectures at the medical college in the city's university. He may have been one of the many dropouts and occasional students who drifted in and out of the universities of early modern Europe. In a letter dated 25 April 1565 and printed in the *Tesoro*, Fioravanti wrote that he had been "continuously practicing" medicine for thirty-two years.[3] This would seem to indicate that he was apprenticed to a surgeon or empiric about 1533, at age sixteen. Although Fioravanti claimed he left Bologna *adottorato*, with a medical doctorate, no record of his matriculation at the University of Bologna has turned up. As we will see, the university finally awarded him a doctorate in 1568, when he was fifty years old.

In any case, Fioravanti's education did not really begin until he left his native city to go searching for "experiences." In a pattern typical of the professors of secrets, Fioravanti became disenchanted with traditional medicine and looked elsewhere, outside of the university, to learn the secrets of the art. His search took him wandering all over the Mediterranean world, "by land and by sea," learning secrets from empirics, distillers, shepherds, farmers, and surgeons. Fioravanti traveled extensively in the years after 1548, "always exercising the art where I traveled, never growing weary of study or of searching for the choicest experiments, whether from the most learned doctors of medicine, or even from empirics, and every other sort of folk, whether peasants, shepherds, soldiers, religious people, country women, and every other quality."[4]

Fioravanti's journeys took him first to Palermo, in Sicily, where he learned the art of distillation. According to the *Tesoro*, he astounded the entire city with his discoveries. He healed leprosy, scrofula, and syphilis, causing the people to marvel that he must have been from some "distant country" to know such secrets.[5] He developed a technique for removing cataracts and at every opportunity tried out his *balsamo artificiato* for treating wounds. Most marvelous of all was an event of 1549. Years later

Fioravanti would still remember it with amazement. An army captain called Matteo Greco came to him begging him to treat his young wife Marulla, who suffered grievously from a hypertrophied spleen (*opilatione della milza*). The physicians who examined her advised the captain that the spleen had to be removed. They assured him the operation was a simple matter, without danger, although none of them was willing to do it. When Fioravanti finally visited the woman, her spleen "had grown so much in her body that you could not imagine it greater. It caused both of her legs to be horribly ulcerated, so that the poor woman didn't even want to live."[6] In desperation, the woman begged her husband to find someone who could remove the spleen, whatever the risk. The captain, having heard of Fioravanti's reputation, asked him if he "had the courage" to take out the spleen. This is Fioravanti's version of what happened next:

> "Of course," I answered cheerfully, even though up to that time I'd never taken out anything. . . . So I promised him I'd do it, and the man, being promised, begged me every day to do it. But to tell the truth, even though I promised him, I really didn't want to, because I didn't want to make a blunder of it. So I sent for a certain old surgeon I knew from the city of Palo, in the kingdom of Naples. This old man, who was called Andreano Zaccarello, operated in that city by cutting (*di taglio*), removing cataracts and similar things, and was very expert in that profession. The old man hurried to my house. I asked him, "Dear Master Andriano, a strange wish has come to the wife of Captain Matteo the Greek. She wants to have her spleen taken out! Can this be done safely?"
>
> The old man answered, "Certainly sir, it can be done. I've done it many times."
>
> "Do you have the courage to go ahead with it?" I urged. He replied that he would do it with me, but otherwise not. So we resolved do it together. I went back to arrange things with the woman and her husband, and having arranged things, I went to the authorities to give her up as dead, as is the custom. Having such license, we went one morning to the woman's house. The good old man took a razor and cut the body of the woman above the spleen. When it was cut, the spleen popped out of the body, and we proceeded to separate it from the tissue until we took it completely out. Then we sewed up the body, leaving only a tiny air hole. I treated the wound with composited hypericon oil, powder of incense, mastic, myrrh with sarcocolla, and made her drink water boiled with a dried apple, hypericon, betony, and *cardo santo*. And so I continued to care for her in such a way that the poor woman in twenty-four days was healed. And she went to mass at the Madonna of the Miracles, according to her obligation, and was healed and saved.[7]

The woman's spleen weighed thirty-two ounces: Fioravanti weighed it. He brought the marvelous thing to the loggia of the merchants, where it was displayed for three days. "The glory of this experiment was given to me," he gloated, "and because of this the people gathered about me, as to an oracle."[8]

It would seem incredible, on the face of it, that such drastic surgery could be successfully accomplished in the sixteenth century without antisepsis or anesthesia. Yet it is now well known that the spleen is unnecessary to sustain life, a fact first proved experimentally in the seventeenth century by the Italian surgeon Giuseppe Zambeccari.[9] Moreover, there is evidence to suggest that splenectomies were not infrequently performed by sixteenth-century surgeons. Hypertrophied spleen was a common condition in Mediterranean countries, where malaria was prevalent. The ancient Romans were also familiar with the operation. Pliny reported that because a swollen spleen impeded running, athletes sometimes had the spleen removed.[10] He also noted that splenectomy could be done on animals without deleterious effects. Sixteenth-century visitors to Turkey reported that Turkish surgeons splenectomized mail carriers to facilitate their running.[11] Meanwhile, academics vigorously debated the spleen's function.[12] Although anatomists all agreed that the organ performed some necessary physiological function, there was little agreement over what function it served. According to the principle *natura non facit saltum*, they reasoned that since the body has a spleen, the organ must have a function.[13] Fioravanti appears to have been completely unaware of these academic debates. To him, the fact that a surgeon could remove one before his eyes was an empirical truth and something positively wonderful. "To tell the truth," he concluded from the incident, "there is no better way to learn than to go out into the world, because every day there are new things to be seen, and various and diverse important secrets to be learned."[14]

Whatever his successes were as a surgeon and empirical doctor—and undoubtedly there were many of them—Fioravanti was even more skillful at advertising them. A few weeks after the famous splenectomy, the Sicilian viceroy's wife appointed him to care for the patients in the Ospedale degli Incurabili (Hospital of the Incurables) in Palermo, many of whom suffered from syphilis. Fioravanti claimed he healed them all. Soon after, the viceroy invited the surgeon to his court in Messina, where Fioravanti dedicated himself to studying the art of distilling with an alchemist he took into his lodgings.[15] He also trained under Matteo Guaruccio, an old surgeon who taught him many secrets "with which he divinely treated wounds of all kinds."[16]

In 1549, Fioravanti went to Tropea, in Calabria, to learn the art of rhinoplasty from two famous surgeons, Pietro and Paolo Vianeo. The

Vianeo brothers' ingenious technique for rebuilding severed noses drew many wealthy patients to Calabria. In that violent age of ferocious tempers, Italian gentlemen not infrequently had their noses severed or bitten off in duels and street brawls.[17] Pretending to be a Bolognese nobleman who wanted to observe the operation firsthand in order to arrange for the treatment of a relative who had lost his nose in a fight, Fioravanti was allowed to watch the operation: "and thus I saw the whole secret, from beginning to end, and learned it."[18] From Tropea Fioravanti went to Naples to continue his practice. In 1551, he was commissioned by the Spanish viceroy, Don Pedro of Toledo, as *protomedico* and personal physician to Toledo's son, Don Garcia, who was then serving in the Spanish Armada.[19] Fioravanti accompanied Don Garcia to Africa, where he witnessed the seige of Monastir. On the voyage, he was called aboard a Florentine galley to attend the wounds of an officer who had taken a beating in a quarrel with another officer, an event that gave Fioravanti another chance to put his famous balsam to the test.[20] During the African campaign Fioravanti also had an opportunity to try out, in his own rough-and-ready way, the skills he learned from the Vianeo brothers:

> A Spanish gentleman named Signor Andres Gutiero, twenty-nine years of age, was strolling through the camp one day and came to words with a soldier. They drew weapons and with a backhand stroke the soldier cut off Signor Andres's nose, which fell in the sand, and I saw it because we were together. The quarrel ended and the poor gentleman remained without a nose. So holding it in my hand, all full of sand, I pissed on it, and having washed it with urine I attached it to him and sewed it on very firmly, medicated it with balsam, and bandaged it. And I had him remain thus for eight days, believing the nose was going to rot. However, when I untied it I found it was very well attached again. I medicated it only once more and he was healthy and free, and all Naples marveled at it. This was indeed the truth; and Signor Andres can tell of it because he is still alive and healthy.[21]

In 1555, Fioravanti moved on to Rome, where he hoped to practice at the papal court. But instead of finding a patron, he had the first of an ongoing series of confrontations with the medical establishment, which earned him the reputation of an unscrupulous "vendor of secrets" that was to haunt him the rest of his life. Although he had a license from the *protomedico* to practice in Rome, his cures were so popular among princes and papal legates that the local physicians felt threatened.[22] In 1557 the physicians, who according to Fioravanti feared losing their own clientele at the papal court, conspired to remove him from the court and to prevent him from selling his remedies in Rome.[23] The conspirators got supporting testimony from various local physicians, including the anatomist Realdo Colombo, who was then teaching at the Roman Sapienza—Fioravanti contemptuously called him *Palombo*, implying he was the leader of the pack of hounds who drove him from Rome. Although Fioravanti lost the battle, he later claimed the victory in the dispute: "Blessed God, who wants the infamous and the damned to be separated from the company of the good, caused a great miracle to happen. Before a year had passed, he called Giovan da Auricola [one of the conspirators] and Realdo Palombo to him to know the truth of the affair. Thus both are dead, and have gone to where the Lord will pass judgment on the good works done in the world."[24]

In 1558, Fioravanti's career took a new turn. Already famous—not to say notorious—as a surgeon and vendor of miraculous cures, he decided to take up a new profession, one that promised to make his name universally known, and perhaps secure for himself a permanent patron: he would become a writer and publish his secrets. Above all, Fioravanti wanted the fame he thought he deserved: "Only those who are written up in books stay alive forever, and their names will never die."[25] To pursue this new ambition, he went to Venice to join the many other adventurers of the pen who had found employment there as professional writers.[26]

Fioravanti *Poligrafo*

For most authors, writing for the popular press was a precarious existence. With few effective copyright provisions, books were easily plagiarized, and authors were not paid when their works were reprinted by other publishers. Consequently, most writers had to seek patrons to support their literary efforts or find some other means to supplement their income. Many, such as Ruscelli, found employment as editors, translators, or proofreaders in the larger publishing houses. Fioravanti was perhaps more fortunate than other *poligrafi*, since he could still practice surgery and dispense his cures. He earned as much as 150 scudi for a single cataract surgery, when a salary of 400 scudi per year for a professor of medicine was considered excellent.[27] Fioravanti took a house near the church of San Luca, where he set up a laboratory, concocted his medicaments, and engaged two local pharmacists to sell them. He used his books to advertise his remedies: "Anyone who would like to avail himself of my remedies," he announced in the *Secreti rationali*, one of his first published works, "will find me in Venice at San Luca, where I will always be ready to the service of everyone."[28] For those from other parts of Italy who wished to consult him, he advised them to write to his publisher, Ludovico Avanzo, and promised a prompt reply. By 1564 he was engaging Sabbà di Franceschi, the proprietor of the Orso pharmacy, to prepare and sell his nostrums.[29]

Fioravanti's entry into the Venetian publishing world was not unlike that of other new writers. It came through his association with small printing houses that specialized in publishing vernacular literature. In 1558 Ludovico Avanzo, a small publisher of vernacular medical and scientific works, printed Fioravanti's first book, the *Secreti medicinali*, a compilation of the empiric's favorite remedies. Avanzo hired Fioravanti to do a new edition of a surgical textbook by Pietro and Ludovico Rostini, to which Fioravanti added a brief surgical tract of his own.[30] Avanzo also printed Fioravanti's famous *Capricci medicinali* (1561), which contained the recipes for his balsam, his "imperial electuary," and other new medicaments he had invented. The "medical caprices" must have returned a sizable profit to Avanzo: the printer brought out new editions of the work in 1564, 1568, and 1573. Fioravanti continued to maintain a close relationship with the Avanzo firm even while publishing his works with other printing houses. Andrea Ravenoldo, another small printer, published Fioravanti's *Regimento della peste* in 1565, along with an edition of his *Specchio universale*. Within a few years the popularity of Fioravanti's works enabled him to approach two of Venice's largest printers. In 1564 the Valgrisi firm brought out a new edition of his *Secreti rationali* and his *Dello specchio di scientia universale*. The Sessa printing house published his *Cirurgia* (1570), *Tesoro della vita humana* (1570), *Della fisica* (1582), and new editions of his other works.

Fioravanti became a familiar part of the circle of *poligrafi* living in Venice in the 1560s. He wrote warmly of literary companions such as Francesco Sansovino and Girolamo Ruscelli, whom Fioravanti said everyone revered "as a true miracle of nature." In his book on military practice, Ruscelli raved about Fioravanti's new methods for treating gunshot wounds and marveled at his military inventions, which Fioravanti demonstrated in Ruscelli's house.[31] Fioravanti submitted drafts of his books to the professional writers and proofreaders working in the larger printing establishments, and in turn doctored them when they fell ill.[32] Fioravanti's closest friend among the Venetian writers was Dionigi Atanagi (1504–1573), a celebrated poet and humanist whom Fioravanti "miraculously resurrected from death to life" after Atanagi received a brutal gunshot wound to the head. Atanagi dedicated a sonnet to the "stupendous and illustrious operations which the singular doctor and surgeon Leonardo Fioravanti produces every day for the world's benefit, with the new and unheard-of healing powers of his precious, vivifying liquors."[33]

The other world in which Fioravanti moved was that of the Venetian pharmacies.[34] With its unrivaled commercial relations with Constantinople, Syria, and Egypt, the Venetian republic had a flourishing trade in pharmaceuticals. When Fioravanti arrived there, Venice had around seventy pharmacies. The pharmacy was in many respects the center of medical practice in the community. In addition to making house calls, physicians frequently consulted patients in the pharmacies, while the pharmacist stocked his shelves according to the advice of the doctors. Often physicians entered into formal contracts with pharmacists, seeing patients in the pharmacy and regularly ordering drugs from the pharmacist who operated the shop. Thus the pharmacy, like the printing house, was a center of discussion and innovation as well as a place of manufacture and trade. Physicians and humanists who wrote on materia medica relied on advice from pharmacists, while pharmacists made it their business to be knowledgeable about new drugs introduced into the marketplace by innovators and empirics such as Fioravanti. The Bolognese empiric maintained close relations with at least two pharmacies: the Felice at San Luca, where he resided, and the Orso at Santa Maria Formosa, which was run by Sabbà di Franceschi.[35] It was in the pharmacies and printing houses, among surgeons, empirics, and *poligrafi*, not in school, or from the physicians, that Fioravanti developed his ideas.

Although as a writer Fioravanti specialized in medical tracts promoting his own treatments, his success as an author depended on his ability to write for a broad audience. He spiced his books with social criticism, moral platitudes, fables, witticisms, and above all with anecdotes recounting his spectacular successes with his medical secrets—all meant to entertain and to edify as well as to provide medical information. One of the novellas included in his *Dello specchio di scientia universale* (Mirror of universal science), a commentary on vain ambition, was about a poor peasant, who with a gratuity from his master, "built his castle in the sky, and soon went to ruin."[36] He also promised a book on the "good and evil effects of language," but never published it.[37] Fioravanti's most ambitious nonmedical work was the *Specchio*, a survey of the major arts, sciences, and trades of the day. Fioravanti included in this popular work a series of philosophical essays on everything from the behavior of princes to the present state of Christendom. There were reflections on friendship, women, marriage, and the fragility of life. The essays reveal Fioravanti the moralist, inveighing against social injustice, ambition, and human folly. Especially vehement were his denunciations of the Italian nobility. Death's "cruel tyranny," he thought, ought to move people to change their wicked ways. But he was pessimistic about the likelihood of such an occurrence. Fioravanti summed up his views in an essay, "On the World":

I don't know how our world can be more deservedly or fitly characterized than to call it a cage of crazies, a house of asses, a camp of mischief-makers, a kitchen full of smoke, a perpetual travail. If you want to prove this to yourself, just look around at all the great diversity and contradiction in the world. It's a stupendous thing to see. For everyone who can read is loaded down with weighty thoughts, everyone boasts that he is more learned than the next, and you won't find a single one who wants to be second to anyone. Even the poor man brags that he eats with a bigger appetite than a nobleman but isn't bothered by all the worries great princes have.[38]

It is not difficult to see why the *Specchio* was one of Fioravanti's most popular books. His moral indictments of highbrow culture and his colorful images of a world in chaos resonated with the sentiment widespread among Italians that they had lost control over their lives. Despite Fioravanti's pessimistic worldview, the *Specchio* also had its moments of good-natured humor. Readers could easily laugh along with Fioravanti as he pointed out the folly of human behavior. In his rich colloquial style, Fioravanti confirmed contemporary beliefs about many subjects. The *Specchio* was quite influential in Italian letters. An early attempt to describe the entire social world of sixteenth-century Italy, the work made an impression upon Tommaso Garzoni, who modeled his *Piazza universale* after it.[39] Garzoni's history of the professions was a truly chaotic piazza, comprehensive but without any apparent taxonomy. The *Specchio*, on the other hand, reflected its author's philosophical primitivism: Fioravanti organized the work according to a historical scheme listing the professions in what he believed was their order of discovery. Thus husbandry, taught to Adam by God, ranked first among the professions, even though "today it is much despised, and reduced to the hands of rustic people and idiots, who practice it solely with a little practical knowledge, without having the slightest knowledge of its science."[40] After agriculture came medicine, which men learned not from the gods (as the ancients said), but by observing how animals healed themselves. Following these came the arts of surgery, weaving, navigation, commerce, painting, shoemaking, gardening, and scores of other trades. Last in the scheme came law, dyeing, distilling, the making of musical instruments, and dancing. Fioravanti said nothing to upset the balance that put the more courtly arts last, comfortably distant from the first. But he did not hide his conviction that the lowly manual arts were closer to the source of true wisdom: the voice of God.

The popularity of Fioravanti's books brought him fame in Italy and abroad. In 1564, the Holy Roman Emperor invited him to become a military surgeon in the imperial army in its campaign against the Turks in Hungary.[41] Fioravanti turned the offer down because he had already embarked on a lucrative business venture in Venice. In 1562, he received a commission from the Serenissima to oversee the reclamation of the ancient Istrian town of Pola, which had gone to ruin as a result of war and pestilence.[42] Fioravanti's partner in the scheme was none other than Sabbà di Franceschi, the proprietor of the Orso pharmacy, the main center for the distribution of Fioravanti's drugs.[43] Evidently the venture was a success; anyway Fioravanti boasted of it in his book on the plague and offered his methods to other Italian cities. Yet his unconventional practice and his contempt for the physicians caused continual friction with the medical establishment. In 1568, a group of Venetian physicians and pharmacists attempted to prohibit him from practicing medicine and the local pharmacists from dispensing his remedies. They charged him with practicing medicine illegally, without a medical degree. Although Fioravanti claimed he had received his doctorate twenty years earlier and was already licensed to practice medicine in Rome and Naples, he could not produce his credentials.[44] So he returned to Bologna in March 1568 and presented himself to the medical college as a candidate for the doctorate of medicine and philosophy.

In seeking the degree, Fioravanti had a choice: he could apply for the degree *alla Bolognese*, as a native citizen of Bologna, and pay full fees; or he could take it *amore Dei*, as a Bolognese citizen unable to pay any fees; or, even though he was a native of Bologna, he could apply *alla forestiera*, as a nonnative, with the payment of smaller fees. Applying as a native citizen would qualify him for becoming a member of the medical college, although admittance into that elite group was not automatic but by vote of the college. It would also confer upon him the title of Golden Knight and Palatine Count, a privilege the emperor Charles V granted the college in 1530. If he applied as a nonnative, on the other hand, he would not qualify for admittance into the college, nor would he receive the knighthood.[45] Fioravanti apparently had no aspirations to teach. All he wanted was a degree to legitimate his claim of being a real doctor. Hence he applied as a nonnative, the fees for which were considerably lower than those for a native Bolognese.[46] However, the day after filing his application, he changed his mind and decided to apply as a native. The reason for his decision soon became apparent, when he began routinely to display the coveted title of nobility on the frontispieces of his books. The medical faculty granted Fioravanti's request on the condition that he pay the remaining fees within one year and promise not to apply for admission to the medical college. Since there is no evidence Fioravanti ever actually matriculated at the university, it is difficult to know quite what to make of this affair. The documents pertaining to the incident tell us only that the medical college faculty unanimously judged Fioravanti "suitable and adequate" for promotion.[47] It is possible that Fioravanti won the degree on the basis of his lengthy publication record. The author of six popular books and the inventor of more than a dozen new drugs, he made an imposing candidate. The faculty of the Bologna medical college would have known Fioravanti's worldwide reputation as a surgeon and healer. Perhaps it granted the degree to take credit for its famous native son's growing fame. If so, it suggests that writing for the popular press was coming into use as an adjunct to traditional academic procedures.

Immediately after receiving the degree, Fioravanti returned to Venice and presented his credentials before the Venetian College of Physicians. But even though he had won the degree, his troubles with the physicians did not end. In October 1568, just seven months later, the medical faculty at Bologna received a letter from the Venetian College of Physicians complaining of Fioravanti's incompetence and questioning the authenticity of his degree. He did not even know the rudiments of grammar, the letter charged. Not only was he a vagabond, he was also a murderer, having caused the deaths of several of his patients with his unorthodox potions. If he did not get the degree by some fraud, the letter intimated, how could the Bologna medical college have promoted such a charlatan, even giving him the title of knight?[48] The Bologna medical college, affronted by this attack on its integrity, appointed a special commission to investigate the charges. After two weeks of deliberation, the commission determined the letter was fraudulent. It did not in fact originate in the Venetian College of Physicians but was part of a "conspiracy" by a group of local physicians and pharmacists. The investigation revealed that the conspiracy was hatched in the Golden Head pharmacy, which was famous for its theriac. Fioravanti had been a vehement critic of pharmacists who sold adulterated theriac at inflated prices.

The Bologna medical college repudiated the charges and reaffirmed the authenticity of Fioravanti's degree. Fioravanti never forgot the college's loyalty: in all of the books he published after the event, he lavished praise upon Bologna and its university. In the *Tesoro della vita humana* (1570), the first book to display his newly won titles, he wrote two dedications to the university, one praising the college of arts, another paying homage to the medical faculty and recommending Bologna to all who would seek a medical degree.[49] The *Tesoro* was Fioravanti's most self-confident work. It displayed his portrait, published his autobiography, and paraded before readers dozens of testimonials to his miraculous cures. Fioravanti had arrived.

With success and fame came disciples who followed his doctrine. Former apprentices—including barbers, surgeons, and empirical "doctors" from Nicosia (Cyprus) to Venice—defended his doctrine in public debate and proselytized for his "new way."[50] But wherever Fioravanti prac-

LEONARDO FIORAVANTI CAV

6.   Portrait of Leonardo Fioravanti. From *Tesoro della vita humana* (Venice, 1570).

ticed, he got into trouble with the physicians. A holograph letter dated 22 April 1573 discloses another battle, this one with the Milanese physicians, who had him thrown into prison for malpractice. Writing from his prison cell, Fioravanti complained to the viceroy of justice that he had been jailed for eight days at the instance of the physicians, who charged him with not "medicating in the canonical way, as they do." The physicians also accused him of killing several of his patients. Fioravanti insisted that the real reason for his incarceration was that the physicians were afraid of competition, "seeing that my name continues to grow." To prove the worth of his doctrine, he issued a challenge: "that there be consigned to me alone twenty or twenty-five sick people with diverse ailments, and an equal number with the same infirmities to all the physicians of Milan, and if I do not cure mine faster and better than they do theirs, I am willing to be banished forever from this city."[51] Unfortunately, we do not know whether (as is unlikely) the physicians took up

Fioravanti's challenge. But within a few days the court set him free. Despite the physicians, he continued to practice and to propagate his medical gospel.

Nor did Fioravanti refrain from giving unsolicited advice to princes. In November 1570, emboldened by the rightness of his worldview, he wrote to Alfonso II, the duke of Ferrara (to whom he had dedicated his *Capricci medicinali*) expressing the "great sadness and sorrow" he felt over "the ruin of your most noble city of Ferrara, and over your personal troubles." He was moved to write the letter by news of a series of violent earthquakes that struck Ferrara in November–December 1570, causing massive destruction and killing hundreds of people.[52] Fioravanti was convinced such terrible events did not happen by chance; they were God's punishment for a wicked world.[53] And who in late sixteenth-century Italy would have doubted the misery and misfortune of that unhappy land? Alfonso's court had been a place of dazzling splendor. The duke lavished his patronage on some of Italy's leading poets, musicians, artists, and humanists. Yet to sustain this magnificent show of wealth, he ground the people down with oppressive taxes, monopolies on grain and salt, and callous game laws.[54] It was plain to Fioravanti that the earthquakes were the visible sign of God's wrath. Unable to contain his moral outrage over the rotten state of Ferrara, he warned the duke that princes are often "followed by many flatterers who make them see black for white":

> To speak plainly, they say in these parts that Your Most Illustrious Lordship has let himself be bribed by the evil men with whom he has involved himself, and in doing so every year has allowed many heavy taxes to be levied on the poor people in his magnificent city; besides this, that Ferrara has become a Sodom and Gomorrah of public sodomy, and that in joking about this, everybody scorns God with blasphemy and filthy words, sins so great that God's majesty cannot suffer them. For all this they blame Your Most Illustrious Lordship, as prince and patron, saying you should remedy the whole situation with the arm of your justice. And so, My Illustrious Lord, I have taken it upon myself to make all these things known to you, because I am certain that as a Christian and Catholic prince you will turn to God for help and counsel, and will see to all these things, so that our Lord God will calm his anger and return your city to its pristine state, as he did to the city of Nineveh. And if it is Your Illustrious Lordship's wish, after all this has come to pass, I will show him the true way to quickly restore and augment the city.[55]

It is fascinating to see Fioravanti posturing as a modern-day Jonah, admonishing the duke to drive out the "flatterers" from his court and to return the city to its "pristine state." Otherwise, Fioravanti warned, God would destroy the city, as he threatened to destroy the corrupt city of Nineveh before it reformed its sinful ways. Just as God sent Jonah to warn Nineveh, he was now sending Fioravanti to help restore the ruined Ferrara.

In 1575, the plague broke out in Italy. By the following year the disease had reached Venice. The return of the plague confirmed Fioravanti's cosmic pessimism and his belief that God was angry with the world. In his work on the plague, *Della peste* (1565), he wrote that "the plague never comes when the Divine Majesty is pleased; he sends it to us in order to chastise us for our enormous sins." As proof, God never sends the plague without first sending certain "ambassadors" of warning, such as famine, disease, flood, war, and social discord. No one in sixteenth-century Italy would have had any doubt that the unfortunate land had been visited by all these "ambassadors." Italy's misery was evident everywhere. Before visiting a people with plague, Fioravanti believed, God also sent a prophet to admonish people of their wickedness, "just as it was done when he wanted to flood Nineveh, first he sent the prophet Jonah to warn them."[56] There is no doubt Fioravanti believed he was a modern-day Jonah; only unlike the biblical prophet's, his advice was not heeded. It was possibly because of his disappointment over this that in 1576, he made a pilgrimage to Spain to visit that most holy Christian monarch, King Philip II, in whom of all princes Fioravanti placed his most ardent hope.[57]

Philip welcomed the famous old empiric into his court. Fioravanti, the king remembered, had rendered distinguished service as a surgeon in the Spanish army in Italy. But even though Fioravanti earned the king's respect, he was attacked by the royal physicians. Just three months after his arrival in Madrid, the king's prosecutor brought charges against him of malpractice and of using unorthodox and unapproved medicines.[58] According to the charges, Fioravanti had obtained a license to practice medicine at the court on the basis of falsified documents. Furthermore, he was ignorant of Latin, gave his patients drugs solely of his own invention, and prescribed poisons, killing several of his patients. In his defense, Fioravanti explained that his "new way" of healing was drawn from experience and from ancient Hippocratic doctrine, which was completely misunderstood by the modern physicians. Although he admitted that one of his patients had died, it was not on account of his extreme remedies. For death is natural; when God ordains that a man's hour has come, no physician can help him. As for being ignorant of Latin, had not Hippocrates, Galen, and Avicenna written in their own mother tongue and not in Latin? Evidently Fioravanti's defense succeeded. He remained at Philip's court for about a year, continuing his studies and his practice. He experimented on tobacco and other

New World plants, his curiosity piqued by Monardes's *Historia de las Indias* (Seville, 1571), and studied Ramon Lull's alchemical theories.[59] Fioravanti's brief residence in the Spanish court was the apex of his career. He earned the friendship of one of the king's physicians, Diego de Olivares, and he left behind one of his disciples, the Bolognese empiric and "alchemista terribilissimo" Anzolo di Santini, who continued to practice at the Spanish court.[60]

In 1576, Fioravanti returned to Venice to supervise the publication of his latest works. In 1580, Melchiore Sessa brought out a new edition of the *Cirurgia*. *Della fisica*, Fioravanti's final work, appeared in 1582. It was his last medical testament, and it presented the fullest statement of his medical philosophy. He dedicated the work, appropriately, to King Philip II. After 1582, except for numerous historical references to him (usually as a charlatan and vendor of secrets), Fioravanti completely disappears from the historical record. Tradition has it he returned to Bologna and died there in 1588.[61]

Fioravanti's Medical Primitivism

In his defense against charges of malpractice at the Spanish court, Fioravanti maintained that his "new way of curing" (*nuebo modo de curar*) was in reality a return to the rules of nature and to the precepts of the ancient physicians, which had been misunderstood by the "moderns"— by which Fioravanti meant everyone from Galen to the physicians of his own day.[62] All diseases, he explained, were caused by corruption, whether by bad food, unhealthy climate, or poisoned wounds. For the body to be restored to its "pristine health," these corruptions had to be expelled from the body with powerful vomitives and purges. Open wounds had to be closed up and treated with "solidifying" unguents made from nature, because "the intention of nature is to consolidate and join together." Above all, the body had to be treated with medicaments that "fortify and strengthen nature."[63] Fioravanti developed these medical principles in part from his extensive reading, which included not only the standard medical authorities but also works on alchemy, history, religion, and moral philosophy. But he read these works through the lens of his own experience as a surgeon, clinician, experimenter, and observer of the social and political world of sixteenth-century Italy. His medical system was conditioned by his religious beliefs and by the cynical pessimism of the Counter-Reformation era. For Fioravanti, corruption and purification were metaphors that applied not only to the body, but also to society, religion, and politics. Just as in healing the physicians' goal was to restore the body to its "pristine" health by purging corruption, so in the practice and teaching of medicine Fioravanti advocated purging medicine of doctrinal corruptions and returning to "nature's way," the way of the "first physicians." As we have seen, he urged a similar path of reform upon the duke of Ferrara, urging him to purge his court of corrupt advisers. Obsessed by the terrifying image of the prophet Jonah, Fioravanti came as a prophet to warn the world to mend its ways.

Fioravanti's social criticism mirrored the general concerns of the circle of Venetian writers of which he was a part. The *poligrafi* were particularly intolerant of what they regarded as the irrelevancy of academic learning to everyday life. Education, they argued, should be a preparation for the *vita civile*, the active life.[64] But they doubted that learning as it was then understood could accomplish this. Instead of academic learning, many critics extolled practical experience, common sense, and sound judgment, which could not be taught in the schools. Anton Francesco Doni praised the life of "good ignorance" (*ignoranza da bene*), a carefree, nonintellectual approach to life that stressed attending to one's own affairs without interfering in the affairs of others or trying to improve the world.

The criticism of the Italian world also extended to society, politics, and religion. Like their northern European contemporaries, Italians were deeply concerned about clerical abuses and the Church's inability to respond to the spiritual needs of the people. The wave of evangelism that swept Italy in the mid-sixteenth century spawned an intense dislike for abstract theology and a desire among critics to return to a religion of simple piety and moral purity.[65] Although few Italians openly embraced Protestantism, spiritual restlessness and growing concern over moral decline gave rise to numerous utopian visions of a society based upon primitive simplicity.[66] Doni imagined a "star city" in the New World, a society without property, social distinctions, government, or formal education, where marriage was eliminated and unfit infants were dropped into a well. He wanted to return to a golden age in which everyone lived in peace, harmony, and perfect equality, where men, women, and children lived together naked without shame, before property and the arts and letters were introduced to create dissension and to corrupt humanity.[67]

Fioravanti's critique of official medicine was just as sweeping. Like Doni and other utopian writers, he was inspired by a kind of primitivism that recalled images of a golden age when men were uncorrupted by greed and by the conceit of philosophical systems. Fioravanti insisted that he did not intend to follow the practices of Galen, Avicenna, and the other medical philosophers, but meant to follow "only my own judgment—and experience, the mother of all things."[68] Surveying the present state of medicine, he saw nothing but confusion and discord. The physicians all seemed to have different remedies and conflicting systems that they continually argued over: "One says he understands well the Galenic system and will practice medicine along that road. Another wants to comment on the subtleties of Avicenna, and another to read in a professorial chair. And so they all believe themselves to know more than the others, and nobody wants to yield to his companion. And in this way the world always gets so confused that it's impossible to know the truth."[69]

In order to practice "true medicine," Fioravanti maintained, the physician must first forget everything the ancients said.[70] But obviously, even Fioravanti could not do this. When he sat down to codify his medical system in the *Della fisica*, he had no other theoretical guide than tradition, which he tried to simplify and to turn into a therapeutics of active healing as opposed to compliant submission to a regimen imposed by a physician. He accepted the ancient doctrine of the four elements, seasons, complexions, and humors. To it he added his own doctrine of the four medical "operations": "I say there are no more than four operations that make up all of medicine, and they are these: namely, those [symptoms] that are too hot, chill them; those that are too cold, heat them up again; those that are too dry, moisten them; and those that are too wet, dry them out. And in this is reduced the whole philosophy of our medicine."[71] The effects produced by drugs were likewise four, according to Fioravanti: vomiting, purging, sweating, and urinating.[72]

Rejecting what he regarded as the abstruseness and complexity that had beset medicine with the establishment of competing schools, Fioravanti wanted to return to the pristine and simple system of the "first physicians," who "knew no medical system, nor any method at all, but had only good judgment."[73] According to Fioravanti, the "first physicians" learned medicine from the brute animals, who knew by instinct how to cure themselves:

> It's quite true that nature gave all the animals a very great gift, which was that each animal, all by itself, without aid or counsel from anyone, knew how to cure its infirmity. . . . The dog, when it feels sick, goes to the forest and finds there a certain sort of herb, which it recognizes by natural instinct, and eats it, and that herb immediately makes it vomit and evacuate from behind; and it is cured at once. The ox, horse, and mule, when they feel themselves aggravated by some infirmity, bite the end of their tongue until blood flows out, and are healed. Hens, when they are sick, take out a certain membrane under the tongue, and the blood flows from it, and immediately they are healed. And many other animals do similar things to cure various infirmities. . . . The animals therefore really know how to doc-

tor themselves, and haven't previously studied medicine. They don't have it by science, but by experience and the gift of nature. . . . And so each time men saw these things they observed it, and in this way came to know that evacuation and bloodletting were very useful.[74]

From this observation Fioravanti made the following generalization, which became the basis of his entire therapeutic system: all diseases result from two principal causes, the "bad qualities and indisposition" of the stomach, and the "alteration and corruption" of the blood.[75] Accordingly, Fioravanti reduced therapy to two prime means: drugs to purge the stomach of "indispositions," and bloodletting to relieve the body of corruptions. In practice, Fioravanti rarely recommended bloodletting, because blood is our soul (*anima nostra*); it is better to save the blood and purify it of bad qualities. However, for certain ailments he found it efficacious to draw a little from under the tongue, following the example of the animals.[76] In general he preferred robust purges to cleanse the stomach:

> The first cause of all infirmities is the indisposed and corrupt stomach, from which follows the corruption of the entire body, and by reason of this cause the blood along with all the interior parts suffers, and for this reason it follows that to be able to liberate the body from all kinds of infirmities it is necessary to evacuate it of these corrupt humors, whether by vomiting or by purgation. And the truth of this is verified every day by experience, which shows that those medicines which provoke vomiting, evacuating a great deal, cause much better effects than any other for the health of the sick body.[77]

Rejecting the physicians' reliance on diet and weak medications, he advocated a therapeutics of "direct action" in the form of violent emetics and purgatives to "remove the sickness and return the body to its pristine health."

Fioravanti blamed the abuses of medicine on the rise of the schools. A spokesman for the empirical practitioners, he contended that in the beginning of time all physicians were empirics. But with the rise of philosophical medicine, the physicians "usurped" medicine: "they kept it in such a way that everyone else was deprived of it when they conferred degrees upon themselves."[78] When the physicians convinced the authorities to prohibit anyone from practicing medicine who did not have a doctorate, Fioravanti maintained, their numbers decreased and their incomes grew, but they neglected the needs of the people. Like Nicholas of Poland and a host of empirics throughout history, he rejected the physicians' claim that diseases cannot be treated without knowledge of their causes:

Now [the physicians] maintain that those first empirics could not have known the cause of the infirmity. But I don't know any doctor who treats the cause. As for me, I've never seen anyone treat the cause, but I have certainly medicated, have seen others medicate, the disease, which is the effect. For the cause is always first, and the effect follows after. Thus if the cause is never cured, but only the effect is, why should we have to know about this damned "cause," when it's never cured? To treat a wound you have to know what is important to the wound, not why the man was wounded.[79]

Fioravanti maintained that every branch of science and medicine had profited more from unlettered experience than from all the books and debates of the philosophers: "I have seen a great many shepherds, peasants, artisans, citizens, gentlemen, and lords, who without having even the slightest knowledge of medical methodology have understood so many pretty secrets and experiences in medicine and also in surgery, and have had infinite experiences in various and diverse sorts of infirmities."[80] According to Fioravanti, surgery was discovered by shepherds and other "experimenters in natural things."[81] Formal anatomy was "very pretty," he thought, but quite useless: if you want to know anatomy, it would be better to study agriculture.[82] Physicians, guided only by theory, are like the navigator who gazes at maps in his study and, when he goes out to sea, gets lost or shipwrecked:

> The physicians will study a very pretty theory; they will find out the causes of infirmities and remedies to cure them, and then when they come upon some difficult case, they won't know how to bring about a cure; . . . and then truly some old experienced hag will come along, who with the rules of life and an enema will make the fever cease, with an unction will make the pain go away, and with some fomentation will make the patient sleep; and in so proceeding, the old hag will know more than the physician.[83]

"As for me," Fioravanti concluded, "I believe more in a little experience than in all the theories of the world taken together."[84]

Fioravanti based his entire medical system on this primitivist, anti-intellectual outlook. It was the philosophical foundation of his stance that experience was the "mother of all things," and of his insistence that the physician must know how to prepare medicaments as well as administer them.[85] He justified his own remedies with the same argument. Discussing the philosopher's stone, he rhetorically asked "the great question of the philosophers": is there a single medicine that can cure all diseases? His answer was in the affirmative, according to the following Fioravantian logic:

[First], it is seen that the animals of the earth don't ever treat themselves of any infirmity, except of the stomach, and when they seek to heal themselves, they eat herbs that cause them to vomit. This teaches us that they do not suffer from any other infirmity than the aforesaid. By the experience of animals I prove that illnesses have their causes in the stomach. [Second], all medicines that the philosopher's stone enters, as soon as they arrive in the stomach, attract to themselves all the evil humors of the stomach, and of the entire body, and embrace them together, and nature condemns them to be vomited out in succession. Thus the stomach is emptied of all such material, and the body remains free of every impediment of infirmity. So for this reason I say that the philosopher's stone of our invention can heal all sorts of infirmities.[86]

Fioravanti's *pietra filosofia*, whose active ingredient was mercury, may well have acted as a powerful vomitive when prescribed in careful doses.

The Politics of Purgation

Fioravanti claimed that his therapeutical system was a return to the "natural way" of healing, a methodology discovered by the earliest physicians but corrupted by theory. In combing the cities and countryside of Sicily, he discovered that these natural methods still survived among the common people as unwritten "rules of life." Indeed, they were the secret of longevity. In Messina, he met a 104-year-old man who reported that the only medicine he ever took was some soldanella in the springtime. "Every time I take it," said the old man, "it makes me vomit thoroughly and leaves my stomach so clean that for a year I cannot fall ill."[87]

It seems as if Fioravanti was obsessively concerned with purifying the body of putrid and corrupting substances. Violent emetics and purgatives played a prominent role in his therapeutics. He had an imposing armory of emetics and purgatives whose active agents included hellebore, veratrum, antimony, and mercury. He gave them catchy trade names like "angelic electuary," "magistral syrup," "blessed oil," and his powerful and trusted standby, *dia aromatica*, the "fragrant goddess" he prescribed as the first course of action against almost every ailment he encountered. Not only did Fioravanti believe passionately in the efficacy of vomiting and purging. Equally striking was his nearly obsessive fascination with the strange and revolting matter that spewed forth from the bodies of his patients. In 1558, he attended a woman who suffered from terrible attacks of indigestion. After he gave her two ounces of *dia aromatica*, "she began to vomit, . . . and spent the whole night vomiting the rubbish that lay in her belly. And amongst the other things that she brought up, there was an object like a uterine tumor, but round and hairy in form, and alive, which filled me with wonder, because I had never seen the like before. I washed it and placed it in a box in cotton wadding so that I might show people this portentous object; but [later] upon inspection, I found that it had dissolved and that so small a quantity remained that it had no shape. Nonetheless, when the woman vomited it, it was large and wondrous."[88] Another woman vomited "a great quantity of putrid matter including a great tumor as large as a hand, alive, which lived in tepid water for another two days."[89]

Nor were emetics and purges always sufficient to cleanse the body of its pollution. Fioravanti often combined them with diuretics, expectorants, and sudorifics. In the summer of 1569, while he was living in Messina, he was asked to treat the inhabitants of the neighboring villages, who were dying one by one from some sort of "putrid fever": "The first thing I did was to give them a bolus that made them vomit greatly. After this I gave them every morning for three to four days a syrup solution which moved their bowels strongly, and followed this with a cupping-glass, whilst I anointed their bodies with oil of hypericum [to make them sweat]. . . . Of the three thousand I treated, only three died, and they died of old age."[90] One of Fioravanti's patients had to endure vomiting, bloodletting from underneath the tongue, fomentations, and eight consecutive days of purging. Falling sickness, not surprisingly, called for extreme measures. Fioravanti purged one poor epileptic for ten days and then made him follow "the rules of life, but no diet."[91]

Fioravanti's obsessive concern with purgation was by no means unique. It was a therapeutic approach advocated by many of the popular healers of the day. We see it, for example, in the writings of Tommaso Zefiriele Bovio of Verona (1521–1609), a lawyer turned medical practitioner who attacked the orthodox physicians in books with titles like *Scourge of the Rational Physicians* and *Thunderbolt against the Supposed Rational Doctors*.[92] Bovio claimed the physicians killed their patients with their strict diets and weak medicines. Like Fioravanti, he preferred strong vomits and purges. He especially recommended his "Hercules," which he reported made one of his patients "vomit a catarrh as big as a goose's liver, and emit from above and below loathsome excrement." Bovio urged the same course of therapy for every internal ailment: "chase away the evil, then maintain nature," that is, use robust purges followed by healthy foods.[93] According to the conception of disease accepted by Fioravanti, Bovio, and many other popular healers, illness was not some benign "imbalance" of humors that could be rectified by diet and regimen. It was an invasion of the body by "corruptions" that had to be forcefully expelled with potent drugs. Only then could the body be restored to its "pristine health." In the struggle between sickness and health that these popular healers waged, therapeutic intervention necessarily took on heroic dimensions.

The conception of disease as a contamination of the body by worms, foul-smelling creatures, and other strange corrupting agents resonated with the tense political and religious climate of post-Tridentine Italy. Piero Camporesi has observed that the therapeutics of purgation, the use of vomitory and expulsive techniques, reached a peak in popularity during the second half of the sixteenth century: "The decades which knew the highest rate of diabolism, the golden age of witch-hunts, the Tridentine 'reconquest,' Catholic supremacy, the hegemony of theocracy, happened to coincide with the age of . . . superpurges, of ostentatious and vehement cleansings of the sullied flesh, of obsession with individual catharsis and collective purification."[94] Camporesi's idea of a "medico-ecclesiastical ideology of supercatharthis" provides a valuable framework for looking at Fioravanti as a medical practitioner. First of all, it should be noted that exorcisms were more widely practiced during the Counter-Reformation than ever before.[95] Often publicly performed, exorcisms demonstrated in a dramatic fashion the Church's jurisdiction over supernatural forces, which made them effective instruments of anti-Protestant propaganda. The exorcist, who protected society from a pollution within, held an office of considerable spiritual power. Yet considerable ambiguity surrounded the office of the exorcist, since not only priests but also laymen could perform the rite. Girolamo Menghi, who wrote one of the standard texts on the subject, allowed that "certain devout persons with or without [the office of] exorcism can undo maleficial infirmities and chase away demons from tormented bodies."[96]

Since the line between demonic possession and diseases of natural origin was often difficult to draw, exorcists and physicians vied over contested jurisdictions. This ambiguity of social roles and professional identities made it possibile for popular healers to appropriate the role of exorcist, and for exorcists to play the role of the doctor. Adding to the confusion, there was a tendency to physicalize demonic possession. Hellebore, highly praised and much used by Fioravanti and Bovio, was a staple antimalefic agent. It was recommended by authorities such as Menghi, who reported having seen all manner of maleficent objects "emitted by the body in vomit or from beneath," during exorcism. Menghi also noted that there were exorcists who abused the art by indulging in carnivalesque, self-aggrandizing displays of power over demons. He compared them to the quacks and charlatans who promised to cure anyone. "From being honest, straightforward exorcists," he wrote, "many become doctors and charlatans, seeking this or that herb against the demons, offering medicines, powders, and similar things to the possessed, temerariously usurping to themselves the office of physicians."[97] Scipione Mercurio, a physician, angrily denounced exorcists who "prescribe purging medicines and very strong solvents in order to purge the spiritual body of wicked humors, not being familiar with the quality of these medicines nor, much less, the patient's temperament."[98] In that world, where it was often impossible to tell whether diseases were of natural or demonic origin, the figure of the exorcist and the charlatan merged.

I do not think it is going too far to suggest that in cleansing the stomach and driving out its polluting sickness, Fioravanti was performing a kind of physiological exorcism. His treatments, acts of purification, mimicked the exorcist's rite of "chasing away demons from tormented bodies." Rarely did Fioravanti merely heal his patients; he nearly always "healed and saved" them. Nor, perhaps, is it surprising that he should have projected the metaphor of catharsis onto the body politic, urging the rulers of the day to expel the "flatterers" from their courts, to restore the moral order of ancient times, and to return society to its "pristine" condition of equity and justice. According to the anthropologist Mary Douglas, the physical body is a microcosm of the social body: "The physical experience of the body, always modified by the social categories through which it is known, sustains a particular view of society. There is a continual exchange of meanings between the two kinds of bodily experience so that each reinforces the categories of the other."[99] Fioravanti believed that the cause of Italy's moral and political decline was an internal pollution that began in the courts and spread throughout the commonwealth. As the "bad quality of the stomach" spreads its contagion to all the body's organs, so corrupt rulers and their fawning courtiers ruined the whole body politic. He chastised princes for giving themselves over to flatterers who made them "see black for white." He reminded them that the duties of princes were similar to those of physicians, to care for their subjects with compassion and love. By abandoning the people, were they not behaving like the physicians? How different were modern princes, he lamented, from those ancient rulers, like Octavian, who loved his subjects and treated them with justice, liberality, and clemency.[100] As his letter to the duke of Ferrara suggests, as in medicine so in politics: rid the body politic of wicked ministers, greed, and corruption, then restore it with virtue and justice. "The operations of this our medical system all consist in two things: purgation and restoration. Purgations are evacuations, which get rid of the disease; restorations are the things that give nutrients to the body after purgation."[101]

Despite his radical critique of theoretical medicine, Fioravanti still believed that "method" was necessary for the perfection of the medical art.

The proper way to reach an understanding of nature was not to give oneself over completely to experience, but to find a method to guide experience toward truth. "Now, if experience by itself is sufficient to make pretty things," he asserted, "one is obliged to hand the work over more to practice than to theory, since the latter by itself has never accomplished any work. Therefore it might be possible to say that method is not necessary. But in truth it is not so, because method or theory, as we want to say, is the light, and the road to walk down to true experience, because the road of experiment alone is perilous."[102] It is not at all clear from Fioravanti's writings what this method should be. He called it a "theory of experience," but he did not formulate it in a systematic way. It almost seems more of a visceral reaction against the confines of scholastic philosophy than a coherent methodology.

Was Fioravanti a Paracelsian, as he is often labeled?[103] Like many empirics, he was enthusiastic about chemically prepared drugs. The great majority of his remedies were "quintessences" of animal, vegetable, and mineral substances. An avid student and practitioner of alchemy, he was full of admiration for Paracelsus, Arnald of Villanova, Ramon Lull, Phillip Ulstad, John of Rupescissa, and the iatrochemists. However, it would be wrong simply to label him a Paracelsian. He gave no special place to Paracelsus in his teachings, and while he praised Paracelsus along with other iatrochemists, his writings show little evidence of adherence to a "chemical philosophy." Unlike the Paracelsians, he rarely made reference to the microcosm-macrocosm analogy, and the concepts of the astra, the archeus, and the doctrine of signatures played no role in his medical system. It was Fioravanti's appreciation of Paracelsus's general approach to therapeutics, his iconoclasm, and his enthusiasm for spagyric drugs, rather than sympathy with Paracelsus's philosophy, that links Fioravanti with the Paracelsian tradition. Like Paracelsus, he vigorously opposed Galenic therapy based upon regimen, which he insisted prolonged diseases rather than cured them. His preference was for active intervention with strong vomitives and purges. The Paracelsian doctrine of the stomach's centrality in disease was consistent with Fioravanti's outlook, but there is little evidence that he was in other respects a disciple of the German medical reformer.

Fioravanti's bricolated natural philosophy, cobbled out of sources as diverse as Hippocrates, Ramon Lull, the Vianeo brothers, and the Venetian pharmacists, was not a system that can be identified with any sixteenth-century medical "school." It was, as many of his contemporaries recognized, a vernacular medical system unique to Fioravanti. To be sure, Fioravantism had elements in common with the views of other pharmacists, surgeons, and empirics in sixteenth-century Italy. Its deeply religious and moralistic tone mirrored the Counter-Reformation.

But in other respects Fioravanti was radically unconventional in his views. The single most dominant characteristic of his thought was its antiauthoritarianism and its anti-"modernism." Indeed, Fioravanti did not claim to be advancing a new medical philosophy. Instead, he wanted to return to the pristine and simple practice he imagined existed in ancient times, before medicine was corrupted by academic disputes and arcane theories. He was a firm advocate of treatment by simples, an admirer of the botanist Pietro Andrea Matthioli, the translator of Dioscorides and a severe critic of the practice of "adulterating" drugs with complex and contradictory ingredients. But he saw no contradiction between this position and the nostrums he prescribed to his patients, which he insisted were according to nature. He believed that the animals, which cured themselves by using simples, had "true medicine" by instinct; but he also believed that humans, endowed with reason and science, could do better. Medical scholasticism, he insisted, had resulted in a separation of theory from practice, the elevation of medicine above surgery, and obfuscation of the true principles of therapeutics.[104] Reversing the scholastic relation between theory and experience, Fioravanti believed that theory must be proven by practice and experience, not vice versa. It was the bedrock certainty upon which he grounded his entire medical philosophy.[105]

Fioravanti fondly called his therapeutic system the "natural way" (*via naturale*), contrasting it to the "indirect way" of the physicians. Perhaps the rather simplistic manner in which he rejected academic medicine does not display a very sophisticated understanding of contemporary medical theory. But we should not let this obscure the fact that he was an immensely popular writer. As a polemicist he knew exactly what to say to the largely popular audience whom he addressed in his books. When he insisted that "nature is the master of all things," and that the physician was only a "helper of nature, not its master," he intended to discredit the pretentious theories the orthodox physicians hid behind. "Let the sick person eat, and leave the blood in his veins," Fioravanti urged, dispatching in one stroke theories about regimen and phlebotomy. Fioravanti's angry indictment of orthodox medicine had a profoundly moral intent: his attack was above all against the arrogance and avarice of the physicians. Sixteenth-century Italy was a nursery ground for a host of moralists. In the sphere of medicine, Fioravanti was chief among them. Italian readers loved it, and he was too sensible to spoil his style by schooling it.

It is probably impossible to find consistency in a system such as the one Fioravanti constructed. Precisely because it was a bricolage, it lacked systematics. But Fioravanti's methods were convincing enough to gain many adherents. His works were translated and cited down through the seventeenth century. His remedies were prepared and sold throughout Italy and as far away as London.[106] Fioravanti's polemical writings also provided ammunition for the seventeenth-century assault on Galenic physiology. The famous splenectomy he performed with the surgeon Zaccarello was frequently cited in surgical and anatomical literature. Fioravanti's account of the operation in the *Tesoro* played an important role in renewing interest in splenectomy among anatomists and surgeons.[107] Between the early sixteenth and mid-seventeenth centuries, despite heated academic debate over the function of the spleen, almost nothing was introduced to change the Galenic account, which held that the organ was necessary to sustain life. Of course, Fioravanti showed empirically that this assumption was false. But typically, he did not see any relevance in his experience for academic anatomy. For him, the two-pound spleen he extracted from the Greek woman's body was an object of wonder, to be displayed in the piazza. It was not until 1657 that the Helmontian physician George Thomson performed experimental splenectomy specifically to test the Galenic theory. By successfully extracting the spleen from a dog, Thomson proved it was possible for an animal to live without the organ, thus rendering vacuous current disputes over the spleen's function.[108]

Fioravanti claimed not to have invented a new medical doctrine but to have rediscovered an old one, a doctrine that was radically different from the "canonical way." Imitating Hippocrates, he even wrote a series of aphorisms to guide clinicians who wished to follow his principles.[109] However, more than a doctrine, Fioravanti found a voice that convinced others to challenge the authority of academic medicine. He prided himself as a teacher and exposer of the physicians' fraud: "Being, as I am, an author, my office is to teach my doctrine and experience to those who do not have it, and who wish to know it."[110] He claimed many pupils who followed his *via naturale*. In the *Tesoro*, he mentioned with special fondness a disciple from Calabria, one Giovanpaolo de Cuglielmi, of whom Fioravanti proudly wrote, "He follows my doctrine in every way and continues to argue for it; he has disputed in Naples, Padua, Venice, and other places, where he has acquired great fame."[111] Fioravanti wrote many books that were consulted by scores of readers. Unfortunately, as far as we know Giovanpaolo de Cuglielmi wrote nothing at all.

SIX

NATURAL MAGIC AND THE
SECRETS OF NATURE

THE professors of secrets put little faith in theory. They rarely asked why particular recipes worked. Nor did they use experiments to test theories. For them experiments were tests of the efficacy of recipes or attempts to improve upon them. They were guided more by the trial-and-error methodology of the inventor than by the theoretical bent of the experimental scientist. Moreover, they were distrustful of theory, which struck them as vague, abstract, and dubious. They knew that in empirical matters, explanations do not necessarily contribute to successful results. Following a rule tested by an experienced practitioner is safer than attempting to devise new techniques based on general principles, as every user of a cookbook knows. The professors of secrets were convinced that experience was a more reliable guide to truth than theory.

Nevertheless, they inherited a coherent view of the natural world and carried out their investigations within an intellectual framework that most of their readers shared: otherwise their books would not have sold so well. The philosophical statement of this outlook was natural magic, the science that attempted to give rational, naturalistic explanations of the occult forces of nature. The basic assumption of natural magic was that nature teemed with hidden forces and powers that could be imitated, improved upon, and exploited for human gain. As Isabella Cortese expressed it, "man is not content with investigation, but strives to put everything into works, to make himself the ape of nature, indeed to supersede nature, as he tries to do what to nature is impossible. And that this might be true, he is able to dig up secrets that every day are seen being put into execution."[1]

The expanding Renaissance interest in magic has been amply documented.[2] However, modern scholarship has tended to treat natural magic as a strictly intellectual problem, while ignoring its political and social dimensions. Nor have historians looked seriously at the relationship of natural magic to popular values and attitudes. Yet natural magic was an ideology as much as it was a natural philosophy. Its rise to prominence in the sixteenth century and the campaign waged by the Church against it cannot be understood in isolation from the momentous Counter-Reformation debate over the future of Catholicism. Indeed, the controversy over natural magic was essentially a political and religious dispute. At issue were doctrinal uniformity and the Church's jurisdiction over supernatural forces. As John Bossy has pointed out, the aim of the Tridentine reform was to enforce a code of uniform parochial practice.[3] Among the various measures it took to enforce parochial uniformity, the Tridentine Church was particularly vigorous in its attempt to eradicate the "errors and superstitions" of popular religion. Besides teetering dangerously close to demonic magic, natural magic was anathema to the Church because it smacked of pagan superstition. Moreover, because it claimed to make "miracles" natural, it encroached upon the Church's jurisdiction over supernatural forces. In its attempt to protect the faithful from the superstitions of supernatural magic, the Church condemned all magical activity as heretical. In the heat of the Reformation conflict, the net grew too large. Natural magic was caught up along with popular superstitions, witchcraft, and consort with demons.[4]

Giambattista Della Porta,
Perscrutatore dei secreti naturali

The professors of secrets' "research program" consisted of an aggressive search for the "secrets of nature," which they believed were hidden underneath nature's exterior appearances. But that quest shaded dangerously into impious curiosity about demonic forces, and worse, into heretical attempts to control them. Traditional exhortations against forbidden knowledge were still widespread in religious and academic circles, particularly against looking into the secrets of nature, the secrets of God, and the secrets of the state. Carlo Ginzburg has stressed the ideological meaning of this triple exhortation. "It tended to maintain the existing social and political hierarchy by condemning subversive political thinkers who tried to penetrate the mysteries of the State," he writes. "It tended to reinforce the power of the Church (or churches), subtracting traditional dogmas from the intellectual curiosity of heretics. As a side effect of some importance, it tended to discourage independent thinkers who would have dared to question the time-honoured image of the cosmos."[5] Nevertheless, many of the traditional limitations imposed on the acquisition and dissemination of knowledge were being broken down in the sixteenth century. Niccolò Machiavelli's *Prince*, a "how-to" book on how to gain and keep power, exposed the secrets of statecraft to the entire community of European intellectuals. Within a few decades, Protestant propaganda was hammering out a profound religious secret, announcing to a popular audience the possibility of gaining salvation on one's own, without the intervention of the priesthood.

In the realm of natural philosophy, the assault upon the concept of forbidden knowledge came from a new front: in the growing number of apologetics for natural magic as a legitimate scientific pursuit, entirely distinct from the demonic magic so vigorously opposed by the churches, both reformed and Catholic.[6] Yet despite attempts by thinkers such as Marsilio Ficino (1433–1499) to create an occult philosophy based on purely nondemonic principles, it was practically impossible to find a clear distinction between the impersonal planetary spirits that were supposedly the basis of natural magic and the personal spirits (i.e., demons) that resided in the planetary spheres and performed exactly the same function as the celestial spirits. If the only difference between demons and celestial spirits was that demons were intelligent souls, how was the magical operator to tell them apart? Any planetary effect might be caused by a demon. Moreover, demons were notoriously deceptive. How could the magus be sure he was not invoking evil demons when they were always lying in wait for the opportunity to deceive those who tried to make contact with celestial spirits? If by his attempt to purify magic Ficino tacitly reopened the door leading to demonic magic, Heinrich Cornelius Agrippa von Nettesheim (1486–1535) recklessly rushed through it.[7] In his *De occulta philosophia*, Agrippa defended invocation of demons so long as they are good demons (and hence the agents of divine miracles), and openly discussed methods for compelling the assistance of evil demons in magical operations. In attempting to distinguish between truly religious ceremonies, which are based on faith, and superstitious practices, which are based on credulity, he was forced to conclude that the effects of credulity and faith are about the same. The outcome of this line of reasoning was to explain religious miracles in terms of the psychological effects of ceremonies, a stance that was very dangerous to religion.[8] Agrippa anguished over the atheistical implications of magic: although he completed his work on magic in 1510, he did not have it printed until 1533, several years after publishing a retraction of his views (*De vanitate scientiarum*, 1530).

Unlike Agrippa, the Neapolitan philosopher and magus Giambattista Della Porta never disavowed his commitment to magic, despite continual harassment by the Inquisition. From his first juvenile effort to his last unpublished work, Della Porta dedicated his life to establishing natural magic as a legitimate empirical science. He wrote more than a dozen books on various aspects of natural magic and was acknowledged as Europe's foremost authority on the subject. Regarded by many contemporaries as the age's "most diligent scrutinizer of the secrets of nature,"

Della Porta articulated the theoretical underpinnings of the professors of secrets' research program.[9]

Della Porta was born in 1535, the second son of Nardo Antonio Della Porta, a Neapolitan nobleman.[10] Like most of the Italian nobility, he and his two brothers, Giovan Vincenzo and Giovan Ferrante, were taught by private tutors. Their education consisted not only of classical letters, but also of "chivalric excercises" such as music, dancing, riding, and gymnastics.[11] None of the Della Porta brothers obtained a university degree or pursued an academic career. Yet the courtly atmosphere of the Della Porta household, with its informal cultivation of aristocratic *virtuosità*, provided the brothers with a sort of connoisseur's familiarity with the arts, crafts, and letters. Giovan Ferrante, the youngest, had a large collection of crystals and geological specimens, which he left to his brothers at his premature death.[12] The eldest, Giovan Vincenzo, was an authority on antiquities and an avid collector of marble statues, busts, and medals. Giambattista, who according to Nicholas-Claude Fabri de Peiresc revered his older brother as a father, used Giovan Vincenzo's collection in doing the research for his books on physiognomy.[13] In addition to his books on natural magic, Giambattista wrote on alchemy, astrology, physiognomy, cryptography, the art of memory, agriculture, optics, geometry, pneumatics, and munitions. He was the most accomplished comic playwright of his generation.[14] And though reportedly tone-deaf, all three of the Della Porta brothers studied musical theory at the Scuola di Pitagora, an exclusive academy for musicians.[15]

This courtly, amplitudinous style of education helped shape the development of Della Porta's interests in natural magic. It freed him from the rigorous constraints of scholarship that were the usual prerequisite for the study of natural philosophy, and it stimulated his somewhat dilettantish interests in "experiments" designed to exhibit *meraviglia*. Even his dramatic works displayed an exaggerated preoccupation with the marvelous: "When he turned from the secrets of nature to write a play, he could not leave off hunting miracles."[16]

Equally important for the development of Della Porta's interests was the cultural milieu of mid-sixteenth-century Naples. Nicola Badaloni has suggested that the circle of natural philosophers known to have influenced the Della Portas was motivated by the ideal of a comprehensive moral, religious, and medical reformation.[17] Matteo da Solito, the Della Portas' astrology teacher, reportedly held radically anticlerical views. According to Nicolò Franco's testimony before the Roman Inquisition in 1568, Solito "frankly and menacingly denied the power of the papacy, declaring it to be null and vain, and he said the same of the Sacrifice of the Mass and of the communion and all the other articles of the faith."[18] Giovan Abioso da Bagnola, another Della Porta tutor, thought that the conjunction of 1524 portended the possibility of an immanent reformation of human manners, a social reformation, and a restoration of human dominion over the material world.[19] To accomplish this *instauratio magno*, Abioso proclaimed it would be necessary to turn away from the books of the ancients and to "hunt for new secrets of nature" (*venari nova naturae secreta*). Abioso urged the study of the habits of animals as a way to learn how to act directly on the primeval forces of nature in order to use them for human benefit. Alchemy, he thought, provided a tool with which to recover the primordial *substantia vitae*, thus producing a universal medicine that could cure all diseases and restore human vitality. The quintessence, separated off from the dross of organisms through distillation, was for Abioso a metaphor for the reformation of society. Entirely consistent with this "primitivist" outlook was the avid interest among many Neapolitan intellectuals in the idea of the golden age, their fascination with American Indians, and their feverish desire to restore society to its pristine simplicity. As for the latter, some intellectuals advocated the overthrow of Spanish rule in Naples and the restoration of the aristocracy.[20] There can be little doubt that these radical ideas attracted Tommaso Campanella to the Della Porta circle on his first visit to Naples in 1589.[21]

This complex medley of ideas and political ambitions found a sympathetic hearing in the informal literary and philosophical academies that flourished in Naples—until they were closed by the Spanish viceroy in 1547. As we have already seen, the style of scientific research motivated by Abioso's dictum *venari nova naturae secreta* came to a fruition in the Accademia Segreta that Girolamo Ruscelli helped to organize in the 1540s. Although Della Porta would have been too young to take an active part in the academy's deliberations, he is likely to have been aware of its existence. Badaloni speculates that the precocious Della Porta may even have attended some of the Segreti's meetings.[22] Although there is no direct evidence for this conjecture, circumstantial evidence suggests a link between Della Porta and the Ruscelli group. Historians have generally expressed incredulity over Della Porta's claim to have written his famous *Magia naturalis* (1558) at the tender age of fifteen. Generally they have dismissed the claim as Della Porta's attempt to pass himself off to the public as a wunderkind.[23] But if Della Porta attended meetings of the Segreti (possibly in the company of his older brother), he would have heard frequent discussions of the subjects he later wrote about in the *Magia naturalis*. His book on magic may have been inspired by the Accademia Segreta.

Additional circumstantial evidence suggests a connection between the youthful Della Porta and the Segreti. I have already presented evidence to suggest that the Accademia Segreta existed under the patronage of Ferrante Sanseverino, the prince of Salerno, and that it disbanded following Sanseverino's unsuccessful rebellion against the Spanish government. The Della Portas were also a noble family from Salerno. Giambattista's father, Nardo Antonio, owned an estate in Vico Equense on the Sorrentine Peninsula, in the principality of Salerno, where Giambattista was born.[24] Nardo Antonio was thus a subject of the prince of Salerno and also held noble privileges in Naples. Evidently some members of the Della Porta family supported Sanseverino's rebellion, a fact Giambattista bitterly lamented.[25] Although Della Porta's father must have remained loyal to the Spanish viceroy during this affair (for he did not lose his estates or the office he held in Naples), the involvement of relatives in the rebellion cast suspicion on the entire family. About this time the family's fortune dwindled somewhat, necessitating the sale of Giovan Ferrante's collection of crystals and geological specimens.[26] Events may also have forced Giambattista to leave Naples, the occasion of his first trip to northern Europe. Years later, Della Porta would recall, "From my parental heritage I suffered some adverse and very melancholy things; suffered exile and persecution."[27]

A final, more telling piece of information suggests that Della Porta was familiar with the activities of the Accademia Segreta. It is well known that Della Porta later organized his own experimental academy at his home in Naples.[28] What has not been generally noticed is that his academy was remarkably similar to the one described by Ruscelli. Even the name Della Porta gave it, the Accademia dei Secreti, brings to mind Ruscelli's academy. Moreover, Della Porta's first published work, *Magia naturalis* (1558)—consisting largely of recipes and experiments in medicine, the crafts, optics, and other "secrets of nature"—is so similar to Ruscelli's *Secreti nuovi* and to the *Secreti* of "Alessio Piemontese" that there can be no mistaking its close relationship to the general aims of experimental science articulated by Ruscelli. Indeed, the *Magia naturalis* reads like a manifesto for a new scientific methodology: that of science as a *venatio*, a hunt for "new secrets of nature." The nearly identical names of the two academies, their proximity in time and place, and the similarity of their experimental methodologies, was surely no coincidence.

Tantalizingly little is known about Della Porta's Accademia dei Secreti. Della Porta himself mentioned the academy only once, in the preface to the second edition of his *Magia naturalis* (1589), where he wrote, "I never wanted also at my House an Academy of curious Men, who for the trying of these Experiments, chearfully disbursed their Moneys, and employed their utmost Endeavours, in assisting me to Compile and Enlarge this Volume, which with so great Charge, Labour, and Study, I had long before provided."[29] From this passage it appears that the academy was formed with the express purpose of trying out the experiments Della Porta had proposed in the first edition of his *Magia naturalis*, and of expanding the scope of that earlier work with the addition of new experiments. The *Magia naturalis* may thus be read as an extension of the research program set forth in the Accademia Segreta and continued in Della Porta's academy. This interpretation is corroborated by the testimonies of two of Della Porta's younger contemporaries, which provide additional details about the academy's activities. In an *éloge* of 1640, Giovanni Imperiali reported that the academy met at Della Porta's house, and that no one was admitted to it "unless he had discovered some new secret of nature useful in medicine or the mechanical arts and beyond the level of ordinary comprehension."[30] Somewhat later, Lorenzo Crasso referred to Della Porta's "famous academy called *de' Secreti*."[31] Pompeo Sarnelli, who owned Della Porta's letters, gave the fullest account of the academy:

> Not content with his own intelligence, he submitted his opinions to some of the more learned men, for whom he had erected in his house an academy with the title of *Secreti*. And these men vied with one another to add new discoveries to his researches, which, being well examined in the academy, they were pleased to see established afterwards. To accomplish this he embarked on a pilgrimage, and traveled (as he himself said) through all of Italy, France, and Spain, visiting very learned men and famous libraries, in order to find out new things; and returning home, he examined all the opinions in his academy, registering only those that had been proved as true.[32]

Della Porta's informal academy devoted itself to a research program almost identical to the one described by Ruscelli: to seek out "secrets" from books and from other savants, to put them to the test of experiment, and to "register only those proved true."

It is not known when the Accademia dei Secreti began its meetings. According to Sarnelli, the society already existed before Della Porta left Naples for an extended tour of Italy, France, and Spain, about 1563.[33] When he returned to Naples around 1566, Della Porta reconvened the group in order to discuss and experiment on the information he had collected during his travels. Although no official roll for the academy exists, we can be reasonably certain about the general character of its membership. Besides Della Porta and his brother Giovan Vincenzo, the group of "curious men" who "cheerfully disbursed their monies" on the academy's experiments was in all probability composed mainly of other local noblemen.[34] As we have seen, aristocratic Neapolitan social life revolved around the academies. In addition to such unnamed individuals, several of Della Porta's close intellectual companions can be identified.

7.   Portrait of Giambattista Della Porta at the age of fifty.
From *Magia naturalis* (Naples, 1589).

Domenico Pizzimenti, a classicist, alchemist, and one of Della Porta's teachers, may also have been a member of the academy.[35] Another likely candidate is Donato Antonio Altomare, a local physician, close friend of Della Porta, and a founder of the Altomare, an academy devoted to the study of medicine.[36] Giovan Antonio Pisano, a professor of practical medicine and anatomy at the Studio of Naples and another of Della Porta's teachers, might also have participated in the society's meetings. Pisano had been a member of the short-lived Euboli, which was closed down by the viceroy.[37] Della Porta also reported that he engaged craftsmen, including a distiller and an herbalist, to help with the experiments undertaken in the academy.[38]

In his biography of Della Porta, Sarnelli rather cautiously and tentatively mentions another event in connection with the activities of the Secreti: some unnamed Neapolitans denounced Della Porta to the Inquisition for the group's magical activities. Sarnelli goes on to say that when he was brought before the Inquisition, Della Porta proved to everyone's satisfaction that all of his secrets were natural, and defended himself so well that he was given an official commendation.[39] This was a complete whitewash, one of many pious lies invented by Della Porta's family and friends to save his reputation. In fact, Della Porta was brought before the Holy Office on at least two occasions.[40] His first appearance before the Inquisition occurred in 1574, when he was arrested and sent to Rome to answer for "things concerning the faith."[41] As a result of the event, he was forced to disband the Accademia dei Secreti.[42] Della Porta's case was reopened in February 1580, when he was brought before the Neapolitan inquisitors.[43] Again the charges related to his magical activities.[44] According to the eighteenth-century historian Giuseppe Valletta, who had the Inquisitorial documents before him, Della Porta was charged with "having written about the marvels and secrets of nature."[45]

But the *Magia naturalis* was published in 1558, and since that time Della Porta had published nothing on magic. What caused the Holy Office to reopen his case twenty-two years later? The answer is that he was implicated in a famous international dispute over witchcraft between the French jurist Jean Bodin (1529–1596) and the German physician Johann Wier (1515–1588). In his treatise on witchcraft, *De prestigiis daemonum* (1564), Wier argued against the persecution of witches, citing Della Porta's experiment demonstrating that the "witch's salve," supposedly used to transport witches into flight, could be understood according to naturalistic principles. Della Porta maintained that the witch's salve was in reality a sleep-inducing hallucinogenic drug, which when rubbed on the body caused supposed witches to fantasize their nocturnal flights.[46] Attacking Wier in his *Démonomanie des sorciers* (1580), Bodin brought Della Porta into the dispute, damning him as "un grand Sorcier Neapolitain" and accusing him of propagating demonic magic. Natural magic, experimental science, or sorcery: it was all the same to Bodin, who was absolutely convinced about the reality of witchcraft.[47]

Although Wier denied the reality of witches, he did not deny the existence of demons. Nor did he accept any attempt (including Della Porta's) to naturalize magic. A Lutheran, he condemned all magic, but especially that of the Roman Catholic church, as diabolical.[48] Wier argued that the operations of magic and witchcraft were delusions induced by demons.[49] He attacked the witchcraft persecutions on the grounds that supposed witches were not heretics who had made a pact with Satan, but innocent persons whose imaginations had been deranged by evil demons. Since Della Porta had concluded from his experiment with the witch's salve that the ingredients in the unguent simply caused credulous old women to experience bad dreams, his argument nicely suited Wier's purpose.

Bodin easily dismantled Wier's position. It was a contradiction, he pointed out, to admit that demons could cause witches' delusions, and yet maintain that such deeds were due solely to demons and never to the individuals who evoked them. If the devil can traffic with humans to wreak physical effects, as Wier constantly admitted, why cannot humans equally traffic with demons? To make his position even more impossible, Wier did not deny the latter possibility. While the supposed witches suffered only from melancholic delusions and did not consort with demons, he argued, the magicians did traffic with demons and should be punished accordingly! As Bodin pointed out, such "capricious and sophistic" reasoning denied the whole relationship between cause and effect, and served only to absolve witches of responsibility for their *maleficia*.[50] Such reasoning must have come from either ignorance or wickedness. Since Wier was obviously not ignorant, he must be in league with the devil and must have written his work with the aim of teaching the black arts. Turning to Della Porta, Bodin pointed out that by publishing an account of how to make the witch's salve, the "Neapolitan sorcerer," like Wier, was actually an accomplice in the devil's design to subvert the social order.

Della Porta defended himself against Bodin in the second edition of the *Magia naturalis* (1589), insinuating that Bodin was a heretic. Nevertheless, he deleted the account of the witch's unguent from that work, and for the rest of his life he struggled with the Inquisition to gain approval for the publication of his books.[51] In 1592, when he tried to publish an Italian edition of his work on human physiognomy, the Inquisition stepped in. On orders from the Holy Office in Rome, the Venetian Inquisition halted the work's publication and forbade Della Porta, under pain of excommunication and a fine of five hundred ducats, to publish anything without the express permission of the Roman High Tribunal.[52]

Magic and Superstition: The Politics of Occult Qualities

It would be a mistake to conclude from these events that Della Porta's troubles with the Inquisition were the result simply of the "witchcraft hysteria." As I have already indicated, his case was first opened in 1574, six years before the Bodin-Wier dispute took place. It was his work on natural magic, not suspicion of practicing witchcraft, that brought him before the Holy Office. The debate over magic in the late sixteenth century makes better sense if understood within the broader context of post-Tridentine politics. The Church, determined to consolidate an ecclesiastical monopoly over access to supernatural forces, saw any attempt to utilize occult powers as a threat to its jurisdiction over the miraculous. One of the Church's principal targets in its struggle to defend this territory was popular superstition. Della Porta believed that natural magic would also eradicate superstitions, not by suppressing them, but by giving them rational, scientific foundations. The only thing superstitious about popular magic, he maintained, was the belief that demons were responsible for it.

The history of Inquisitorial processes in the sixteenth century confirms the Church's growing concern over popular superstitions. Until about 1580, the Roman Inquisition busied itself with combating the Protestant heresy.[53] Having succeeded in stamping out Protestantism in Italy, or at least driving it underground, the Inquisition turned its attention to eradicating popular magic. Although the statistics vary somewhat from city to city, the pattern of persecution was everywhere the same: illicit magic replaced doctrinal heresy as the most common charge brought before the local tribunals of the Holy Office. In Naples, where Protestantism was considered to be less of a problem for the Inquisition than it was elsewhere, magic became the single most common crime after 1570 and remained so down through the seventeenth century. The overwhelming majority of such cases of "superstitious errors" concerned magical healing, love magic, and divination. That is, the accused were charged with using charms, incantations, and magical devices to heal various physical ailments, to prevent disease, to detect thieves, to find stolen objects and buried treasure, or to incite sexual passion. In contrast to northern Europe, the focus of the witch craze, popular "superstitions" rather than satanic witchcraft constituted the bulk of the crimes prosecuted by the local tribunals of the Italian Inquisition.[54]

Historians have argued that these activities were part of a systematic campaign by the Tridentine Church to reform popular culture, to eradicate popular superstitions, and to "divert all streams of popular religion into a single parochial channel."[55] With that as the Church's principal objective, it is not difficult to see why popular magic came under attack. While Protestantism emphasized the need to accept conditions as they were, Catholicism offered a religion that was "remedial" in concrete ways.[56] Popular devotional life revolved around such practices as appealing to saints for their intercession in human affairs, requesting aid from holy relics and religious amulets, and using prayers to seek divine assistance. Such practices were in principle entirely orthodox, but the distinction between acceptable practice and superstition was repeatedly blurred. Although ordinary people were encouraged to pray for divine aid, many of the prayers that circulated among the people were more like spells or magical secrets than supplications. Extrapolated from conventional prayers, these "spiritual recipes," which tacitly guaranteed success, implied a mechanical means of manipulating divine power.[57] Put simply, magical remedies competed with clerical ones. The Church was "a repository of supernatural power which could be dispensed to the faithful to help them in their daily problems."[58] The problem lay in the irresistible temptation luring unauthorized persons to try to get at this power.

The Tridentine Church regarded such unauthorized attempts to gain access to supernatural powers as "superstitions." This represented a significant change in the Church's definition of magic and witchcraft. Prior to the fifteenth century, Inquisitorial jurisdiction limited itself to cases involving explicit pacts with demons.[59] The most common cases of popular sorcery, *maleficia* (causing physical harm through magic), were treated like other crimes involving physical harm to persons, crops, or animals without regard to the magical means employed. Fifteenth-century legislation shifted the focus to the means employed, as opposed to the ends. All magical activity, whether harmful or beneficial, came under suspicion as involving, implicitly or explicitly, a pact with demons.[60] In 1607, the Council of Malines drew up a definition of supersition that summarized the general practice followed by the Roman Inquisition in the second half of the sixteenth century: "It is superstitious to expect any effect from anything when such an effect cannot be produced by natural causes, by divine institutions, or by the ordination and approval of the Church."[61] Magic, even without directly invoking demons, drew on forces not controlled or sanctioned by the Church, and hence was superstitious and presumptively diabolical. In elaborating the "typical" figure of the witch during the fifteenth century, theologians conflated activities traditionally regarded as stemming from mere ignorance or credulity with stereotypes drawn from learned magic and the underworld of heresy. As modern studies of witchcraft have shown, the identification of the ordinary village sorcerer with the magician who knowingly enters into a pact with Satan dramatically extended the scope and severity of witchcraft persecutions.[62]

In light of these changing juridical practices, it is easy to see why natural magic came under attack. It was not just that churchmen suspected natural magic of involving consort with demons; worse, they saw it as an arrogant attempt to usurp the Church's prerogative in controlling access to supernatural forces. The dangerous tendencies of naturalism were made glaringly clear earlier in the century, when the Paduan philosopher Pietro Pomponazzi (1462–1525) eliminated all nonnatural agency from physical causation, thus denying the existence of angels and demons and rejecting even the miracles attested in Scripture.[63] The violent controversy that erupted over Pomponazzi's views brought the issue of miracles into sharp focus and put the Church on guard against naturalism. In the tense political climate of the post-Tridentine period, the Church was incapable of accepting naturalistic interpretations of such practices as chiromancy or healing by sympathetic magic, both of which the proponents of natural magic considered natural. Nor was the Church able to condone astrological forecasting, because it threatened to usurp its guardianship over scriptural prophecy.[64]

Della Porta wrote all of his works in Latin, hence not for a popular audience. Yet his books were in steady demand. His *Magia naturalis* appeared in more than twenty Latin editions and was translated into Italian, French, German, English, and Dutch. More than fifty editions of the work appeared in the sixteenth and seventeenth centuries. After his trial, however, he had great difficulty getting his books published. His Latin treatise on physiognomy, *De humana physiognomia*, was printed in 1586, after a three-year delay. That same year, Pope Sixtus V issued a bull banning works on astrology and divination. Della Porta had to wait ten years for approval of his petition to publish an Italian translation of the work. He complained bitterly that it took longer to get his books published than to write them.[65] But from the Church's point of view, censorship of magical books was a necessary step toward maintaining its control over occult forces.

In order to understand why the Church considered Della Porta's works to be dangerous, we must take a closer look at Della Porta's views on magic, sorcery, and "superstitions." In his youthful *Magia naturalis*, Della Porta made a traditional distinction between black and white magic, pointing out that whereas sorcery "is infamous, and unhappie, because it hath to do with foul spirits, and consists of Inchantments and wicked Curiosity," the other magic "is natural, which all excellent wise men do admit and embrace, and worship with great applause; neither is there any thing more highly esteemed, or better thought of, by men of learning."[66] But this was not to be Della Porta's last word on demonology. He returned to the theme in a later, unpublished work entitled the *Criptologia*. This small treatise was to make up a book of his planned magnum opus, the *Taumatologia*, which the Inquisition steadfastly refused him permission to have printed.[67] The *Criptologia* was a head-on challenge to the Church's conception of popular magic: "In this book are treated the most hidden secrets that are buried in the intimate bosom of nature, for which neither natural principles nor probable explanations can be given—but are not for that reason superstitions." There were great truths in popular magic, Della Porta concluded. But these truths were distorted by popular superstitions and by learned de-

monology. On the one hand, the common people were extremely foolish to believe that the effects of natural magic were caused or enhanced
by the invocation of supernatural aid. Witches and cunning men produced their effects by using natural forces; the ceremonies, rites, and
spells connected with their practices were useless and blasphemous. On
the other hand, the scholastic philosophers and, following them, the
Church were just as credulous in attributing these effects to demonic
agencies. According to Della Porta, Aristotelian natural philosophy was
hopelessly inadequate to account for "superstitions" and hence contributed to their growth: "The ignorant philosophers, when they cannot
give reasons to these things according to the principles of Aristotle (as if
he knew all things), judge them as superstitions. But learned men know
well that of the infinite number of things seen in this great machine of
the world that they want to know, they know scarcely a particle of
them."[68] Della Porta believed that popular and learned views about
witchcraft were equally superstitious. He wanted to eradicate both kinds
(*omni explosa superstitione*).[69]

Della Porta went on to give a critical but unusually positive evaluation
of demonology. He thought that demons, or fallen angels, although deprived of God's grace, "did not because of that lose their natural ability
to know the virtue of the heavens, metals, stones, plants, and animals."[70]
Like angels, demons were superior natural philosophers, although they
used their superior intelligence to deceive humans in order to "pull
them from paradise to the eternal darkness of damnation." In reality,
demonic magic is nothing more than natural magic, to which many superstitions, magical rituals, ceremonies, and secret words had been adjoined. Moreover, "these secrets would have been very difficult things
for man to uncover" had they not been revealed (*manifestati*) by demons.[71] Undaunted by the prospect of losing his soul, Della Porta resolved to seriously study these phenomena, and to prove that they are
natural by putting them to the test of experiment: "I began to look over
the books of the physicians, printed and manuscript, that were full of
magical rituals and worthless words, and I discovered that the effects
derived from natural causes; and by simply experimenting on these
things, and arriving at the truth, I unmasked (*detexi*) the frauds and the
diabolical trickery."[72] Thus demonic magic was based on the mistaken
belief that the rites, words, and ceremonies attached to magical events
were the causes of those events, when in reality they had purely natural
causes.

Della Porta's claim to have "experimented" on the effects of demonic
magic should not be taken literally. His method of research in the *Criptologia* was primarily textual and historical. In investigating popular superstitions, he examined the "empirical" books of Marcellus, Pseudo-

Apuleius, Pliny, and others, and discovered that the ancients employed many of the same practices village healers used, but without adding to them any magical formulas or sacred words. Della Porta applied this research method to various forms of love magic employed by the village *streghe* (witches), such as touching the flesh with a white magnet while reciting certain incantations to provoke amorous feelings.[73] Della Porta discovered that Marbod and Kyranides affirmed the natural aphroditic agency of white magnets, thus "exploding" the popular superstition that demons were behind it. In another literary "experiment," he exploded a superstition of his own. When he was much younger, he used a certain "mystery of nature" as an antidote against poisons. He had the patient place his bare foot on the ground, and then, after tracing the footprint in the dirt, he inscribed inside the footprint the words *Caro, caruze sanum reduze, reputa sanum, Emanuel, paraclitus.* Della Porta used this technique successfully on several occasions, once healing some harvesters bitten by tarantulas. But he began to have doubts about the procedure's orthodoxy. After consulting a cleric and learning that the incantation was sacrilegious, he stopped using it. Later, upon observing an acquaintance's success with the same technique, but with an entirely different incantation, Della Porta "immediately discovered the devil's fraud, because a certain blasphemy was inherent in the words, and the same thing can be produced without appealing to these formulas. Having experimented on it, I obtained the same results." Then, clinching the argument, he determined that Marcellus Empiricus and other ancient authors recommended similar remedies to be administered on bare feet.[74]

   With this method Della Porta discovered that many popular "superstitions" were in reality instances of profound natural magic. He ridiculed the practice of certain "imposters" who carved magical characters on a forked stick, which they used as a divining rod for finding buried treasure. The technique, he maintained, was in reality a natural method used by miners to find veins of ore, the enchantment of the stick having no effect whatsoever upon the stick's natural attraction to precious metals.[75] He observed that many psychosomatic ailments only appeared to be treated magically. In reality they are treated naturally with herbs and other hidden devices, and "cured 'marvelously' with the secure conviction on the part of the afflicted ones that they will be healed."[76] The amulets the *streghe* used as love charms, the enchanted herbs with which they exorcised demons, and their supposed magical remedies against impotency were all natural, as we can plainly see by reading Ovid, Pliny, and Dioscorides. Such "superstitions" were all corroborated by the ancients, who said nothing whatsoever about needing incantations to make them work. Philological research even enabled Della Porta to discover the fish that the biblical Tobias used to cure Sarah of demonic possession.[77] Della Porta did not doubt the effects produced by these magical techniques. He only disputed the efficacy of the incantations and rituals associated with them in the popular (and learned) imagination.

But the *Criptologia* was no defense of the people's wisdom. It was a critique of their credulity. Although Della Porta did not deny the existence or potency of demons, he maintained that their powers were purely natural. All the rituals added by sorcerers and witches to these powers were nothing more than useless words, superstitions stemming from credulity and ignorance. Della Porta maintained that humans can exploit the same powers without adding to them the superstitions of sorcery. Yet his research led him to believe things even stranger than the people's magic. He maintained, for example, that people could tap the natural force in the witch's salve, which caused old women to be seized by visions and hallucinations, to transmit their imaginations, thus enabling them to communicate across distances. Instead of being seized by "demonic" hallucinations, the natural magician could control the forces that caused hallucinations in order to telegraph his thoughts.[78]

By classifying demonic magic as natural, Della Porta opened up to experimental investigation events that the Church condemned as diabolical. Although the "superstitions" he attacked were the same as those the Inquisition hunted, his position on the status of demonic forces was radically different. Since demonic forces are natural, he maintained, it is legitimate to investigate and even employ them, as long as one avoids the rituals adjoined to them. Indeed, he claimed that the natural magic originally taught by demons could be used to eradicate demons. He believed that the research program of the *Criptologia* would lead to a "smashing of superstitions"—not by persecution, but by exploding the false beliefs surrounding the real, natural phenomena. He also thought it would lead to a recovery of the true, natural magic practiced by the ancient magi.[79] The most convincing way to expose the fraud of demons, thought Della Porta, was to discover the natural truth behind the superstitions of popular magic, and to demonstrate these by producing marvels naturally.

Obviously, such a radical position on diabolism could never gain the Church's approval. Della Porta's position was essentially that of Pomponazzi, whose *De incantationibus* was condemned by the Inquisition. Not only did Della Porta's view amount to an assault on the Church's jurisdiction over the marvelous, it directly challenged the official mechanisms against demons sponsored by the Church, such as exorcism. Just as the Church was intensifying its campaign against popular superstitions, Della Porta was developing a theory that would undermine its entire conception of superstition. But the Church needed demons as much as it needed Aristotle. Witchcraft was a necessary component of the sixteenth-century moral economy. As Stuart Clark has argued, it was one of its "conventions of discourse."[80] In a world predisposed toward seeing things in terms of binary opposition, such polar opposites as good/evil, rule/misrule, and chaos/order were universal principles of intelligibility as well as configurations of the real world. A well-ordered commonwealth implied the existence of a world upside-down.[81] Witches, "conventional manifestations of disorder," confirmed the legitimacy of established authority because, according to the operative tautology, "the Devil's tyranny was an affront to all well-governed commonwealths [and] to every state of moral equipoise." Thus witchcraft legitimized established authority and the instruments by which it exercised its power: "Each detailed manifestation of demonism presupposed the orderliness and legitimacy of its direct opposite, just as, conversely, the effectiveness of exorcism, judicial process and even a royal presence in actually nullifying magical powers confirmed the grounds of authority of the priest, judge or prince as well as the felicity of his ritual performance."[82] The *Criptologia*, by attempting to naturalize the miraculous, presumed to spell out the instruments of occult power and threatened to put them into the hands of anyone who could read. Reasonably enough, the Church could never allow such a work to be published.

Although the *Criptologia* represents an extreme version of Della Porta's general theory of natural magic, his temerity in venturing into such forbidden territory makes it easier to understand why the Inquisition found it necessary to summon him to answer questions "concerning the faith." Not only did Della Porta think it was legitimate to investigate demonology, he argued strenuously for the benefits of doing so. "I rejoice more in having found out this fraud of demons," he declared, "than in whatever else I have done in my life of seventy years, having thereby opened the way for investigators to recover other things that are manifested in the inexhaustible treasure of God."[83] Serious, experimental study of necromancy, he thought, would lead to knowledge of "the most hidden secrets" of nature; it would eradicate superstitions; and by recovering a lost ancient science, it would clarify once and for all the real distinction between superstition and natural magic.

The *Magia naturalis*

According to Della Porta, natural magic was a science of the extraordinary. In contrast to Aristotelian natural philosophy, whose aim was to explain the normal, everyday aspects of nature, natural magic explained the exceptional, the unusual, and the "miraculous."[84] The overarching aim of Della Porta's major and best-known work, the *Magia naturalis* (1558 and 1589), was to develop naturalistic explanations for the supposedly "marvelous" phenomena recorded in the great medieval encyclopedia of secrets: as he put it, to "reduce all those secrets into their proper places."[85] The task he took on was already a familiar one. To some extent it was the same as that undertaken in the thirteenth century by the author of the pseudo-Albertine *De mirabilibus mundi*, a work Della Porta knew well. In devising a metaphysics to account for nature's "secrets," he made generous use of the standard authorities on the occult tradition. But he also drew on local influences and synthesized the diverse influences that made up the Neapolitan intellectual tradition into a distinctly south-Italian philosophy of nature.

Della Porta's natural philosophy was a blend of Aristotelian physics, Renaissance Neoplatonism, naturalistic metaphysics in the tradition of Telesio, and a poetic fancy mainly his own.[86] Although he subscribed to the Aristotelian doctrine of substance as a composite of matter and form, he believed that the scholastic interpretation of the doctrine was incomplete. In particular, it was unable to account for the special "properties" (*proprietates*) of things, the unique, insensible qualities that give rise to magical operations. Following Aristotle, he explained that the four elements of matter (earth, air, fire, and water) contain within them the common, "primary" qualities (hot, cold, dry, and moist). When the elements are mixed, they give rise to "secondary" and "tertiary" qualities, such as hardness, density, malleability, and smoothness. Yet in addition to such uniform material qualities, every object also has its own peculiar "properties" that cannot be accounted for in terms of the qualities of the four elements alone. The power of the lodestone to attract iron, for example, is peculiar to it and not to other stones or earthy substances. Nor does the ability of certain drugs to expel poisons seem to arise from mixtures of the elements. The Peripatetics attributed such properties to the power of essential form. But in Della Porta's view, this vague formulation did not go far enough. It could not explain the origin of occult properties. What was needed, he believed, was a closer analysis of the union of form with matter, the intersection where occult properties arose. In undertaking this analysis of hylomorphism, he discovered that the deepest secrets of nature were all instances of attraction and repulsion, sympathy and antipathy, love and hate, which jointly govern the economy of properties. Concord and discord rule the world, but not willfully: their dominion is determined by form.

Della Porta meant this literally. Nature is alive. Everything inherently possesses the sensations of attraction and repulsion, the sentiments of love and hate. Matter is not altogether without force in determining the distribution of these occult properties, he conceded: the elements and their qualities are the sensible matrix upon which form impresses the properties of concord and discord. But Aristotle's material cause was vastly inferior in potency to formal cause. Indeed, while material qualities are prepared by the elements, all occult properties proceed from form. Della Porta likened form to an artisan who carefully selects the materials appropriate to his work, then shapes and molds them according to his specific design:

> But the Form hath such singular vertue, that whatsoever effects we see, all of them first proceed from thence; and it hath a divine beginning; and being the chiefest and most excellent part, absolute of her self, she useth the rest as her instruments, for the more speedy and convenient dispatch of her actions: and he which is not addicted nor accustomed to such contemplations, supposeth that the temperature and the matter works all things, whereas indeed they are but as it were instruments whereby the form worketh: for a workman that useth a graving Iron in the carving of an Image, doth not use it as though that could work, but for his own furtherance in the quicker and better performance thereof. . . . Wherefore that force which is called the property of a thing, proceeds not from the temperature, but from the very form itself.[87]

Understanding this, the magus, who is also a kind of artisan, discovers ways to manipulate the power of form and makes nature do his bidding.

The superiority of form over matter was the linchpin of Della Porta's theory of natural magic. Form was the immediate cause, or as close as he could come to one, of the generation of occult qualities. Form, he explained, proceeds from God, who communicates it to the celestial intelligences, the spiritual beings intermediate between God and man—in the Christian view angels, in the Neoplatonic view daemons. The intelligences, in turn, communicate with the elements through the heavens. The planets and stars immediately govern the allocation of form. They stamp upon everything the capacity of sensation and the forces of sympathy and antipathy:

> The form, as it is the most excellent part, so it cometh from a most excellent place; even immediately from the highest heavens, they receiving it from the intelligences, and these from God himself; and the same original which the Form hath, consequently the properties also have. . . . For God, as Plato thinks, when by the Almighty power of his Deity he had framed in due measure and order the heavens, the stars, and the very first principles of things the Elements, . . . and he enjoyned inferiour things to be ruled of their superiors, by a set Law, and poured down by heavenly influence

upon every thing his own proper Form, full of much strength and activity: and that there might be a continual encrease amongst them, he commanded all things to bring forth seed, and to propagate and derive their Form wheresoever should be fit matter to receive it.[88]

The celestial intelligences, the stars, and the planets were thus links in a chain of multiple causation in which form, emanating from God, permeates the universe like the rays of the sun emanating from a central source. This doctrine, Aristotelian, Stoic, and Neoplatonic in origin, was widely embraced in Renaissance natural philosophy.[89] "No man doubts but that these inferiour things serve their superiors," wrote Della Porta, "and that the generation and corruption of mutable things, every one in his due course and order, is over-ruled by the power of these heavenly Natures."[90] But the influence of the stars over natural objects was only one instance of the interconnectedness of things. God fashioned the universe and the social hierarchy according to the same principles of proportion. The entire universe was a vast system of interrelated correspondences, a hierarchy in which everything has its assigned place, a world in which everything acts upon everything else. Evoking the image of a well-ordered utopian society, whose members are all "linked together in their rankes and order," inferiors serving superiors and deriving their natures from those above them, Della Porta concluded: "The parts and members of this huge creature the World, I mean all the bodies that are in it, do in good neighborhood as it were, lend and borrow each others Nature; for by reason that they are linked in one common bond, therefore they have love in common; and by force of this common love, there is amongst them a common attraction, or tilling of one of them to another. And this indeed is Magick."[91]

We cannot hope to fully explain the properties arising from the complex web of correspondences that is the universe, Della Porta admitted. They exist because "it is the pleasure of Nature to see it should be so."[92] But we can know occult qualities empirically and can use them for human benefit. Once begun, Della Porta's illustrations of such properties proliferate rapidly. There is a "deadly hatred" between vines and coleworts, which is why coleworts are a good remedy against drunkenness. Rue and hemlock are also enemies: if rue is handled with bare hands, "it will cause Ulcers to arise; but if you do chance to touch it with your bare hand, and so cause it to swell or itch, anoint it with the juice of Hemlock." By analogy, rue is an antidote against hemlock poisoning. A wild bull is tamed by being tied to a fig tree; thus it was discovered that wild fig stalks can be used to tenderize beef. Humans and serpents are such deadly enemies that if a pregnant woman catches the scent of an adder, it will cause her to abort her child. Because of the antipathy between wolves and dogs, "a Wolves skin put upon any one that is bitten of a mad Dog, asswageth the swelling of the humour."[93]

But how is experience to weave its way through this labyrinthine network of correspondences and hidden similitudes? How are nature's "secrets" to be discovered? The answer is that nature puts a mark on things: the outward appearances of things provide clues or signs pointing to the properties that would otherwise be totally hidden from view. These "signatures," or visual likenesses, enable us to know, for example, that the herb *scorpius*, which resembles the scorpion, is a good remedy against the scorpion's sting; that the milky galactites, powdered and sprinkled over the back of a goat, will cause the goat to give milk plentifully to her young; or that the wine-colored amethyst prevents drunkenness.[94] Such signatures were not merely coincidences but were divinely ordained. They were woven into the fabric of nature, giving it meaning and intelligibility. Without signatures, nature would be baffling and impenetrable.

The doctrine of signatures was a cornerstone of Della Porta's theory, "the very root of the greatest part of [the] secret and strange operations" of natural magic.[95] He devoted several independent works to the study of natural signatures, including treatises on chiromancy and physiognomy.[96] His science of physiognomy, like his science of demonology, was aimed at discrediting the "superstitions" of the various charlatans and palm-readers who made prognostications about future events.[97] By giving physiognomy and chiromancy credible theoretical foundations, he believed it would be possible to restore the science to its ancient status.

Della Porta drew his examples of sympathies, antipathies, and correspondences straight from classical and medieval sources. In this respect, he was very much a product of his age. Nothing could be more characteristic of Renaissance natural history than its reverence for classical antiquity and its attachment to philology. As Brian Copenhaver writes of the Renaissance naturalists, "their curiosity about antiquity and their reverence for it stirred them so powerfully that their natural history became as much philological as biological."[98] Printing gave scholars of Della Porta's generation access to ancient texts unimaginable a century earlier. The embarrassment of riches produced by humanist scholarship convinced them that the entire book of nature could be read in the books of the ancients. Swamped as he was with written sources, there is little wonder that Della Porta was better able to record and broadcast ancient testimony about occult powers than to explain or refute it. That such boundless enthusiasm and such earnest devotion should have gone into unearthing ancient reports of occult qualities testifies, more than anything, to the poverty of erudition.

8. Title page of Giambattista Della Porta, *Magia naturalis* (1589), with the motto "Aspicit et Inspicit." At the top of the page a lynx stalks its prey; the border depicts a series of lockets, which are opened by natural magic to reveal the secrets of nature.

Despite his wholehearted approval of philological methods, Della Porta also insisted upon the primacy of experience. Natural magic, he wrote, "is nothing else but the survey of the whole course of Nature."[99] Like Fioravanti, he believed that nature was a great teacher that has long instructed humankind on how to use art to imitate its work. "Knowledge of secret things depends upon the contemplation and view of the face of the whole world," he wrote, "of the motion, state and fashion thereof, as also of the springing up, the growing and the decaying of things: for a diligent searcher of Natures workes, as he seeth how Nature doth generate and corrupt all things, so doth he also learn to do." In a lengthy discussion that reminds us of Fioravanti's primitivism, Della Porta described how the natural magician "learns from living creatures, which, though they have no understanding, yet their senses are far quicker then ours; and by their actions they teach us Physick, Husbandry, the art of Building, the disposing of Houshold affairs, and almost all the Arts and Sciences." The ancients observed, for example, that doves gather bay leaves to protect their young against enchantments. Elephants eat wild olives as an antidote against chameleons, "whence Solinus observes, That the same is a good remedy for men also in the same case." Partridges eat leeks to clear their voices, and "according to their example Nero, to keep his voice clear, ate nothing but oil of leeks certain days in every month." The beasts "have also found out purgations for themselves, and thereby taught us the same."[100]

Such examples from the *Magia naturalis* could be multiplied almost indefinitely. Obviously, Della Porta and his contemporaries adopted empirical standards that were radically different from ours. For the Renaissance naturalists, experience did not necessarily imply direct, personal observation. Indeed, by their very nature the "extraordinary" phenomena that were the subjects of natural magic were not accessible to ordinary experience. They were open only to careful scrutiny, or to some special intuition such as that of the magus. You had to "know" what you were looking for.

How did Della Porta know what to look for? He knew by reading the ancient authorities. For the Renaissance naturalists, the basic topography of nature had already been mapped out in classical literary sources such as Ovid and Pliny, who portrayed nature as full of wonder, variety, and surprise.[101] In completing the map of nature, the sixteenth-century naturalists were essentially fulfilling the promise of classical literature. Della Porta's "empiricism" was not some proto-Baconian inductivism, but history and literature verified by experience. Historical research generated certain expectations about nature that, he presumed, experience would fulfill. Della Porta did not make a clear distinction between history and experience. For him, "reading" nature was like reading a text. The book of nature could be expected to contain many of the same delights and surprises he found in Pliny's *Natural History* or in Ovid's *Metamorphoses*. Not surprisingly, he did find them in nature. In Della Porta's world, fecund nature brought forth myriad creatures "of her own accord": the silt overflowing from the Nile River engendered mice, horses brought forth wasps and hornets, and red toads arose spontaneously out of dirt and menstrual blood.[102] Della Porta barely needed a creator God: nature was sufficient to spontaneously populate the world: "The earth brought forth of its own accord, many living creatures of divers forms, the heat of the Sun enliving those moisture that lay in the tumors of the earth, like fertile seeds in the belly of their mother; for heat and moisture being tempered together, causeth generation. So then, after the deluge, the earth being now moist, the Sun working upon it, divers kinds of creature were brought forth, some like the former, and some of a new shape."[103] Ever changing, ever joking, subtle, ingenious, and prodigious, nature exhibited the same luxuriant style as the best poets of antiquity.[104]

The Magician as Artisan

But if for Della Porta scientific research began in the library, it did not end there. He conceived of natural magic as an empirical, even experimental science. "In our Method I shall observe what our Ancestors have said," he wrote, "then I shall show by my own experience whether they be true or false."[105] To him nature was an "inexhaustible treasury of secrets." Wherever he went he collected specimens of plants, gems, geological materials, and "curiosities," amassing a sizable private natural history museum.[106] At his villa in Vico Equense, he kept a garden in which he cultivated the botanical specimens he collected during his travels. He carried on a lively correspondence with other naturalists, exchanging specimens and information with them.[107] These activities were all channeled toward one aim, which was the ultimate goal of natural magic: to bring to light the hidden secrets of nature and to put them to practical use. "Our task," he wrote, is "to teach the way and method of searching out and applying of secrecies."[108] Natural magic was thus both the "consummation of natural philosophy" and its "practical part." The natural magician was like a superior artisan who possessed an intimate knowledge of materials and the forces acting upon them. Substituting philosophy for the artisan's cunning, he is able to do with nature what nature cannot do on its own:

> Art being as it were Nature's Ape, even in her imitation of Nature, effecteth greater matters than Nature doth. Hence it is that a Magician being furnished with Art, as it were another Nature, searching thoroughly into those works which nature doth accomplish by many secret means and close operations, doth work upon Nature, and partly by that which he sees, and partly by that which he conjects and gathers from thence, takes his sundry advantages of Nature's instruments, and thereby either hastens or hinders her work, making things ripe before or after their natural season, and so indeed makes Nature to be his instrument.[109]

Natural magic does not work against nature but is a minister of nature, supplying by artificial means what nature wants: "The works of Magick are nothing else but the works of Nature, whose dutiful hand-maid Magick is, . . . as in Husbandry, it is Nature that brings forth corn and herbs, but it is Art that prepares and makes way for them."[110]

The *Magia naturalis* is about making marvels naturally. Della Porta began the work, appropriately, with a chapter on natural prodigies and monstrous births, examples of which he found in abundance in ancient and medieval literature. But his real interest in monstrous births was to discover ways to supersede or improve on nature: he wanted to produce "marvels" of his own. The way to do this was to imitate nature. "Whosoever wouldst bring forth any monsters by art," he advised, "must learn by examples, and by such principles be directed."[111] Consider, for example, the powerful force of imagination, which, acting upon the seed during conception forces its imprint upon the offspring. Imitating nature, it is possible to produce animals of a predetermined color by causing the female to gaze upon that color during conception. Della Porta reported that stablemasters hang multicolored tapestries in the stables where mares are bred, "whereby they procure Colts of a bright Bay colour, or of a dapple Gray, or of any one colour, or of sundry colours together."[112] Similarly, a woman can be made to bear beautiful children if pictures or statues of Cupid, Adonis, or Ganymedes are placed in the bedroom where she conceives, a secret Della Porta tested by experience:

> After I had counseled many to use it, there was a woman who had a great desire to be the mother of a fair Son, that heard of it, and put it in practice; for she procured a white boy carved of marble, well proportioned every way; and him she had always before her eyes. . . . And when she lay with her Husband, and likewise afterward, when she was with child, still she would look upon that image, and her eyes and heart were continually fixed upon it, whereby it came to pass that when her breeding time was expired, she brought forth a Son very like in all points to that marble image, but especially in colour, being as pale and as white, as if he had been very marble indeed. And thus the proof of this experiment was manifested and proved.[113]

Nature can also be imitated in other activities, such as gardening, metallurgy, distillation, and the making of cosmetics, perfume, ammunition, and fireworks. You can make strawberries ripen out of season by picking them while the berries are still white, storing them in an airtight container, and setting them out later in the sun: this is natural magic. To have lettuce for a winter salad, bind the leaves to blanch them, and transplant them indoors at the end of the summer. To make roses bloom later in the season, pinch the early buds and let the later ones bloom.[114] Whenever a gardener plants hotbeds indoors in the winter or early spring to extend the growing season, he is practicing natural magic. The natural magician is like a gardener, who neither leaves things to nature nor acts contrary to nature, but imitates nature's methods and uses nature to his advantage.

Domestic economy can also be improved through the techniques of natural magic. Della Porta's instructions for preserving fruits, storing grain, and making oil, wine, and jam were all similar to the techniques described in other books of secrets. He showed how to make bread from cheap grains and nuts, "that not only the Householder may provide for his family with small cost, but when provision is dear he may provide for himself with small pains in Mountains and Deserts."[115] He discussed the causes of putrefaction. Terrestrial organisms decay, he explained, when they come into contact with air, whose temperature varies because of the celestial influences acting upon it. The "active principle" of heat in the air acts upon the "passive principle" of cold in vegetables or fruit and consumes it. "But man is not of such a dull sense, and of such a blockish wit, but that he can tell how to prevent these inconveniences, and to devise sundry kinds of means whereby the soundness of Fruits may be maintained against the harms and dangers both of cold and of heat."[116] Natural magic teaches that one can preserve fruits and vegetables merely by keeping them in airtight containers.

Della Porta dedicated a full chapter of the *Magia naturalis* to alchemy, which he considered to be "an excellent Art discredited and disgraced" by certain practitioners who abused it, "so that nowadays a man cannot handle it without the scorn and obloquy of the world, because of the disgrace and contempt which those idiots have brought upon it."[117] He warned readers against those who attempted to manufacture the philosopher's stone or the elixir of life, "a mere dream." Alchemy promises no such miracles. Della Porta's restrained view of alchemy was similar to the stance taken in the *Rechter Gebrauch d'Alchimei* and the Italian books of secrets: "I do not here promise any golden mountains, as they say, nor yet that Philosopher's Stone, which the world hath so great an opinion of, . . . neither do I promise here that golden liquor." Instead, his recipes for the "transmutation" of metals were designed mainly to improve the quality of metals, to make them "more excellent" or "worthier," or to prepare them for specific processes. Some of the techniques, he advised, were but "sleights" and "counterfeits" (such as the methods for coloring metal surfaces to make them resemble precious metals): "let them be esteemed no better than they deserve." But they were useful to learn because they trained alchemists in necessary skills and in turn led to new discoveries. Alchemy was for Della Porta not unlike other craft procedures, although nobler because it showed the way "to the searching out of the secrecies of Nature." His laboratory was the artisan's workshop, where he observed technical operations firsthand, absorbed the lore of the crafts, and attempted to distinguish between folklore and empirically tested techniques. His chapter on tempering steel exhibits the extent to which he profited from these experiences. He drew attention, for example, to the importance of color tests to guide tempering, a technique not mentioned in the *Kunstbüchlein*:

> When the Iron is sparkling red hot, that it can not be hotter, that it twinkles, they call it Silver; and then it must not be quenched, for it would be consumed. But if it be of a yellow or red color, they call it Gold or Rose color; and then quenched in Liquors, it grows harder. This color requires them to quench it. But observe that if all the Iron be tempered, the colour must be blue or violet color, as the edge of a Sword, Razor, or Lancet; for in these the temper will be lost if they are made hot again. Then you must observe the second colors; namely, when the Iron is quenched, and so plunged in, grows hard. The last is Ash color; and after this if it be quenched, it will be the least of all made hard.[118]

This was an acute observation. As the historian of metallurgy Cyril Stanley Smith pointed out, it led Della Porta to realize the advantages of a two-stage quench over a direct quench, and as a consequence to reject outright some of the exotic quenching baths that had characterized earlier metallurgical literature. He emphasized the necessity of using clear quenching liquids so that the tempering colors could be discerned, and recommended rubbing a blade with soap before heating it, "that it may have a better color from the fire." Della Porta roundly criticized a recipe from Albertus Magnus that had appeared in numerous craft booklets: "Albertus, from whom others have it, says that Iron is made more strong if it be tempered with the juice of Radish, and Water of Earthworms, three or four times. But I, when I had often tempered it with juice of Radish, and Horseradish, and Worms, I found it always softer, till it became like Lead; and it was false, as the rest of his Receipts are."[119]

There are no printed antecedents for these observations. Della Porta learned metallurgy, as he learned other crafts, by observing artisans at work and by experimenting on his own. By putting conventional techniques to the test of experiment, he hoped to find ways of improving existing techniques and to make new inventions. The results were often successful. Investigating the problem of how to make armor musketproof, he noticed that finished armor often contained tiny, almost invisible flaws that caused the iron to break under stress. Observing the process by which smiths made armor, he discovered that they often heaped coals over the iron to make the iron heat up more quickly. "And with this trumpery-dust, there are always mingled small stones, chalk, and other things gathered together in pieces; which, when they meet in the fire, they cause many knots outwardly, or cavities inwardly, and cracks, that the parts cannot well fasten together." Della Porta's solution: wash the coals in water, then dry them out before using them to pile over the ironwork. The magus was rhapsodic about this discovery: "What a blessing of God this profitable Invention is!" he exclaimed, "for thus men make Swords, Knives, Bucklers, Coats of Mail, and all sorts of Armour so perfect that it were long and tedious to relate: for I have seen Iron breasts that scarce weighed above twelve pounds to be Musketproof."[120]

Della Porta's experimental method did not aim to test general hypotheses. Instead, it attempted to imitate nature in order to produce utilitarian knowledge, and to correct and amplify the written tradition. Based on his experiments, Della Porta could assert that the metalworker can control the degree of hardness more effectively by observing color changes in the heated metal than by adding organic material to the quenching bath; that the attractive force of a magnet is not augmented when it is heated red hot, nor diminished when garlic is rubbed on it; or, refuting Pliny's claim, that seawater can be freshened by distillation but not by straining. Apologizing for the mundaneness of some of his experiments, he pointed out that only from such common and familiar examples can the mind ascend to higher truths. "Sometimes from Things most Known and meanly esteemed, we ascend to Things most Profitable and High, which the Minde can scarce reach unto," he affirmed. "One's Understanding cannot comprehend High and Sublime Things, unless it stand firm on most true Principles. . . . Wherefore I thought it better to Write true Things and Profitable, than false Things that are great. True Things be they never so small, will give occasions to Discover greater things by them."[121] Instead of being grounded in past opinion, natural science was to be a dedicated search after the secrets of nature and the arts that were hidden from the intellect and revealed only to those willing to undertake the laborious search. "You have heard the beginnings," wrote Della Porta; "now search out things, work on them, and put them to the test."[122]

Magic and Court Culture

Della Porta was widely acknowledged as Italy's most distinguished natural philosopher and "scrutinizer of natural secrets" (*perscrutatore dei secreti naturali*).[123] Until Galileo taught a new generation an entirely different way of investigating nature, Della Porta dominated the scene of Italian science. The Antwerp printer Christopher Plantin's superb Latin edition of the *Magia naturalis* (1560) assured the work's wide distribution in scholarly libraries throughout Europe. Translations of the work into Italian, French, Dutch, German, and English established Della Porta's reputation among educated and popular readers alike. It was said that the two greatest tourist attractions of Naples around the year 1600 were the baths of Pozzuoli and Giambattista Della Porta.[124]

However, in Della Porta's day literary fame and status were not measured solely by the number of books a writer might sell, any more than artistic styles were fashioned by popular tastes. More importantly, they were measured by the patronage one received from princes and by the recognition one gained in courts. By this measure, Della Porta succeeded brilliantly. Princes and prelates throughout Italy and Europe recognized him as a master magus and lavished their patronage upon him. The Holy Roman Emperor Rudolf II and the duke of Florence sent embassies, while the duke of Mantua came in person to see the Neapolitan wonder-worker. Toward the end of his life, the old magus recorded that his patrons had given him more than 100,000 ducats for his research, a sum about ten times the annual salary of the Spanish viceroy of Naples.[125] To understand Della Porta's success, we must understand how culture was produced at the courts.

Italy in the late Renaissance was a "honeycomb of princely courts," each a center of autonomous political power and a magnet for ambitious artists and men of letters.[126] The engine that drove the patronage system was the competition among princes and courts for "honor," which took the form of lavish displays of grandeur, wealth, and magnificence. Political power in sixteenth-century Italy rested not just upon military might, but upon the performance of rituals of power: in other words, upon theater and spectacle. The luxurious ostentation of court culture was no mere show; it was a display of the prince's power. "Without such an exhibition, there was somehow no sufficient claim or title to the possession of power."[127] Art and literature were deployed to enhance the self-image of the prince. All public occasions, such as marriages, baptisms, funerals, coronations, and religious processions, gave princes the opportunity to self-glorify and to affirm their titles to rule.[128] The cultivation of self-images made courtly art a form of political propaganda. Since they aimed to enhance the prince's self-image, practically all forms of culture produced by the courts served a propagandistic function.

In a world dominated by courts, princely patronage not only provided monetary support to cultural activity, it was also a mechanism for fashioning social identities. To belong to a court, even to a minor one, was to share in the prestige that went with rendering service to an overlord. Honor—that is, the praise and recognition of a prince—was the reward of servitude and the engine of courtly ambition. Torquato Tasso caught the dynamics of self-fashioning at court in his *Malpiglio*, a dialogue on courtly virtues, wherein one of the interlocutors observes,

"The courtier's end is the reputation and honor of his prince, from which his own reputation and honor flow as a stream from a spring."[129] Patronage was also a mechanism for legitimizing the products of cultural activity, whether works of art, theatrical performances, or scientific books.[130] The style of art or science that won favor in the courts was in large part conditioned by the replication of princely self-images. To be sure, science also served practical needs. Astrologers, engineers, and mathematicians performed important practical services by casting horoscopes, designing bridges and fortifications, and providing expertise on matters such as ballistics. But if in retrospect the prince's primary needs were for technical assistance, from a contemporary perspective even more important was "reputation," since what others thought of him was an important determinant of what he actually was. Machiavelli noted that "a prince ought to show himself a lover of ability (*virtù*), giving employment to able men and honoring those who excel in a particular field." Above all, he wrote, "a prince should endeavor to win the reputation of being a man of outstanding ability (*virtù*)."[131]

Princely *virtù*, rehearsed and extolled ad nauseum in the courtly literature and in dedications of books to patrons, gave rise to a potent cultural ideal, that of the "learned prince." The cult of the learned prince, which deployed the Platonic image of the philosopher-ruler who combines power with wisdom, helped to legitimize the political rule of the Medici dukes and the newly established *condottieri* rulers at Milan, Mantua, Ferrara, and Urbino. *Sapientia/potentia*—wisdom combined with power—became a conventional formula in humanist political rhetoric. As an epigram to Lorenzo de' Medici proclaimed, fate united power with wisdom in the person of the prince: "Because you know everything, O Medici, you are all-powerful" (*Sic sapis, o Medices, omnia sicque potes*).[132]

Among the qualities of the idealized "learned prince" (hence of the ideal courtier), two stand out as being particularly important for their influence upon courtly science: curiosity and virtuosity. The stern, negative medieval attitude toward intellectual curiosity was completely absent from the Renaissance courts.[133] Not only was curiosity considered a virtue worthy of a prince, it was an important symbol of his power. Striking visual demonstrations of the new valorization of curiosity could be seen in the cabinets of curiosities (*Wunderkammern*) that Renaissance princes collected and put on display at the courts.[134] In these fabulous collections, the court projected an aura of the uncanny and the superhuman. Carved gems, watches, antiques, mummies, and mechanical contrivances were displayed side by side with fossils, shells, giants' teeth, unicorns' horns, and exotic specimens from the New World, making up an encyclopedia of the bizarre and the marvelous. Full of strange and exotic *naturalia* and *mirabilia*, they were meant to delight spectators and to provoke wonder rather than to serve as museums for scientific research. Indeed, the curiosity-cabinets were arranged so as to deliberately exclude the normal and the ordinary. Objects were not considered worth collecting unless they were monstrous or had some bizarre peculiarity.[135] As if to mock the Augustinian concept of *curiositas*, these collections invited viewers to gawk at them, to be amused by them, and to be filled with wonder. For that matter, the court itself was a menagerie of bizarre forms, where clowns, dwarfs, and exotic animals mingled amiably with princes and courtiers.

However, the curiosity-cabinets and the living curiosities that inhabited the courts were not mere amusements. They were visual manifestations of the prince's power. In them the prince appropriated and reassembled all reality in miniature, symbolically demonstrating his dominion over the world.[136] Similarly, the intricate courtly botanical gardens, with their complex mathematical designs and their fantastic floral and architectural displays, demonstrated the prince's power to transform nature itself. As a guidebook to one of the gardens expressed it, "in this little theater, as it were a world in miniature, a spectacle will be made of all the marvels of nature." In its design and arrangement, the Renaissance botanical garden was a microcosm reflecting a measured and orderly universe at the prince's command and disposal.[137]

The new attitude toward curiosity, which affirmed the value of inquisitiveness about nature, gave rise to another cultural ideal: virtuosity (*virtuosità*). Put simply, the virtuoso was "the product and fusion of two traditions, of the courtier and the scholar."[138] He was, so to speak, the codification of the "learned gentleman" described by Castiglione at the idealized court of Urbino. Besides being the offspring of court culture, the ideal of virtuosity was also the result of the Renaissance reevaluation of the concept of nobility. According to Castiglione, whose formulation became conventional in the late Renaissance, "true nobility" consisted not of noble birth but of *virtù*, which more than anything else depended upon the acquisition of courtly skills and manners. "The true foundation of nobility is virtue," wrote Girolamo Muzio in his *Gentilhuomo* (1571), the handbook he wrote explaining how to acquire it.[139] As Castiglione's *Il Cortegiano* makes clear, court culture raised the cultivation of *virtù* to a high art. It was this refined, courtly conception of *virtù*, not the traditional Christian concept of virtue, that gave rise to the early modern ideal of virtuosity.

The distinguishing mark of the courtier, according to Castiglione, was *grazia*, or grace, "a seasoning without which all the other properties and good qualities would be of little worth."[140] Essentially identical with elegance, urbanity, and refinement, grace was the highest achievement of culture. *Grazia* may be displayed in any action, but the key to it was an art for which Castiglione coined the term *sprezzatura*, a kind of smoothness and nonchalance that hides the effort that goes into a difficult performance. However, "nonchalance" conveys only part of the meaning of *sprezzatura*. The root of this untranslatable word is the verb *sprezzare*, meaning to scorn or despise. When Castiglione demands that the courtier act with "una certa sprezzatura" toward what is unimportant, he implies acting with an attitude of disdain and scorn for normal human limitations or physical necessities.[141] Castiglione put it down as a "universal rule" of courtly behavior that to achieve gracefulness one must "practice in all things a certain *sprezzatura*, so as to conceal all art and make whatever is done or said appear to be without effort and almost without any thought about it." The more difficult the performance, the greater the possibility of manifesting *sprezzatura*, the art that makes what is difficult seem simple and natural. This is why, when a courtier accomplishes an action with *sprezzatura*, his behavior elicits another characteristic courtly response, *meraviglia*, or wonder: "because everyone knows the difficulty of things that are rare and well done; wherefore facility in such things causes the greatest wonder."[142]

Courtly virtuosity, with its feigned disregard for normal limitations and the high esteem it attached to wonderment, fostered—indeed idealized—a dilettantish approach to intellectual and cultural pursuits. The accomplished courtier did not pursue learning with the diligence of a scholar, nor play the lute like a professional, nor fight like a *condottiere*. He performed everything with *sprezzatura*, which made his actions appear as a pastime, success as a matter of course. Feigning his accomplishments as natural made the courtier seem to be the master of himself, of society's rules, and even of physical laws.[143] Holding himself above the common crowd, disdaining the obvious and the merely useful, he turned his curiosity toward what was obscure, rare, and "marvelous."

If, as Tasso claimed, dissimulation was the essential courtly virtue, natural magic was courtly science par excellence. The magus's essential characteristics—his passionate quest for secrets, his craving for rarities, his cultivation of wonder, and his tendency to view science as a theatrical performance designed to delight and astonish spectators—perfectly fit the courtly manner. To read Della Porta's *Magia naturalis* is to plunge headlong into the world of courtly discourse. The magus assured his readers that his "Catalogue of Rarities" contained nothing base or common, only those exquisite, curious, and ingenious things fitting the taste of a gentleman. Knowledge of such secrets separated "noble minds" from the vulgar, who would only abuse and discredit the art. A certain dissimulation was necessary in order to keep secrets from falling into the hands of rude and ignorant men: "Such as are magnificent and most excellent, I have veiled by the artifice of words, . . . yet not so, but that an ingenious reader may unfold it, and the wit of one that will throughly search may comprehend it."[144] Like the courtier, the magus had to be knowledgeable about many arts. In addition to being a "perfect philosopher," he had to be a physician, an herbalist, and an artisan. He had to be familiar with metals, minerals, gems, and the art of distillation. He had to be an accomplished alchemist and pyrotechnician. He knew physics, mathematics, astrology, and the principles of optics, which taught him how to create visual deceptions. The magus had to know the principles of magic so well that he could artfully conceal the causes of the marvels he produced. For the aim of magic was to provoke wonder in the beholder, which the magus could do only by understanding the causes of marvels well enough to hide them. Imitating the tricks of the trades and theatrical legerdemain confirmed a general rule about natural magic: "If you would have your works appear more wonderful, you must not let the cause be known: for that is a wonder to us, which we see to be done, and yet know not the cause of it. For he that knows the causes of a thing done, doth not so admire the doing of it; and nothing is counted unusual and rare, but only so far forth as the causes thereof are not known."[145]

This was no sober theoretical treatise; it was manual for producing marvels *con sprezzatura*. Like the art of courtly self-fashioning, natural magic used art to bring nature to perfection. "It is nature's part to produce things and give them faculties," wrote Della Porta, "but art may ennoble them when they are produced, and give them many several qualities."[146] The magician was a "servant of nature" who coaxed, cajoled, and extracted from nature what nature could not produce unaided. His art resembled the masterful masquerading of the courtier, whose "artful dissimulation" won the favor of a prince or lady without flattery or force.

Della Porta's apparent fascination with tricks and illusions has made it difficult for modern historians to take the *Magia naturalis* very seriously.[147] He enthusiastically described how to produce half-white, half-black figs, gigantic leeks, nuts without shells, sweet lemons, cucumbers shaped like dragons; how to sire multicolored horses, counterfeit gemstones, or make an artificial egg as big as a man's head. His ardor seems to result more from perverse curiosity than from genuine scientific interest. But to Della Porta and his contemporaries, such amusements were not only prescribed by the rules of courtly discourse, they were considered appropriate categories for the study of natural history. Nature's subtlety and "playfulness" was thought to be an essential part of the world's architecture.[148] Sometimes nature mimics itself, as in the seahorse or the mandrake, while at other times it contorts itself and exhibits monsters, giants, and dwarfs, or it demonstrates its cunning by creating artful imitations of human artifacts. So nature cloaks herself in masks, which it is the task of the naturalist to remove, thereby exposing nature's secrets. The illusions and sleights of hand that Della Porta included in his book were imitations of nature hiding herself. They were part of the unmasking of nature that natural magic aimed to accomplish. Once exposed, nature might be imitated by the art of natural magic. But if nature "plays" for those who understand her secrets, she also deceives those ignorant of causes, as a juggler, magician, or craftsman seems to create marvels in the eyes of amazed onlookers. Illustrating the mechanism underlying cunning, whether human or natural, was an essential part of Della Porta's "science of secrets."

It is significant that besides being Italy's foremost magus, Della Porta was also an accomplished playwright. Between 1590 and 1615, his presence was almost as commanding on the Italian theatrical scene as it was over Italian science.[149] He was a master of the *commedia erudita*, or "learned comedy," which dominated the Italian theater of the Counter-Reformation. The defining characteristic of *commedia erudita*, like that of natural magic, was its delight in *meraviglia*, whether in action, situation, character, or plot. Exaggeration was its prime tool. Tortuous plots and *imbrogli*, characters stylized beyond all pretense to realism, exhibitions of legerdemain, slapstick humor, macaronic language, superfluous disguises, and outlandish caricature were the marks of Della Porta's comic style. Everything was done in an atmosphere of hilarious unreality, with grace, gravity, and *sprezzatura*. His elegant, learned, and ridiculous comedies were expressions of the same courtly virtuosity that produced his natural magic. Both were predicated upon ingenuity, complexity, preciosity, and dexterity. Their trademark was artificiality and invention. Neither merely imitated nature; they surpassed nature, stylizing it into hyperreality.

For Della Porta, the quintessential virtuoso, natural magic no less than comedy was a spectacle. He delighted in exhibiting marvels and took pride in his reputation as a seer and wonder-worker. He dazzled courtly audiences with his prognostications of future events, as when, upon being shown a portrait of Henry IV of France, he predicted the monarch's violent death.[150] But as carefully as he cultivated his reputation as a magus, he nurtured his relations with princes. Natural magic was not only an instrument for fashioning nature according to human desires; it was also an instrument for self-fashioning at the courts. Della Porta told one of his patrons that magic was especially useful to princes because knowledge of the secrets of the universe may be applied to government. The kings of Persia, he noted, studied magic "that by the example of the commonwealth of the whole world, they also might learn to govern their own commonwealth. . . . for as nature governs the world by the mutual agreement and disagreement of the creatures, after the same sort they also might learn to govern the commonwealth committed unto them."[151] An imperious "survey of the whole course of nature," natural magic befitted the image of the prince. Indeed, the magus was in a sense a prince-in-miniature. He could read the secret signs in nature; he understood the physiognomy of things, what their uses were, and what the heavens portended.

Della Porta received offers of patronage from many different princes. In 1579, he entered the service of Cardinal Luigi d'Este and spent several years at the brilliant Estensi court at Ferrara.[152] He wrote plays for the cardinal, sent him reports of his experiments, and dedicated to him the enlarged edition of his *Magia naturalis* (1589). He reported that Este provided over 100,000 ducats to support his research.[153] Della Porta tried to use the cardinal's influence to angle his way into the papal court either as papal engineer or as secretary of ciphers, but was unsuccessful.[154] Other princes also vied for his services. He refused an offer of 150 ducats for one of his secrets from Cardinal Orsini, assuring the solicitous prince that he was being excellently provided for in the Este household.[155] The duke of Mantua came to visit whenever he was in Naples, and stayed at Della Porta's house until three or four in the morning discussing the mysteries of nature. Cosimo II de' Medici, whose avid interest in alchemical and magical secrets made the Tuscan court a center of Paracelsian research, also approached Della Porta. The expectant duke sent the old magus a gift of a gold necklace hoping to entice a few secrets from him.[156]

Potentially Della Porta's most prestigious offer came from Europe's greatest patron of the occult sciences, the Holy Roman Emperor Rudolf II. Although the details of Rudolf's entreaties are not known, more than once he tried to lure the famous Neapolitan magus to his court in Prague. The melancholy emperor, who had turned his court into a laboratory for the study of alchemy, astrology, and the occult sciences, was eager to learn the secret of the philosopher's stone, which Della Porta had supposedly discovered. In 1597, he sent an envoy to Naples to invite Della Porta to Prague.[157] A few years later, in 1604, the emperor sent a letter through his emissary Christian Harmio asking Della Porta to disclose some of his "choicest secrets." Once again he urged the old magus to come to Prague.[158] Della Porta was himself partly responsible for the rumor that he had found the philosopher's stone: years earlier he had boasted to Cardinal d'Este that he was on the verge of discovering the secret, proclaiming it to be "the most beautiful thing on earth."[159] Of course he was unable to comply with the emperor's request. In any case he was nearly seventy years old and was doubtless reluctant to leave

Naples to spend his last years in Bohemia. Instead, he sent the index to his unfinished *Taumatologia*, which was to contain the "quintessence" of his science, promising that the completed work would "open the door to new philosophies, to new speculations, to new and more profound mysteries, which have been for a long time hidden in the most secret recesses of nature."[160]

Della Porta anguished more over the *Taumatologia* than over any of his other writings. In 1609 he reported he was doing experiments on the virtues of numbers for the work. From this point on, the book is mentioned repeatedly in his correspondence. In 1611, he noted that experiments on the five hundred secrets in the work had cost him great pain and expense. In 1612, he wrote to Cardinal Borromeo that he had completed the index, and requested the cardinal's assistance in getting the work printed.[161] Unfortunately, the censors refused to license even the index, with its brief synopsis of the projected book. The manuscript was never published; only fragments of it survive.[162]

The Accademia dei Lincei

In the meantime, Della Porta found a new patron—or rather, a new patron found him. In the spring of 1604, he was visited by Federico Cesi (1585–1630), the eighteen-year-old marchese di Monticello, who was then trying to save from dissolution a recently founded brotherhood of "searchers of the arcane sciences."[163] Motivated by Della Porta's ideal of science as a quest for rare secrets of nature, Cesi had founded the society as an attempt to put these ideals into practice. He was joined in the venture by three companions: Joannes Eck (1577–1620), a Dutch physician; Francesco Stelluti (1577–1651), a mathematician; and Anastasio de Filiis (1577–1608), a student of mechanics. The foursome called their society the Accademia dei Lincei (Academy of Lynxes).[164] In adopting the lynx as their emblem, the Lincei were inspired by the impresa Della Porta had chosen for his 1589 *Magia naturalis* and by the words in the preface describing the natural philosopher, "examining with lynx-like eyes those things which manifest themselves, so that having observed them, he may zealously put them into operation." Explaining the choice of the lynx as the society's emblem, Stelluti wrote that the academy's purpose was to "penetrate into the inside of things in order to know their causes and the operations of nature that work internally, just as it is said of the lynx that it sees not just what is in front of it, but what is hidden inside."[165]

The Lincei dedicated themselves entirely to natural philosophy, vowing never to enter into disputes concerning politics or religion. They pledged not to quarrel among themselves over trivial matters, nor make vain promises, nor boast of their accomplishments.[166] To protect themselves against any interference from Cesi's suspicious father, the duke of Aquasparta, the Lincei kept their deliberations strictly confidential. They adopted secret names and wrote letters to one another using a cipher they learned from Della Porta's work on the subject. Cesi imagined the group as a tightly knit brotherhood of philosophers imbued with the spirit of *lincealità*, a "way of life" (*modo di vivere*) marked by its fervent, almost religious devotion to science. It would be a sort of monastic order devoted to secular rather than religious learning, its members sworn to vows of chastity and completely dedicated to experimental investigation of "the secret miracles of nature."[167] When Eck expressed his desire to marry a woman from his hometown, Cesi absolutely forbade it, sternly reminding him of the principles of *lincealità*.[168]

Such militant devotion to science, inspired in part by the successes of the Jesuit Order, was positively uncourtly.[169] Cesi detested court life. "I loathe the courts and courtiers like the plague," he confided to Stelluti.[170] The Lincean academy was for him a refuge from the intrigue and dissimulation of court circles. It was impossible to be a true philosopher in the courts, he maintained, because the courtier's only aim was to earn the prince's favor. "It is extremely dangerous for the philosopher to fall from his honored station into such a treacherous place full of parasites, buffoons, and flatterers."[171] Cesi spoke from the bitter experience of having seen his beloved academy become the victim of court intrigue. No sooner had the Lincei agreed upon their first projects than his father, suspicious of his son's magical activities, disbanded the academy and drove the young prince's friends from the Cesi household. With the Lincei temporarily scattered, the dejected prince journeyed to Naples to seek the counsel of Della Porta, his intellectual mentor. The famous magus, perhaps remembering his own disbanded Secreti, responded enthusiastically. He encouraged the young marchese's project and dedicated to him his recently completed work on distillation.[172] Writing from Naples, Cesi reported that Della Porta was a "very good friend" of the Lincei, and that he had learned "many wonderful secrets" from him.[173] When Cesi acceded to the dukedom of Aquasparta upon his father's death in 1610, leaving him free to develop his cherished academy, he invited Della Porta to join and installed him as the head of its newly formed Neapolitan branch. The world-famous magus had a name to conjure with; besides, he had a large library, which Cesi hoped he would bequeath to the academy.

The spirit of Della Porta continued to inspire the Lincei until 1611, when Cesi enlisted the Lincei's sixth and soon-to-be most celebrated member, Galileo, who had recently published his sensational *Sidereus Nuncius* (*Sidereal Messenger*). Galileo's electrifying discoveries with the telescope gave the Lincei an entirely new sense of purpose. Almost immediately, Galileo's influence on the society began to eclipse that of Della Porta. The telescope was seen as the perfect symbol of the lynx's keen vision, of the enhanced possibilities of the human intellect, and of the Lincei's scientific objective. Abandoning its dilettantish preoccupation with secrets, the Lincei took up the Galilean program with a renewed sense of mission. Esotericism gave way to an identification with the "republic of letters." Publication became one of the academy's principal aims. Cesi continued to pay his respects to Della Porta, indulging the old magus when he suggested various rituals, vestments, and ceremonies for the society.[174] But when Della Porta attempted to admit into the academy a group of Neapolitan noblemen, the marchese resisted.[175] For while Cesi humored Della Porta, he had become Galileo's disciple. "No one will ever be admitted [to the academy] without your knowledge," he assured the new star of the Lincei, "and those who are will not be slaves of Aristotle or of any other philosopher, but will be noble and free intellects."[176] By the time Della Porta died in 1615, the Lincei were committed to the research program of the Galileans. When the society came under attack for upholding Copernicanism, Cesi protested that "this is not so, as they unanimously claim only freedom in philosophizing about things in nature."[177]

The change in the principles underlying the Lincean research program is clearly discernible in an oration Cesi delivered before the academy's Neapolitan branch in 1616, just a few months after Della Porta's death. In his address "On the Natural Desire for Knowledge," Cesi outlined a new vision for the society. He described the Lincei as an organization whose aim was to support humanity's "innate disposition" for knowledge. It was to be a place where knowledge would be pursued not for profit, honor, or reputation, as in the courts, but for its own sake and for the improvement of humanity's spiritual and material condition. The society, he promised, would endeavor to "remove all the obstacles and impediments to fulfilling this good disposition [for knowledge], setting before itself the keen-sighted lynx as a continual spur and remembrance of that acuteness and penetration of the mind's eyes that is necessary for discovering things, and for examining minutely and diligently, outside and inside, by whatever means, all the objects that exist in this great theater of nature."[178] Turning the academy away from its esotericism and its preoccupation with natural secrets, Cesi dedicated it to "the propagation of knowledge, to the communication and advancement for public utility of our virtuous toil and the results made by them."[179]

There can be little doubt that the main influence behind this new policy was Cesi's new idol, Galileo, whose meteoric rise to scientific prominence was made possible by the publications revealing his telescopic discoveries. Galileo expressed the academy's intentions with these words: "The Lincei are a society of Academicians . . . [who] have as their aim the study of letters, and in particular of philosophy and other sciences profitable thereto, and moreover they expect the more expert [among them] to write and publish their labors, to the benefit of the republic of letters."[180]

Galileo's *Sidereus Nuncius* "was not so much a treatise as an announcement," not only of new discoveries but also of the beginning of a new era in science. "It told the learned community that a new age had begun and that the universe and the way in which it was studied would never be the same."[181] The work proposed "great things for inspection and contemplation by every explorer of Nature. Great . . . because of the excellence of the things themselves, because of their newness, unheard of through the ages, and also because of the instrument with the benefit of which they make themselves manifest to our sight."[182] Ironically, a few months before Galileo's announcement of his discoveries Della Porta dismissed the telescope as being unimportant (*una coglionaria*). He was at the moment preoccupied with other "new things," including perfecting his secret for communicating at a distance, a sort of voodoo-like telegraphy by which pricks in the body would be felt in the same part of the body by another person some distance away.[183] A year later he lamented that his preoccupation with the *Taumatologia* had cost him the credit for the invention of the telescope.[184] Plainly, Galileo was diverting science into a completely different channel from the one in which Della Porta swam.

It was not just that science was passing Della Porta by: the principles governing scientific communication were also changing. Della Porta remained wedded to the tradition of esotericism in science. He believed that the deepest secrets of nature should not be revealed indiscriminately but should be reserved for princes. "Great things are suitable only for great princes," he wrote in his dedication of the *Taumatologia* to Rudolf II. To reveal secrets to the public at large, he warned, would risk profaning them and would possibly result in social calamity. "I am sending [my secrets] to you in manuscript because it does not seem fitting that they should have to die with me," he told the emperor, "nor should secrets of such great price be profaned in the hands of the vilest sort of people that might happen upon them; but Pythagoras would have the king yield to divine sovereignty and make divine mysteries clear to the world."[185] Yet ironically, even as Della Porta urged withholding secrets from the vulgar, he proclaimed himself a great divulger of secrets. And despite his claim that the *Taumatologia* revealed secrets of nature appropriate only for a great prince, he made numerous attempts to have the work printed. The truth is, Della Porta could not get the *Taumatologia* published. It was too hot for the Inquisition to handle, and it was irrelevant to the world of science.

SEVEN

THE SECRETS OF NATURE
IN POPULAR CULTURE

THE flood of books of secrets that streamed from the sixteenth-century presses had two long-term effects upon early modern culture. First, it opened up a burgeoning market for self-help manuals in medicine, the crafts, and the domestic arts. The market for "how-to" books carved out by the professors of secrets was eagerly exploited by a host of surgeons, empirics, and self-styled "experimenters," who toward the century's end began producing dozens of little booklets of secrets for ordinary people. Whereas the professors of secrets had written primarily for a middle- and upper-class audience, the new writers on secrets wrote for the commonest, simplest readers. The *professorini di secreti* (as we might call them) were in fact the mountebanks of Ben Jonson's *Volpone*.[1] They appropriated the language, the empiricist outlook, and many of the recipes from the books of secrets already in circulation. Their pocket-sized pamphlets, consisting of but a single signature folded into octavo size, were widely distributed in the cities and towns, and throughout the countryside of Italy, thus propagating the attitudes and values of the professors of secrets to the people.

Second, the avalanche of books of secrets advanced a new conception of the scientific enterprise: the concept of science as a *venatio*, or a hunt, an aggressive search for the "secrets of nature." Instead of viewing science as consisting solely of logical demonstrations, the professors of secrets tended to think of science as an exploration of unknown territory, a search for "secrets" that lay hidden in the innermost recesses of nature. This conception of science was propagated initially through translations of works by Alessio Piemontese, Leonardo Fioravanti, and Giambattista Della Porta, whose works were known all over Europe. But other writers emulated the precepts laid down by the professors of secrets. As we will see later on, the hunt as a metaphor describing science came into widespread use in the seventeenth century, and was one of the underpinnings of the "new philosophy." Before exploring that territory, however, we must look briefly at how the "secrets of nature" became the common property of the people of Europe.

The Rise of the *Ciarlatani*

A "secret" was supposedly a unique recipe or technique. Often would it bear the name or trademark of its inventor to distinguish it from other, "common" recipes. Fioravanti gave his secrets catchy trade names like *olio angelico* and *siroppo maestra Leonardi*. Ironically, however, by flooding the marketplace with secrets, the professors of secrets made their recipes anything but unique. Livio Agrippa, one of the *professorini di secreti* of the late sixteenth century, declared, "I do not believe the world has ever had such an abundance of secrets as are found in it today."[2] Tommaso Garzoni, always quick to label a fool, never dreamed there would be "such aboundance of them that hunt after strange secrets, insomuch that in Bergamo there starts me up one that vaunted of a secret he had, which would convert the Turk, and would have sold it to a Phisition, and a friende of mine, for a peece of forties if he so pleased, a matter enough (if he shoulde have knowne so much) to have made Fioravanti of Bologna to dispaire in himself, for not placing it among his medicinall toies, under title of Angelicall & divine Elixir Fioravantyne."[3] Alessio Piemontese's worst fears had come true: secrets, once rare and precious, had become "public and common," hence no longer secrets. If "all secrets were equivalent," what was to prevent anyone from making up his own, or from peddling those he lifted from a printed book?

Although printing made the cheapening of secrets possible, it was not the principal cause of it. More important was the commercialization of the economy as a whole.[4] The wheels of commerce turned at all levels of the urban economy, right down to its hub, the piazza. In addition to swarms of consumers and bystanders, the market square drew a motley and colorful assortment of peddlers, petty merchants, pickpockets, cutpurses, and entertainers. It was among such throngs of people, including those gullible enough to buy new secrets and those willing to sell them cheaply, that the recipes in the books of secrets were disseminated.

The piazza was also a focus of popular culture. People went there to watch puppet shows, to see jugglers and fire-eaters, or to listen to ballads, if the singers were not drowned out by the cries of charlatans, chapbook sellers, and chestnut women. They bought pamphlets, pictures, songs, pornography, and the latest pasquinades.[5] Garzoni gave a vivid description of the lively crowd of hawkers and entertainers one might expect to encounter almost any day in the piazzas of the major Italian cities:

Intartenimento che danó ogni giórno li Ciarlatani in Piazza di S. Marco al Populo, d'ogni natione che matina e serà. ordinariamente, ui concore Giacomo Francò Forma con Priuilegio

9. *Ciarlatani* on the Piazza San Marco, Venice. On the stage in the foreground, standing to the left of the lute player, are two *pauliani*, or snake handlers, one of whom is selling a vial of St. Paul's earth. On the lute player's right is Pantalone and a Zanni, while on the stage in the upper right, il Dottore Graziano doffs his hat.

In one corner of the piazza you will see our gallant Fortunato with Fritata entertaining the crowd every evening from four till six spinning yarns, inventing stories, telling fables, making up dialogues, singing extempore, squabbling, making up, dying with laughter, quarreling again, falling on the stage, arguing, and at last passing around the hat to see how much money they can raise with all their polished chatter. In another corner Burattino yelling as if the executioner were flogging him, with his porter's sack on his shoulders and wearing his rogue's cap, shouts to the crowd at the top of his voice. . . . Nor does Zan della Vigna fail to make an appearance with his crowd of clowns and jugglers who, after showing off their art, act like performing monkeys and baboons. There is Catullo with his lyre and the Mantuan dressed like a Zanni, while Zottino sings a sonnet about the French pox with a pretty Sicilian rope-dancer. [Then] Tamburino takes up his post and, after sizing up the crowd like a general inspecting his troops, tries to spin an egg on a stick.[6]

Garzoni, who went on for several pages in this manner, was describing the crowd of entertainers who accompanied the *ciarlatani* (or *ciurmatori*) and *montimbanchi* who set up their portable scaffolds in the Piazza San Marco in Venice, where they performed for throngs of bystanders and sold their salves and nostrums. Fynes Moryson, an English traveler to Italy in the late sixteenth century, described these marginal healers and vendors of secrets:

Italy hath a generation of empiricks who frequently and by swarmes goe from citty to citty and haunt their markett places. They are called *montibanchi* or mounting banckes or litle scaffolds, and also *ciarletani* of prating. They proclaime their wares upon these scaffolds, and to drawe concourse of people they have a *zani* or foole with a visard on his face, and sometymes a woman, to make comicall sporte. The people cast their handkerchers with mony to them, and they cast them backe with wares tyed in them. . . . The wares they sell are commonly distilled waters and divers oyntments for burning aches and stitches and the like, but espetially for the itch and scabbs, more vendible than the rest. Some carry serpents about them and sell remedyes for their stinging, which they call the grace of St. Paule, because the viper could not hurt him. Others sell Angelica of Misnia at twelve pence English the ounce, naming (as I thincke) a remote Country to make the price greater, for otherwise that colde Country shoulde not yealde excelent herbes. Many of them have some very good secrets, but generally they are all cheaters.[7]

Cheats or not, the *ciarlatani* attracted large crowds. The abundance of contemporary references to the mountebanks leaves little doubt they were endless sources of entertainment to the people. Foreign travelers to

Italy were greatly amused by the *ciarlatani*. No visit to Venice would have been considered complete without a walk through the Piazza San Marco to watch their performances. Thomas Coryat, an Englishman who visited Venice in 1608, recorded that the mountebanks "oftentimes ministred infinite pleasures unto me." Each day, he reported, they set up in St. Mark's,

> in which, twice a day, that is, in the morning and in the afternoone, you may see five or six severall stages erected for them. . . . These Mounte-banks at one end of their stage place their trunke, which is replenished with a world of new-fangled trumperies. After the whole rabble of them is got-ten up to the stage, whereof some weare visards being disguised like fooles in a play, some that are women (for there are divers women also amongst them) are attyred with habits according to that person that they sustain; after (I say) they are all upon the stage, the musicke begins. Sometimes vocall, sometimes instrumentall, and sometimes both together. This mu-sicke is a preamble and introduction to the ensuing matter: in the meane time while the musicke playes, the principall Mountebanke which is the Captaine and ring-leader of all the rest, opens his truncke, and sets abroach his wares; after the musicke hath ceased, he maketh an oration to the audi-ence of halfe an houre long, or almost an houre. Wherein he doth most hyperbolically extoll the vertue of his drugs and confections. . . . After the chiefest Mountebankes first speech is ended, he delivereth out his com-modoties by little and little, the jester still playing his part, and the musi-tians singing and playing upon their instruments. The principall things that they sell are oyles, soveraigne waters, amorous songs printed, Apothecary drugs, and a Commonweale of other trifles. The head Mountebanke at every time that he delivereth out any thing, maketh an extemporall speech, which he doth eftsoones intermingle with such savory jests (but spiced now and then with singular scurrility) that they minister passing mirth and laughter to the whole company, which perhaps may consist of a thousand people that flocke together about their stages.[8]

As the antics of the *ciarlatani* suggest, theatrical elements were essen-tial in the marketplace defined by the piazza. In order to attract the throngs of people that became the buyers of their nostrums, the moun-tebanks put on a sort of slapstick comedy, using the characters, devices, and gigs of what would later be called (in politer circles) the commedia dell'arte.[9] Indeed, Garzoni's description of the charlatans clearly sug-gests that the commedia dell'arte was born in the piazza, not in the court (as is often argued). Contemporary descriptions of the *ciarlatani* are practically indistinguishable from those of the more "sophisticated" commedia troupes:

From time to time a company of these gallant men [*ciarlatani*] come into a city with their women, without whom they are given little applause, and spread the word that they will serve the public by selling excellent secrets and presenting a free comedy. They choose a place in the public square, where they put up a scaffold. . . . Every day at an appropriate hour a Zanni or similar character begins strumming an instrument or singing to attract an audience. In a moment another actor appears, then another, and often a woman joins the show. They all perform tricks and mix it up with one another and with the audience. Then comes the head charlatan (*Archiciar-latano*), the seller of secrets, and with a fine manner introduces the great and incomparable glory of his marvelous remedies. Once the money is collected, another charlatan makes his pitch, if he has not gone first, and then *la Signora* sells her dainties or whatever delicacies she has. Finally the head charlatan cries, "Bring on the comedy! Let the comedy begin!" The boxes and trunks are packed, the bench changed into a scene, every charlatan becomes a comedian, and there begins a comic performance that will last around two hours, filling the people with laughter and delight.[10]

More can be learned about the *ciarlatani*'s "secrets" from the medical chapbooks many of them sold, along with their nostrums, in the public squares.[11] Dozens of such pamphlets were printed and distributed for a few soldi apiece. Usually the *professorini di secreti* had their booklets printed up by someone like Prospero Malatesta of Milan (who printed many of them), and licensed for sale in several major cities. The authors of the tracts often identified themselves by their "stage" names, all stock commedia dell'arte characters. Tomaso Maiorini, for example, played Policinella on the mountebank's stage and sold a pamphlet entitled *Frutti soavi colti nel giardino* (Delicate fruits cultivated in the garden). A certain Francesco, who took the part of Biscottino, published a tract called *Giardino di varii secreti* (Garden of various secrets). Pietro Maria Mutii, "il Zanni bolognese," sold a tract called *Nuovo lucidario de secreti* (New illumination of secrets). Another *ciarlatano*, who called himself "the little marquis of Este" (*il Marchesino d'Este*), wrote *Il Medico de' poveri, o sia il gran stupere de' medici* (The physician of the poor, or the amazement of the physicians). Lodovico Monte of Bologna, the author of a tract called "The Little Garland Blossoming with Various Lovely Secrets," played the guitar (*della Chittariglia*), while "Il Pesarino" was a sleight-of-hand artist (*gran Giocator di mano*) who sold a chapbook called *Secreti utilissimi e nuovi* (Very useful and new secrets). "Fioravanti Cortese," the pseudonym used by the author of the *Giardino et fioretto di secreti* (Garden and bouquet of secrets), was obviously a play on the names of two famous professors of secrets. Similarly, "Il gran Piemontese," the author of another medical tract, may have been

10. Popular Italian booklets of secrets:
"Treasuries of Secrets" by Lorenzo Leandro
(above) and Biasio "Il Fagadet" (opposite).
Such booklets were sold in the piazzas of the
major Italian cities by *ciarlatani*, popular
healers, and self-styled "experimenters."

a takeoff on Alessio Piemontese. Biagio, "Il Figadet," wrote a booklet of
secrets "collected from various clever men," while Giovanni Cosson, a
French mountebank who called himself "il Bontempo Francese," wrote
a booklet of practical jokes that "have greatly delighted the French,
Spanish, and Italian princes and gentlemen."[12]

"Dottor Gratiano Pagliarizzo da Bologna," the author of a work on
"new and rare secrets," was undoubtedly modeled on the commedia
mask of Graziano, the quack doctor and astrologer. One can imagine
him in his tight knee-breeches, ruffled doublet, and cloak, holding forth
with his learned platitudes and his ridiculous malapropisms. "He who is
sick cannot be said to be well," he would expound in a mock-serious
tone, parodying the physicians, and he would prove it on the analogy

that he who walks cannot be said to stand still. In another commedia dell'arte scenario Graziano advises a patient with a toothache, "Hold a ripe apple in your mouth and put your head in the oven; before the apple is cooked your toothache will be gone."[13] Another *ciarlatano* styled himself "the son of the great physician to the king of France" (*Giovanni Battista figliola del gran medico del Re de Francia*). The thought of a charlatan parodying a quack doctor in a comedy, then peddling his own quack nostrums to the audience, may at first glance seem ludicrously ironic. Doubtless, the irony was intended. Yet the stark contrast between the physician's elegant but meaningless prattle and the charlatan's instant, surefire remedies struck a responsive chord in the audience gathered around the mountebank. The empiric/charlatan was more deeply connected to the social realities of the people than the official Galenic physician, whose humoral theories were far removed from the rules and beliefs by which most of his patients lived.[14] Then as now, the people wanted action, not an intellectual understanding of their ailments' causes.

The title pages of these pamphlets fairly ring with the cries of the mountebanks extolling to the skies the virtues of their marvelous remedies. Some boasted exotic secrets from distant lands. Benedetto "il Persiano" translated his "marvelous occult secrets of nature" from the Persian language, while "Americano" wrote about "A True and Natural Fountain, from which flows forth a living water fountain of miracles and health-giving secrets, in the reading of which you will draw such desired living water of health. The same fountain reserves other secrets for him of great worth, and offers them to whoever values virtue and is generous with his money when the situation merits it."[15] Many of the chapbooks advertised the wares of empirics like Giovanni Vitrario, called "il Tramontano" (the Northerner), who described himself as a "surgeon and distiller in Rome in the Piazza Navon, at the sign of the Phoenix." Andrea Fontana, another surgeon and distiller, made cosmetics and facial waters for "honored ladies" and offered to teach the art of distillation to anyone interested, exchanging secret for secret. Gulielmo Germerio also offered to teach the art of distillation. He invited interested parties to his house in Venice, where they could see his cabinet of curiosities, which included "ten very stupendous monsters, marvelous to see, among which there are seven newborn animals, six alive and one dead, and three embalmed female infants."[16]

Despite the hype generated by this carnival atmosphere, the booklets contained little that was new. Most of the recipes were lifted from books of secrets already in circulation or else were fabricated on models from the books of secrets. Doubtless, a number of "experimented" folk remedies also crept in, although the nature of the "experiments" supposedly made on them remains a mystery. Most of the recipes were prescriptions for self-cure of common ailments such as headaches, nosebleed, burns, toothache, intestinal worms, bladder stone, ringworm, and various eye ailments. Recipes for "itch and scab" (*rogna*), as Moryson observed, were also popular among the *ciarlatani*'s unwashed customers. Livio Agrippa wrote a whole pamphlet on antidotes to poisons, while Francesco Scarioni published a recipe for Fioravanti's famous balsam. Most of the booklets offered at least one or two abortifacients, a potion to cure impotence, an oil or powder to stanch the flow of blood from a wound, helps for women in labor, and several hair-dyeing recipes. Practically all of the *professorini di secreti* felt obliged to include a recipe for killing bedbugs, for preventing venereal disease after visiting a prostitute, and for predicting whether a pregnant woman will bear a male or female child. Simplified rules of physiognomy adapted from Della Porta and other masters of the science were also popular. Although the majority of the recipes appealed to practical needs, the booklets also described various parlor tricks, practical jokes, and simplified feats of "natural magic," such as how to make a ring dance on a table or a candle burn under water. Popular entertainers like "Il Pesarino" appropriated tricks from works like Della Porta's *Natural Magic* and purveyed them to ordinary people in the piazzas. Giovanni Buger, who styled himself "the great magician (*Giocatore di mano*) to the king of France," called his joke book *Virtus Occulta Perit* (Occult virtue vanishes) and described in it various practical jokes and sleight-of-hand tricks similar to those found in the works on natural magic. Many of these pranks, which were perhaps also part of the mountebank's *burle*, unveil a sense of humor quite foreign to ours. A favorite trick, repeated in many of the booklets, was to make someone fall asleep at the dining table. Others described how to make cooked meat seem full of worms or to make dogs urinate on someone's shoe. Even for the people, magic was being exposed as entertaining hocus-pocus.

Not only did the *ciarlatani* appropriate recipes from earlier books of secrets, they also popularized the values and philosophical views of the professors of secrets. Thus Domenico Fedele (the "little Mantuan") copied word-for-word the dedication from Timotheo Rossello's *Summa de' secreti universali* (1561), extolling humanity's "natural desire" to search for secrets, and used it as a preface for his *Con il Poco farete assai* (With a little you will do a lot). But he left off at the point where Rossello began addressing his exalted patron, and concluded with the words, "I am compelled to present to you, O people! this little garden of lovely flowers, in their variety pretty to look at, but in substance beneficial to the human body."[17] Giacomo Trabia of Bologna also appropriated the language of the books of secrets. The preface to his *L'Esperienza vincitrice, epilogata in diversi secreti* (Experience the Conqueror, summed up in various secrets) was obviously influenced by, among others, Alessio Piemontese's famous autobiography:

The common desire that motivates everyone to want to help the world drives some people to find out new inventions, and others to improve the arts and to add new ways and forms for their use. Thus everyone goes off imagining useful inventions. Being among those who know no better way to demonstrate this desire than to give my fellow man such help that I can, I resolved, submitting to disasters, to changes in climate, and (what is worse) to the censure of malicious men, to take up residence in voluntary exile, solely to investigate and to experiment on secrets deemed most worthy, which could serve both to acquire health and to maintain it. Wherefore having experimented on them many times, and having tried them out with very successful results, I have made a selection of my own inventions and of those of the most prudent and virtuous men, having also made of them many experiences, and of those I certified, I took a selection to the printer;

. . . and the ones that are worthy, easy to make, and cheap, these I dedicate and give to you.[18]

The rise of the *ciarlatani* illustrates several characteristics of popular culture in early modern Europe. The first was its variety and originality. Popular culture was always derivative but was never a mere stamping of elite culture upon a blank slate. It was always a kind of intellectual brico-lage, a construction built up of the bits and pieces of available materials. The *ciarlatani* took recipes from the printed books of secrets and appro-priated the social role of the folk healer. They played on the middle class's fascination with secrets, and they advertised their wares with the improvised antics of the street entertainer. The *ciarlatani*'s gigs be-came, in turn, the masks and *scenari* of the commedia dell'arte.

The emergence of the *ciarlatani* also sheds light on the mechanisms of cultural change, specifically on the transformation of the popular healer in an age of commerce. In the culture of the piazza, the popular healer doubled as an entertainer and a salesman. The mountebank's portable stage was his storefront, his medical theater a form of advertis-ing that was well suited to a society that was becoming more and more commercialized. However, the connection between healing and dra-matic performances was a traditional one. That is how shamans oper-ate.[19] Thomas Coryat reported seeing a charlatan

hold a viper in his hand, and play with his sting a quarter of an houre to-gether, and yet receive no hurt; though another man should have beene presently stung to death with it. He made us all beleeve that the same viper was linealy descended from the generation of that viper that lept out of the fire upon S. Pauls hand, in the Island of Melita now called Malta, and did him no hurt; and told us moreover that it would sting some, and not oth-ers. Also I have seene a Mountebanke hackle and gash his naked arme with a knife most pittifully to beholde, so that the blood hath streamed out in great abundance, and by and by after he hath applied a certain oyle unto it, wherewith he hath incontinent both stanched the blood, and so throughly healed the woundes and gashes, that when he hath afterward shewed us his arme againe, we could not possibly perceive the least token of a gash.[20]

Garzoni observed similar performances, including one by "Mastro Paolo da Arezzo, who appears on the piazza with a great long standard unfurled, on which you can see St. Paul with a sword in one hand and in the other a swarm of hissing snakes." Master Paolo was one of the wan-dering *pauliani*, healers who claimed descent from St. Paul, referring to his experience in Malta, when "there came a viper out of the heat, and fastened on his hand . . . and he shook off the beast into the fire and felt no harm" (Acts 28:3–5). This passage gave rise to the folk belief that Maltese earth was an antidote to the bite of a poisonous serpent. The *pauliani* sold it under the trade name "St. Paul's grace" (*la gratia di S. Paolo*). Yet it was certainly not the folk tradition alone that convinced people of the efficacy of Maltese earth. The theatrical part of Paolo da Arezzo's profession was equally if not more important. "He struck such fear in the crowd," Garzoni related, "that the people trembled, and did not want to leave the city gates without taking some of the powder with them."[21]

The boundaries between public performances, advertising stunts, and traditional healing rituals are not easy to draw in the context of early modern culture. Certainly snake handling was a good stunt; but it was also a tradition associated with the *pauliani*. Self-wounding of the sort Coryat described was another stunt; but it may also have been associated with shamans and their rituals of purification. It is not at all clear how much of this was ritual, how much it involved local healing traditions, and how much was plain ballyhoo. Apparently the quasi-religious elements of folk medicine were gradually becoming commercialized and transformed into advertising gimmicks. In the *ciarlatano*, the folk healer emerges as a "commercialized shaman."[22]

The *ciarlatani* were also, to some extent, isolated individuals. Drifting from city to city in troupes whose composition changed whenever they repacked their trunks, they were relatively free of the normal controls and constraints imposed by society. As vagabonds, they had no legal status. They were accompanied by women of dubious reputation, and in their parodies and jests, they cocked a snook at the establishment. No wonder the authorities regarded the *comici ciarlatani* as dangerous to the moral fabric of the community. Cardinal Paleotti of Bologna declared that the mountebank's comedies were excuses for youths to play truant and to steal from their parents and employers. In times of heresy and pestilence, he warned, the crowding on the piazzas to watch the spectacles was particularly undesirable.[23] In 1565 the governor of Milan prohibited all "Masters and players of comedies, herb-sellers, charlatans, buffoons, Zanni, and mountebanks . . . who are used to mount their platforms and to draw a crowd around them" to play on church feast days or in Lent or to erect their stages near the church.[24] Cardinal Borromeo enacted stern measures in an attempt to censor the performances, but all were unsuccessful. As Paleotti observed, it was practically impossible to censor improvised comedy: "They use the most lascivious and indecent gestures that are not indicated, they introduce women of the town and quite shamelessly they will suddenly devise such things as it would never occur to one to forbid."[25]

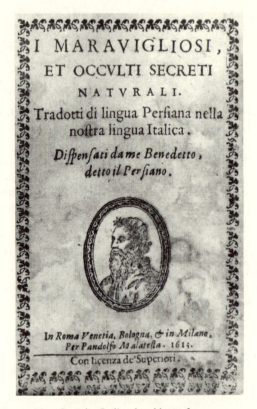

11. Popular Italian booklets of secrets:
The "Marvelous and Occult Secrets" of Benedetto
"The Persian" (above) and the "Garden of
Very Rare Secrets" of Francesco Ricci
(opposite).

These official reactions to charlatanism stemmed in part, no doubt, from economic considerations. To a certain extent the *ciarlatani* competed with the regular physicians. There were already too many doctors, complained the physicians, without unlicensed charlatans peddling their pills and powders in the public square. Some cities required the mountebanks to obtain privileges to sell their nostrums. In 1609, for example, Girolamo di Ferrante, "l'Orvietano," obtained a license from the town government in Florence to "mount a bank with his company and sell a certain secret against poisons in the city square."[26] The remedy, *Orvietano*, was touted as a theriac substitute. Enforcement of such ordinances was fairly lax at the beginnning of the seventeenth century, although as

we shall see, the medical establishment soon intensified its efforts to control the charlatans.

A more important motive for regulating the *ciarlatani* was fear of crowds, of novelty, of strangers, and, perhaps, of laughter itself. Because their profession obliged them to wander, the *ciarlatani* saw things others did not see. They traded in satire, ridicule, and burlesque. Nothing was sacred to them, not doctors, nor priests, nor government authorities. They sold aids to licentious behavior, including treatments for venereal disease and drunkenness, ways to drink large quantities of wine without getting drunk, and remedies "to enable one to go to a prostitute without catching any kind of disease."[27] Their loose moral behavior made a mockery of society's norms. As Mikhail Bakhtin has shown, laughter was an expression of freedom, however ephemeral. Elemental and spontaneous, it threatened order and gave people a momentary victory over the fear inspired by religion and power. While seriousness terrorized, demanded, and forbade, laughter liberated the soul from fear.[28]

Laughter was the stock-in-trade of the *ciarlatani*. They were heralds of truths expressed in curses, satire, and abusive words. Embodiments of criticism and change, they sometimes professed radical or heretical views. As the bizarre case of Costantino Saccardino suggests, the culture of the piazza gave expression to their discontent.

Saccardino, a Roman charlatan, distiller, and sometime court jester, was arrested in Bologna in 1622, along with his son and two other accomplices, and charged with heresy. The four conspirators were accused of having desecrated the sacred statues of the city, soiling the images with excrement and affixing to them placards full of blasphemies and vague threats against the political and religious authorities.[29] The sensational trial that followed revealed the fullness of Saccardino's shockingly heretical beliefs.[30] He scorned the Holy Scripture as a book full of lies and false promises. He denied that Jesus was a true God, and believed that Christ was born of an adulterous relationship: his placards called the Virgin a whore. He ridiculed indulgences, relics, and the cult of saints. He denied that the pope had any spiritual power, and maintained that religion was a lie perpetrated by politicians. He denied free will and asserted that everything occurs by necessity. Most shocking of all were "the strange and ridiculous opinions he held in his confused and corrupt mind about the origin of man (denying that Adam and Eve were the first of all created beings), . . . maintaining that man was born like so many frogs or August toads from the fat earth with the help of the sun's rays."[31]

Before setting up shop as a distiller, Saccardino had been employed as a court jester (*buffone*) in the service of the grand dukes of Tuscany and of the *Anziani* (elders) in Bologna. He played the role of Graziano the charlatan-doctor, playing his guitar, singing, and improvising comedy for the court.[32] He also had a notable reputation as a healer. He was the author of a medical tract entitled *Libro nominato la verità di diverse cose, quale minutamente tratta di molte salutifere operationi spagiriche, et chimiche; con alcuni veri discorse delle cagioni delle lunghe infirmatà, & come si devono sanar con brevità* (Book named the truth of various things, which minutely treats many salutary spagyrical and chemical operations; with some true discourses on the reasons for long-lasting illnesses, and how they can be cured quickly). There were no hard-and-fast lines separating Saccardino's various professional roles as charlatan, healer, court jester, and actor. The one thing that underscores them was his persistent ridicule of the medical establishment. In his *Libro*, he railed against the "idle modern physicians who run away from experimental labors and attend only to learning soothing logical formulas and other rhetorical discourses so they can pretend to give reasons for great invisible, impalpable, and incomprehensible things . . . which can't ever be seen or known." How different are modern physicians from the "wise philosophers" of the "good old days" who, "full of love and charity, used to visit the languishing sick and disinterestedly bring them medical relief made by their own hands; without ambition or pride, not desiring pomp or greatness, with humility they visited so many poor and rich people alike. They did not enter into disputes as some physicians do now, so that often the poor patient succumbs before such bickering and dies in their very presence."[33]

Saccardino's medical philosophy does not easily reduce to "precursors." The "spagyrical" doctrines of the Paracelsians are evident in his tract but are blended with the materialistic metaphysics of Italian naturalism and with elements of popular charlatanism.[34] He cited Mattioli as an authority when he criticized the apothecaries for adulterating drugs. But his main inspiration was Leonardo Fioravanti, from whose works he borrowed repeatedly. He accepted Fioravanti's doctrine that the cause of all diseases was the "bad qualities and intemperance of the stomach." Yet he took the theory much further than Fioravanti had intended it to be taken. Fioravanti's doctrine convinced him that it was possible, contrary to the opinions of the ancient and modern physicians, to find a single cure for all diseases and to preserve the health of all patients, whatever their humoral disposition. For Saccardino, the doctrine of the four humors was the Great Lie of the physicians. It enabled them to sustain the orthodox view that proper treatment had to be based upon a subtle understanding of physiological differences among patients, the kind of theoretical understanding available only to trained physicians. But if all illnesses stemmed from a single cause, the "corruption of the stomach," then all could be cured by a single "universal medicine."[35] It was a theory that perfectly matched the *ciarlatani*'s promise of cheap, fast cures.

Saccardino had been brought before the Inquisition twice prior to his arrest in Bologna: once in Ferrara (1616) and again in Bologna (1616). From the records of these trials we learn that he wandered from city to city proselytizing among artisans, teaching that religion, especially the notion of hell, was pure fakery. "You're baboons if you believe it," he would say. "Princes want us to believe it, because they want to do as they please. But now, at last, the whole dovecote has opened its eyes."[36] As it turns out, this was but one of Saccardino's many borrowings from Fioravanti. In *Dello specchio di scientia universale*, Fioravanti had written of the ancient physicians:

> At that time they got people to believe whatever they wanted, because in those days there was a great shortage of books, and whenever anyone could discourse even a little about *bus* or about *bas* he was revered as a prophet,

and whatever he said was believed. But ever since the blessed printing press came into being, books have multiplied so that anyone can study, especially because the majority of them are published in our mother tongue. And thus the kittens have opened their eyes.[37]

Saccardino took the short step that Fioravanti prudently avoided, transposing the metaphor from the sphere of medicine to that of religion and politics. The crude lies of religion, he was saying, kept the people in check, just as Fioravanti had insisted that the lies of the physicians endangered the people's health and kept the physicians on top. Fioravanti and the other professors of secrets wrote for a safe audience; but when hawked in the piazza by heretics like Saccardino, their ideas posed a threat to the political order. Saccardino the court buffoon put his ideology into action by throwing excrement at religious images and posting placards denouncing the establishment. To the authorities, it only proved that the riotously funny could threaten to turn into a riot. Saccardino ended his life, as he lived it, on the piazza. In 1622 he, his son Bernardino, and two coconspirators were hanged in the public square before a great crowd. Their bodies were burned, the Inquisition's usual method of purging society of incorrigible heretics.[38]

The Diffusion of the
Books of Secrets, 1570–1600

Saccardino's metaphysics, a mosaic of elements from oral and written cultures, is not readily reducible to "antecedents." Nor is it possible, practically, to identify all the sources for the thousands of recipes in the early modern books of secrets. As we know from Alessio's preface and from Ruscelli's history of the Accademia Segreta, some came from books and manuscripts, many of which, doubtless, are now lost. Yet we also know that the professors of secrets learned recipes directly from the common people. Alessio's claim to have traveled the world over to learn secrets from empirics, peasants, and women is reflected in the entire books-of-secrets tradition.[39] Then, almost as soon as the recipes were put into circulation, they were appropriated by empirics and popular writers, who redistributed them in booklets aimed at the common people. In what amounted to a reshuffling and recycling of recipes and "experiments," the oral and written traditions merged in popular culture.

In addition to the *ciarlatani*, a host of surgeons, distillers, and self-proclaimed "experimenters" perpetuated the tradition of writers like Alessio and Fioravanti. In penny-pamphlets and broadsheets, these empirics self-consciously followed in the footsteps of the famous sixteenth-century professors of secrets. Andrea Fontana, a surgeon and distiller from Brescia, produced one of these tracts, calling it "A Fountain Where Water of Secrets Flows" (*Fontana dove n'esce fuori acqua di secreti*). The booklet published a selection of Fontana's medical recipes, an advertisement for his distilled waters, and his "marvelously practiced" cure for cataracts.[40] Fontana claimed he had discovered a facial water that would make a person look ten years younger. He offered to teach the art of distillation (or "occult philosophy," as he termed it) to "honored ladies" who wanted to learn how to make cosmetic waters. To other experimenters, he proposed exchanging secret for secret.[41] Besides being commodities, the secrets-pamphlets advertised the practice of empirics. Vincentio Lauro, another "experimenter in secrets," published a booklet of secrets he had "experimented on and proved" in his private practice in Rome.[42] A number of self-styled "experimenters" published booklets on natural magic. The *Tesoro di varij Secreti* (Treasury of various secrets) by the Venetian experimenter Lorenzo Leandro supposedly contained experiments he had "tried out with great effort and expense of time and money." In reality, he got most of them from existing books of secrets.[43]

The Italian tradition of books of secrets was spread to northern Europe through editions and translations of works by Alessio Piemontese, Fioravanti, and Della Porta, the three most famous Italian professors of secrets. By 1700, twenty-four Latin editions of Della Porta's *Magia naturalis* had appeared, establishing his reputation in the learned community as Europe's leading authority on natural magic. Della Porta's publishers included two of Europe's most prominent printing houses, Antwerp's Plantin press and the Wechel press at Frankfurt. The *Magia* was also translated into French (thirteen editions), English (two editions), German, and Dutch (one edition each). Fioravanti's works were also popular in northern Europe, especially in England. But more than any other, it was the legendary name of Alessio Piemontese and the popularity of his works that insured the widespread diffusion of the Italian books of secrets.

Alessio's *Secreti* was a spectacular best-seller. Within just four years of its appearance, seventeen editions of the work had been published, including translations into French, Latin, English, and Dutch (see table 3). In the 1560s, German, Spanish, and Polish translations appeared. By the end of the century, seventy editions of the work had been printed; another thirty-four appeared in the seventeenth century. The *Secreti* was excerpted by almost every encyclopedist of the early modern period, from Antoine Mizauld to Johann Heinrich Alsted. By the seventeenth century, Alessio's recipes had become such common currency that it is difficult to find a medical or technical recipe book that does *not* include

TABLE 3
Alessio Piemontese, *Secreti*: Editions and Translations,
1555–1699, by Language

	1555–1599	*1600–1699*	*Total*
French	17	11	28
Italian	17	7	24
German	9	6	15
English	12	2	14
Latin	7	3	10
Dutch	4	3	7
Spanish	3	0	3
Polish	1	1	2
Danish	0	1	1
Totals	70	34	104

at least a few of them.[44] Although the work appealed principally to middle-class readers, Pierre Bayle reported that in France the colporteurs sold copies of Alessio's secrets at the village fairs "along with their other little blue-covered books."[45]

The popularity of the original *Secreti* quickly led to the appearance of three additional compilations that went under Alessio's name. The provenance of these works is anyone's guess; mine is that they were commissioned by printers who hoped to capitalize on Alessio's name and reputation.[46] The recipes in the new Alexian corpus are nearly identical in character with those of the original Alessio: most are medical, but others relate to gardening, making cosmetics, preserving fruits, eradicating insects, ink making, cleaning fabrics, and metalworking. What is distinctive about the new Alessio (especially part 2) is the predominance of magical recipes in comparison to the original *Secreti*. Included, for example, are recipes for making marvelous lamps that make things appear green or black, or that ward off wild beasts and protect a house from enchantments. According to the new Alessio, the skin of a hyena or crocodile hung over a gate keeps away thunder and lightning, while the heart of an ape put under the pillow at night causes one to dream of wild beasts. To stop a woman from menstruating, "Take a toad and bind him with a little band, and hang it about the woman's necke that hath the infirmitie, and in a few daies she shalbe cleered of it."[47] Doubtless many of these recipes were copied out of various medieval manuscript collections of magical secrets, such as that of Pseudo-Albertus. Whatever their sources, they are eloquent testimony to the survival of medieval popular magic.

By the time the last part of *Secreti* came out, in 1568, Alessio Piemontese was something of a legend in popular culture, more a trade name synonymous with medical and technical secrets than a real historical figure. In the preface to the last part of the *Secreti*, Alessio gave another account of how he came to publish his secrets. Although this version was somewhat different from the original (nothing, for example, is said about his guilt over causing the artisan's death), it deployed the familiar topoi associated with the literary figure of the professor of secrets. "Alessio" described his lifelong search for secrets, his travels to foreign lands, his consultations with "divers sortes of people," and his dedication to experimenting, "by reason whereof in short space, . . . I became so famous, that I obscured the names of the most expert Phisitions." He told again of his youthful aversion to revealing his secrets, "so much was I blinded in worldly glorie." But now, being old, he wrote, "I coulde no longer absteyne from wryting and publishing them to adorne therewith the worlde, bicause otherwise dying, I shoulde worthily have beene accompted not inferiour unto an homicide." Compared to the final part, which supposedly contained his deepest secrets, Alessio's earlier publications were "trifling things."[48]

Alessio's conversion to the ethics of openness was a central part of the Alessio legend. To many readers, his renunciation of niggardliness and his decision to disclose his secrets contrasted starkly with the attitude of the physicians. Whereas Alessio resolved to freely publish "for the benefit of the world," the physicians, it was said, deliberately withheld knowledge from the people in order to prolong illnesses and to maintain a monopoly over medical practice.[49] These qualities gave Alessio's *Secrets* a kind of moral force that made it especially attractive to middle-class readers. Alessio's English translator, the Puritan physician William Warde, found much that was morally edifying in the work.[50] In dedicating the translation to Francis Russell, the earl of Bedford and a prominent Puritan sympathizer, Warde noted that besides benefiting the people's health, Alessio's work demonstrated deep religious truths. It showed that God, as creator and absolute master of the universe, had implanted in things "certaine secrete vertues, whiche be manifeste signes of goddes love and favoure towardes man." Warde believed this knowledge might serve to combat the resurgence of pagan naturalism, which held that nature alone, not God, was responsible for the secret virtues in things.[51] Warde's *Secretes of the reverende maister Alexis of Piemount* was one of the most popular medical books in sixteenth-century England. Some twenty editions were published between 1558 and 1615, and in 1657 the work was still being listed in William London's *Catalogue of the most vendable books in England*.[52] The work's practical content accounts in large measure for Alessio's popularity, but the moral lesson to be drawn from the Alessio legend could only have enhanced its utilitarian value.[53]

Although Fioravanti was not as famous outside of Italy as Alessio, his reputation as Italy's foremost "spagyrical" philosopher made his books popular among the Paracelsians. French editions of the *Capricci medicinali* appeared in 1586 and 1598.[54] The work also appeared in a seventeenth-century German edition.[55] Since numerous German works by Paracelsus were in circulation, along with the continued publication of Ryff's works, Fioravanti did not have many German followers. In England, however, his works found an enthusiastic translator and advocate in John Hester (fl. 1576–1593), a "spagiricke distiller" who owned an apothecary's shop on St. Paul's Wharf in London. Although Hester is remembered as England's foremost Paracelsian, it would be more accurate to describe him as an English Fioravantian.[56] In all probability he learned his Paracelsus secondhand through some compiler such as Philip Hermanni. Nor was he particularly devoted to the Paracelsian philosophy. As he confessed, he read the chemical philosophers "not for their method, which I meddle not with, but for their medicines, which I usually make."[57] But he was an avowed disciple of Fioravanti. In his style, his practice, and his emphasis upon experience as the "mother of all things," Hester was directly influenced by the Italian empiric. While he expressed no interest in translating the Paracelsian corpus into English, Hester announced his intention to translate all of Fioravanti's works, "in the which," he wrote, "are comprehended the whole art of Phisicke and Chyrurgery."[58]

An autobiographical passage in the Epistle to the Reader of Hester's *Key of Philosophy* gives a vivid account of the making, not of an English Paracelsian, but of an English professor of secrets. After a period of youthful dissipation, Hester related, he thought it "high time to set downe a surer compasse." He told of his eagerness to earn a reputation, "driven at last by a greedy kinde of jelousie, to envie the store that I saw in others, in respect of mine owne penurie: and there withall I fell into consideration how I might become one of the small number of those, whom the greatest number Wondered at." At first he considered going to Oxford or Cambridge, but the seven-year road to the M.A. degree seemed too long. Finally he encountered the two "minions" to whom he would dedicate his life: the Mistress of Mines and Minerals, and the Lady of All That Grows upon the Face of the Earth. "Divers and sundry their affaires have they imployed mee in," Hester wrote, "in the which I have faithfully, painefully, and chargeably applied myselfe, and attained by their instructions (to mine owne destruction almost) manie their hidden secretes as well in Mettalles and Mineralles, as in Hearbes and

Spices."[59] Hester's quest for fame, his abandonment of philosophy, and his devotion to nature and experience were well-worn insignia of the professor of secrets.

Hester's apothecary shop at Paul's Wharf was well established by 1576, when the surgeon George Baker recommended him as London's most skillfull preparer of the new chemical drugs.[60] Fioravanti's oils and balsams were part of Hester's regular stock-in-trade, and he used the Fioravanti translations to advertise his business.[61] In his scathing attacks on the orthodox physicians, Hester proved himself a worthy pupil of Fioravanti. The physician who doesn't know distillation, he charged, "is little or nothing worth, although they take the pacients mony."[62] He praised Fioravanti's discoveries in surgery, which enabled him to cure wounds "in halfe the time which is or hath bene used heretofore, by either ignoraunt or arrogant Professors and Practicioners of that noble and profounde Science, which as they more esteeme a great gaine to them-selves, then a little ease to their pacientes, and a long protracting of the cure for a large payment."[63]

Hester's translations were not, strictly speaking, popular works. They were written for a relatively small audience that included mainly surgeons, empirics, druggists, and distillers like Hester himself. But the Continental books of secrets supplied English popular-science writers with a proven model and a plentiful supply of material for their works. By the middle of the sixteenth century, the production of cheap handbooks of science and technology had become a profitable enterprise. Scientific and technical "secrets" were the staple of the popular-science writers. Leonard Mascall (1546?–1589), a clerk in the household of the archbishop of Canterbury, was an avid collector and compiler of recipes. His publications included works on grafting and the care of orchards,[64] on poultry raising,[65] and on the care and management of cattle.[66] He also produced works on the domestic arts, including a translation of the Dutch *Bouck van Wondre*.[67] In the preface to his book on grafting, Mascall gave expression to the rising sentiment for "hands-on" empiricism in science. He praised the natural philosophy of the husbandman, by which "not only we may see with our eyes, but also feele with our handes in the secret works of nature." To see, smell, and touch nature is to understand nature. What greater pleasure or profit can there be, he wrote, "than to smell the sweete odour of herbes, trees, and fruites, and to beholde the goodly colour of the same, which in certain tymes of the yeare commeth foorth of the wombe of their mother and nourse, and so to understande the secrete operation in the same."[68]

Mascall's popular works were among the first of many household recipe books to appear in England during the last decades of the sixteenth century. Ranging from cookbooks to collections of technical recipes, these works obviously filled a need for practical recipe books in Elizabethan households. Among the most familiar of such works were John Partridge's *Treasurie of commodious conceits, and hidden Secretes* (1584) and *The Widdowes Treasure* (1599), which went through a combined sixteen editions between 1584 and 1629. The two works were in fact the same book (rearranged and published under different titles). In addition to the medical recipes that made up the bulk of Partridge's compilation, the work included recipes for cosmetics, cooking, preserving, dyeing, ink making, and stain removing. Partridge's recipes were not of his own invention but were "gathered out of sundry Experiments, lately practised by men of great knowledge."[69] He took many from the *Secretes* of Alexis and obtained others from contemporary physicians. "Dr. Stevens's water," a cordial found in nearly all the seventeenth- and eighteenth-century English home remedy books, appeared for the first time in Partridge. The recipe, wrote Partridge, was kept secret by Dr. Stevens "till of late, a litle before his death, Doctor Parker, late Archbishop of Canterburie, did get it in writing of him."[70] Other remedies "proved" by leading physicians included "Doctor *Barties* medicine for Aches, with swellings" and a remedy "for the heat in the back, approved by Doctor Huick."[71]

In popular literature, "secrets" shaded into marvels, and marvels into magic. Nature revealed her secrets in a redundant display of prodigies and wondrous events; humans, imitating nature, can produce their own wonders through the art of natural magic. Although magic had always played an important role in folk culture, its efficacy was generally thought to depend upon the "cunning" of individuals long experienced in esoteric practices. Magic was the preserve of wizards who were endowed with special knowledge and had unusual powers. Every town and village had its cunning men and wise women who "could do a thing or two."[72] Magical practices were rites performed at specified times and places, with specific tools, materials, and agents. In short, traditional magic did not consist merely of actions, but of ritual actions performed by extraordinary individuals.

What impact might the publication of magical "secrets" have had upon traditional magic? We might speculate, first of all, that by itemizing magical practices in lists of natural curiosities, the books of marvels presented magical practices as techniques rather than as rites. By being "spelled out," magic lost some of its quasi-religious character. Moreover, exhibiting magic in itemized lists of the "marvels of the world" displayed magical practices as instances of nature, not cunning. In a rudimentary fashion, the books of marvels exposed the underlying "principles" of magic, showing that the effects of natural magic were the results of knowledge about nature (which can be learned from books), not of the peculiar gift of a wizard. Books of marvels made magic accessible to anyone. But if anyone can make magic, is it really magic any longer?

Books of marvels and magical experiments were quite popular in Elizabethan England. The essential character of these works is captured in the opening sentence of Thomas Johnson's *Cornucopiæ, or divers secrets* (1595), which declares, "Manie are the woonders & marvailes in this world, and almost incredible, were it not that experience teacheth the contrarie: for who could bee perswaded to beleeve that the Owstridge could eate or devoure cold & hard Iron, or that hote burning Iron could not hurt her stomacke, were it not that it hath and is daylie seene and knowne."[73] In a manner characteristic of the genre, this declaration is followed by a lengthy list of natural marvels and magical secrets:

> The Satyres have heads like unto men, and bodies like unto goats, and are capable of reason and speech, which is both strange & wonderful.
> The stone found in an Eagles neast, bound to the left arme of a woman with child preserveth from abortion, but bound to her thigh in her travaile, causeth easie and speedie deliverance.
> The hart of the Crowe or a Batte borne upon one suffereth not the partie to sleepe till it bee taken awaie.[74]

And so on. Thomas Hill (ca. 1528–1576), a prolific popular-science writer, mixed curiosities, natural magic recipes, and "conceits" in his *Naturall and Artificial conclusions* (1581), a compendium he compiled "for the recreation of wittes at vacant times."[75] The work included many of the standard magic tricks recorded by Della Porta, Alessio Piemontese, and Pseudo-Albertus Magnus. Although most were designed merely to amaze one's friends, a few, such as the metalworking recipes and the technique for making hens lay through the winter, had practical utility. The anonymous *Booke of Pretty Conceits* (1586) published recipes from various manuscripts and printed books, including many taken directly from Hill. These "very merry, very pleasant, conceits" included such tricks as making fire by directing the sun's rays through a glass of water onto some inflammable material, which, the author wrote, "is very strange to beholde, the rather because Fire which as a hot and dry element, is procured out of Water, which is cold and a moist Element."[76] Examples of sympathetic magic abound in works of this sort. "The tongue of a water Frogge put on the head of one sleeping causeth him to speake in his sleepe," Thomas Johnson attested.[77] Putting a dog's tongue in your shoe is said to keep dogs from barking at you, while anointing the eyes with bat's blood enables one to see at night.[78] To determine if a woman is a virgin, "Burne Mother-Wort, and let her take the smoak at her nose, and if she be corrupt she shall pisse, or else not."[79] Such odds and ends of learning must have fascinated Elizabethan readers. The curiosity-books contained seemingly endless lists of them.

The most remarkable Elizabethan compilation of magical lore was Thomas Lupton's *Thousand notable things of sundrie sorts* (1579), a fantastic miscellany of folklore and popular magic.[80] Lupton, a hack writer, gathered his material from Pliny, Pseudo-Albertus, Della Porta, Alessio Piemontese, Mizauld, Lemnius, Cardano, Ryff, and a dozen or so other authors. He collected from unnamed "old books" and from oral sources, and dished up his "notable things" in bite-sized morsels. Although the work contained a few items of practical information, its overall emphasis was on the magical. Lupton recommended rubbing garlic on the soles of the feet as a remedy for a toothache and prescribed powdered mouse for incontinence. He told of a poet who wore lead shoes to prevent himself from being blown away by the wind, "his body was so light and so lytle." He related numerous other marvels, such as the story of an Italian woman who vomited nails and the butcher's wife who, "as she was stirring the blood of a Beast newly killed, a little thereof did chaunce to sprinkle or spirt on her face: which she with her hand suddainly wiped of, and then wiped the same on her left thigh: Who after being brought to bed of a boy, the same boy had & hath the like mark, or blood spot on the left thigh."[81]

It is difficult to imagine how seriously contemporary readers took these works. Hill characterized his recipes as "conceits," as did the anonymous author of the *Booke of Pretty Conceits*, which seems to suggest they were intended as conversation pieces or as amusements rather than as "scientific" subjects. Yet if the secrets-tradition teaches anything, it is that during its heyday in the sixteenth century, the line between "conceit" and serious science was not clearly drawn. Wonder was a perfectly appropriate response to nature's prodigious variety. Lupton presented magical material in a serious, or at least believing, tone. He assured readers that his reports were "proved," although it is fairly certain he did not himself actually test the recipes. There is little doubt that many of these magical formulas were widely practiced not only in the countryside but also in urban and educated circles. As late as the 1660s, a controversy broke out over the methods of the Irish occult healer Valentine Greatrakes, whose apparent ability to heal by sympathetic touch attracted the attention of many of England's leading intellectuals. The *fact* that Greatrakes cured by touch was not at issue. What was at issue was whether his cures were natural or miraculous.[82] Lupton's magical recipes were fundamentally no more outrageous than Greatrakes's healing methods. Indeed, they were pretty much the same as those found in Della Porta's *Magia naturalis*, a work that was highly regarded among sixteenth-century intellectuals. In matters scientific, the intellectual space separating common readers from the learned was not very wide in the late sixteenth century.

The Elizabethan books of secrets underscored a problem that would be thrust to the forefront of scientific culture in the seventeenth century. The "grain-and-chaff problem," as we might call it, was that of distinguishing between authentic empirical data and spurious testimony. Is it possible to find reliable criteria by which to make this distinction? What constitutes reliable empirical knowledge? These questions became increasingly urgent in the seventeenth century, as natural philosophy gradually dislodged itself from scholastic suppositions and moved toward more experimental methodologies. The books of secrets presented mounds of "data" in the form of recipes, which in effect said, "If you do such-and-so, these results will follow." A "secret" was both a recipe and a prescription for an experiment that, if successful, might reveal some "secret of nature." The results of such "trials" were confirmed by the author's testimony, "I've done it, it happened" (*probatum est*). However, for a variety of reasons such testimonies only obscured the issue. First of all, since the books of secrets were a popular genre, they catered to popular credulity and to the people's fascination with marvels. They made no distinction between magical and technical recipes. They mixed the marvelous with the natural. Moreover, the authors of the books of secrets were suspect. They came from nowhere, were usually without academic credentials, and were often defiant in their criticism of established doctrines and methods. Their view of what constituted "scientific" knowledge was radically unorthodox. Hence the problem of finding reliable empirical knowledge became increasingly one of determining the reliability of the testimony for such knowledge claims. As we will see in the next section, separating the scientific grain from the superstitious chaff was inextricably bound up with the problem of what to do with popular culture.

Popular Errors

The idea that the common people possessed "secrets" making up a body of knowledge unknown to the savants was a popular notion in the sixteenth century. As we have seen, Leonardo Fioravanti believed that the common people's empirical knowledge about the "rules of life" was superior to the medical learning of the schools. Giovanni Battista Zapata reported that the poor people had simple remedies they learned from experience, for which they spent only a few pennies, yet they received the same medicinal benefits as rich men who spent hundreds of ducats for their exotic cures.[83] Della Porta thought popular "superstitions" concealed profound truths about nature, while the Danish Paracelsian Peter Severinus urged studying the "astronomy and terrestrial philosophy of the peasantry." The French jurist René Choppin, in the course of describing a suit brought by the Paris Faculty of Medicine against a peasant woman practicing medicine, commented "how many savants in Medicine have been outdone by a simple old peasant woman, who with a single plant or herb has found a remedy for illnesses despaired of by physicians."[84] Such sentiments were widely shared among early modern intellectuals. As Natalie Zemon Davis has shown, the sense that the people possessed a common wisdom stimulated great interest in popular proverbs and songs, which many writers praised for their "naïveté"—a kind of naturalness they thought should be emulated in prose.[85] Montaigne thought the villanelles of his native Gascony were more beautiful than poetry perfected by art.[86]

The raw empiricism of the "hand-in-the-wound" school elicited two diametrically opposed reactions, one of sympathy and one of extreme hostility. On the one hand skeptics like Montaigne used it to support their assault on scholastic pedantry. "I do distrust the inventions of our mind," wrote Montaigne, "in favor of which we have abandoned nature and her rules."[87] The English antiquarian John Aubrey (1626–1697) believed that folk wisdom contained useful information and perhaps profound truths. "Old customes, and old wives-fables are grosse things," he wrote, "but yet ought not to be quite rejected: there may some truth and usefulness be elicited out of them."[88] Aubrey collected hundreds of examples of folk customs in the belief they might yield useful information. He thought the shepherd's weather prognostications were worth examining, that popular medical recipes contained much good, and that certain folk customs were "relique[s] of Naturall Magick." Citing the methodology of Sir Francis Bacon, who asserted that "the fables of the poets are the mysteries of the philosophers," Aubrey observed that "Proverbs are drawn from the experience and Observations of many Ages: and are the ancient Natural Philosophy of the Vulgar." He collected them, he said, to provide "Instantiae Crucis, for our curious moderne Philosophers to examine, and give διότι to their ὅτις (i.e., justification to their statements)." Investigating one, he found it to be "the profoundest natural Magick, that ever I met with in all my Life."[89]

Seventeenth-century intellectuals took a more critical view of the people's wisdom. Rural customs could hardly stand up to the test of philosophical rationalism or endure the onslaught on "superstition" and credulity. And needless to say, the university-trained physicians did not remain silent in the face of the ridicule aimed at them by the mountebanks and empirical healers. In a series of works on "popular errors" published in the late sixteenth and early seventeenth centuries, the physicians rallied to the defense of medicine, lashing out against "popular superstitions," folklore, and the "errors of the people" in medicine and health. This line of writings was originated by Laurent Joubert, the chancellor of the Faculty of Medicine at Montpellier, who in 1578 published the first volume of a projected series of works on *Erreurs populaires*. In Joubert's judgment, popular culture was shot through with error, ignorance, and superstition. Even when the people did get something right from a medical point of view—in one of their proverbs, for instance—it was always for the wrong reasons. They used bad logic and did not understand rational causes. They dosed themselves with whatever remedies some midwife, empiric, or grandam recommended, so that in the end, "their poor bodies are altered and mixed up by a chaos of remedies, and their minds tossed about by hope and despair."[90]

The main problem, Joubert concluded, was that "everybody makes medicine his business." There were too many "meddlers" who tried to "cut in on a portion of the profession" with their quack remedies and panaceas. He reserved special contempt for midwives, who made extravagant but foolish claims for "secrets" they alone supposedly knew. "What disgusts me," he wrote, "is how these women share among themselves a few small remedies, which, after all, are not even of their own invention but were taken at some time or other from physicians and later passed around among themselves. For women have never invented a single remedy; they all come from our domain or from that of our predecessors. They are very ignorant to think we do not know about these remedies and to think they know more about them than we do." As Joubert's diatribe suggests, in the struggle between orthodox medicine and empirical practice, women healers and midwives became symbolic of the ignorant intruder into the domain of official medicine. The hierarchy of medicine demanded that midwives, surgeons, and barbers be instructed and overseen by physicians. "It is most fitting that [the physician] be present everywhere," wrote Joubert, "in case some complication arises. For all illnesses are within his knowledge and under his jurisdiction. All those who meddle in treating any illness are underlings with respect to physicians, such as surgeons, who have middling jurisdiction, and midwives, who have the lowest."[91]

Joubert's work had considerable impact within the learned community. It went through several French editions and was translated into Latin and Italian. The work was widely imitated. In 1603 the Roman physician Scipione Mercurio (d. 1615) opened up the Italian front in the war against popular errors in his *De gli errori popolari d'Italia*. Mercurio wrote the work in response to Joubert's invitation to physicians everywhere to record popular errors concerning medicine and health

262 SECRETS IN THE AGE OF PRINTING

and add them to his catalog: the physicians declared total war on "superstitions." Like Joubert, Mercurio was especially alarmed about errors committed by women healers and midwives, "because most errors are committed by women, who intrude too much in medical matters."[92] The people also exhibited their foolishness by playing tricks on the physicians, as, for example, when they brought him the urine of an ass or a horse for diagnosis, pretending it came from a patient.[93] Mercurio lashed out against "errors committed in the piazza" by empirics and charlatans, who endangered the public with their ridiculous and often poisonous drugs. He was amazed that people could be so foolish as to credit such remedies as those "made of useless junk and sold in the piazza to the imprudent public, authorized by the presence of a vagabond dressed in velvet and wearing a gold tricorne, approved by a clown, registered by the doctrine of Dr. Graziano, proved by an unbridled whore, sealed by Burattino's jokes, confirmed by a thousand false testimonies, and accompanied by as many lies." Besides being vagabonds, he wrote, the *ciarlatani* have no understanding of the causes of diseases. They imagine that practically all ailments are caused by worms, which they claim their potions will quickly eradicate. In reality, however, diseases have complex causes relating to humoral imbalances. "Since a medication cannot take into account all these things unless it is composed by a very learned physician," he pronounced, "the *ciarlatani*, who are very ignorant, cannot compose them safely."[94]

Joining Mercurio in the assault on the *ciarlatani* was a throng of priests and moralists. The ecclesiastical denunciation of the *ciarlatani* was part of the Counter-Reformation moral crusade against popular culture. The crusade was directed in particular at popular festivities, such as carnival, and at popular activities such as comedy. The carnival, churchmen argued, was unchristian because it contained traces of ancient paganism, because it distracted the people from the Church, and because it encouraged them to licentious behavior.[95] In a series of diatribes, sermons, and proclamations at Milan, Cardinal Carlo Borromeo incessantly denounced comedy for its obscenity, charging that it encouraged lechery, drunkenness, and disrespect for the authorities.[96] Others linked comedy with diabolism.[97] The Roman priest Cesare Franciotti set out with ponderous scholastic argumentation to prove that those who watched comedies committed a mortal sin, an opinion shared by the religious authorities in general.[98] One gains the distinct impresssion that popular comedy bore the brunt of the attack against theater. When the Jesuit Giovanni Ottonelli attacked actors, he carefully distinguished between the *commedianti virtuosi*, the professional actors who played in private houses for the elite, and the *ciarlatani del banco*, who performed from their makeshift stages in the piazza.[99] Despite Ottonelli's moderate stance, the identification of the *ciarlatani* with actors lent powerful moral force to the condemnation of popular healers, who were now singled out as targets of the Counter-Reformation polemic against the vices of the age.[100]

That there was also an economic issue at stake in the moral crusade against popular culture is easy to see. The writers on medical errors all singled out, as particularly objectionable, the unauthorized selling of medical "secrets." The nostrums vended in the open marketplace competed directly with conventional remedies and cut into the physicians' monopoly over the medical marketplace. Mercurio regarded "secrets" as charlatanism's greatest fraud, because they deceived the public and endangered patients. He reported, for example, that the *ciarlatani* made a powder for intestinal worms out of ground coral, which they bought for a lira and sold for more than twenty.[101] But it was not only the ignorant and credulous common people who fell for such frauds. The English physician James Primrose (1592–1659) knew a gentleman who paid twenty pounds for a "secret" he could have bought from an apothecary for a fraction of that amount: "Others I know, who have the *pils of amber, aqua mirabilis,* and many other such remedies, which are to be had in every Apothecaries Shop, and yet they account them as great secrets."[102] In a section of his *Popular Errours* dedicated to exposing "them who are thought to have some secrets," Primrose observed that the empirics often represented common remedies as "great secrets, which they will reveale to no man." This was wisely done, he wrote, because in reality "they have nothing that is worthy the name of a secret." Primrose did not unequivocally deny the value of supposed secrets; instead, like Joubert, he appropriated them, claiming they were invented by physicians in the first place. "Hence it appeares how much many, both men and women here in England, are beguiled, where all do busie themselves in gathering, receits (as they call them) when oftentimes those remedies . . . did at the first come from some physician, who himselfe had nothing that was secret." As long as the physician rationally understood the causes of diseases and the methods for treating them, he had no need for secrets, "for he is able to prescribe as good remedies, yea perhaps better then those secrets are, which are so highly boasted of." According to Primrose, science abolishes the need for secrets. "Those remedies are the best which are no secrets," he insisted, "but best knowne, as being confirmed with more certaine experience."[103]

The most comprehensive seventeenth-century attack on "popular errors" was delivered by the English physician Sir Thomas Browne (1605–1682). In his *Pseudodoxia Epidemica* of 1646, Browne exposed errors not only in medicine and health, but in natural knowledge generally. In writing the work, he was probably influenced by Sir Francis Bacon's appeal for the compilation of a *"calender of popular errors*, . . . that man's knowledge be not weakened nor imbased by such dross and vanity."[104] The *Pseudodoxia Epidemica* takes us away from the heat of the campaign against popular culture and into the quiet study of one whose knowledge of the subject came almost entirely from books. For Browne, popular culture was hopelessly irremediable. He addressed his work not to the people ("whom Books do not redress, and are this way incapable of reduction"), but to the "knowing and leading part of Learning." Like Bacon, he was less concerned with correcting the people's errors than with extirpating the errors that inhibited the advancement of science among the learned.

In the opening pages of the *Pseudodoxia*, Browne discussed in detail the causes of errors. Foremost among them was original sin, which made people naturally credulous and easily deceived. Credulity, "a weakness in the understanding . . . whereby men often swallow falsities for truth," was a condition common to humanity. Skepticism also leads to error, wrote Browne, because it discourages the intellect from inquiry into unknown or uncertain things. Another cause of error is supinity or neglect of inquiry: it is easier to believe what we are told than to investigate for ourselves. Plain laziness accounted for the Scholastics' tendency to record what they had read rather than to inquire into causes for themselves, "and this is one reason why, though Universities be full of men, they are oftentimes empty of learning." But the "mortallest enemy unto knowledge" was blind adherence to authority, especially the tendency to establish beliefs on the "dictates of Antiquity." "There is scarce any tradition or popular error but stands also delivered by some good Author," Browne observed.[105] As prime examples, he singled out some of the most famous professors of secrets, including Girolamo Cardano, Alessio Piemontese, Antoine Mizauld, and Giambattista Della Porta.[106] Cardano, he thought, was "a great Enquirer of Truth, but too greedy Receiver of it." While Della Porta was an excellent experimenter, his *Natural Magic* merely dazzled readers instead of encouraging them to experiment: "Which containing various and delectable subjects, withall promising wondrous and easie effects, they are entertained by Readers at all hands; whereof the major part sit down in his authority, and thereby omit not onely the certainty of Truth, but the pleasure of its Experiments."[107]

Great as the imperfections of the mind were in individuals, thought Browne, they were multiplied a hundredfold in the masses. The common people were "the most deceptable part of Mankind and ready with open armes to receive the encroachments of Error." Feebleminded, illiterate, and governed by the passions and not reason, the people not only fell into error but swarmed with every conceivable vice because of it.

They would sooner believe in their foolish proverbs than accept a reasonable argument. They were incapable of thinking abstractly and thus degraded God himself into physical images. They readily fell for the deceptions of charlatans, quacksalvers, fortune-tellers, and jugglers. Hopelessly deluded, they were even mistaken in their judgments as to sense perception, being convinced, for example, that the moon is bigger than the sun. To Browne, popular culture teemed with superstition, ignorance, and perversion:

> Their individual imperfections being great, they are moreover enlarged by their aggregation; and being erroneous in their single numbers, once huddled together, they will be a farraginous concurrence of all conditions, tempers, sexes, and ages; it is but natural if their determinations be monstrous, and many waies inconsistent with Truth. And therefore wise men have alwaies applauded their own judgment, in the contradiction of that of the People; and their soberest adversaries, have ever afforded them the stile of fools and mad men; and, to speak impartially, their actions have made good these Epithets.[108]

The diatribes against popular errors exposed deep-seated cultural conflicts, and bespoke a widening gulf between learned and popular cultures.[109] It was not just the "errors of the people" that the elite rejected, but their entire culture. Popular errors were seen as instances of the people's delusion and of their tendency to be easily led astray into confusion, heresy, and anarchy. Untrained and ruled by their passions instead of being guided by reason, the common people were like wild animals that threatened the social order: when they could not be tamed, they would have to be caged. The fable of Orpheus taught Bacon that one of the main purposes of philosophy was to convince the vulgar to "forget their ungoverned appetites" and to listen to reason and obey the law. But he held out little hope for it. In times of social unrest, he observed, men would always return to "the depraved conditions of their nature."[110] The argument for the social utility of science did not die out, as we will see below. Increasingly, however, it was framed in the context of the growing split between elite and popular cultures, and by the fear and paranoia that elite culture exhibited toward the recalcitrant masses. Thus Galileo argued that Copernicanism supported traditional biblical exegesis because it presumed the need for a deeper understanding of scriptural passages that were set down in common language for the vulgar, "who are rude and unlearned." Since only theologians were capable of making such subtle interpretations, the new cosmology preserved the distinction between the wise and the multitude. "For the sake of those who deserve to be separated from the herd," he argued, "it is necessary that wise expositors should produce the true sense of such passages, to-

266 SECRETS IN THE AGE OF PRINTING

gether with the special reasons for which they were set down in these words."[111] Nor should scientific discoveries be freely communicated to the general public: "Even if the stability of heaven and the motion of the earth be more than certain in the minds of the wise, it would still be necessary to assert the contrary for the preservation of belief among the all-too-numerous vulgar."[112] In a similar fashion, seventeenth-century English spokesmen for the new philosophy defended experimental science on the grounds that it tempered religious passions, combated sectarianism, and promoted the new political economy.[113]

The books of popular errors also brought to the forefront the age-old question of the reliability of empirical knowledge. The validity of empirical knowledge depends to a large extent upon the reliability of the testimony adduced for it. No individual investigator can personally witness all the empirical or experimental facts that fall into the domain of a complex research problem. Many, perhaps the majority, of those facts are accepted with the understanding that they have been witnessed by other qualified, reliable observers. The question therefore becomes: who qualifies as a reliable observer? In the sixteenth century numerous claims were made that "naive" observers who were unbiased by philosophical opinions were the best witnesses. Montaigne, for example, made this point in his essay "Of Cannibals." Montaigne's informant regarding the strange customs of the New World Indians was "a simple, crude fellow" who had lived for some time in Brazil: "a character fit to bear true witness." In order to get reliable reports about things we haven't seen for ourselves, wrote Montaigne, "we need a man either very honest, or so simple that he has not the stuff to build up false inventions and give them plausibility."[114] Such views contrasted sharply with the new philosophy, whose validity rested on the claim that "naive" empirical knowledge was inherently unreliable. Thus Galileo contrasted the "eyes of an idiot" with those of "a careful and practiced anatomist or philosopher."[115] The sublime mysteries of the universe were beyond the capacities of the common people. Similarly, the books of popular errors raised doubts about the trustworthiness of popular testimony with regard to empirical knowledge. Indeed, not only did the writers on popular errors question the credibility of popular *testimony*, they doubted whether the common people were capable of reliable *seeing*. Thus Browne wrote that because their senses were so dulled, the people "are farther indisposed ever to attain unto truth; as commonly proceeding in those wayes, which have reference unto sense, and wherein there lyeth most notable and popular delusion."[116] By invalidating their testimony, the new philosophy disqualified the people from the arenas where experimental knowledge would be generated.

PART THREE

THE "NEW PHILOSOPHY"

EIGHT

SCIENCE AS A *VENATIO*

THE professors of secrets affirmed the superiority of experience over reason in the search for scientific knowledge. They believed that nature was permeated with "secrets" and occult forces that lay hidden underneath the exterior appearances of things. Neither reason nor authority, nor any of the traditional instruments of inquiry, they insisted, were capable of gaining access to the occult interior of nature. Some new way had to be found to penetrate nature and to capture its secrets.

The "new" scientific epistemology advanced by the professors of secrets was in reality one of the most ancient epistemologies of all: that of the hunter. The hunter of nature's secrets experiences nature as a dense woods in which theory offers a poor guide. Just as the hunter tracks his hidden prey following its spoor, the hunter of secrets looks for traces, signs, and clues that will lead to the discovery of nature's hidden causes. This "art of indications" (*ars indicii*), as Sir Francis Bacon called his experimental methodology, demanded entirely different skills from those traditionally required of natural philosophers.[1] A sharp eye, intuition, good judgment, and sagacity—not theoretical knowledge or facility at logic—were the marks of the scientist as hunter. The "secrets of nature," which were inaccessible to the intellect, could be found out only by long experience in the ways of nature.

The conception of science as a hunt for the secrets of nature was by no means peculiar to the writers of the books of secrets. It was widely shared by sixteenth- and seventeenth-century natural philosophers. "In these centuries," observed Paolo Rossi, "there was continuous discussion, with an insistence that bordered on monotony, about a logic of discovery conceived as a *venatio*, a hunt—as an attempt to penetrate territories never known or explored before."[2] One implication of this idea was that nature was a great uncharted unknown, and that science had to begin anew. Another was that new methods and guides had to be found to help the intellect weave its way through the labyrinth of experience. This chapter looks at the emergence of the conception of science as a hunt for the "secrets of nature," and at the epistemology it generated. In the idea of science as a *venatio* we will trace the roots of the "new philosophy" of the seventeenth century.

New Worlds of Secrets

The metaphor of science as a hunt occurs repeatedly in the scientific literature of the early modern period. In the 1520s, the Neapolitan astrologer Giovanni Abioso urged natural philosophers to turn away from the books of antiquity and to "hunt for new secrets of nature." His motto, *venari nova secreta naturae*, helped shape the ideology underlying the research program of the Neapolitan "academies of secrets."[3] Not content merely to gaze upon nature's exterior, Giambattista Della Porta proclaimed that the goal of the Accademia dei Secreti was to discover occult secrets "locked up in the bosom of nature."[4] His use of the lynx, the keen-sighted predator, as the emblem for his book on natural magic inspired the name of the Accademia dei Lincei. According to one of the society's early members, the Lincei sought to "penetrate into the inside of things in order to know their causes and the operations of nature that work internally."[5] Another academy, the short-lived Accademia Cacciatore (Academy of Hunters) founded at Venice in 1596, adopted as its device the symbol of a dog pursuing a hare.[6]

Not surprisingly, images of the hunt turn up frequently in the Renaissance courts. Hunting was the signorial sport par excellence, "a true pastime for great lords," according to Castiglione.[7] The conception of science as a *venatio* permeated courtly science. The Renaissance princes quested passionately after natural "secrets," especially those pertaining to alchemy and magic. The emperor Rudolf II, Europe's most famous patron of the occult sciences, was an avid collector of secrets. A Venetian observer at the imperial court reported that Rudolf "delights in hearing secrets about things both natural and artificial, and whoever is able to deal in such matters will always find the ear of the Emperor ready."[8] The Medici grand dukes were also zealous seekers of alchemical, magical, and technical secrets. Francesco I de' Medici became famous for his preoccupation with alchemical and technological experiments, which he carried out in a laboratory at the ducal palace.[9] Arriving at the laboratory early in the morning and staying until late at night, Francesco experimented with porcelain, enamel, and majolica; distilled medicinal waters; raised silkworms; and made incendiaries, counterfeit jewels, and china.[10] Francesco's son Don Antonio, a devotee of alchemy, personally conducted alchemical experiments in the *fonderia* he had built at the palace. Don Antonio recorded his experiments in a huge four-volume manuscript containing "secrets" on everything from the transmutation of metals to chiromancy, and from astrology to ballistics. Agnolo della Casa, one of the alchemists active at Don Antonio's laboratory, recorded another eighteen volumes of alchemical experiments.[11] The same fervent interest in secrets consumed Galileo's patron, Cosimo II de' Medici, whose court at Florence was a magnet for every sort of "experimenter" who claimed to have some new "secret."[12] Galileo, eager to win Cosimo's patronage, assured the grand duke's secretary that he too had "particular secrets, as useful as they are curious and admirable, . . . in great plenty. Their very abundance has worked to my disadvantage, . . . for had I but a single one of these I should esteem it highly, and with that incentive I could have interested some great ruler, which I have not hitherto done or attempted."[13]

What do all these these "secrets" and experiments signify? To some extent they attest to an interest in applied science. Many of the secrets recorded experimental attempts to improve artistic or technological processes. But the preoccupation with secrets at the courts also had a political purpose, in that it represented the prince as a repository of preternatural, superhuman secrets, and as the heir to a tradition of esoteric wisdom. In this respect the hunt was a particularly suitable metaphor for courtly science. For just as hunting demonstrated in a spectacular fashion that the goods of the earth existed first and foremost for the prince, so science carried out as a hunt—that is, as a capturing of rare secrets—demonstrated that nature's occult forces existed for the use and delight of the prince.

The logic of discovery implied by the hunt metaphor also influenced the fashioning of professional identities at the Renaissance courts. When a prospective client approached a prince to make a bid for patronage, he was expected to present an appropriate gift, and the patron was bound to reciprocate. Gift giving was not primarily a form of economic exchange, but a statement of status and identity, a process of "self-fashioning" at court.[14] Gifts were tokens of the *amicizia* that sealed the relationship between patrons and clients. According to the ritual of gift-exchange, natural philosophers who hoped to find a place at a court had to invent or discover things they could present as gifts to their patrons. Such discoveries did not have to have high intrinsic value; more important, they had to be unusual or rare: only something novel, exotic, or surprising suited a prince. Rare specimens of plants and animals, mysteriously behaving objects, alchemical recipes, *mirabilia*, exotica, and secrets of any kind were all more appropriate gifts to give princes than practical, technological devices. One of the most spectacular scientific gifts of the era was that of the Medicean stars, which Galileo dedicated to his patron Cosimo de' Medici.

What is significant about these examples is that all were premised on the idea of scientific inquiry conceived as the discovery of *new* things rather than as attempts to demonstrate the known. The theme of novelty appears repeatedly in the scientific literature of the early modern period. Lynn Thorndike observed that words such as *novus* and "un-heard-of" recur in the titles of hundreds of scientific books of the seventeenth century.[15] The *Ne plus ultra* inscribed on the ancient pillars of Hercules became a favorite device to illustrate the tyranny of ancient philosophy over creative thought.[16] The growing awareness that superstitious reverence for antiquity hampered progress in learning magnified the importance of new discoveries and the value of novelty for its own sake.

One of the most important events contributing to Europe's heightened consciousness of novelty was the discovery of the New World—or rather, of new worlds, since the geographical explorations of the era brought back accounts of discoveries made in Asia and Africa as well as in America. News of the discoveries, which revealed regions completely unknown to the ancients, raised Europe's awareness of the sheer immenseness of the world. The explorers brought back specimens of exotic plants and animals, hair-raising tales of adventure, and accounts of completely new cultures. Above all, the discoveries demonstrated that ancient philosophy and science were not necessarily eternal verities. The relations of the voyagers to distant parts of the world seemed to confirm that, as one explorer wrote, "experience is contrary to philosophy."[17] In 1514, the Italian physician Giovanni Manardi pronounced that the problem of whether life was possible in the torrid zone had been put to rest by the Portuguese voyages to those parts. "If anyone prefers the testimony of Aristotle and Averroes to that of men who have been there," wrote Manardi, "there is no way of arguing with them other than that by which Aristotle himself disputed with those who denied that fire was really hot, namely for such a one to navigate with astrolabe and abacus to seek out the matter for himself."[18] Sir Thomas Browne thought ancient philosophy and science were so fraught with error that "the America and untravelled parts of Truth" still awaited discovery.[19]

The New World explorer provided natural philosophers with a heroic self-image. Writing in the late seventeenth century, the Italian surgeon Giuseppe Zambeccari compared his physiological experiments to the explorations of the New World: "I courageously embarked upon this exactly in the same way as the discoverers of the New World, who under the mercy of fortune first entrusted themselves to the sea, without knowing (so to speak) what they were doing and where they were going."[20] The explorer was singled out as the model empiricist, unpretentious in his learning and skeptical of the opinions of the Schoolmen. "The simple sailors of today have learned the opposite of the opinion of the philosophers by true experience," asserted Jacques Cartier.[21] Such pronouncements turn up again and again in the relations of the explorers. The brilliant achievements of the modern navigators brought into sharp relief the limited intellectual horizons of the ancients and demonstrated the superiority of empirical over bookish knowledge. The French explorer Jean de Léry contrasted an experienced seaman's knowledge to that of a scholar: "I have seen one of our pilots, named Jean de Meun of Harfleur, who, although he did not know his ABCs, nevertheless made such progress in the art of navigation through long experience with his charts, his astrolabes, and his Jacob's staff that he silenced at every turn an educated man who had embarked with us and who discoursed about theory with a haughty air."[22]

The "search for secrets" in unknown regions of nature was an image that appeared in the period's scientific literature with monotonous regularity. The promise of revealing "secrets" hidden even from the eyes of the ancients made excellent copy for works aimed at middle-class readers, whose curiosity about novelty was aroused by the revival of magic, the discovery of new worlds, and the revelation of new and bizarre natural phenomena. At the end of the sixteenth century, the French historian Louis Le Roy was positively giddy with excitement as he contemplated the marvels of the new age: "All the mysteries of God and secrets of nature are not discovered at one time. . . . How many have bin first knowen and found out in this age? I say, new lands, new seas, new formes of men, manners, lawes, and customes; new diseases, and new remedies; new waies of the Heaven, and of the Ocean, never before found out; and new starres seen? Yea, and how many remaine to be known by our posteritie? That which is now hidden, with time will come to light; and our successours will wonder that wee were ignorant of them."[23] In the mid-seventeenth century, the English virtuoso Joseph Glanvill (1636–1680) still envisioned the opening up of an "America of secrets and an unknown Peru of nature." As long as we stick to Aristotle, wrote Glanvill, "we are not likely to reach the Treasures on the other side of the *Atlantick*, the directing of the World the way to which, is the noble end of true Philosophy."[24] Like the New World, nature stood before investigators as uncharted territory.

The Encyclopedias of Secrets

One indicator of the widespread acceptance of this view of nature and the epistemology it implied was the publication of numerous encyclopedias of secrets in the second half of the sixteenth century. The early modern period witnessed a rebirth of the polymath. Encyclopedism reached new heights and branched out in a wealth of new directions. The period saw the publication of compendiums of almost every specialized field of knowledge as well as comprehensive works that attempted to unite the various branches of knowledge.[25] Compiling "secrets" from printed books and manuscripts became a respectable occupation for minor writers, whose encyclopedias and miscellanies confirmed the fecundity and diversity of nature and proclaimed the wonders of art.

Antoine Mizauld (ca. 1520–1578), a French physician in the service of Princess Margaret of Valois, was one of the most prolific of these polymaths.[26] Mizauld's voluminous writings included a series of anthologies of arcana and memorabilia bearing titles like *The Little Forest of the Arcana of Nature*.[27] In writing these popular compilations, Mizauld drew encyclopedically from classical and medieval authors as well as from the books of secrets. His works were probably read as much for entertainment as for practical instruction. Nevertheless, they were extremely popular among middlebrow readers and were widely excerpted by vernacular writers of the sixteenth century. Doubtless, his writings were more effective in disseminating the ideas of natural magic in France than the works of Della Porta or Agrippa. No one could read Mizauld without gaining the impression that all of nature was a maze of *mirabilia*.

While Mizauld's works were frankly intended to entertain readers, the Flemish physician Levinus Lemnius (1505–1568) had in mind a more serious purpose in publishing his widely read *Occulta naturae miracula* (*Secret Miracles of Nature*, 1559). Lemnius wanted to preserve the presence of God in nature against contemporary philosophy's tendency to naturalize miracles. But instead of condemning philosophers for attempting to find natural reasons for occult qualities, he praised the attempt and asserted his intention to "search out the demonstrations and causes of [occult] things." Such an effort, he thought, would exhibit all the more clearly God's wonders. Although most natural phenomena could be explained in terms of natural causes, he wrote, "Yet I cannot deny, or gainsay, but there are many hidden and secret things in nature, of an hidden and unknown effect, that it would be undiscreet to attempt to declare the reason and cause of such things." Such phenomena, "which we cannot attain to by the Reason and Judgement of the mind, we cast . . . into the hidden essence and secret properties, and by such refuge we do escape and deliver ourselves out of that Labyrinth."[28]

Yet if there were extraordinary powers in nature, producing a multitude of natural miracles, it was not because of the intrinsic force of nature. Everything proceeds from God, Lemnius proclaimed, "For in the smallest works of Nature the Diety shines forth, and all things are good and beautiful."[29] Like Augustine, Lemnius wanted to preserve nature's divinity and to rescue it from the naturalism that characterized the Italian school of natural magic.[30] Nature is not autonomous; it is the mind of God:

Nature, in which the prints of Divinity do shine forth apparently, is the beginning of every thing, whereby all things consist. Nature is the Mind or Divine Reason, the Efficient of natural works, and the preserver of things that be. Which power can be ascribed to none else but God, and to Christ who is inseparably united with him . . . ; and in him is the life and vigour of things created, and from him is there a living quality infused into all things; That is, by him everything subsists in its natural force, by an inset faculty propagates and maintains itself: there is nothing in so great an Universe that is barren or idle, nothing was made rashly, or by chance, or in vain.[31]

The *Occulta naturae miracula* was an extraordinary compendium of traditional occult phenomena, natural prodigies, herbal lore, and folk beliefs, all deployed to prove that "in the smallest works of Nature the Diety shines forth." To Lemnius, natural wonders also demonstrated moral and political truths. Thus he attributed monstrous births to "unseasonable venery" (copulation during the woman's period), which, he thought, "should teach all men and women to use all decency, and orderly proceedings in their mutual embracings, lest Nature should be wronged thereby." Moreover, just as the wonders of nature showed that God's virtue is diffused through all things, nature proved the reasonableness of the political system in which wise rulers govern through magistrates, "who in executing their offices and publike charges, take great care and pains, whereby they may hold all men in their duties, and all things may be kept peaceably, and the Commonwealth not rent by any Civil broils or seditions."[32]

With Lemnius's deeply pious work, the encyclopedia of secrets and wonders shaded almost imperceptibly into the encyclopedia of prodigies and monsters, examples of which are legion in the sixteenth century. In analyzing these works, Katharine Park and Lorraine Daston observed a shift in the interpretation of monsters in the second half of the sixteenth century. Whereas in the early part of the century monsters were generally seen as portents or signs of God's wrath, later sixteenth-century natural philosophers tended to view monsters as purely natural wonders.[33] The shift toward more secular interpretations of monsters can be illustrated through a comparison between the French and English versions of Pierre Boaistuau's popular *Histoire prodigieuses* (1560). Boaistuau advanced a traditional view, affirming that monsters "do for the most part discover unto us the secret judgement and scourge of the ire of God, . . . which maketh us to feele his marvellous justice so sharpe, that we be constrained to enter into oure selves, to knocke with the hammer of our conscience, to examin our offences, and have in horrour our misdeedes." His English translator, Edward Fenton, on the other hand, regarded monsters as supernatural only in the sense that "they come by the permission or speciall appointment of God." Rather than seeing monsters as instances of God's wrath, Fenton thought they revealed man's ignorance of natural causation. "Touching things supernaturall or above nature," he wrote, "we are to think they are not so cald in respect of nature, as though she had made ought by chaunce, wherof she was not able to yeld a reason, but rather having regard to us, whose weake understanding cannot conceive hir secrete meanes in working."[34] Similarly, Ambrose Paré acknowledged that some monstrous births are evidence of God's wrath; but most, he felt, proceed from such natural causes as too much or too little seed, narrowness of the womb, illness, the imagination, or heredity.[35]

Inevitably, efforts were made to organize this proliferating archive of secrets and marvels. The polymaths not only assembled secrets in compendiums and encyclopedias, they also assimilated them into the contemporary intellectual framework by incorporating them within a tradition of secrets-literature dating back to antiquity. Traditional Renaissance encyclopedias defined the circle of learning (*orbis doctrinae*) and unified the knowledge contained within its parameters; the encyclopedias of secrets expanded the circle of learning beyond the normal, quotidian aspects of nature (which were the subject of traditional *scientia*) to include the extraordinary and the marvelous.[36] They "domesticated" secrets and brought them under the cloak of the familiar.

The German polymath Johann Jacob Wecker (1528–1586) produced one of the most famous of these compilations. His work, *De secretis libri XVII* (1582), was a huge Latin compilation of recipes, experiments, and observations that Wecker extracted from classical authors and from the books of secrets.[37] Wecker's list of 129 authorities was one of the most unusual assortments of authors ever assembled in a learned work. In addition to such familiar classical names as Aristotle, Pliny, Galen, Ovid, and Virgil, it included several famous professors of secrets: Alessio Piemontese, Ruscelli, Fioravanti, Falloppio, Della Porta, and Cortese. The work took recipes from the *Kunstbüchlein* and from the magical books of Hermes, Damigeron, Zoroaster, and Pseudo–Albertus Magnus. Modern authors such as Lemnius, Cardano, and Paracelsus appeared alongside the secrets of Democritus, Orpheus, and Apollo's oracle. In Wecker's list of authors, history seemed to demonstrate that the revelation of natural secrets was a never-ending process. "The treasury of nature is inexhaustible," he wrote, "and undoubtedly much that lies hidden will with the efforts of sagacious men of succeeding times be uprooted."[38] was a manifesto of the program of science as a *venatio* and an encyclopedia of its discoveries.

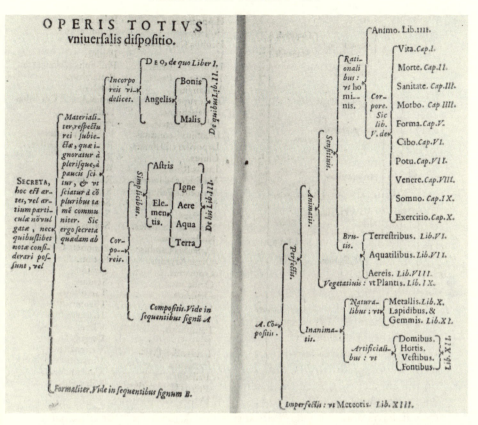

12. Secrets in space: Ramist table of secrets from Johann Jacob Wecker, *De secretis libri XVII* (Basel, 1582). In such works, academic culture appropriated and domesticated popular secrets.

Wecker, a former professor of logic at the University of Basel, was not content merely to list recipes as other writers did. He organized his work according to the principles of Ramist logic, ranking secrets on a descending scale of importance, from the secrets of God to the secrets of games and pastimes.[39] Wecker defined secrets as "arts, or particulars of arts, not common (*vulgatae*) or known by everyone." In Wecker's scheme, secrets fell into two broad categories, depending upon whether they were considered materially, according to the bodies or elements concerned, or formally, according to the arts or sciences concerned. In the first category Wecker treated immaterial bodies (God, angels, demons) and material bodies (the physical sciences). The second branch of secrets included all the arts, from the trivium to carpentry. The subdivisions seem endless, and while Wecker's mushrooming dichotomies doubtless had a ring of authority, the rationale behind his distinctions probably eluded most readers. Encyclopedic in scope, the *De secretis* contained the "secrets" of God, man, and nature; of plants, animals, metals, and machines; of economics, mathematics, and politics; of war and peace; of natural magic and the mechanical arts. Obviously, Wecker's "secrets" were by this time anything but secret, having appeared in dozens of printed books. He paraded before his learned audience many of the familiar antidotes, love charms, medical remedies, gardeners' helps, parlor tricks, and feats of natural magic that had entranced readers of the vernacular books of secrets for half a century. The same tricks to prevent drunkenness that the *ciarlatani* peddled in the city squares were now offered, attired in respectable Latin, to learned readers. Despite the familiarity of its subject matter, the *De secretis* went through another four Latin editions before the century's end and was translated into German, French, and English. The work found a place in the libraries of physicians, natural philosophers, and literary figures throughout Europe.[40] Because it aspired to universal coverage, the work was widely regarded as "an encyclopedia of arts and sciences," as its English translator characterized it.[41]

The conception of science as a *venatio*, a search for the secrets of nature and the arts, was obviously more than just a figure of speech. For many natural philosophers, it was the methodology of discovery at the very core of science. Just how seriously sixteenth-century intellectuals were preoccupied with the discovery of secrets can be gauged by a work on the subject by one of the leading Italian natural philosophers of the day, Girolamo Cardano (1501–1576).[42] A physician who taught medicine at Pavia and Bologna, Cardano is best known for his mathematics and for his massive encyclopedias of the sciences, *De subtilitate* (1550) and *De rerum varietate* (1557). Among his voluminous writings (which fill six volumes in the *Opera* of 1663) is a short tract entitled *De secretis* (On secrets), a sort of theoretical prospectus for science conceived as a *venatio*. First published in 1562 along with Cardano's treatise on dreams, the tract became Tommaso Garzoni's model for his characterization of the "professors of secrets."[43]

Cardano's *De secretis* was essentially an attempt to construct a taxonomy of the various kinds of secrets and of the methods by which they are discovered. What are secrets? According to Cardano, they were of three kinds: unknown things that may eventually come to light; things known to but a few and hence dear; and common things whose causes are unknown.[44] Some secrets are purely speculative and can only be contemplated (*in contemplatione sola consistunt*). Secrets such as the nature of God or the substance of the heavens, for example, exist solely for the delight of the intellect (*sola scientia delectant*). Others, although speculative and not fully intelligible, can nevertheless be put to useful work.

Such occult secrets include magnetism and the virtues of plants and stones. Some secrets can be known directly, while others can be detected only by their effects. There are great secrets of nature as well as mediocre and trivial ones (*magna, mediocria, & levia*). Some are "perfect" (*perfecta*) in that they always produce an expected outcome, while others work only most of the time or rarely. Some secrets are completely illusory, like the tricks of jugglers and ventriloquists.

Having defined a *secretum*, Cardano then went on to analyze the means by which the secrets of nature are discovered. Some are found out by deductive reasoning, others by analogy with things known. We learn many secrets from the authorities; others come by revelation or appear to us in dreams. Some, especially secrets of the mechanical arts, are discovered purely by accident.[45] Cardano was sure of one thing: there were secrets of practically every conceivable subject. He made an ambitious proposal to write the definitive work on secrets, which he envisioned as a massive encyclopedia comprising a hundred books on the secrets of all branches of knowledge. The first would be *De secretis* itself, laying the theoretical foundations of the project. Then would come books devoted to the secrets of the heavens and the earth; of air, wind, and water, and of machines powered by them; of sounds, odors, flavors, and of things "marvelous to the touch" (*tactus miraculis*); of stones, herbs, fish, serpents, and other creatures; of agriculture, medicine, mining, metallurgy, glassmaking, architecture, navigation, distillation, weaving, and other mechanical arts. The project would encompass the secrets of mathematics, rhetoric, the art of memory, and the differing customs and mentalities of foreign peoples. Cardano even envisioned books on the secrets of love and marriage, the care of the family, the education of children, and a book on secrets "of whose kinds we hitherto are ignorant" (*De secretis quorum genera adhuc ignota sunt*). Every conceivable subject had secrets unknown to the ancients and the moderns, and which awaited discovery. Every branch of knowledge was new, unexplored territory. Knowledge was not given; it had to be found out through a great hunt that would mobilize the intellectual energy of generations.

For Cardano, the discovery of secrets was the accomplishment of what he called "subtlety" (*subtilitas*), a concept he articulated, with hundreds of examples, in his massive *De subtilitate* (1550).[46] Subtlety was "a certain intellectual process (*ratio*) whereby sensible things are perceived with the senses and intelligible things are comprehended by the intellect, but with difficulty."[47] Only by *subtilitas* are we able to discover or comprehend secrets. In addition to being a special intellectual faculty, *subtilitas* referred to the nature of things discovered by it. In general, *subtilitas* had to do with the occult causes of phenomena, which are only with great difficulty discovered. According to Cardano, subtlety exists in substances, in accidents, and in representations. In material substances, it included such characteristics as thinness, smallness of quantity, fluidity, and divisibility. In the exploration of incorporeal being it included the secrets of God and the order of the universe. In accidents, having external causes, subtlety was more varied and included the invention of ingenious mechanical inventions or cunningly devised objects. In representations, it included such things as riddles, acrostics, and mathematical conundrums. Subtlety lay at the very edge of perceptibility and intelligibility. It included inquiry into anything from the unusual and unexpected to the difficult or implausible.

The concept of subtlety was supposed to make natural secrets more intelligible. Yet it only raised more difficult epistemological problems, a point the Aristotelian scholar Julius Caesar Scaliger (1484–1558) drove home in a refutation of Cardano's *De subtilitate* published in 1577.[48] Scaliger ridiculed Cardano for his linguistic obscurity and his inability to distinguish between mere words and real things. A notorious polemicist, Scaliger struck a blow at the heart of Cardano's methodology. When treating subtlety, he asked, how can we be sure we are dealing with something real and not simply with ambiguity? Indeed, "secrets" must be the most subtle things of all, for once discovered they become something else. In the final analysis, a "secret" could never really be defined and hence could not in any conventional sense be an object of knowledge. By the very nature of its status as obscure or hidden, a secret defied exact formulation. So long as it remained secret, it could not be known; once made public, it was no longer a "secret."

For a man who published so much, one who so flaunted his discoveries before the public, Cardano was obsessed with secrecy.[49] He was acutely aware that in the age of printing, publication was the route to the fame he coveted.[50] Nevertheless, he embraced the esotericism of the Renaissance magical tradition, of which he was a principal spokesman. "If secrets are divulged and made common, they lose their beauty and dignity," he wrote. "Hence it is permissible to preserve knowledge within oneself, for those [secrets] are more precious, none of whose parts are commonly known. Therefore those that are contained in one person only are more perfect, because one is better than many. . . . Thus the ancient physicians hid their compositions, even giving them beautiful names, so they might possess them as secrets. For a secret is not a secret because it is hidden; it is a secret because it is worthy of hiding."[51]

Cardano betrayed his fascination with secrets in his intense interest in divination. He wrote extensively on the subject, including books on chiromancy, astrology, geomancy, and a work of thirteen books containing studies of "metoposcopy," or physiognomy. He also wrote a work on the interpretation of dreams. In these writings Cardano elaborated a methodology for discovering "secrets." It involved the examination and classification of tiny clues that would lead to the comprehension of a deeper, otherwise unattainable reality. He believed that dreams contained secrets hidden within a maze of apparently irrelevant data. The outward "data" of the dreams—the actual events of the dream—were really coded messages, which, if properly read, would lead to the dream's true meaning. Dreams fascinated Cardano; he thought they held important clues illuminating character and signaling future events.[52] Similarly lines, marks, and warts on a person's hands or face were signs of his inner character. Although this methodology seems rather dubious when applied to a "science" such as metoposcopy, as we will see in the next section, it shared some essential similarities with the "venatic" paradigm that emerged in early modern experimental science.

The Epistemology of the Hunt

In an important essay (which has not received the attention among historians of science it deserves), Carlo Ginzburg discovered in the hunter's methodology the roots of an "evidential paradigm" that Ginzburg contrasted to the Platonic (or Galilean) model.[53] The hunter, who follows clues that lead to an unseen quarry, uses a conjectural methodology. His lore is characterized by the ability to make the leap from apparently insignificant facts, which can be observed, to a complex reality that cannot be experienced directly. "In the course of countless chases [the hunter] learned to reconstruct the shapes and movements of his invisible prey from tracks on the ground, broken branches, excrement, tufts of hair, entangled feathers, stagnating odors. He learned to sniff out, record, interpret, and classify such infinitesimal traces as trails of spittle."[54] The "venatic methodology" was essentially the same as that of the ancient diviner, who used clues found in the guts of animals, in the heavens, or in the flight of birds to look into the future.[55] It was the forerunner of psychoanalytic methodology, which, as Freud put it, "is accustomed to divine secret and concealed things from unconsidered or unnoticed details, from the rubbish heap, as it were, of our observations."[56] Conan Doyle's fictional detective Sherlock Holmes used a similar method of interpretation when he looked for clues in marginal data to reconstruct the circumstances of a crime.

The Greeks called this type of knowledge *mêtis*, by which they meant a kind of practical intelligence based upon acquired skill, experience, subtle wit, and quick judgment: in short, cunning.[57] Mêtis, or cunning intelligence, was entirely different from philosophical knowledge. It applied in transient, shifting, and ambiguous situations that did not lend themselves to precise measurement or rigorous logic. Its stratagems were especially applicable to situations of conflict requiring foresight, quickness, and trickery: hunting and warfare, for example. The scheming intelligence of the sophist and the shrewdness of the politician were varieties of *métis*. So was the knowledge employed by craftsmen, diviners, navigators, and physicians. All were guided not by a fixed method but by the circumstances themselves, and by observable signs. Paracelsus noted that woodworkers "have to understand their wood by chiromancy of it, what it is good and apt for."[58] The navigator conjectures his route through the raging seas by following signs in the stars, winds, and water. The physician looks for signs in the fluctuating sea of symptoms in order to make a diagnosis of the illness he will treat. Primed with knowledge acquired through long experience, like the pilot at the tiller he makes his way by conjecture and seizes the right moment to apply his art. Wholly oriented toward the world of becoming, *métis* applied in situations that demanded action rather than reflection, where there were no ready-made rules for success, and where agility, intuition, and quick judgment were more valuable intellectual traits than facility with logic.

The venatic paradigm contrasted sharply with philosophical knowledge. Greek philosophy posed a radical dichotomy between the worlds of being and becoming, between the intelligible and the sensible. The sphere of being, the unchanging, was the domain of true and definite knowledge. The sphere of becoming, of the sensible and the unstable, was the world of changing opinion.[59] Mêtis or conjectural knowledge, which is oriented toward the sensible world of becoming rather than the intelligible world of being, can have no place in the realm of philosophy understood this way. Hence the conjectural paradigm was pushed into the background by philosophy, erased from the realm of knowledge and relegated, according to the circumstances, to speculation, opinion, or charlatanry. Plato condemned all knowledge based on experience, conjecture, and practical skills, whether it be in medicine, divination, or the crafts. Such knowledge, he maintained, cannot be the object of exact science, whose domain is the absolute and the measurable. While Aristotelian philosophy rehabilitated some forms of conjectural knowledge (e.g., prudence), it did not accept that such knowledge could become the basis of science.

Whereas the classical concept of truth dominated scholastic discussions of method, the venatic paradigm reemerged in seventeenth-century discussions of scientific method. Many natural philosophers rejected the classical idea that only rigorous logic and exact measurement could provide scientific knowledge. Such methods, they argued, offered few insights into the changing, shifting world of becoming. Only through experiments was it possible to make known the vast regions of unknown nature. The hunt metaphor was used increasingly as natural philosophers attempted to elucidate and to vindicate experimentalism. The French savant Pierre Gassendi (1592–1655) described scientific discovery as a way of "sagaciously" (*sagaciter*) examining nature, looking for clues or signs (*media*) that will lead the searcher to the hidden aspects of nature. Knowledge, he wrote, can be about either things manifest or things hidden. The truth that natural philosophy seeks "is not of manifest things since that is public knowledge; nor is it of totally hidden things (*penitus occultae*) since our ignorance of them is invincible." Rather, it is about things that are either temporarily hidden (*occultae ad tempus*) because of some obstacle such as distance, or that are naturally hidden (*occultae natura*), "which cannot become evident by their own nature, or by themselves, but which we can nevertheless know and understand through something else." Although we cannot know such occult things directly, we can know them by means of signs (*signa*) or indicators (*indicia*). Gassendi compared this method to that of the dog who hunts for hidden game by finding a footprint (*vestigium*) or by picking up the scent (*subodorando*) of the prey, and by sniffing his way to where the game is hiding. "This must be something we already know (*primo notum*) and may be called a sign since it leads us to the knowledge of something hidden in the way that tracks are a sort of sign indicating to a dog which way he should pursue the chase in order to catch the quarry."[60] Although the secrets of nature are hidden from the senses, Gassendi maintained, we can know them by their traces.[61]

The most detailed elaboration of the hunt metaphor came, not too surprisingly, from Sir Francis Bacon (1561–1626). Describing the experimental methodology he called "learned experience," Bacon compared it to "Pan's hunt," after the ancient god of hunting. According to the legend, Pan, while hunting, accidentally discovered the hidden goddess Ceres when all the other higher gods failed in their quest after her. Bacon interpreted the myth to mean that "the discovery of things useful to life . . . is not to be looked for from the abstract philosophies, as it were the greater gods, no not though they devote their whole powers to that special end—but only from Pan; that is from sagacious experience and the universal knowledge of nature, which will often by a kind of accident, and as it were while engaged in hunting, stumble upon such discoveries."[62] As a scientific methodology, Pan's hunt proceeds from one experiment to another, as if being guided by an invisible hand, in the same way a hunter tracks his prey deliberately, step by step, guided by footprints and signs. Bacon called this method "a sagacity and a kind of hunting by scent, rather than a science."[63] The experimental scientist is a hunter of the secrets of nature (*venator naturae*, according to Gassendi),[64] whose "sagacity" and vast experience enable him to see things others cannot see. Instead of "groping in the dark," he patiently reads the minute signs and clues that will lead him to his prey hiding in the dense thicket of experience.

The advent of the hunt metaphor in scientific discourse testifies to the emergence of a new conception of the aims and methods of science. Instead of viewing natural philosophy as a sort of hermeneutics—"natural philosophy without nature," as John Murdoch aptly characterized late-medieval physics—intellectuals of the early modern period tended to think of science as a search for new and unknown facts, or of causes concealed beneath nature's outer appearances.[65] This conception of science rested, in turn, upon a redefinition of what constitutes scientific knowledge. In medieval natural philosophy factual knowledge, or knowledge of individual, isolated events, did not qualify as science unless it could be demonstrated that such facts occurred by logical necessity. Medieval natural philosophers "had not dwelt upon phenomena and objects that did not fit within existing theories."[66] Facts were tucked snugly in under the blanket of *scientia*.

Whereas in the scholastic tradition the aim of science was to demonstrate the known, early modern science began to include among its aims the discovery of new and "curious" phenomena. As Bacon expressed it, Aristotle's practice "was not to seek information from unfettered experiment but to exhibit experience captive and bound. He did not introduce a wide impartial survey of experience to assist his investigation of truth; he brought in a carefully schooled and selected experience to justify his conclusions."[67] In early modern science, particularly in the Baconian tradition, facts in the sense of novel, unexplained data—"nuggets of experience detached from theory"—began to take on powerful significance.[68] In the natural magic tradition, such novel, previously unnoticed facts were signs ("signatures") that might guide investigators to nature's deepest arcana. Such "clues" were often marginal, apparently irrelevant, details. Della Porta wrote, "True things be they never so small, will give occasions to discover greater things by them."[69]

To these contrasting images of science—one as logical demonstration and the other as a hunt—corresponded radically different images of nature. One conceived of nature as a geometrical cosmos, a work of God who framed the world *numero, pondere, et mensura*, a reality whose essential features could be known by reason. The other viewed nature as a dense forest, an uncharted domain, a labyrinth in which method offered but a thin thread to orient the hunter, "as if the divine nature enjoyed the kindly innocence of such hide-and-seek, hiding only in order to be found."[70] Renaissance natural philosophy stressed nature's bounty, variety, and changeableness. "The infinite multitude of things is incomprehensible," wrote Della Porta, "more than a man may be able to contemplate."[71] Bacon compared matter to the ancient god Proteus, who dwelt in an immense cave and who constantly changed his shape to avoid capture.[72] Galileo thought nature's effects were so myriad that many existed "not only unknown but unimaginable."[73] Noting that experiments often lead to unexpected and surprising results, Galileo observed "how conclusions that are true may seem improbable at first glance, and yet when only some small thing is pointed out, they cast off their concealing cloaks and, thus naked and simple, gladly show off their secrets."[74]

According to the epistemology of the hunt, since nature's secrets were hidden beyond the reach of ordinary sense perception, they had to be sought out by extraordinary means. Instruments had to be made, for example, which would enable researchers to "look out at and look into" (*auspicit et inspicit*) nature, as the motto of the Lincean Academy expressed it. Experiments had to be devised that would enable researchers to penetrate nature's interior, "twisting the lion's tail" to make her cry out her secrets. As Bacon expressed it, nature, like Proteus, had to be constrained by experiments that forced it out of its natural condition, for "the secrets of nature reveal themselves more readily under the vexations of art than when they go their own way." Finally, new methods of reasoning had to be found to take the place of scholastic logic, which according to the early moderns was incapable of reaching the inner recesses of nature and laying bare its secrets. "Before we can reach the remoter and more hidden parts of nature," Bacon wrote, "it is necessary that a more perfect use and application of the human mind and intellect be introduced."[75]

Bacon and the Hunt of Pan

In the *Novum Organum*, Bacon described the difficulty of knowing nature in the following way:

> The universe to the eye of the human understanding is framed like a labyrinth, presenting as it does on every side so many ambiguities of way, such deceitful resemblances of objects and signs, natures too irregular in their lines and so knotted and entangled. And then the way is still to be made by the uncertain light of the sense, sometimes shining out, sometimes clouded over, through the woods of experience and particulars; while those who offer themselves for guides are (as was said) themselves also puzzled, and increase the number of errors and wanderers.[76]

Nature's opaqueness was a source of profound discouragement to natural philosophers, thought Bacon. For whenever they try to weave their way through the labyrinth of nature, they "fall to complaints about the subtlety of nature, the hiding places of truth, the obscurity of things, the entanglement of causes, the weakness of the human mind."[77] Too easily they bow to authorities instead of striking out on their own. The empirics, on the other hand, commit themselves to the "waves of experience" and proceed without any method at all. Natural philosophy was in the same situation as navigation before the invention of the compass:

> As in former ages, when men sailed only by observation of the stars, they could indeed coast along the shores of the old continent or cross a few small and Mediterranean seas; but before the ocean could be traversed and the new world discovered, the use of the mariner's needle had to be found out; in like manner the discoveries which have been hitherto made in the arts and sciences . . . lay near to the senses and immediately beneath common notions; but before we can reach the remoter and more hidden parts of nature, it is necessary that a more perfect use and application of the human mind and intellect be introduced.[78]

In circumstances so difficult we cannot hope for success by relying upon authorities, Bacon insisted, nor by simply experimenting in a random fashion. "No excellence of wit, no repetition of chance experiments, can overcome such difficulties as these," he wrote. "Our steps must be guided by a clue, and the whole way from the very first perception of the senses must be laid out upon a sure plan."[79]

Bacon devoted the bulk of his work *De augmentis scientiarum* to elaborating such a methodology of discovery. He referred to his methodology as "learned experience" (*experientia literata*) or, using a favorite simile, the "Hunt of Pan." The method was meant to give order and direction to experimentation, so that scientists would not have to waste time "groping in the dark."[80] Learned experience proceeds by analogy from experiments that produce known results to unknown situations in which similar experiments might prove equally fruitful. According to this method, "new knowledge is discovered by ingenious adaptation of existing knowledge, rather than by formal inference from fundamental principles."[81]

Learned experience extends experimental knowledge in eight different ways: by "the Variation, or the Production, or the Translation, or the Inversion, or the Compulsion, or the Application, or the Conjunction, or finally the Chances, of experiment."[82] The variation of experiments (*variatio experimenti*) produces new knowledge by varying the materials, the efficient cause, or the proportions of the materials in an experiment. Thus while paper is normally made of linen, experiments might be tried making it with hair, wool, or cotton. Amber when rubbed attracts straw; varying the efficient cause, will it also attract when heated? If a one-pound leaden ball dropped from a tower reaches the ground in ten seconds, will a two-pound ball reach the ground in five seconds? Experimental knowledge may also be extended by the production of experiments (*productio experimenti*)—that is, by making repeated applications of the same experimental procedure to different situations. Another way was by the translation of experiments (*translatio experimenti*), such as by reproducing artificially situations observed in nature, or by transferring experiments from one art or practice to another. Cider-makers, for example, having observed that grapes in clusters ripen fast, imitate nature by storing apples in heaps to accelerate their ripening.

Inversion of experiments (*inversio experimenti*) takes place when an experiment is devised to explore the opposite effect of the one originally intended. For instance, heat is generated by a magnifying glass; can a type of glass be made to generate cold? Other experiments test the limits of the known properties of bodies by the compulsion of experiments (*compulsio experimenti*). "In the other hunts the prey is only caught," wrote Bacon, "but in this it is killed." As an example of such an experiment, he suggested treating a magnet in such a way that it loses its power to attract iron (say by burning the magnet or by steeping it in aquafortis). The results of certain experiments might also be extended to other experimental situations by the application of experiments (*applicatio experimenti*). Thus, having found out that flesh putrefies slower in some cellars than others, we might devise experiments to discover airs more or less healthy to live in. Or a series of experiments having similar effects might be performed in a chain of applications to produce even greater effects by the conjunction of experiments (*copulatio experimenti*). We can make roses bloom late either by plucking off early buds or by exposing their roots in the spring; by applying both techniques, might we make them bloom even later?

The final method of extending experimental knowledge was by the chances of experiments (*sortes experimenti*). Determined to leave no stone in nature unturned, Bacon urged trying experiments so outlandish that no one has ever attempted them before. Such were the typically "Baconian" experiments that "try anything" and see what happens to nature: "This form of experimenting is merely irrational and as it were mad, when you have a mind to try something, not because reason or some other experiment leads you to it, but simply because such a thing has never been attempted before. . . . For the *magnalia* of nature generally lie out of the common roads and beaten paths, so that the very absurdity of the thing may sometimes prove of service."[83]

Although Bacon characterized learned experience as an inductive methodology, he did not mean induction as it was commonly understood, which was to him "a puerile thing [that] leads to no result."[84] The traditional interpretation of Baconian induction—a mechanical procedure advancing by minute degrees from the observation of particulars to the formulation of general laws—is far from the method Bacon actually recommended. The methodology he had in mind worked not by simple enumeration of particulars, but by analogy and conjecture. The creative imagination played an important role in this process.[85] Learned experience relies on the use of analogies to relate the known results of previous experiments to other experiments and fields of inquiry. Having assembled natural and experimental histories, the investigator then searches through them for clues—that is, instances that stand out in one way or another to guide the researcher to the discovery of how nature works. Such clues, which Bacon calls "prerogative instances," provide shortcuts to the discovery of the nature of things.[86] In recognizing prerogative instances, the investigator has to rely on the imagination to fill in the gaps in existing experimental histories. Sometimes he must make intuitive leaps from the seen to the unseen. Analogy is employed, for example, "when things not directly perceptible are brought within reach of the sense, not by perceptible operations of the imperceptible body itself, but by observation of some cognate body which is perceptible."[87] This does not mean that "anything goes," or that the imagination is given free reign. "The understanding must not . . . be allowed to jump and fly from particulars to axioms remote and of almost the highest generality." Rather, it should proceed gradually by successive steps from particulars to lesser axioms, and from lesser axioms to generalities. Nor was learned experience intended merely to produce more experiments, but to lead "by an unbroken route through woods of experience to the open ground of axioms."[88]

Bacon thought existing experimental methods were "blind and stupid." Some experimenters concentrated on too narrow a range of experiments. The alchemists, for example, wasted their time trying to make gold, while William Gilbert (1544–1603) wrote an entire experimental treatise on the magnet (*De magnete*, 1600). "No one successfully investigates the nature of a thing in the thing itself," Bacon observed; "the inquiry must be enlarged, so as to become more general." Other experimenters proceeded without any method at all. "Wandering and straying as they do with no settled course, taking counsel only from things as they fall out, they fetch a wide circuit and meet with many matters, but make little progress."[89] Without mentioning specific authors, Bacon aimed this criticism at the natural magicians and the professors of secrets. Although he praised natural magic for its attention to experiment and endorsed its operative aim, Bacon condemned natural magic on ethical grounds. He accused its practitioners of fraud and megalomania, and attacked their preoccupation with provoking wonder. The natural magicians, he wrote, persisted in seeing themselves as privileged illuminati whose knowledge was the product of a cunning beyond the reach of ordinary intelligence. Hence they deliberately obscured their discoveries in enigmatic language. Even when it produced some genuine effect, natural magic did so only to excite wonder at its novelty rather than to serve a useful purpose. "The mark of genuine science is that its explanations take the mystery out of things," he wrote. "Imposture dresses things up to seem more wonderful than they would be without the dress."[90]

Bacon threw the professors of secrets into the same batch of "talkers and dreamers" that included the natural magicians, the alchemists, the astrologers, and the diviners. By devoting themselves exclusively to rarities and conceits, "those who promise to reveal secrets" neglected experiments on common, ordinary things. Like the Aristotelians, they took ordinary things for granted. "What in some things is accounted a secret has in others a manifest and well-known nature," he observed, "which will never be recognized as long as the experiments and thoughts of men are engaged in the former only." Bacon also took the professors of secrets to task for failing to understand the proper purpose of experimentation, which was to lead to the discovery of laws of nature. "My course and method," he insisted, "is . . . not to extract works from works or experiments from experiments (as an empiric), but from works and experiments to extract causes and axioms." Bacon denounced the practice, characteristic of the books of secrets, of randomly compiling experiments in books. He insisted that experiments must be properly reported and arranged in tables, or "histories," designed to lead investigators to axioms or laws of nature. Moreover, he redefined the purpose of experimentation. Experiments, he thought, were of two kinds: "experiments of fruit" (*experimenta fructifera*), recipes designed to produce some particular effect or useful purpose in the mechanical arts, and "experiments of light" (*experimenta lucifera*), which are of no use in themselves but simply serve to discover the natural causes of some effects. He insisted that experiments of light were far more important in natural philosophy than experiments of fruit. Whereas the professors of secrets conceived of experiments as tests of whether a recipe worked, for Bacon an experiment was an attempt to find an entrance into "the secrets of nature's workshop."[91]

In an important sense, Baconian induction was an attempt to translate *mêtis*—whether it be the artisan's cunning or the natural magician's intuition—into a method. As we have seen, *mêtis* was a kind of knowledge for which no recognized method existed in the philosophical tradition. For the ancient philosophers, *mêtis*, or conjecture, was nothing more than guesswork, which could not possibly lead to certainty.

Bacon's learned experience, by contrast, was essentially an attempt to define a rigorous methodology for conjecturing from the seen to the unseen aspects of nature, and from effects to causes. Bacon condemned divination and natural magic, which supposedly conjectured by a kind of intuition or cunning that was beyond ordinary intelligence. He wanted to reduce cunning to a rule, and to provide an orderly and systematic way of proceeding from particulars to axioms. This required extensive experience with nature, which could be gained only through the combined efforts of many researchers working together. Such efforts, Bacon believed, would enable researchers to discover in experiments clues that would lead them to nature's interior, just as Pan, having long experience in the ways of nature, fell upon Ceres apparently by accident, but in reality by a kind of "sagacity." This is why Bacon called his method "a sagacity and a kind of hunting by scent, rather than a science."[92] Only in Bacon's scheme Pan's "sagacity" would be replaced by learned experience, an orderly method that began with the compilation of experiments and observations, then proceeded to the discovery of prerogative instances, and ended with the eduction of axioms and laws of nature.

In his utopian tract *New Atlantis*, Bacon described an imaginary research institute called "Solomon's House," which embodied the ideals of his program.[93] In certain respects Solomon's House resembled the Italian academies of secrets. It was equipped with various laboratories, furnaces, and workshops where experiments could be undertaken. It had a well-stocked pharmacy, an herb garden, and a distillery. It employed various servants and workers to carry out the purely manual operations involved in its research. The research program at Solomon's House bore a striking similarity to those of the Italian academies. "The End of our Foundation," he wrote, "is the knowledge of Causes, and secret motions of things (*et motuum, ac virtutum interiorum in Natura*); and the enlarging of the bounds of Human Empire, to the effecting of all things possible."[94] The Solomonians' goal of understanding the "secret motions" and the interior virtues of nature brings to mind the aims of Ruscelli's Accademia Segreta and of the early Lincei, while their utilitarian motive recalls Isabella Cortese's assertion that humans naturally try to supersede nature. Finally, like the Accademia Segreta, the fellows of Solomon's House swore an oath of secrecy and consulted among themselves over which experiments to publish and which to keep secret.

Despite these similarities, Solomon's House represents a fundamental departure from the academies of secrets. Its methods were radically different from those of the earlier academies. Solomon's House was a fictional model for the implementation of the Baconian program. Its teams of researchers were organized according to a hierarchy reflecting the ascending stages of Baconian induction. At the bottom of the scale the Merchants of Light gathered the books of secrets from which the society drew its experiments. At the next level, the Depredators selected experiments from the books of secrets, while the Mystery-Men (or *Venatores*, hunters) collected experiments from the mechanical arts. The Pioneers or Miners tried out new experiments. The Compilers recorded them into tables and prepared them for induction. Next, the Dowry-men or Benefactors (actually practitioners of learned experience) drew new experiments out of existing ones. Then the Lamps, who were searchers after prerogative instances, directed new experiments. The Inoculators executed these experiments. Finally, at the highest level, the Interpreters of Nature "raise the former discoveries by experiments into greater observations, axioms, and aphorisms."[95]

The *New Atlantis* was one of Bacon's most popular works. It appeared in eight editions between 1626 and 1658. The English public was probably more familiar with Bacon's utopian tract than with any of his more philosophical works. The *New Atlantis* provided a model for numerous schemes for creating a national scientific institution and had an important influence on the foundation of the Royal Society of London. One possible reason for the work's impact was that its way was paved, so to speak, by the popular experimentalism of the books of secrets. These works, which were avidly read by the virtuosi, familiarized Bacon's readers with experimental practices and gave them a sense of the possible kinds of experiments that might be done. Yet many of the virtuosi became convinced that random experimentation of the kind found in the books of secrets led nowhere. Some way or method had to be found to mobilize experimental activities toward the common goal of discovering nature's hidden causes. Natural philosophy needed a method that would lead, as Bacon put it, "through the woods of experience to the open ground of axioms." The methodology the *New Atlantis* so effectively depicted seemed to answer this need. As we will see in the next section, the discovery of the "secrets of nature" by experiment became the main objective of the seventeenth century's "new philosophy."

The Secrets of Nature in the New Philosophies

The early modern natural philosophers often distinguished themselves from the medieval Scholastics by insisting that instead of basing their science upon commonsense observations, they attempted to discover the "hidden causes of things." Thus Bacon criticized the Scholastics for ending their investigations "where sight ceases. . . . Hence all the working of the spirits enclosed in tangible bodies lies hid and unobserved. . . . And yet unless these . . . things . . . be searched out and brought to light, nothing great can be achieved in nature."[96] Francesco Stelluti declared that the Lincei wanted to understand "the operations of nature that work internally, just as the lynx is said to do with its look, seeing not only that which is on the outside, but also that which is hidden inside."[97] Sir Walter Raleigh (ca. 1552–1618) pointed out that "the schoolmen were rather curious in the nature of terms, and more subtile in the distinguishing upon the parts of doctrine already laid down, than discoverers of any thing hidden."[98] Logic was unhelpful for finding out new things, Raleigh asserted. Experiment alone (by which Raleigh meant natural magic), gave access to nature's interior and revealed the "secrets of nature." Raleigh contrasted the "brabblings of the Aristotelians" with the methods of natural magic, "which bringeth to light the inmost virtues, and draweth them out of Nature's hidden bosome to humane use."[99]

As these passages suggest, the epistemology of science as a *venatio* rested upon a distinction between knowledge of nature gained by common sense, which revealed only nature's outer appearances, and knowledge of the inner causes of phenomena. Early modern natural philosophers understood this difference in terms of the distinction between manifest and occult qualities, a problem that was the focus of heated controversy in the sixteenth and seventeenth centuries. Keith Hutchison has pointed out that the program of the new philosophies of the seventeenth century was to explain occult qualities in mechanical terms rather than banishing them from the dominion of science, as the new philosophers claimed scholastic science had done. In other words, seventeenth-century natural philosophers sought mechanical explanations for phenomena that, they insisted, the Scholastics conceded were in principle unknowable.[100] While Aristotelian physics offered explanations for "manifest qualities" such as tastes and colors, argued the new philosophers, it could not coherently explain insensible qualities such as magnetic force. Such "occult qualities" were taken to be beyond the boundaries of scholastic *scientia*. Seventeenth-century natural philosophers, on the other hand, fully accepted the reality of occult qualities. Moroever, the new philosophers "saw their acceptance of such occult qualities as one of the marks of the superiority of their new philosophy over then-orthodox systems of thought. They saw Aristotelianism as unable to handle occult qualities because it placed too much emphasis on the importance of sensation, and failed to solve the central epistemological paradox posed by occult qualities: How can a science based on sense perception handle agencies which by very definition are insensible?"[101]

Far from banishing occult qualities, if anything the new philosophers banished manifest qualities. All qualities are occult, they argued, but are nevertheless knowable.[102] The English atomist Walter Charleton (1620–1707) maintained that to the casual observer "all the Operations of Nature are meer Secrets."[103] Joseph Glanvill made precisely the same point. Attacking Aristotelianism for its reliance upon common sense instead of experiment, Glanvill asserted that nature always works through invisible causes. "The most common phenomena may be neither known, nor improved, without insight into the more *hidden* frame," he wrote, "For *Nature* works by an *Invisible Hand* in all things."[104] We never directly observe the causes of phenomena, the new philosophers maintained. Nature's effects alone are observable, from which we must conjecture probable causes. "Many things, which are most obvious and open to the Sense, as to their *Effects*," wrote Charleton, "may yet be remote and in the dark to the Understanding, as to their *Causes*." Charleton attacked the Aristotelian doctrine of occult qualities, which he claimed the Scholastics used as a refuge to hide their ignorance of causes. "That ill-contrived Sanctuary of Ignorance, called OCCULT QUALITIES," he argued, prevented the Scholastics from investigating nature's hidden workings: "Instead of setting their Curiosity on work to investigate the Causes [of a problem], they lay it in a deep sleep, with that infatuating opium of Ignote Qualities; and yet expect that men should believe them to know all that is to be known, and to have spoken like Oracles . . . ; though at the same instant, they do as much confess, that indeed they know nothing at all of its Nature and Causes. For, what difference is there, whether we say, that such a thing is Occult; or that we know nothing of it."[105]

In the new philosophies, the concept of occult qualities was not an ending point but a beginning of inquiry. However, the new philosophers maintained that the opaqueness of nature prevented reason from plumbing its depths. "Nature is an immense Ocean, wherein are no Shallows, but all Depths," wrote Charleton, "and those ingenious Persons, who have but once attempted her with the sounding line of Reason, will soon confess their despair of profounding her."[106] Similarly Glanvill pointed out that only nature's grosser ways of working are sensible. Her machinery is forever hidden from the unaided senses: "Nature is set a going by the most *subtil* and *hidden* Instruments; which it may be have nothing *obvious* which resembles them. Hence judging by visible appearances, we are discouraged by supposed *Impossibilities* which to *Nature* are none, but within her Spear of Action. And therefore what shews only the outside, and sensible structure of Nature; is not likely to help us in finding out the *Magnalia*."[107] But if occult qualities were in principle knowable, as the new philosophers claimed, by what means might they be known? If reason and the unaided senses were incapable of unlocking the door to "nature's workshop," where was the key to that locked door to be found?

Seventeenth-century natural philosophers were in general agreement that they could gain access to nature's secrets only by adopting a two-fold strategy that consisted of right method combined with instruments to aid the senses. The English Paracelsian John Webster (1610–1682) expressed a view of Peripatetic philosophy that most proponents of the new philosophy would have accepted. The principal defect of scholastic natural philosophy was its superficiality, Webster argued, for it "is only conversant about the shell, and husk [of nature], handling the accidental, external and recollacious qualities of things, confusedly, and continually tumbling over obscure, ambiguous, general and equivocal terms, . . . but in no way consorting or sympathizing with nature it self." Getting at nature itself was possible only by experiment, "the only certain means . . . to discover and anatomize nature's occult and central operations." Although Webster's Paracelsianism and his vigorous defense of natural magic were somewhat out of step with the new philosophy, many contemporaries concurred in his assessment of the Aristotelians, who, Webster observed, "think they can argue Dame Nature out of her secrets, and that they need no other key but *Syllogisms* to unlock her Cabinet."[108]

Instead of expecting nature to "follow us into our Chambers, and there in idlenesse communicate her secrets unto us," the new philosophers insisted that nature had to be interrogated by methodical inquiry and "anatomized" by experiment. In the 1680s, Robert Hooke (1635–1703), the Royal Society of London's curator of experiments, formulated a detailed set of research principles to guide the society's virtuosi. Hooke's *General Scheme, or Idea of the Present State of Natural Philosophy* embodies many of the ideals characterizing the new experimental philosophy of the seventeenth century. In spite of numerous improvements over Scholasticism, Hooke asserted, natural philosophy was still defective and in need of reform. In attempting to account for these defects, Hooke noted that humans gain information about nature solely by sensations. The senses, however, are deficient because they give us information only about nature's exterior. "Man is not indued with an intuitive Faculty, to see farther into the Nature of things at first, than the Superficies and out-sides, and so must go a long way about before he can be able to behold the Internal Nature of things." Hooke maintained that logic was useful for some purposes, but "as to the Inquiry into Natural Operations, what are the Kinds of secret and subtile Actors, and what the abstruse and hidden Instruments and Engines there made use of, may be; It seems not, to me, as yet at all adapted and wholly deficient."[109]

Hooke contended that to overcome these deficiencies natural philosophy needed a new method—as he put it, a "Philosophical Algebra, or an Art of directing the Mind in the search after Philosophical Truths," which would make it possible "for the meanest Capacity acting by that Method to compleat and perfect it."[110] Hooke's "philosophical algebra" would consist of two main parts: helps to the senses and new rules of reasoning. Since the foundation of natural knowledge is sensation, he argued, we must first ensure that the mind is supplied with accurate sense information. The natural defects of the senses first had to be identified, and then ways of overcoming those defects had to be found. Hooke proposed making a detailed examination of the sensual faculties so that natural philosophers would have a better understanding of how empirical knowledge was limited and how these limitations might be corrected. He then outlined a method for collecting empirical data, arranging them into tables and histories, and preparing them for the induction of axioms. Not only are the unaided senses inadequate, Hooke observed, but nature itself is deceptive. Hence even when our perceptions are accurate, they cannot always be taken at face value. Experiments would help surmount this problem. In the hunt for the secrets of nature, wrote Hooke, "the footsteps of Nature are to be trac'd, not only in her *ordinary course*, but when she seems to be put to her shifts, to make many *doublings* and *turnings*, and to use some kind of art in indeavouring to avoid our discovery."[111] The experimental methodology Hooke proposed was essentially the same as Bacon's learned experience, although his explication of the method was informed by an intimate familiarity with the nuances and hazards of experimention.[112]

In addition to proper experimental methods, Hooke maintained that the senses had to be armed with instruments to enhance their powers of perception. Foremost among such instruments were the telescope and the microscope. In his important treatise on the microscope, the *Micrographia* (1665), Hooke argued that with the aid of such instruments, "the subtilty of the composition of Bodies, the structure of their parts, the various texture of their matter, the instruments and manner of their inward motions, and all the other possible appearances of things, may come to be more fully discovered; all which the antient *Peripateticks* were content to comprehend in two general and . . . useless words of *Matter* and *Form*." The microscope and the telescope revealed worlds completely unknown to Aristotle and the Scholastics. Every new improvement revealed things never known before, Hooke rhapsodized, "producing new Worlds and *Terra-Incognita's* to our view." Hooke even supposed that such instruments would enable humankind to regain the original acuteness and perfection of the senses that Adam lost with the Fall, and thus reverse the effects of original sin: "By the addition of such *artificial Instruments* and *methods*, there may be, in some manner, a reparation made for the mischiefs, and imperfection, mankind has drawn upon it self, by negligence, and intemperance, and a wilful and superstitious deserting the Prescripts and Rules of Nature, whereby every man, both from a deriv'd corruption, innate and born with him, and from his breeding, and converse with men, is very subject to slip into all sorts of errors."[113]

Hooke was optimistic that the microscope would reveal the mechanical causes of phenomena, enabling natural philosophers to "discern all the secret workings of Nature, almost in the same manner as we do those that are the productions of Art, and are manag'd by Wheels, and Engines, and Springs, that were devised by humane Wit." By exposing the interior "machinery" of nature, the microscope would also confirm the principles of the mechanical philosophy. This, he believed, would go a long way toward resolving the ancient problem of occult qualities. By means of instruments to aid the senses, Hooke wrote, the virtuosi "find some reason to suspect, that those effects of Bodies, which have been commonly attributed to *Qualities*, and those confess'd to be *occult*, are perform'd by the small *Machines* of Nature, which are not to be discern'd without these helps, seeming the meer products of *Motion, Figure*, and *Magnitude*."[114] Henry Power (1623–1668), another champion of the microscope, pointed out that the invisible aspects of nature, though the least known, are "the things that govern Nature principally." Power believed that the microscope would enable the virtuosi to see "what the illustrious wits of the Atomical and Corpuscularian Philosophers durst but imagine, even the very Atoms and their reputed Indivisibles and least realities of Matter, . . . and those infinite, insensible Corpuscles which daily produce those prodigious (though common) effects amongst us." Armed with the microscope, Power asked, "who can set a *non-ultra* to [natural philosophy's] endevours?"[115]

Like Bacon, Hooke attempted to elucidate a method whereby discovery would not have to depend upon the genius or cunning of a few investigators, but would rely instead upon the dedicated work of teams of researchers. With the new tools of natural philosophy, he asserted, discovery "will not be so much the Effect of acute Wit, as of a serious and industrious Prosecution" of method. The philosophy of the schools worked in the opposite manner, Hooke asserted. The Scholastics grounded their doctrines on a few observations and studied "more to gain Applause and make themselves admired, or the Head of some Sect, . . . than to perfect their Knowledge, or to discern the Secrets of Nature."[116] In order to overcome the danger of such dogmatism, the new philosophers insisted upon basing natural philosophy upon experimental "matters of fact." Glanvill praised the Royal Society for rejecting preconceived theories and opinions and grounding its research upon experiments: "So that the relations of your *Tryals* may be received as undoubted *Records* of *certain events*, and as securely be depended on, as the *Propositions* of *Euclide*."[117] Not only was experiment the safest and surest way of understanding nature, it was also the most satisfying. Glanvill stressed the sheer delight of doing experiments: "Yea, whether they succeed to the answering the particular *aim* of the *Naturalist* or not; tis however a *pleasant* spectacle to behold the *shifts, windings* and *unexpected Caprichios* of distressed *Nature*, when pursued by a *close* and *well managed Experiment*."[118]

The repeated references to the "secrets of nature" in seventeenth-century scientific literature should not be dismissed as mere rhetoric. Far from being a mere hackneyed metaphor, the continual appearance of that well-worn phrase indicates a fundamental shift in the direction of natural philosophy. The concept of nature's "secrets"—that is, the idea that the mechanisms of nature were hidden beneath the exterior appearances of things—was the foundation of the new philosophy's skeptical outlook, and of its insistence upon getting to the bottom of things through active experimentation and disciplined observation. The Scholastics had been too trusting of their senses, the new philosophers asserted. Their naive empiricism was responsible for the erroneous belief that nature exhibits her true character on the outside. Moreover, scholastic philosophy's inclination to accept at face value whatever the unaided senses revealed had given rise to the sterile science of elements and qualities. In reality, the new philosophers declared, nature's workings are hidden. The unaided, undisciplined senses do not reveal reliable information about what makes nature tick any more than observing the hands of a clock reveals how the clock works.[119] All the dogmatic pronouncements of scholastic philosophy were but chimeras based upon unreliable empirical foundations.

The rejection of Aristotelian natural philosophy's naive empiricism— its assumption that nature could be immediately observed—was an essential precondition of the Scientific Revolution. The concept of the "secrets of nature" was one of the premises underlying the "Baconian sciences," which according to Thomas Kuhn emerged simultaneously with the revolution in the classical sciences.[120] This new cluster of scientific interests, Kuhn argued, was characterized by the accumulation of observations and experiments rather than the revolutionary overthrow of existing paradigms. To the extent that these sciences existed at all prior to the Scientific Revolution, he maintained, they were little more than interesting classes of phenomena. There was no governing paradigm to explain them or to give them cohesion. Thus Kuhn argued that the Baconian sciences emerged essentially without precedents during the period of the Scientific Revolution. In contrast to the traditional, mathematical disciplines of the university curriculum, whose paradigms were completely overthrown, the Baconian sciences emerged from a "preparadigm" condition.

Kuhn's thesis, which is based largely upon a "Copernican Revolution" model of the Scientific Revolution, seems to me to diminish the revolutionary significance of the Baconian sciences. For the Aristotelian *philosophy of nature* was no less entrenched in the scholastic curriculum than were the "quadrivial," or mathematical, disciplines, and was no less a part of the traditional natural philosophy paradigm than Ptolemaic astronomy. Aristotelianism precluded inquiry into occult qualities. Insofar as paradigms define the boundaries of research, this exclusion, I have argued, was one of the essential features of the Aristotelian paradigm. In contrast to scholastic natural philosophy, which accepted the visible qualities of nature as real and excluded occult qualities from the domain of science, the new philosophy made explanation of occult qualities one of its principal goals. The inner "secrets of nature," not nature's outer appearances, became the object of the new science. The thrust of the new philosophies was to explain all physical change in terms of the motions of insensible particles. Manifest qualities became "secondary qualities," hence occult qualities in the sense that they were considered to be the effects of occult causes. However, even if the causes of such phenomena remained elusive, it was nevertheless possible to establish natural effects experimentally. In spite of its skepticism, the experimental philosophy could establish "matters of fact."[121] The new philosophers did not consider such experimental "matters of fact" as ends in themselves, but, provided the proper method were followed, as the "traces" of nature that would lead to hidden causes.

These considerations may give us a better understanding of the "Baconian" research programs of the Royal Society and other seventeenth-century scientific organizations. It is misleading to regard the Royal Society's activities as involving merely random experimentation and indiscriminate collection of "curious" facts. Many of the apparently random experiments performed by the virtuosi were actually directed at the investigation of occult qualities—in Baconian jargon, the "hidden causes of things." The premise of the society's research program was that passive observation was incapable of seeing beyond the surface of things. Only by experiment and disciplined observation was it possible to arrive at a knowledge of causes. When the new philosophers attacked occult qualities, they were attacking the scholastic *doctrine* of occult qualities, which they regarded as a refuge for ignorance that ended investigation: it was the ne plus ultra of Scholasticism. For the Baconians, however, occult qualities were the beginning point of experimental research. The fervent interest in alchemical and "hermetic" subjects in the early Royal Society, to which recent research has called attention, was not an eccentric deviation from the society's "scientific" research.[122] Such interests were linked to the society's concern with unveiling the "secrets of nature" and were part and parcel of the Baconian program. As John Henry has shown, "the investigation and demonstration of occult qualities in matter . . . were major factors in establishing experimentalism as the safest and surest way to truth in natural philosophy."[123]

The Royal Society's fascination with monsters, "singularities," and "curiosities" sheds additional light on the emergence of the Baconian sciences. A casual perusal of any of the early volumes of the society's *Philosophical Transactions* will turn up dozens of reports of bizarre occurrences and curiosities. This omnivorous fact gathering, notwithstanding the pleasure it reportedly brought the virtuosi, was a serious part of the Baconian approach to finding out the "secrets of nature." According to Bacon, nature exists in three states: in her ordinary course (the liberty of nature); forced out of her ordinary course by the "perversity and insubordination of matter," as in the production of monsters (the errors of nature); and constrained and molded by art (the bonds of nature). Natural history was therefore threefold. In addition to amassing histories of nature's normal productions, Bacon urged the compilation of histories of the mechanical arts and of nature's anomalies and monsters.[124] All the bizarre objects and rarities that had fascinated and delighted visitors to the Renaissance curiosity-cabinets became urgently relevant to the Baconian scientific enterprise.[125]

What have ostrich eggs, two-headed calves, and Indian featherwork to do with the Scientific Revolution? Such curiosities were an essential part of the Baconian natural histories. They take us directly back to the evidential paradigm that characterized Baconian epistemology. Such oddities were the visible clues and out-of-the-ordinary events that the Baconians believed would lead to a knowledge of hidden causes. Among the "prerogative instances" Bacon encouraged assembling were those he called "singular instances." Bacon was contemptuous of those (like Lemnius) who would "go no further than to pronounce such things the secrets and mighty works of nature, things as it were causeless, and exceptions to the general rules." He was confident that such "miracles of nature" might be understood in terms of universal causes: "For we are not to give up the investigation until the properties and qualities found in such things as may be taken for miracles of nature be reduced and comprehended under some form or fixed law."[126] Bacon also urged collecting "deviating instances," that is, the "errors, vagaries, and prodigies of nature." Like the former, they correct the erroneous impressions we have gotten of nature by basing our generalizations upon common phenomena and too few instances. Such anomalies were the "traces" that, if properly read, would lead to the discovery of hidden causes.

The new philosophy redrew the boundaries of science: it brought the secrets of nature back into the picture. But the reformed doctrine of occult qualities was no throwback to medieval esotericism. As we shall see in the following chapters, the Baconian idea of the secrets of nature paved the way toward the conception of science as public knowledge.

NINE

THE VIRTUOSI AND THE
SECRETS OF NATURE

T HE renewed interest in nature's "secrets," the exotic, and the marvelous was part and parcel of a new sensibility that emerged in seventeenth-century polite society. Originally an Italian tradition, virtuosity (*virtuosità*) had a pronounced influence upon upper-class tastes throughout Europe. It is to the English tradition of the movement, however, that we must look to find virtuosity's strongest and most visible alliance with natural philosophy. The virtuoso movement has not figured conspicuously among the various explanations that have been advanced for the flowering of experimental science in seventeenth-century England. Outside of a classic article by Walter J. Houghton, little has been written on the subject.[1] Yet for several reasons, I believe, virtuosity should be given a more prominent place in accounts of the origins of experimental science.[2] For when seventeenth-century English natural philosophers called themselves "virtuosi," they did so for quite specific reasons. First of all, the label identified a community of "new philosophers" who contrasted themselves to the scholastic "pedants." Virtuosity thus gave the new philosophers a badly needed sense of identity and cohesiveness. The sensibility of virtuosity—above all its courtly, genteel manners—also described a way of doing natural philosophy. The virtuosi insisted that the new philosophy was marked by "civility" in contrast to the "litigiousness" and pugnacity of scholastic debate. They did not defend abstract systems as if their reputations or livelihood depended on it; such, they insisted, was not their manner. They settled disputes "civilly," being guided by the conventions of courtly discourse.[3] The virtuoso, in other words, was not only a scholar but also a gentleman. He wore virtuosity as a badge of honor that distinguished him both from the scholastic "pedant" and from the "vulgar sort." Moreover, the virtuoso was also a curioso. He categorically denied the traditional view that natural curiosity endangered the soul. As I will argue in this chapter, the seventeenth-century rehabilitation of curiosity was in large measure a product of the virtuoso sensibility. Like any cultural ideal, that of virtuosity recommended certain activities, interests, and tastes above others. The virtuoso's curiosity was particularly drawn to the rare, unusual, and "extravagant" phenomena that might entertain and delight as well as instruct. The "secrets" of nature—in the broad, seventeenth-century sense of that term—fascinated and delighted him to the point that the virtuoso became the butt of endless jokes about the dilettante who dabbles in trifles but understands nothing of himself. Trifles or not, the virtuoso's "secrets" were the stuff of early modern experimental science.

The Emergence of Virtuosity

The virtuoso sensibility was to a large extent a symptom of the ailing fortunes of the nobility. Between 1560 and 1640, the English aristocracy suffered a steep decline in prestige and economic condition. The reasons for this slump have been spelled out at length by Lawrence Stone: the shrinkage of the aristocracy's territorial possessions, the granting of titles for cash rather than merit, the rise of individualism, and the demands by the state for a competent administrative elite irrespective of rank.[4] The decay affected not only the aristocracy's social status, but also its self-image. Traditionally the aristocracy conceived of itself as a class whose principal role was that of state service. The Elizabethan educational treatises reflected this view.[5] Sir Thomas Elyot, in his influential *The Governour*, encouraged studies that would most benefit the gentleman's career in public office, such as law, history, mathematics, and moral philosophy.[6] But as the peerage became increasingly dependent upon the Crown for its livelihood, and as their local power and wealth diminished relative to those of the gentry, the thirst for office became not merely a patriotic duty or an embellishment but an urgent necessity. The court became glutted with once-powerful local potentates transformed into fawning courtiers.

Eventually, however, the funds to support this vast patronage system dried up. The Crown simply could not afford to employ all those who aspired to office. The situation worsened as a result of James I's policy of granting titles for cash. As the number of unemployed nobles multiplied, contempt for the aristocracy intensified. The foppish nobleman became a trope, and the nobility itself came under attack as an institution unworthy of the respect it once commanded. By the early seventeenth century, the basic characteristics of the transformed aristocracy were established: those who managed to find employment in government service became tame state pensionaries; the less fortunate, who were by far the majority, were doomed to lives of boredom and loneliness on their country estates.[7]

The cultural ideal of the virtuoso was born out of the crisis of the aristocracy. The virtuosi were drawn mainly from the ranks of unemployed gentlemen with too much time on their hands. As opportunities for an active civic life declined, gentlemen tended to devote their leisure hours to activities and studies that set them apart from vulgar "upstarts." A growing emphasis upon the pleasures as opposed to the practical uses of knowledge began to appear in treatises on gentility in the early seventeenth century. The ideal of education as essential training for state service gave way to the conception of learning as an ornament. Henry Peacham's *Compleat Gentleman*, which in the seventeenth century replaced Elyot's *Governour* as the model for the training of the elite, expressed the new ideal. "Learning is an essential part of Nobilitie," wrote Peacham, hence "it followeth, that who is nobly borne, and a Scholler withall, deserveth double Honour, . . . for hereby as an Ensigne of the fairest colours, hee is afarre off discerned, and winneth to himselfe both love and admiration; heigthing with skill his Image to the life, making it pretious, and lasting to posteritie."[8] Yet while the gentleman might also be a "scholar," he was at all costs to avoid being a pedant. The disputatious, hectoring, uncivil manner of the Scholastic was considered unbecoming of a gentleman; moreover, his subjects of study were unsuitable for polite conversation.[9] A popular seventeenth-century college tutor's manual contained a special reading list designed for young gentlemen who wished to acquire only "such learning as may serve for delight and ornament and such as the want wherof would speake a defect in breeding." The list recommended no heavy scholastic tomes, only books like Della Porta's *Magia naturalis*, Estienne's *Wonder of Wonders*, and other works befitting *Studia Leviora*.[10]

In his description of the attributes of the gentleman, Peacham echoed Castiglione and the Italian courtesy books, which had a major influence in defining the English concept of virtuosity. Indeed, as Walter Houghton pointed out, the movement rested above all upon the spread of Italian influence in England.[11] Italy was the ultimate destination of every Grand Tour to the Continent, and no tour of Italy was complete without visits to the cabinets of curiosities belonging to aristocratic collectors. Nor did the English gentleman have to travel to Italy to absorb Italian culture: it was all around him. Italian dress, Italian manners, Italian music, Italian comedy, even Italian cooking were all in vogue. Books about Italy and in Italian were regularly printed in England. Published diaries and itineraries, such as those of Moryson and Coryat, also brought Italian tastes to the attention of the less adventurous. Italian treatises on conduct and civility, including works by Stefano Guazzo, Giovanni Della Casa, and Romei, implanted the ideals of virtuosity. Indeed, in the introduction to his translation of Castiglione's *Cortegiano*, Sir Thomas Hoby opined that the book "advisedly read and diligently followed but one year at home in England, would do a young gentleman more good, I wis [know], than three years' travel abroad spent in Italy."[12] It is significant but hardly surprising that the English gentry, in seeking to locate themselves in the fluctuating social landscape of the Jacobean age, found the idealized Italian courtier a perfect model for their own self-fashioning.

If virtuosity was a symptom of aristocratic defensiveness, it also owed its existence to boredom and its characteristic malady, melancholy, a disease for which the nobility seemed most at risk.[13] Such, at least, was the diagnosis of Robert Burton, whose *Anatomy of Melancholy* provided the fullest index of the virtuoso's sensibilities. Burton observed that melancholy particularly afflicted the gentleman who "for want of employment knows not how to spend his time": "Idleness is an appendix to nobility, they count it a disgrace to work, and spend all their days in sports, recreations, and pastimes, and will therefore take no pains, be of no vocations: . . . and thence their bodies become full of gross humours, wind, crudities, their minds disquieted, dull, heavy, etc."[14]

One of the remedies Burton prescribed for melancholy was "study." The subjects he recommended read like a curriculum for the training of a virtuoso: "to read, walk, and see maps, pictures, statues, jewels, marbles"; to visit cabinets of curiosities and to gaze upon "such variety of attires, faces, so many, so rare, and such exquisite pieces, of men, birds, beasts, etc., . . . Indian pictures made of feathers, China works, frames, thaumaturgical motions, exotic toys, etc."; to page through scientific books that depicted "herbs, trees, flowers, plants, all vegetals expressed in their proper colours to the life, . . . birds, beasts, and fishes of the sea, spiders, gnats, serpents, flies, etc."[15] The virtuoso, according to Burton, cultivated an interest in "rarities" both to alleviate boredom and to distinguish himself from his social inferiors. Such activities were suitable indoor "exercises" that gave pleasure and had therapeutic value, but were not recommended as ends in themselves. The idle melancholic might "distract his cogitations" with them, but "let him take heed he do not overstretch his wits." Antiquities, natural and artificial curiosities, conceits, and mechanical gadgets—all typically collected by the virtuoso—were like ornaments betokening the man of means. "The possession of such Rarities," wrote Peacham, "by reason of their dead costlinesse, doth properly belong to Princes, or rather to princely minds." Indeed, as Peacham informs us the term "virtuoso" originally meant one who was devoted to the collection of antiquities and curiosities. "Such as are skilled in them," Peacham wrote, "are by the *Italians* termed *Virtuosi* as if others that either neglect or despise them, were idiots or rakehels."[16]

The hold of this tradition of virtuosity over the gentry lasted well into the seventeenth century. In his travels on the Continent in the 1640s,

John Evelyn (1620–1706) recorded his delight touring the gardens at the palazzo d'Este in Tivoli, "feasting our Curiositie with [its] artificial Miracles." He was wide-eyed with wonder upon witnessing the Venetian nobleman Carlo Rugini's cabinet of rarities, noting its "divers curious shells, & two faire *Pearles* in 2 of them," petrified "Eggs, in which the *Yealk* rattled, . . . an whole hedgehog, a plaice on a Wooden Trencher turned into stone, & perfect, . . . a *Diamond* which had growing in it a very fare *Rubie*, . . . divers pieces of *Amber* wherein were several *Insects* intomb'd, in particular one cut like an heart, that contain'd [in] it a *Salamander*."[17] In a letter of 1657 to Benjamin Maddox, a young gentleman about to make a Grand Tour of the Continent, Evelyn advised him to "seek therefore after nature, and contemplate that great volume of the creatures whilst you have no distractions: procure to see experiments, furnish yourself with receipts, models, and things which are rare."[18] This image of virtuosity became so well established in literary and popular circles that in 1676 Thomas Shadwell used it to create his caricature of the virtuoso, Sir Nicholas Gimcrack, in his comedy *The Virtuoso*. "We virtuosos never find out anything of use," exclaims Sir Nicholas, " 'tis not our way."[19] Writing at the end of the century, Mary Astell defined the virtuoso as "one that has sold an Estate in Land to purchase one in *Scallop, Conch, Muscle, Cockle Shells, Periwinkles, Sea Shrubs, Weeds, Mosses, Sponges, Coralls, Corallines, Sea Fans, Pebbles, Marchasites* and *Flint Stones;* and has Abandoned the Society of Men for that of *Insects, Worms, Grubbs, Maggots, Flies, Moths, Locusts, Beetles, Spiders, Grashoppers, Snails, Lizards,* and *Tortoises*."[20] In spite of the completely different sense of the term employed by the Royal Society— that of the serious experimental scientist—the caricatures suggest that in the popular mind at least, the virtuoso was still a dilettante and a dabbler.

Conceits, Curiosities, and "Extravagants"

The gentry's preoccupation with rarities was nourished by numerous books of "conceits" published in the early seventeenth century. Most of these simple and browsable works resembled the series of pamphlets first published by Thomas Johnson in the 1590s and reprinted in the 1630s.[21] Johnson's *Dainty conceits* (1630), with its "rare and witty inuentions . . . made and inuented for honest recreation, to passe away idle houres," was evidently written for the newly arrived gentleman: it began with "directions for choosing a country estate" and proceeded to card tricks, the care of oxen, making beer and ale, metallurgy, and grafting fruit trees.[22] Johnson's *New Booke of new Conceits* (London, 1630) served up more novelties, "some profitable, some necessary, some strange, none hurtful, and all delectable."[23] In addition to recipes quarried from the books of secrets, Johnson recorded various magical tricks (many of which go back to Pseudo-Albertus), physiognomic exercises, and an ample dose of medieval plant and animal lore. His works were prototypes of a host of curiosity-books bearing titles like *Newe Recreations, or The Mindes release and Solacing* (1631), *Hocus Pocus Junior* (1638), and *Sports and Pastimes: or, Sports for the City, and Pastime for the Country* (1676).[24] These pamphlets reveal the beginnings of a transformation of the popular image of magic. By being exposed to the mechanics of conjuring tricks, people were less likely to take magic seriously, more likely to view it simply as "hocus-pocus."

Practical utility was obviously not the main impetus that drew the virtuosi to the books of secrets. While they occasionally delved into utilitarian subjects, they rarely did so *because* of their utilitarian value. The virtuosi gathered "secrets" not because they found them useful but because they regarded them as pleasant diversions and as ornaments of gentility. Evelyn typified the attitude of the virtuosi in his letter to Mr. Maddox, where he recommended collecting "many excellent receipts to make perfumes, sweet powders, pomanders, antidotes, and divers such curiousities":

> Though they are indeed by trifles in comparison of more solid things, yet if ever you should affect to live a retired life hereafter, you will take more pleasures in those recreations than you can now imagine. And really gentlemen despising those vulgar things, deprive themselves of many advantages to improve their time, and do service to the desiderants of philosophy; which is the only part of learning best illustrated by experiments. . . . Commonly indeed persons of mean condition possess them because their necessity renders them industrious; but if men of quality made it their delight also, arts could not but receive infinite advantages, because they have both means and leisure to improve and cultivate them; and, as I said before, there is nothing by which a good man may more sweetly pass his time.[25]

Closely related to these pastimes was the virtuosi's fascination with what they termed "mathematical magic," a grab bag of subjects that included mechanical gadgets, optical illusions, tricks, problems, and various forms of recreational mathematics. Burton recommended such mathematical diversions as a cure for the melancholic. "Let him demonstrate a Proposition in Euclid," he advised, "extract a square root, or study algebra." Better yet, let him "demonstrate with Archimedes" and pursue applied mathematics: "What so intricate, and pleasing withal, as to peruse and practise Hero Alexandrinus' works, . . . Those rare instruments and mechanical inventions of Jac. Bessonus and Cardan to this purpose, with many such experiments intimated long since by Roger Bacon, in his Tract *de secretis artis et naturae*, as to make a chariot to move *sine animali*, diving boats, to walk on the water by art and to fly in the air, to make several cranes and pullies, . . . and such thaumaturgical works."[26]

John Bate, a London instrument-maker, reported that he was often asked by the virtuosi to demonstrate various "engines" and mathematical "experiments." So great was the virtuosi's interest in such subjects, said Bate, that to avoid the inconveniences of their entreaties he was forced to write a book describing the various gadgets and experiments that had amused his fashionable clients. Bate's *Mysteries of Nature and Art* (1634) was essentially a handbook of diversions and mechanical "conceits" for the virtuosi.[27] In one section of the work he gave instructions for making various hydraulic devices employing pumps, siphons, and steam power. Most were mechanical toys and gadgets in the tradition of Hero of Alexandria: fountains with chirping mechanical birds, water clocks, sound machines, and water-powered automata. In another section Bate gave instructions for making fireworks, including rockets, flares, "raining fire," and firecrackers. A third section explained the mysteries of the arts of drawing, painting, and engraving. The final part, much like a traditional book of secrets, contained a mixture of home remedies, technical recipes for making ink and dye, instructions for soldering, tempering, gilding, glassmaking, and for making bait to catch fish and fowl.

Bate's *Mysteries* and works like it were quite fashionable in seventeenth-century England.[28] The audience for such writings was large enough in 1633 to warrant the translation of Heinrich Van Etten's *Récréations mathématiques* (1624), another compilation based largely upon Hero of Alexandria.[29] Van Etten said he wrote the work "to satisfie the curious, who delight themselves in these pleasant studies, knowing well that the *Nobillitie* and *Gentrie* rather studie *Mathematicall Arts* to content and satisfie their affections in the speculation of such admirable Experiments as are extracted from them, than in hope of gaine to fill their *Purses*." In addition to number games and card tricks, the work demonstrated mechanical problems and described various machines, optical illusions, and pyrotechnical experiments. Van Etten proposed such subjects to "imploy the minde in usefull knowledge, rather than to be busied in vaine *Pamphlets*, *Play-bookes*, fruitlesse *Legends*, and prodigious *Histories* that are invented out of fancie."

In 1651, John White published *A Rich Cabinet, With Variety of Inventions*, the first edition of his popular series on "conceits" and "rarities." White, who styled himself "a lover of artificiall conclusions," advertised his "receipts and conceipts" as having been "unlock'd and opened, for the recreation of Ingenious Spirits at their vacant hours."[30] Addressed to "all lovers of ingenious and artificiall conclusions," his compilations included recipes for dyeing silk and leather, for removing spots and stains, for making perfume, soap, and invisible ink; instructions for making sundials, wax carvings, fireworks, and mechanical toys; bait for catching fish; weather lore; and number games—in other words, the sort of "trifles" Evelyn recommended to Mr. Maddox. Truly a "virtuoso's handbook," the work appeared in seven editions in the seventeenth century. White shuffled his material into similar collections with new titles, including *The Rich Cabinet of modern curiousities, Hocus-Pocus, or a Rich Cabinet of Legerdemain Curiosities, Arts Treasury or rareties and curious inventions,* and *Arts Masterpiece.*

Although these works responded directly to the cultural ideal of virtuosity, indirectly they helped shape the scientific interests of a broad spectrum of English readers. First of all, they familiarized the public at large with many of the instruments that would eventually become an integral part of experimental science. Bate and Van Etten described how to make thermometers and gave detailed instructions for their use. Less than a decade after the instrument's invention, the virtuosi were being shown various experiments with it.[31] Van Etten related tricks that could be done with the camera obscura and recorded experiments with various types of optical lenses. Moreover, although they were intended mainly for entertainment, these popular books of "scientific magic" provided a modicum of genuine scientific instruction. Bate observed that everyone who came to him for instruction desired "to know the reasons and causes of those things they were desirous to be informed in." Accordingly, he prefaced his treatment of hydrostatics with the observation that nature's abhorrence of a vacuum "is the ground and foundation of divers excellent experiments." Similarly, he opened the section on fireworks with "certayne Praecognita or Principles, wherein are contayned the causes and reasons" of pyrotechnics.[32] Van Etten wrote for "*Schollers, Students,* and *Gentlemen,* that desire to know the *Philosophicall* cause of many admirable Conclusions." His work, he hoped, would both entertain the virtuosi and "acuate and stirre them up to the search of further knowledge."[33] He concluded many of his experiments with examinations of the scientific principles involved. Optical tricks illustrated the laws of reflection and refraction, while feats accomplished with levers and pulleys demonstrated basic principles of mechanics. If nothing else, the books of conceits demonstrated that science could be fun.

If these works suggest that the scientific interests of the virtuosi were rather frivolous, it should be pointed out that the same activities occupied the minds of many of the seventeenth century's leading scientific personalities. John Wilkins (1614–1672), one of the founders of the Royal Society, published a work similar to those of Bate and Van Etten. Wilkins titled his work *Mathematical Magick* (1648) "in allusion to vulgar opinion, which doth commonly attribute all such strange operations unto the power of Magick." Intentionally a work of popular science, the book was really about the principles of mechanics, which Wilkins illustrated by explaining the operation of various machines and devices. Wilkins was one of the founding members of the Oxford Experimental Philosophy Club, one of several informal natural philosophy groups that emerged during the Interregnum and culminated in the establishment of the Royal Society of London.[34] Walter Charleton's description of the Oxford club's activities makes it clear that "mathematical magic" was an important part of the group's interests: "It is their usual recreation, to practise all Delusions of the sight, in the figures, Magnitudes, Motions, Colours, Distances, and Multiplications of Objects: And, were you there, you might be entertained with such admirable Curiosities, both Dioptrical and Catoptrical, as former ages would have been startled at, and believed to have been magical. . . . Were Friar Bacon alive again, he would with amazement confesse, that he was canonized a Conjurer, for effecting far lesse, than these men frequently exhibit to their friends in sport."[35] When Evelyn visited Wilkins at Oxford in 1654, he was intrigued by the great variety of machines Wilkins had constructed, including "an hollow Statue which gave a Voice, & utterd words, by a long & concealed pipe which went to its mouth, whilst one spake thro it, at a good distance, & which at first was very Surprizing: He had above in his Gallery & Lodgings variety of *Shadows*, Dyals, Perspectives, . . . & many other artificial, mathematical, Magical curiosities: A Way-Wiser, a *Thermometer*; a monstrous *Magnes*, *Conic* & other *Sections*, a Balance on a demie Circle."[36]

Books on "mathematical magic" played an important role in popularizing mechanics and in stimulating interest in science. We know that Sir Isaac Newton's scientific curiosity was awakened in part by such works. As a boy, Newton was fascinated by mechanical toys. Books on "mathematical magic" were among his favorites. An avid reader of Wilkins's *Mathematical Magick* and Bate's *Mysteries*, he took extensive notes on both works. He reproduced Wilkins's mechanical experiments and copied lengthy excerpts on the mixing of paints and on sketching from Bate's *Mysteries* into his school notebook.[37] In a memoir of Newton of 1752, William Stukely tried to establish a connection between Newton's youthful activities and his later experimental research: "It seems to me likely enough that Sir Isaac's early use and expertness at his mechanical tools, and his faculty of drawing and designing, were of service to him, in his experimental way of philosophy. . . . A mechanical knack, and skill in drawing, very much assists in making experiments. . . . For want of this handycraft, how many philosophers quietly sit down in their studys and invent an *hypothesis*; but Sir Isaacs way was by dint of experiments to find out *quid Natura faciat aut ferat* [what nature might do and suffer]."[38]

How might a "mechanical knack" and skill at drawing be useful for experimentation? First of all, it is obvious that experimental facts do not occur "by nature" or without special aids to the senses. They are purposefully made in a laboratory, often with the use of instruments; increasingly in seventeenth-century science, matters of fact were machine-made.[39] Instruments such as the telescope and the microscope made things seen that the naked eye could not see, while the air pump created spaces that, according to traditional natural philosophy, nature "abhorred." The thermometer and the barometer provided standard, quantified measures for subtle meteorological changes that often could not be felt with the unaided senses, and could be reported only impressionistically. Mechanical skills were essential for doing the kind of experiments that were characteristic of the Baconian sciences.

Moreover, the validity of experimental situations rests to a large extent upon acceptance of the notion that the natural and the artificial do not differ in kind, but only in "efficient cause." This view reverses the traditional Aristotelian conception of the relation between nature and art. According to Aristotle, art imitates the work of nature, but its mode of operation is completely different. Hence the productions of art cannot teach us anything about how nature operates.[40] Bacon categorically rejected this distinction. "The artificial does not differ from the natural in form or essence, but only in the efficient," he asserted, "nor matters it, provided things are put in the way to produce an effect, whether it be done by human means or otherwise."[41] Bacon argued that the classical view of the relation between art and nature had bred a premature despair in human enterprises. He denounced those who judged the more difficult and subtle works of nature as "*magnalia* or marvels of nature, and by man not imitable."[42]

Much of Bacon's scientific program was predicated on the assumption that by imitating nature, we come closer to understanding natural processes. This is why in Solomon's House, Bacon's utopian research institute, projects involving the imitation of natural processes were given high priority. There the researchers have "artificial wells and fountains, made in imitation of the natural sources and baths"; houses where they "imitate and demonstrate meteors; as snow, hail, rain"; orchards where they make trees and flowers bear out of season or "make them also by art greater much than their nature, and their fruit greater and sweeter and of differing taste, smell, colour, and figure, from their nature"; furnaces that "have heats in imitation of the sun's and heavenly bodies' heats"; perfume houses where they imitate smells; "perspective-houses" where they create optical illusions and make artifical rainbows; and, finally, houses of "deceits of the senses," where they "represent all manner of juggling, false apparitions, imposture, and illusions."[43] Solomon's House was "an artificial world carefully fashioned and crafted in imitation of the natural world."[44] Methodical imitation of nature was central to Bacon's philosophy of science. Nature, he thought, was like a closed workshop that produced its effects by hidden means. Since nature's modes of operation were no different from those of art (except in the efficient cause), reproducing nature's effects by artificial means was a guarantee of man's knowledge of nature.[45] In this sense, the imitation of nature, such as that attempted in Solomon's House, was the realized goal of Baconian science, not merely a tool of science.

The literature of "scientific magic" made the imitative wonders of Solomon's House commonplace among the virtuosi and also disseminated them in popular culture. Van Etten demonstrated how any schoolchild might make an artificial rainbow by simply standing with his back to the sun and blowing water into the air.[46] Bate gave directions for mechanically imitating the voice of birds, and for making artificial stars with fireworks.[47] Wilkins described a metal spider that moved "as if it were alive."[48] Further evidence for Atlantean experiments centers on Cornelis Drebbel (1572–1633), a Hollander who emigrated to England in 1605 and became attached to the court of Prince Henry of Wales. Drebbel is reported to have demonstrated instruments that made artificial rain, lightning, and thunder, and to have carried out experiments in refrigeration, once making the Great Hall at Westminster so cold on a summer's day that the court was forced outside.[49] Such experiments, some real and some only imagined, demonstrated the "Baconian" principle that there was no essential difference in the classical sense between nature and art.

Another side of scientific virtuosity can be seen in the published and unpublished works of the Elizabethan professor of secrets Sir Hugh Plat (1552–1608).[50] Although Plat was perhaps too practical to qualify as a typical virtuoso, his writings nevertheless illustrate the breadth of the movement's interests. The son of a wealthy London merchant, Plat graduated from Cambridge and then dedicated his entire leisured life to his experiments and inventions in agriculture, horticulture, and chemistry. He corresponded with gardeners and farmers, gathering information about agricultural and horticultural practices from various parts of England. He conducted horticultural experiments on his estate at Bethnal Green and maintained laboratories there and at his home in London, where he did alchemical experiments and worked on his inventions. In

1592, Plat exhibited to a select circle of privy councillors and leading London citizens a précis of his secrets, which he promised to offer to the public upon reasonable compensation.[51] Two years later he fulfilled his promise with the publication of his *Jewell House of Art and Nature* (1594), a collection of "divers new and conceited experiments, from the which there may be sundry both pleasing and profitable uses drawne." Most of Plat's inventions were designed for improvement rather than for novelty. He maintained that "the trew end of all our privat labors and studies ought to bee the beginning of the publike and common good of our country."[52] He offered instructions for preserving meat for a month in any weather; a cheaper way to feed fowl; a chafing dish to keep food hot without coals; a way to brew beer without using hops; methods to keep moths from garments, for getting rid of wasps, and for killing rats; and recipes for making inexpensive wine, candles, perfume, and oil. Yet while Plat was concerned mainly with useful inventions, he was too much of a virtuoso to omit such curiosities as an alchemical experiment "to engender strange forms in a glass." This novelty, he wrote, was useful as a "Philosophical work" worth trying "if it were to no other end, then to put us in mind of *Democritus* his *Atoms*, which concurring together, at length engender bodies."[53] He also showed how to hold a bar of red-hot iron in the hand without getting burned, to write secret letters, to communicate in sign language, and to make an egg stand on its end. Plat's most popular work, a household recipe book entitled *Delights for ladies, to adorne their persons, tables, closets, and distillatories* (1602), described methods for preserving fruits and vegetables, for making perfumes and oils; it provided cooking recipes and household medical recipes, including a version of "Dr. Stevens's water."[54] Plat's *Floraes Paradise* (1608), a work on gardening, exhibited a similar combination of the utilitarian and the curious. While most of Plat's experiments were useful, some were written merely "to the pleasuring of others, who delight to see a raritie spring out of their owne labours, and to provoke Nature to play, and to shew some of her pleasing varieties, when she hath met with a stirring workman." Plat derided the Scholastics, who had written "many large and methodicall volumes on this subject (whose labours have greatly furnished our studies and Libraries, but little or nothing altered or graced our Gardens and Orchards)." By contrast, his more than two hundred experiments were "not written at adventure or by any imaginary conceit in a Schollars private Studie, but wrunge out of the earth, but by the painfull hand of experience."[55]

An indefatigable experimenter and an avid reader of the experimental literature of the day, Plat was constantly testing and improving upon secrets he discovered in the works of Palissy, Quercitanus, Wecker, Della

Porta, Agrippa, and Cardano. He queried artisans, housewives, and other virtuosi, and spoke of exchanging secrets with "*John Hester*, one of the most ancient chimists of my time in London."[56] Plat tried to counteract the impression (which he thought was widespread) that the professors of secrets had exhausted the possibilities of experimental knowledge, thus discouraging new "professors of inventions" from experimenting. His critique of the professors of secrets and their books is revealing: "Because neither *Albertus Magnus, Alexis of Piemont, Cardanus, Mizaldus, Baptista Porta, Firovanta*, nor the rest of that Magical crew, nor yet *Wickerus*, that painefull gatherer and disposer of them all, should here be objected as a matter of terror unto al new professors of rare & profitable inventions, I have thought good (for the better satisfaction of the wiser sort) in a word or two to set downe some of the wants of those ancient Authors . . . which no doubt are left for the *Neoterici* and later writers of our age to supply & furnish." In the first place, Plat pointed out, most of these authors had written in foreign languages and hence were inaccessible to many readers. Second, many of their secrets were untrue or untried. Third, many of them recorded secrets gathered only from theory or speculation, but when they are "tried in the glowing forge of *Vulcan*, they vanish into smoake." Finally, they often wrote assuming too much knowledge of the reader, "as that no man, without a manual maister that may even lead him by the hand through all their riddles," could duplicate their experiments.[57] Plat vowed to "outpractice" the professors of secrets, and to explain his recipes so clearly that anyone could repeat them.

In addition to his eleven published books advertising his secrets and inventions, Plat left several manuscript notebooks recording his ongoing experimental work. The notebooks, which are now in the British Library, leave the unmistakable impression that he was constantly testing and developing new inventions.[58] The manuscripts record Plat's experiments in beer and wine making, horticulture, waterproofing, removing spots and stains from garments, and making ink, dye, soap, perfume, and cosmetics—in short, the staples of the traditional book of secrets. However, his notebooks shed a completely new light on the books-of-secrets tradition. They demonstrate that, in Plat's case at least, his published works were not just compilations drawn from previous books and manuscripts, but were the result of his ongoing experimental research. A notebook in which Plat recorded his dyeing experiments, for example, shows that he worked systematically to improve the colors and fastness of the dyes. He kept a record of each experiment and its results. In July and August 1598, for example, he was at work on a series of experiments in dyeing silk. On 2 August 1598, Plat recorded experiment number fifteen of the series: "I did a Tawney, the galled silke first steeped in a stronger allomed liquor of a lb. of allom to one gallon of water, for 2 houres." On the following day, he reported the results of experiment number sixteen: "I did not like the tawney supra no. 15 and therefore I layd it 7 or 8 howrs in a stronge allome water, & then I did it againe as before in fresh brasill Liquor, . . . and I made a reasonable good tawny. 3 August: 98."[59] Another notebook contains numerous headings for experiments followed by blank spaces, suggesting that Plat laid out ambitious research plans in advance but did not always follow through.[60]

The Rehabilitation of Curiosity

Hugh Plat's inexhaustible curiosity bore witness to one of virtuosity's most important contributions to early modern science: the rehabilitation of intellectual curiosity.[61] Instead of viewing curiosity as a grave vice that endangered the soul, the virtuosi considered curiosity to be an outright virtue and an essential quality of the cultivated gentleman. Indeed, curiosity was the hallmark of the virtuoso: in the seventeenth century, "virtuoso" was synonymous with "curioso."[62] The virtuoso's enthusiastic cultivation of the "secrets of nature and art" embodied attitudes that were the antithesis of traditional evaluations of curiosity. When he gave his undivided attention to the subtle and intricate workmanship of nature, or to ingenious works of art, he did not expect to discover something useful, but merely to satisfy a longing or fancy. The virtuosi *endorsed* curiosity for the very reasons Augustine had condemned it. Instead of resisting the longing to know the trivial, they championed it.

In the seventeenth century, "curious" referred at once to a personal attribute or quality, to an attribute of things, to a state of mind, and to a way of seeing.[63] As a personal quality, curiosity referred first of all to the desire to know things, especially to know the unusual, the novel, and the exotic. Curiosity about such matters prepared the appetite for feasting on deeper truths. Evelyn described his visit to the famous gardens at the Este palace at Tivoli quite literally as an appetizer: "Having feasted our Curiositie with these artificial Miracles, & din'd," they went to see the cascading Aniene River, the ruins of an ancient Roman temple, the emperor Hadrian's villa, and numerous other sights.[64] In its preference for the novel and rare, virtuosity ran counter to the Aristotelian program, whose focus was upon the everyday, normal, and "vulgar" works of nature. At the same time, curiosity meant proficiency in artistic endeavors and ingenuity in matters mechanical. Expert craftsmen displayed curiosity in their workmanship. And as many a failed experiment proved, the new science also required it: Henry Power noted "How many ticklish Curiosities" are required to perform certain experiments.[65] In the seventeenth-century sensibility, curiosity, whether scientific or artistic, was really a form of connoisseurship.

"Curious" was also an attribute of things. Curiosities were always described as rare, novel, exotic, unusual, "conceited," or "extravagant." Out-of-the-ordinary things fascinated the virtuosi. Evelyn visited several shops that sold such curios to collectors. In Paris, at a shop called Noah's Ark, you could purchase "all the Curiosities naturall or artificial imaginable, Indian or European, for luxury or Use, as Cabinets, Shells, Ivorys, Purselan, Dried fishes, rare Insects, Birds, Pictures, & a thousand exotic extravagances." He reported that Rouen "abounds in workemen that make and sell curiosities of Ivory and Tortoise shells, . . . and make many rare toyes; & indeed whatever the East Indys afford of Cabinets, Purselan, natural & exotic rarities are here to be had with aboundant choyce." As these references suggest, "curious" also denoted delicacy and intricacy of workmanship. At a Haarlem church, Evelyn marveled at the "curiosity of the Workmanship" of a candlestick.[66]

"Curious" applied not only to works of art, but also, in a favorite seventeenth-century metaphor, to nature's workmanship. The virtuosi acknowledged a parallel between the curious works of art and nature, and were connoisseurs of both. Hooke observed that the scales of the dogfish were "curiously ridg'd, and very neatly carved," and marveled at "curiously figur'd" snowflakes. Natural objects were often described as curious by virtue of their smallness, exquisiteness of workmanship being exhibited more strikingly in miniature. Hooke noted that when examined under a microscope, the most "curious" works of art appear crude, "whereas in *natural* forms there are some so small, and so curious, and their design'd business so far remov'd beyond the reach of our sight, that the more we magnify the object, the more excellencies and mysteries do appear; And the more we discover the imperfections of our senses, and the Omnipotency and Infinite perfections of the great Creatour."[67]

If exquisite workmanship made objects of art worthy of inquiry (and of acquisition), the subtle and intricate secrets of nature were the most curious of all possible objects of interest. In the early modern period, Lorraine Daston observed, "curiosity, in both its natural and social forms, had a powerful affinity for delving and prying into secrets."[68] Hooke openly acknowledged the voyeurism of the new philosophy. Far from condemning it, he extolled it. Hooke contrasted the microscope's ability to peek at nature without being noticed to the more violent methods of dissection. Instead of "pry[ing] into her secrets by breaking open the doors upon her," with a microscope the observer can "quietly peep in at the windows, without frighting her out of her usual byas."[69] To the almost exclusively male company of virtuosi, nature's secrets were as wonderful and mysterious as those of woman. As nature was feminine, natural philosophy was "a Male Virtu" whose "curious sight" followed nature "into the privatest recess of her imperceptible Littleness."[70]

Such views echoed Bacon's call for a "masculine" science that would be active, virile, and generative in contrast to "passive" and "sterile" Scholasticism. The goal of the Royal Society of London, according to one of its spokesmen, was to "raise a masculine philosophy" of nature.[71] The exact meaning of this phrase is not entirely clear. However, certainly one important attribute of the "masculine" Baconian scientist was his curiosity. Both as a state of mind and as a way of seeing, curiosity was the beginning of science. For one thing, the curious virtuoso looked at things differently from those who relied upon common sense. Instead of gazing contentedly upon nature like the common observer (or, for that matter, like the Aristotelians) he examined it up and down with the lynx's eyes. Hooke contrasted the "Eye of the Vulgar" to that of the "diligent Naturalist," who "may perhaps discover that to be the richest Ambergrease, which another takes for Grease fit for naught but to noint his Shoes."[72] Even when looking at the perfectly ordinary, the virtuoso, examining nature with a discerning and "curious" vision, or through the lens of a microscope, saw something entirely new.

Openness to novelty and rarity, the virtuosi believed, was the starting point of scientific inquiry. For the strange and new provoked wonder, and wonder in turn provoked curiosity about causes. "Wonder is the child of rarity," Bacon said. "By rare and extraordinary works of nature the understanding is excited and raised to the investigation and discovery of Forms."[73] Thomas Hobbes (1588–1679), though a vigorous opponent of the new experimental science, identified curiosity as the "engine" that drives intellect in its desire to know causes: "This hope and expectation of future knowledge from anything that happeneth new and strange, is that passion which we commonly call ADMIRATION: and the same considered as appetite, is called CURIOSITY, which is appetite of knowledge. . . . And from this passion of admiration and curiosity . . . is derived all philosophy."[74]

Although wonder may be the beginning of philosophical inquiry, it was generally acknowledged that the goal of science was to make marvels end by explaining their causes. As Bacon pointed out, something is marvelous only so long as its cause is unknown, whereas "an explanation of the causes removes the marvel."[75] In his *Experimental History of*

Colours, the English virtuoso Robert Boyle (1627–1691) addressed the problem of wonder in experimental philosophy. Boyle eschewed experiments that aimed merely to astonish observers. "It is fitter for mountebanks than naturalists to desire to have their discoveries rather admired than understood," he wrote; "for my part I had much rather deserve the thanks of the ingenious, than enjoy the applause of the ignorant." However, Boyle acknowledged that experiments that provoked wonder were useful in recruiting people to experimental philosophy, and might stimulate philosophical inquiry among those who would "scarce admit philosophy, if it approached them in another dress." He thought the sensible changes exhibited in his color experiments were often so surprising and dramatic "that scarce any will be displeased to see them, and those that are any thing curious will scarce be able to see them, without finding themselves excited to make reflections upon them." He hoped the experiments "may both gratify and excite the curious, and lay perhaps a foundation, whereon either others or myself may in time superstruct a substantial theory of colours." When the techniques for exhibiting such "wonderful" sensible effects are made public, Boyle wrote, they will "quickly lose all, that their being rarities, and their being thought mysteries, contributed to recommend them."[76]

The principles of public knowledge on which Boyle's comments rested underscore one of the paradoxes of the virtuoso sensibility. Evelyn's devotion to the new philosophy could scarcely conceal his regret that science had taken the wonder out of the world. Long after the event, he remembered seeing an ingenious and "curiously" wrought candlestick that spouted streams of water instead of flames: "This seemed then, and was a rarity," he recalled, "before the Philosophy of Compress'd aire made it intelligible."[77] Whether knowledge of causes should be made public or kept secret was still an issue that concerned, and eventually divided, the virtuosi. Concealing the causes of nature's secrets served purposes unrelated to science. Van Etten noted, for example, that mathematical recreations are more entertaining if the tricks are concealed, "for that which doth ravish the spirits is, an admirable effect, whose cause is unknowne, which if it were discovered, halfe the pleasure is lost."[78]

More seriously, knowledge of hidden causes served to distinguish the cognoscenti from the vulgar. This was another reason why, ironically, the new philosophers insisted so strenuously upon the reality of occult qualities. As we have seen, it was not occult qualities that they rejected, but the scholastic doctrine of occult qualities as a refuge for philosophical ignorance. It is well known that the "new philosophers," including those who made up the Royal Society of London, were men whose strictly *scholarly* (that is to say, philosophical) credentials were questionable at best. Knowing secrets was thus a badge of social identity for a heterogeneous group of virtuosi who had a driving need to legitimize their newfangled approach to natural philosophy.[79] In other words, esotericism might serve the new philosophy as effectively as it had served the old.

TEN

FROM THE SECRETS OF NATURE
TO PUBLIC KNOWLEDGE

THE debate over secrecy versus openness in science reached a climax in the seventeenth century, when esotericist practices came under sustained attack. Natural philosophers condemned esotericism in alchemy and the crafts as obstacles to the growth of knowledge. They denounced the preciosity of the virtuoso sensibility, which, to many, seemed particularly harmful to the advancement of knowledge. The political implications of esotericism became the subject of widespread debate. The terms of the debate were in large part defined by the philosophy of Sir Francis Bacon, who put forth a powerful argument for the idea that knowledge would be most efficiently advanced by cooperative endeavor. Since science had to be grounded upon a new foundation of reliable factual information, Bacon argued, it needed the ordinary talents of many investigators rather than the uncommon genius of a few. The project of assembling the great Baconian natural history required a spirit of openness among investigators. If individual discoverers jealously withheld their "secrets" from the public, that collaborative enterprise would certainly fail.

However, the seventeenth-century debate over secrecy versus openness was not simply an argument over the most efficient means by which to advance scientific knowledge. It also had religious, political, and institutional dimensions. The Baconian ideology of openness appealed to the English Puritans because of its ethical component. The gifts of inventors and discoverers were God-given, they argued, and should therefore be used to benefit everyone. The Puritans also used Baconianism to attack monopolies over "esoteric" knowledge, such as medicine and law. Finally, Baconianism gave rise to spirited discussions concerning the organization of science, culminating in the formation of the Royal Society of London in 1661. The issues of secrecy and esotericism played a central role in these debates. Nevertheless, the ideology of openness did not penetrate all layers of the seventeenth-century English scientific community equally. Officially the Royal Society endorsed the Baconian ideal of science as public knowledge; in reality, its commitment to that ideal was only partial. The development of professionalism in science and the pressure of advancing specialization conflicted with the "democratic" Baconianism of the Interregnum. Restoration science was not, after all, to be modeled upon the egalitarian Baconian ideal championed by the Puritans. In practical terms, the doors to Solomon's House would be opened only to those worthy of the scientific calling.

The Critique of Forbidden Knowledge

Bacon opened the debate over esotericism with an onslaught upon traditional theological injunctions against investigating forbidden knowledge, which still had considerable force in the early seventeenth century. In the *Great Instauration*, Bacon wrote that one of the principal obstacles to the advancement of science was the error of thinking "that the inquisition of nature is in any part indicted or forbidden." Whereas God forbade inquiry into the precepts of morality and religion, which are to be accepted on faith, he argued, inquiry into nature's secrets was not forbidden: "For it was not that pure and uncorrupted natural knowledge whereby Adam gave names to the creatures according to their propriety, which gave occasion to the fall. It was the ambitious and proud desire of moral knowledge to judge of good and evil, to the end that man may revolt from God and give laws to himself, which was the form and manner of the temptation. Whereas of the sciences which regard nature, the divine philosopher declares that 'it is the glory of God to conceal a thing, but it is the glory of the King to find a thing out.'"[1]

Bacon went on to describe the adverse consequences of applying to natural philosophy an injunction intended strictly for religion. One result was that it made theologians fearful lest inquiring into the secrets of nature "should transgress the permitted limits of sober-mindedness" in religion. Others feared that "in the investigation of nature something may be found to subvert or at least shake the authority of religion, especially with the unlearned." To these apprehensions Bacon answered, "Natural philosophy is, after the word of God, at once the surest medicine against superstition and the most approved nourishment for faith."[2] There is no danger in curiosity about nature provided that human knowledge is confined to its "true bounds and limitations"—namely, that we do not become so charmed by it as to forget our mortality, that we apply knowledge to the betterment of human welfare, and that "we do not presume by the contemplation of nature to attain to the mysteries of God."[3] Bacon elaborated on the theme of forbidden knowledge in an unfinished fragment entitled *Valerius Terminus of the Interpretation of Nature*, written about 1603. Although the "summary law of nature God should still reserve within his own curtain," he wrote, "yet many and noble are the inferior and secondary operations which are within man's sounding."[4] Overturning the patristic view that the Creator purposely hid nature's secrets from man's prying eyes, Bacon maintained that God had not intended them to be mysteries but rather "took delight to hide his works, to the end to have them found out." Quoting Solomon, he wrote, "*It was the glory of God to conceal a thing, the glory of a King to find it out*; as if the divine nature enjoyed the kindly innocence of such hide-and-seek, hiding only in order to be found, and with characteristic indulgence desired the human mind to join Him in this sport."[5] While the appropriate response to the mysteries of God was humility and wonder, the secrets of nature beckoned. God's game of hide-and-seek *invited* inquiry into nature.

The detachment of the secrets of nature from the traditional domain of forbidden knowledge was an essential part of the Baconian program. Far from being enjoined by the precepts of "true religion," Bacon maintained, natural philosophy had a serious religious purpose. The "true end of knowledge," he insisted, was not curiosity, nor honor, nor material wealth, but "a restitution and reinvesting of man to the sovereignty and power which he had in his first creation."[6] The idea of the Great Instauration—of regaining the dominion over nature that man had lost during the Fall—had a profound influence upon seventeenth-century English science.[7] Being linked with the all-pervasive millennial expectations of the age gave the Baconian program a special urgency and sense of purpose. Moreover, it legitimized the new, positive evaluation of intellectual curiosity. As Hooke wrote of the new philosophy, "And as at first, mankind fell by *tasting* of the forbidden Tree of Knowledge, so we, their Posterity, may be in part *restor'd* by the same way, not only by *beholding* and *contemplating*, but by *tasting* too those fruits of Natural knowledge, that were never yet forbidden."[8]

The Interregnum unleashed a torrent of radical political movements and sectarian religious beliefs.[9] One result of the growth of religious heterodoxy was to bring to the forefront of debate the question of the threat natural philosophy posed to religion. The growth of atheism and materialism alarmed many contemporaries. The archroyalist Walter Charleton said in 1652 that the present age in England had produced more swarms of "atheistical monsters" than any age or nation.[10] Charleton, an atomist, was himself the target of such a charge. The new philosophy was particularly vulnerable to the suspicion of atheism.[11] Closely associated with materialism and moral libertinism, the new philosophy's rise to popularity during the Interregnum coincided with the flourishing of heterodox religious views. Thus apprehensions about science's unorthodoxy were not just the hysterical reactions of religious conservatives; they had real grounds. The potentially atheistic implications of the mechanical philosophy were apparent to all intelligent observers. There was no obvious dividing line between harmless devotion to natural philosophy and the views of notorious materialists such as Hobbes and Spinoza. As one divine put the question, what could be more "absurd and impertinent" than the spectacle of "a Man, who has so great a Concern upon his Hands as the Preparing for Eternity, all busy and taken up with *Quadrants*, and *Telescopes, Furnaces, Syphons,* and *Air-Pumps*?"[12] Given the obvious danger, would it not be wisest to forbid natural philosophy altogether?

The virtuosi, obviously anxious about the growing suspicions of science, rallied to the defense of the new philosophy. Robert Boyle, one of the new philosophy's leading proponents, dedicated several works to defending experimental science against the charge of atheism. In one of his earliest works on the subject, *Some Considerations Touching the Usefulness of Experimental Natural Philosophy*, Boyle wrote, "several divines . . . out of a holy jealousy (as they think) for religion, labour to deter men from addicting themselves to serious and thorough inquiries into nature, as from a study unsafe for a Christian, and likely to end in atheism, by making it possible for men . . . to bring themselves such an account of all the wonders of nature, by the single knowledge of second causes, as may bring them to disbelieve the necessity of a first." Such fears were completely unwarranted, Boyle maintained. God did not intend to make the secrets of nature mysteries, forever concealing them from humanity. Indeed, "whatever God himself has been pleased to think worthy of his making, its fellow-creature man should not think unworthy of his knowing." Far from diminishing God's glory, as some claimed, natural knowledge increased His glory. For we cannot render the praise due the Creator if we remain ignorant of the details of his "curious workmanship." Boyle maintained that "the universal experience of all ages manifests, that the contemplation of the world has been much more prevalent to make those, that have addicted themselves to it, believers, than deniers of a Deity." Developing a standard argument of natural theology, Boyle argued that God's attributes, including his power, his wisdom, and his goodness, were manifested in the creation: "For the works of God are not like the tricks of jugglers, or the pageants, that entertain princes, where concealment is requisite to wonder; but the knowledge of the works of God proportions our admiration of them, . . . that the further we contemplate them, the more foot-steps and impressions we discover of the perfections of their Creator."[13] Natural theology proposed a new *venatio* for science: natural philosophy as a hunt not only for secrets of nature, but even for the secrets of God.

The Baconian Intelligencers

Denouncing the tyranny of philosophical systems, Bacon visualized a scientific community in which efficient organization and effective communication, rather than genius, would be the guarantee of progress: "The course I propose for the discovery of sciences is such as leaves but little to the acuteness and strength of wits," he wrote, "but places all wits and understandings nearly on a level." Bacon's ideal of the organization of scientific inquiry drew its inspiration from the mechanical arts rather than philosophy. Whereas philosophical systems flourish at the hands of the first author, he wrote, afterwards they "stand like statues, worshipped and celebrated, but not moved or advanced." The mechanical arts, by contrast, leave no room for the dictatorial pronouncements of individuals but, "as having in them some breath of life, are continually growing and becoming more perfect."[14] The key to progress in the mechanical arts, Bacon believed, was collaboration rather than "dictatorial" rule by the founders of philosophical systems:

> In arts mechanical the first deviser comes shortest, and time addeth and perfecteth; but in sciences the first author goeth furthest, and time lesseth and corrupteth. So we see, artillery, sailing, printing, and the like, were grossly managed at the first, and by time accommodated and refined; but contrariwise the philosophies and sciences of Aristotle, Plato, Democritus, Hippocrates, Euclides, Archimedes, of most vigour at the first, and by time degenerate and imbased; whereof the reason is no other, but that in the former many wits and industries have contributed in one; and in the latter many wits and industries have been spent about the wit of some one, whom many times they have rather depraved than illustrated.[15]

Such was the paradigm of progress Bacon urged for science. Yet his model scientific society, Solomon's House of the *New Atlantis*, while promoting collaborative research, was far from democratic. In Solomon's House, experimenters were divided into research groups assigned to specific tasks and subjects, working together in an atmosphere that reminds one more of Girolamo Ruscelli's Accademia Segreta than of a modern scientific society. The research groups formed a hierarchy ranging from the lowly Depredators, who collected experiments from books, to the Interpreters of Nature, who interpreted the experiments and created axioms. Like the Segreti, the Solomonians consulted among themselves to decide "which of our inventions and experiences which we have discovered shall be published, and which not; and take all an oath of secrecy for the concealing of those which we think fit to keep secret;

though some of those we do reveal sometimes to the state, and some not."[16] The scientists of Bensalem, in effect members of an international scientific freemasonry, placed the advancement of knowledge above even the interests of the state.

Like all of Bacon's works, the *New Atlantis* was conceived within the framework of an all-embracing plan for religious reform whose aim, as we have seen, was the restitution of human dominion over the natural world. Moreover, Bacon proposed a new model of the scientific investigator as one dedicated to the pursuit of knowledge for the public good. No longer would science exist merely for the pleasure and comfort of the privileged few; it was to be used for the advancement of the commonwealth in general.[17] Although his works had little impact upon contemporaries, with the Puritan ascendancy his reputation underwent a complete reversal. Since Bacon had developed his philosophy within the context of speculation about humanity's Fall, the Puritans found his views to be particularly congenial to their reform program. In the Puritan version of history, humans lost their dominion over creatures with Adam's fall. The life of ease in the Garden of Eden was replaced by a life of endless labor and struggle. However, the Fall was not irreversible: the Puritans believed that with the reform of the church to its primitive state of purity, the New Eden would become a reality. Humanity would again reign as master of the earth under the personal rule of Jesus Christ.[18]

The Puritans also endorsed Bacon's insistence that the true end of knowledge was the improvement of human welfare. Bacon insisted that traditional natural philosophy had dissolved into useless speculation. Like the spider that works on its own web, traditional philosophy "is endless, and brings forth indeed cobwebs of learning, admirable for the fineness of thread and work, but of no substance or profit."[19] The problem was that philosophy had lost sight of the proper criterion of truth. As in the spiritual life, where the inner light of God's grace must be tested and proved by works, philosophical method must be proven by its utility. Since only a true method is capable of producing useful things, works are the proof of its truth. "Truth therefore and utility are here the very same things: and works themselves are of greater value as pledges of truth than as contributing to the comforts of life."[20] Drawing a parallel between the reformation of learning and the spiritual renewal of humanity, Bacon wrote, "The rule in religion that man should demonstrate his faith in works applies in natural philosophy as well; knowledge should be proved by its works. For truth is revealed and established by the evidence of works rather than by disputation or even by sense. Hence the human intellect and human conditions are improved by one and the same means."[21]

The fall of the English monarchy presented the Puritans with the prospect of carrying out their own version of Bacon's philosophy. When the Puritans gained control of Parliament in 1640, they considered dozens of proposals for educational and social reform. Many of these projects emanated from the circle of reformers whose nucleus was composed of three foreigners: Samuel Hartlib (d. 1662), a Palatinate refugee; John Dury (1596–1680), the son of a Scottish minister exiled in Germany; and the Czech reformer Jan Amos Comenius (1592–1670). Hartlib had settled in England in 1628. In the summer of 1641, at Hartlib's invitation offered on behalf of Parliament, Dury and Comenius followed. All had come with high hopes of finding fertile soil in which to transplant their vision of "Pansophia," or universal knowledge, an ideal that emerged out of Protestant Pietism in Germany and eastern Europe.[22] The brainchild of Johann Valentin Andrae (1586–1654), a German Lutheran and the supposed founder of the Rosicrucians, Pansophia essentially meant the unification of all scientific, philosophical, political, moral, and religious knowledge into one all-embracing harmonious worldview.[23] The concept of Pansophia inspired numerous educational reform tracts. It underlay such works, for example, as Dury's *Reformed School* (1650), his *Proposalls Toward the Advancement of Learning* (1653), and Hartlib's *True and Ready Way to Learne the Latin Tongue* (1654).[24] One of the most comprehensive statements of the pansophic ideal was Comenius's *Via Lucis*, written about 1641 but not published until 1668. The metaphor of the struggle between light and darkness, signifying knowledge and ignorance, pervaded the work and gave it a sense of urgency. True knowledge is one, wrote Comenius. It is Light, "a brightness which spreads itself over things and manifests and reveals them," whereas darkness or ignorance "is an obscurity which wraps up and hides things, with the result that we cannot see what surrounds us." In the unceasing struggle, ignorance has gained the advantage, so that in the present age, "darkness covers the earth and gross darkness the people." The remedy for this state of affairs, wrote Comenius, was to restore the Universal Light through a sweeping educational reform that would lead mankind back to "Universal Notions, original and innate." Comenius foresaw the dawning of a new age, the "day of Universal Light," in which the world, after centuries of darkness, will once again be illumined by the universal truths that God revealed at the beginning of creation. He took encouragement from the advent of printing, which spread the light of knowledge, and from the discovery of new "secrets" in science and religion: "With greater skill and ever-increasing effort, and also with happier results, many men are shrewdly investigating the secrets of Nature, the deep places of the Scripture and all the inmost regions of the human heart, and from those treasuries are daily bringing forth into the light more and greater things; and all this comes with a notable increase of human wisdom and the clearest evidence that the divine Providence is now bringing to birth that which it foretold long ago."[25]

With the outbreak of the Thirty Years War, the pansophic enterprise collapsed on the European continent. But its demise did not signal the end of Hartlib's dream of universal restoration. For in the meantime, he had become acquainted with the works of Bacon, whose description of the utopian community of Bensalem in the *New Atlantis* (1626) bore unmistakable similarities to Christianopolis, the fictional utopian community that inspired the pansophic ideal.[26] Bacon's ideas made a deep impression upon the Hartlib circle, which became a pressure group for the promotion of Baconian projects. Whatever else may be said about Hartlib's contributions to science, his enthusiasm inspired an ardent sense of optimism about the possibilities of applying science for social and moral improvement. Largely as a result of Hartlib's propaganda, Baconianism became incorporated into a comprehensive ideology that conceived of science as an instrument to promote Christian unity. The members of the Hartlib circle were passionately committed to the idea of collaboration among scientists and inventors. Besides reconciling warring religious factions, collaboration offered the hope of improving human welfare by promoting useful inventions. The Puritan commitment to the Baconian program thus gave Bacon's plea for scientific collaboration a potent moral force. Hartlib's goal—to build a society in which natural philosophers, physicians, and craftsmen would be encouraged to publish their secrets for the general good of the commonwealth—became one of the major aspirations of the social reformers under the Protectorate. The Revolution offered them the opportunity to realize Bacon's *New Atlantis* in England.

In 1641, the Long Parliament was presented with an anonymous tract entitled *A description of the Famous Kingdome of Macaria*, a utopian work modeled upon Bacon's *New Atlantis*. The author of the tract was Gabriel Plattes (ca. 1595–1644), a craftsman and inventor whose cause was taken up by Hartlib.[27] An intimate member of Hartlib's circle, Plattes had already published works on mining and agricultural improvement, hoping, as he put it, "to turne Plowmen into Philosophers, and to make them to excell their predecessors, even as a learned Physician excelleth an Empericke."[28] With *Macaria*, he offered nothing less than a blueprint for the Great Instauration in the English commonwealth. In many respects, the utopian kingdom of Macaria resembled England. It was governed by a king and parliament; its economic base was a mixture of trade, agriculture, and fishing. Unlike England, however, Macaria was free of religious turmoil and its inhabitants enjoyed prosperity, happiness, and good health. The principal difference between the two kingdoms, according to Plattes, was that the Macarians placed the public good above private gain. Above all, the Macarians valued the universal dissemination of useful knowledge. They erected a "college of experience," which, like Solomon's House, collected and publicized "any experiment for the health or wealth of men."[29] Plattes believed that scientists and inventors had a moral obligation to communicate knowledge of their discoveries and inventions to the public. Attacking the monopolistic and secretive habits of projectors and inventors, he insisted that "knowledge that concerneth the publick good, ought not to be concealed in the brests of a few."[30] In one of his agricultural works, he proposed erecting a "Colledge for Inventions in Husbandrie," presumably modeled upon the College of Experience in *Macaria*.[31] Knowledge, wrote Plattes, "bee worth all the merchandize in the kingdome." Echoing a theme familiar since the time of Fioravanti, he expressed the hope that the spread of knowledge made possible by printing would bring tyranny to an end: "The Art of Printing will so spread knowledge, that the common people, knowing their own rights and liberties, will not be governed by way of oppression and so, by little and little, all Kingdomes will be like to *Macaria*."[32]

Macaria was but the first of many Interregnum proposals aimed at bringing knowledge of science and inventions into the public domain.[33] One of Hartlib's most ambitious schemes was his attempt to establish a national Office of Address modeled after Théophraste Renaudot's (1584–1653) Bureau d'Adresse, which began weekly meetings in Paris in 1633.[34] In addition to being a registry and clearinghouse for goods and services, Renaudot's pioneering agency held regular conferences on scientific and technical subjects. All topics were admitted for discussion, with two important exceptions: religion and affairs of state. The bureau was a place where *civility* governed discussion: "Not only is slander banished, but for fear of irritating minds easily upset by problems of Religion, all such concerns are referred to the Sorbonne. The mystery of affairs of state, partaking of the nature of divine things of which those who have the most to say, say the least, we refer them to the Conseil [du Roi] from where they proceed. All the rest are here, to give free play to your imagination."[35] The Bureau d'Adresse, which was public, not private, and which sought publicity rather than shunning it, broke all the rules governing the elitist, exclusive French academies.

The aims of Hartlib's proposed Office of Address were almost identical to those of Renaudot's bureau, with one notable exception: religion was not banished from discussion. Hartlib's goal was nothing less than to ensure that "all that which is good and desirable in a whole Kingdome may be by this means Communicated unto any one that stands in need thereof." The Office of Address was to have two separate departments. The Office for Address of Accommodations, like Renaudot's bureau, was an agency for the exchange of labor, services, and commodities. The other department, the Office of Address for Communications, was to serve as a clearinghouse of international correspondence for the advancement of religious, scientific, and technical knowledge. Its aims were threefold: first, to rectify mistakes in religion and to "stirre up and waken the sense and love of Piety"; second, "to put in Practice the Lord *Verulams* Designations, *De augmentis scientiarum*, amongst the Learned"; and third, "to offer the most profitable Inventions which [it] should gaine, unto the benefit of the State, that they might be Publikely made use of."[36] Although the scheme was never realized, it generated a great deal of discussion over the idea of creating a national scientific society. Charles Webster concluded, "Perhaps more than any other factor it was responsible for publicizing among scientific intellectuals, the merits of a national scientific institution."[37]

Almost immediately following the announcement of Hartlib's Office of Address in 1647, offers of advice for its implementation poured in. One of the most enthusiastic supporters of Hartlib's proposal was William Petty (1623–1687), a surveyor, inventor, and projector who later went on to earn a medical degree at Oxford. Petty believed that the chief problem with the present state of learning was "want of an union" among intellectuals. In his *Advice of W.P. to Mr. Samuel Hartlib* (1647), he endorsed the Office of Address as a "great Design, whereby the wants and desires of all may bee made known to all, where men may know what is already done in the businesse of Learning; What is at present in doing, and what is intended to be done: to the end that by such a generall communication of designes, and mutuall assistance, the wits and endevours of the world may no longer be as so many scattered coals or firebrands, which for want of union are soon quenched, whereas being but layd together they would have yeelded a comfortable light and heat."[38] The son of a clothier, Petty was primarily interested in promoting trade and manufacturing in the realm. Hence he reoriented the concept of the Office of Address, completing ignoring the religious and pansophic foundations of the enterprise and making its principal goal the compilation of Baconian histories of trades. His proposal called for the building of a *gymnasium mechanicum*, or college of tradesmen, a vocational school for craftsmen that would also serve as a vehicle for compiling histories of the trades. In addition to the gymnasium, there would be a *nosocomium academicum*, or medical research institute, and a *theatrum botanicum* with cages for exotic animals and birds, ponds for fish, a repository for rarities, and a library. Petty gave detailed instructions for compiling the history of trades. He proposed to put readers to work perusing books and recording mechanical inventions. Like the Accademia Segreta, Petty's scholars would "re-experiment the experiments" they found in books, sift the "real or experimental learning" from the false, and compile "one book, or great work."[39]

Numerous other schemes for institutionalizing Baconian science on a national scale emanated from the Hartlib circle. A proposal by Dury suggested converting the royal ordnance at Vauxhall, recently confiscated by Parliament, into a college of mechanics similar to that proposed by Petty. Vauxhall College would bring together artisans and mechanics from England and abroad, and would provide a place "To make Experiments and trials of profitable Inventions."[40] In a rather bizarre scheme, the engineer Thomas Bushell, claiming to be Bacon's chosen disciple for the creation of Solomon's House, proposed establishing Solomon's House at Bushell's estate in Lambeth Marsh.[41] In 1659, John Evelyn discussed with Robert Boyle still another proposal for setting up a college along the lines of Solomon's House. Evelyn's scheme is particularly interesting for its resemblance to the Italian academies of secrets, which he may have learned about during his extended travels abroad. Noting the failure of earlier attempts to establish Solomon's House, he proposed founding a society on a thirty- or forty-acre site outside of London, where a small group of gentlemen might "join together in society, and resolve upon some order and œconomy, to be mutually observed, such as shall best become the end of their union, [and to] resign themselves to live profitably and sweetly together." Evelyn's quasi-monastic society would comprise nine fellows who would be provided with chambers "after the manner of the *Carthusians*." Their experimental work would be conducted in a central house furnished with a laboratory, a library, and a repository for "rarities and things of nature." The house would be surrounded by grounds with an aviary, a garden, and an orchard. The society was to be governed by a strict rule, as in a monastic order, which specified "the promotion of experimental knowledge, as the principal end of the institution." The rule designated periods for reading, discussion, and experiments, and stipulated that "No stranger [be] easily admitted to visit any of the Society, but on certain days weekly."[42]

It is not known how Boyle reacted to Evelyn's proposal. However, judging from an early published work, he would not have been sympathetic to the exclusive, secretive, almost monastic character of Evelyn's imaginary society. Boyle's tract *An Epistolical Discourse . . . inviting all true lovers of Vertue and Mankind, to a free and generous Communication of their Secrets and Receits in Physick* was printed in a collection of works edited by Hartlib.[43] Boyle first met Hartlib in 1646 through his sister, Lady Ranelagh, who was then Hartlib's patroness. Only twenty when he met Hartlib, Boyle immediately fell under the reformer's spell and took up the cause of Pansophia.[44] Boyle came to believe, like Hartlib, that the goal of creating an English New Atlantis required free and open communication of scientific and technical knowledge. One of the main obstacles blocking the way to that goal, he insisted, was the tradition of secrecy and esotericism that inhibited inventors from revealing their secrets and frustrated attempts to penetrate esoteric disciplines like alchemy.

In the *Epistolical Discourse*, Boyle articulated a powerful argument against secrecy and esotericism in science. Written in the form of a letter from Philaretus (Boyle), a natural philosopher, to Empyricus, an empirical chemist famed for his medicinal compositions but caught up in traditional habits of secrecy, the work was an impassioned plea, on moral, religious, and practical grounds, for openness. Philaretus argued, first, that Empyricus had a religious duty to reveal his secrets. "Our Saviour assureth us, that it is more blessed to give than to receive," he argued, "And therefore, the more diffused, and the less selfish, and mercinary our good actions are, the more we elevate our selves above our own, and the neerer we make our approximation to perfections of the Divine nature." Charity also demanded open disclosure of beneficial knowledge. Boyle chastised the "secretists" who withheld their rare recipes from the public, comparing them to usurers who hoard up their goods out of vanity and avarice. "The avarice of profitable secrets, is by so much worse than that of money, by how much the buried Treasure is more excellent," Boyle wrote. "How universally should it be execrated, that in a scarcity would keep his Barnes cram'd, whiles he beholds his pining neighbours starving for want of bread? And yet the censured Miser cannot bestow his corn without losing it; whereas receipts, like Torches, that in the lightning of others, do not wast themselves, may be imparted without the least diminution."[45]

Boyle went on to answer typical objections to divulging secrets, most of which are by now familiar. To those who argued that revealing their secrets would threaten their livelihood, Boyle answered that physicians and empirics ought to place the health of their patients above material gain. Those who withheld secrets out of a desire for fame he entreated "to consider whether it be not a greater glory to oblige (and have many wear one Livery) than to gain the reputation of having buried hoards." One of the "noblest Prerogatives" of secrets, Boyle pointed out, is to be "diffusive of it self." Besides, if the original inventors of secrets had all concealed their knowledge, humankind would have no knowledge at all. Others objected that revealing secrets profanes them, like pearls cast before swine. To this Boyle responded, "Those secrets that were intended

for our use, are not at all profaned by being made to reach their end. . . .
[God] intendeth them for the good of all Mankind, and to make that
Almoner to whom he trusteth them, not the grace but the steward of his
graces." Besides, Boyle argued, divulging secrets would lead to public
examination of them and would ultimately serve to sift the grain from
the chaff, for in the marketplace of ideas, where everything is tested and
only the genuine survives, " 'tis very unlikely, there should be then more
false receipts believed, when there are more true ones extant to confute
them."

Although Boyle's "Invitation" was addressed to an imaginary medical
empiric, his target audience was wider and closer to home. In addition
to attacking secretive alchemists and empirics, he denounced the cult of
curiosities that preoccupied many of the virtuosi, distracting them from
what Boyle considered to be the true aims of science. Indeed, Empyricus, the "cozening" empiric of Boyle's tract, was perhaps more of a
trope constructed out of literary figures such as Alessio Piemontese than
a figure modeled on any real-life character. But even if he was a kind of
stock character, he represented cultural values that were widely held in
the virtuoso community. Since collecting "rarities" was restricted to
those who could afford it and who had the leisure to indulge in it, the
cult of curiosities was a mark of status setting the virtuosi above the
common crowd. In addition to rare objects and exotic specimens, "curiosities" also included alchemical formulas, medicinal recipes, and recipes
for perfumes, cosmetics, scented oils, and other luxury items.[46] The superior tone of the gentleman-scholar emerges continually in the correspondence of virtuosi such as John Evelyn. Despite his professed adherence to the Baconian program and his ardent support of the history of
trades project, Evelyn's own contributions were confined to histories of
"aristocratic" arts such as engraving, oil painting, miniature painting,
annealing, enameling, and marbling paper. In the outline for a history of
trades he presented to the Royal Society, Evelyn preserved a hierarchical
ranking of the trades, beginning with the "Usefull and purely Me-
chanic," ascending to the "Polite and More Liberall," then to the "Curious," and ending finally with "Exotick, and very rare Seacretts."[47]
Considerations of status also reinforced the feeling that secrets should
not be spread abroad or debased by being put to mechanical or practical
uses. Thus Evelyn opted against publishing his findings because he
feared it might "debase much of their esteem by prostituting them to
the vulgar."[48] He finally abandoned the project altogether because of
the necessity of "conversing with mechanical and capricious persons."[49]

Bacon condemned the cult of curiosities and inveighed against the
virtuosi's preoccupation with rarities to the neglect of "experiments familiar and vulgar": "For it is esteemed a kind of dishonour unto learning

to descend to inquiry or meditation upon matters mechanical, except they be such as may be thought secrets, rarities, and special subtleties."[50] Hartlib followed in Bacon's path, urging the virtuosi to communicate their secrets to the public at large. Like his mentor, Boyle too denounced the cult of curiosities and the ethic of exclusiveness that characterized the virtuoso movement. Knowledge should not exist for the chosen few, he maintained, but should be put to use for everyone's benefit. The virtuosi's unwillingness to reveal their secrets militated against the aim championed by the Hartlib circle, of putting knowledge to the service of humankind.

Building Solomon's House

The belief that esotericism had frustrated the growth of knowledge underlay all of the reform projects that emanated from the Hartlib circle. It is thus unhelpful to dismiss Hartlib's projects, as A. R. Hall does, on the grounds that they were infused with the mystical ideals of the "German school of practitioners of esoteric or Hermetic arts."[51] Hall's assertion is not borne out by contemporary impressions of Hartlib. Although the reformer was deeply influenced by the Hermetic currents of the day, he denounced those who withheld secrets out of a desire for profit.[52] Moreover, contemporary descriptions of Hartlib do not convey the impression of a mystic or an occultist, but of a utopian and philanthropist who advocated and practiced open communication of knowledge. Visiting the projector in November 1655, Evelyn described Hartlib's house as a general storehouse of information. "This Gent[leman] was Master of innumerable Curiosities, & very communicative," Evelyn wrote.[53] Hartlib was a prolific correspondent whose contacts included Boyle, Petty, Evelyn, John Beale (1603–1683), Christopher Wren (1632–1723), and others. His large network of correspondence on the continent made him, prior to the establishment of the Royal Society, England's "intelligencer" to the European scientific community.[54] Hartlib seemed to know everyone and where anyone might be found. He came to be regarded as an agent and go-between on behalf of individuals and for projects he thought would advance the public good. His correspondence was staggering in its extent. He received letters, books, and information about scientific work going on all over the world. John Winthrop, the governor of Connecticut, called him "the great Intelligencer of Europe."[55] When Henry Oldenburg (1618–1677) first visited England in 1653, he joined the Hartlib circle and added to it his own contacts with Continental scholars. Oldenburg's correspondence before 1660 is dominated by his exchanges with the Hartlib group. Thus when he was appointed to supervise the correspondence of the Royal Society as the group's first secretary, Oldenburg "had effectively taken over from Hartlib as the major English correspondent of continental natural philosophers."[56]

In addition to inspiring social reform projects, Baconianism stimulated informal associations among individuals interested in promoting experimental science. A number of informal philosophical clubs emerged during the Interregnum. One of these groups began meeting in London in 1645 and later moved to Oxford.[57] Another group, which Boyle called the "Invisible College," was essentially an informal gathering of Anglo-Irish intellectuals patronized by Boyle's sister, Lady Ranelagh.[58] Attempts to find in one or another of these groups the "precursor" to the Royal Society of London have not so far been successful.[59] Nor has it been possible to discover in these various groups a homogeneous religious and political ideology. In particular, attempts to establish a direct connection between Puritanism and science—most recently in the powerful and learned work of Charles Webster—have foundered, largely against the plain fact that "many of the most active scientists of the Interregnum were neither Puritans nor collaborators with the regime."[60] If anything united the scientists of the Interregnum and linked them to the Restoration scientific movement, it was their reaction against sectarianism, their revulsion against extremes in politics and religion, and their aversion to philosophical and religious dogmatism. In other words, the "ideological" foundations of seventeenth-century English natural philosophy were religious and political moderation, not sectarianism or extremism. Science tended to transcend political parties and religious sects. Barbara Shapiro has convincingly argued that "latitudinarianism," as this moderate stance was called, best describes the ideology of seventeenth-century English science.[61] The latitudinarians saw science as an instrument to calm religious dissension and to promote unity. Baconianism appealed to them because of its stand against dogmatism, its distrust of speculation, and its stress upon collaborative research. These ideals were articulated in the official apologia Thomas Sprat wrote for the Royal Society. Experimental philosophy bred a race of men "invincibly arm'd against all the inchantments of *Enthusiasm*," Sprat contended. Straining to disassociate the Royal Society from the "passions and madness of that dismal Age," the Interregnum, he contrasted the society's "moderation" to the views of the "modern zealots." "Seeing we cast *Enthusiasm* out of *Divinity* it self," he wrote, "we shall hardly sure be perswaded, to admit it into Philosophy." By concentrating its energies upon matters of fact, experimental philosophy avoided the rancor of academic philosophy: it "proceeds on *Trials*, and not on *Arguments*." Besides promoting "rational religion" instead of enthusiasm, "*Experimental Philosophy* will prevent mens spending the strength of their thoughts about *Disputes*, by turning them to *Works*. . . . it will free them from *perversity*, by not permitting them to be too peremptory in their *Conclusions.*"[62] Sprat and others envisioned the Royal Society as an organization composed of open-minded, moderate men of diverse professions and social backgrounds, who would arrive at a philosophy of nature that would satisfy all reasonable men.

The Royal Society of London received its charter from Charles II in 1662. With its establishment, the principle of science as public knowledge gained institutional endorsement. The society brought together virtuosi of diverse social backgrounds and philosophical outlooks into a single organization. It also gave prominence and legal standing to the virtuosi's purpose of advancing natural knowledge along the path mapped out by Bacon. Although the society lacked unanimity on theoretical issues, its fellows were united on two basic principles: the ideal of cooperative research and the goal of establishing "matters of fact." Joseph Glanvill reported that the Royal Society was the fulfillment of Bacon's "mighty Design": "To the carrying of it on, It was necessary there should be many *Heads* and many *Hands*; and *Those* formed into an *Assembly*, that might *intercommunicate* their *Tryals* and *Observations*, that might joyntly *work*, and joyntly *consider*; that so the *improvable* and *luciferous Phænomena*, that lie scatter'd up and down in the *vast Campaign* of *Nature*, might be *aggregated* and brought into a *common Store.*"[63] Sprat celebrated the company's diversity, noting that "they have freely admitted Men of different Religions, Countries, and Professions of Life."[64] Despite this professed ideal, the Royal Society's social diversity was more theoretical than real. The overwhelming majority of its members were drawn from the aristocracy and gentry. It was more of a gentleman's club than an organization based on merit.[65] Indeed, Sprat admitted, "though the *Society* entertains very many men of *particular Professions*, yet the farr greater Number are *Gentlemen.*"[66]

In order to put its ideal of cooperative research into practice, the Royal Society committed itself at the outset to communicating the results of its research to the international scientific community. Bacon, noting that memory alone was insufficient to retain the vast number of experiments necessary for the reconstruction of science, had urged that experience be made to "learn her letters": "No course of invention can be satisfactory unless it be carried on in writing," he declared.[67] Accordingly, at their very first meeting, the society resolved to take detailed notes on all of their deliberations, and to preserve the notes in a permanent record book. The fellows appointed Henry Oldenburg as the society's first secretary, charging him with maintaining the organization's correspondence with members in England and abroad.[68] Oldenburg was an ideal candidate for the job. Well educated, widely traveled, and fluent in all the major European languages, he threw himself into the task assigned him with an almost religious devotion. Largely on his own initiative, he vastly expanded the society's contacts with the European virtuosi. Within a few years, the Royal Society of London had become an international center for the exchange of scientific and technical information.

In the Renaissance, the private letter, because of its flexibility and immediacy, had been an important medium for the communication and exchange of natural knowledge. Conrad Gesner (1516–1565) maintained a wide correspondence with leading European naturalists, who sent him specimens of plants, animals, and gems, and who visited him at his home in Zurich.[69] From Naples, Ulisse Aldrovandi exchanged letters and specimens with virtuosi throughout Italy and abroad.[70] In the early seventeenth century, the French scientific community benefited from the extensive correspondence maintained by Marin Mersenne (1588–1648), who, as Baillet remarked, had epistolary relations with "tous les Scavans et Curieux de l'Europe."[71] These channels of communication, however, were more or less informal and restricted to small circles of personal contacts. Oldenburg, on the other hand, acting officially on behalf of the Royal Society, established systematic correspondence with the European virtuosi and provided a public forum for announcing new discoveries and for arbitrating scientific controversies. In a letter to Johannes Hevelius in February 1662/63, Oldenburg wrote:

> Friendship among learned men is a great aid to the investigation and elucidation of the truth; if friendship could be spread through the whole world of learning, and established among those whose minds are unfettered and above partisan zeal, because of their devotion to truth and human welfare, philosophy would be raised to its greatest heights. This our Fellows are striving for with all their might and for that reason they are developing a wider correspondence with those who philosophize truly. . . . We entertain a bright hope that France, Italy, and the Netherlands, and the peoples of other countries will be brought together in aiding and honoring this purpose with all their might, that they will eagerly seek to attain it, and that whatever blemishes the dignity of philosophy will be cast out from it, while its honor and esteem will be defended as far as human means will allow.[72]

These goals, which Oldenburg communicated to scholars throughout Europe, bound the Royal Society to an ideal of public, corporate science and made the society an international clearinghouse for scientific information. Oldenburg's correspondents were made to understand that the contents of their letters might be read or summarized at the society's meetings, and would be entered in abstract into its register. This provided a means for establishing priority in scientific discoveries and for settling priority disputes. The recognition and status conferred by the leading scientific society in Europe induced many natural philosophers to accept free communication of scientific information as a new norm.

The inauguration of the *Philosophical Transactions of the Royal Society of London* in 1665 was perhaps the single most important factor in establishing the society's reputation as Europe's most prominent scientific institution.[73] Scientists throughout Europe became convinced that the *Philosophical Transactions* were, as John Wallis put it, "a proper place for communicating new discoveries."[74] The society's growing reputation as an authoritative body encouraged the virtuosi to submit their work to the judgment of its fellows. The astronomer Johannes Hevelius (1611–1687), praising the society for its work, expressed his confidence that "thus will hidden secrets be revealed at last and new miracles appear that were formerly concealed in the majesty of nature." Hevelius went on to assure Oldenburg that as soon as his own work on comets was published, "I will make it my first care to submit it to the high judgment and due consideration of the Royal Society."[75] The French astronomer and engineer Pierre Petit (1598–1677) also praised the "celebrated Society, to whose judgment I submit all my ideas."[76] Somewhat reluctantly, the society occasionally found itself adjudicating scientific disputes and thus became de facto an international scientific court of appeal.[77] Hevelius's work on comets, for example, touched off a prolonged controversy with Adrien Auzout (1622–1691) over the accuracy of Hevelius's observations of the comets of 1664–1665. The society's members, acting as impartial judges, concluded that Auzout's observations were more accurate than those of Hevelius. In the midst of the dispute, Hevelius assured Oldenburg, "I am in good spirits so long as my voice can be heard."[78]

The other principle that guided the Royal Society's deliberations was also Baconian. The founders agreed that the company ought to avoid debates over theory and to strive for agreement upon "matters of fact." "Their *first* and *chief* Imployment is, carefully to *seek*, and to *report* how things are *de facto*," Glanvill wrote, "and they continually declare against the *establishment of Theories*, and *Speculative Doctrines*, which they note as one of the most considerable *miscarriages* in the *Philosophy* of the Schools. . . . Their aims are to free *Philosophy* from the vain *Images* and *Compositions* of *Phansie*, by making it *palpable* and bringing it down to the *plain Objects* of the *Senses*."[79] In contrast to the philosophers of the "notional way" (i.e., the Scholastics), the virtuosi eschewed contentious theorizing and stuck to the facts. Sprat wrote that "They have been cautious, to shun the overwheening *dogmatizing* on causes on the one hand; and not to fall into a *speculative Scepticism* on the other." The purpose of the society's assemblies, Sprat reiterated, was "to *judg*, and *resolve* upon the matter of *Fact*."[80] The new philosophers repeatedly insisted that the mechanisms and causes of nature were secret, and that debates over such matters were unresolvable. What could be agreed upon were matters of fact.

But how are matters of fact produced, judged, and "resolved upon"? What constitutes a matter of fact? As Steven Shapin and Simon Schaffer have shown, nothing was more crucial to the success of the new philosophy than defining the boundary between fact and opinion, and between facts and theoretical claims.[81] In their detailed studies of experimentalism in the Royal Society, Shapin and Schaffer argue that the production of experimental knowledge rested upon a set of conventions for generating matters of fact. First of all, in order to be trustworthy, such facts had to be witnessed; they had to be produced in a public space before reliable witnesses. Since the resolution of matters of fact required the unanimous assent of witnesses, experiments had to be performed in a social space. The experimental laboratory envisioned by the Royal Society thus contrasted with the secretive space of the alchemist's closet and with the closed "laboratories" of the academies of secrets. Thus Boyle intended to draw "the chymists doctrine out of their dark and smoky laboratories" and bring it "into the open light."[82] Experimental knowledge had to be *public* knowledge.

However, the spaces where experiments were performed also had to be *disciplined* spaces. Proper manners had to be observed in order to preclude disputes about what was being witnessed. Modesty was to be observed in regard to discussion about causal explanations; dogmatic pronouncements were to be avoided; disputes were to be managed with "civility."[83] Discipline was also necessary to insure that experiments were not performed merely as spectacles, to excite wonder in the beholder or to gratify the senses. Such were the "experiments" of the quacks and mountebanks, with whom the virtuosi were sometimes associated.[84] Evelyn's account of an experiment with phosphorus performed before the Royal Society in 1641 illustrates this danger. Although Evelyn had seen numerous phosphors, none glowed as brightly as the one demonstrated on this occasion. Preparing some of it himself, he rubbed it on his hands and face and appeared in the dark "like some spirit, or strange apparition." Evelyn imagined all kinds of mischief had the papists first discovered the substance: "for had they the *seacret* onely, what a miracle might they make it, supposing them either to rub the Consecrated Wafer with it, or washing the Priests face & hands with it, & doing the feate in some darke Church or Cloyster, proclaime it to the Neighbourhood; I am confident the Imposture would bring thousands to them, & do an infinity of mischiefe, to the establishing of the common error about Transubstantiation; all the world would ring at the miracle."[85] Evelyn knew whereof he spoke: the only phosphor comparable was one he had seen years earlier in the Piazza Navona in Rome, where a mountebank performed tricks with a phosphorescent ring, "and having by this surprizing trick, gotten Company about him, he fell to prating for the vending of his pretended Remedies."[86] Evelyn always regretted he had not purchased the recipe.

Demonstrations of rarities and unusual phenomena were important resources in expanding the public culture for science.[87] But the danger that experiments might simply bedazzle onlookers instead of enlightening them was always present. This was why it was important that the space within which experiments were performed be a *disciplined* and to some extent *restricted* space. In order to insure that experimental scientists would not be mistaken for mountebanks or Gimcracks, as well as simply to put the stamp of authority upon matters of fact, the Royal Society insisted that witnesses to experiments had to be reliable and their testimony creditable. As Shapin and Schaffer stress, they had to observe certain social conventions about how knowledge was produced, about what may be questioned and what may not, and about what counts as evidence and proof.[88] What resulted was "a public space with restricted access." In fact, "the space was restricted to those who gave their assent to the legitimacy of the game being played within its confines."[89] Another means by which the Royal Society attempted to reduce the "wonder" of experiments was to insist that experiments be *replicated*. Replication, it was believed, would make secure experimental facts out of what were commonly perceived as wonders. It would also reinforce the distinction between the true experimental scientist and the dilettante or charlatan whose concern was merely to exhibit novelties. As one of the fellows put it, "an Artist or Experimenter, is not to be taken for a maker of gimbals, nor an observer of Nature for a wonder-monger."[90]

The principles of qualified public science were also exhibited in the literary conventions the society adopted. The validity of experimental knowledge depended upon multiple witnesses. Sprat observed that the society never admitted anything as a matter of fact "till the whole *Company* has been fully satisfi'd of the certainty and constancy, or, on the otherside, of the absolute impossibility of the effect."[91] Obviously, however, not *everyone* could be a witness to the production of experimental knowledge; nor, as we have seen, was it practical or feasible to open up experimental spaces to the public at large. In order to multiply the number of witnesses and thereby strengthen the credibility of experimental facts, the society adopted a form of literary technology that Shapin calls "virtual witnessing."[92] Although Boyle originated the technology of virtual witnessing, as a literary form it came to characterize the style of the Royal Society as a whole. The technology of virtual witnessing involved "the production in a reader's mind of such an image of an experimental scene as obviates the necessity for either its direct witness or its replication." The "prolixity" of descriptions of experiments and the detailed, naturalistic illustrations that went into the society's publications aimed to create the impression of verisimilitude. They were designed to convey not just the *idea* of an experiment but "a vivid impression of the experimental scene."[93] Sprat noted that in reporting their experiments the society shunned ornamental language and strove to "represent *Truth*, cloth'd with Bodies, and to bring *Knowledg* back again to our very senses, from whence it was first deriv'd to our understandings."[94]

Consistent with Shapin's findings, Peter Dear emphasizes the novelty of the conventions the Royal Society adopted for communicating discoveries. Dear points out that with the rejection of ancient authority, a new standard of authority had to be found to support one's knowledge statements. In the conventions adopted by the Royal Society, he suggests, the new standard was the authority of the trustworthy reporter. When a fellow of the Royal Society made a contribution to knowledge, he did not write a commentary on a text (the standard procedure in scholastic science) but instead reported an actual event, normally one he had experienced firsthand. "The actuality of a discrete event was the central point to be established in any contribution to the cooperative philosophy of the Royal Society."[95] Frequently a recipelike format appeared in the society's communications. The reader was given a specific set of instructions for the experimental procedure, which he then could replicate. Generally, however, the virtuosi distrusted recipes unless they were accompanied by trustworthy accounts of actual trials attempted. Thus Boyle warned readers "not to be forward to believe chymical experiments, when they are set down only by way of prescriptions, and not of relations; that is, unless he, that delivers them, mentions his doing it upon his own experience."[96] Communications of "relations," examples of which can be chosen almost at random from the pages of the *Philosophical Transactions*, reported events in such a way that they could be "virtually" witnessed.

By such means, the Royal Society established the authority of experimental matters of fact independent of theoretical considerations. As a result of Oldenburg's widening network of correspondence, the internationalization of the Baconian program continued. Oldenburg stressed the importance of cooperative international endeavor. "What we are about is no task for one nation or another singly," he wrote. "It is needful that the resources, labors, and zeal of all regions, princes, and philosophers be united, so that this task of comprehending nature may be pressed forward by their care and industry."[97] To Georg Stiernhelm of Stockholm Oldenburg reported, "It is our object to bring together the studies, labors, and exertions of all the philosophers wherever they may be in compiling a well-ordered philosophical treasury, upon which at some time a solid and fruitful system of natural science may be constructed."[98] Oldenburg invited naturalists all over Europe to contribute communications and specimens to the great natural history that he indicated the society was in the process of compiling. Thus in 1667, initiating what were to become regular exchanges with the Italian naturalist Marcello Malpighi (1628–1694), Oldenburg invited Malpighi "to a whole-hearted alliance of mind and study, so that, through experimental research into Nature, the foundations of philosophy may be laid more solidly, the ancients may be put to the question, and useful novelties may be discovered and made public." Oldenburg announced that the society was "engaged in compiling a history of nature that shall be true and faithful to it." Therefore, he continued, "we earnestly beg you to be so good as to let us know of all that is noteworthy—of which there is much in your island—concerning plants, or minerals, or animals and insects, especially the silkworm and its productions, and finally concerning meteorology and earthquakes, known to you or to other ingenious men."[99] Responding to Oldenburg's request, Malpighi enthusiastically reported, "I have already begun to solicit my friends living in Italy to collect together each the natural history of his own country."[100] By inviting foreign correspondents to use their personal contacts, Oldenburg further widened the international network of scientific correspondence.

The Continental virtuosi emulated the Royal Society by establishing organizations with research programs modeled after the society's Baconian agenda. The French Académie Royale des Sciences, which was founded at Paris in 1666, was explicitly conceived on Baconian principles. Indeed, the Paris Academy, with its exclusive coterie of professional scientists, may have resembled Solomon's House even more closely than did Royal Society. The academy's objective, according to one of its founding members, was "to banish all prejudices from science, basing everything on experiments, to find in them something certain, to dismiss all chimeras and to open an easy path to truth for those who will continue this practice."[101] Cartesians and Jesuits were banned, not because of doctrinal differences, but because they were considered dogmatic partisans of metaphysical creeds rather than open-minded seekers after truth.

The Baconian model also appealed to Italians. After the closing of the Accademia del Cimento in 1660, Italian natural philosophers increasingly looked to England and the Royal Society for a model for the organization of science. The Baconian design of the Royal Society seemed especially appropriate in the context of the Counter-Reformation, as Italian scientists searched for a theoretical position that would permit experimental research without committing them to controversial metaphysical doctrines. Adopting a Cartesian, Gassendian, or Galilean philosophy would almost certainly have embroiled them in heated disputes over metaphysics and raised suspicions of heterodoxy or atheism. Baconianism appealed to the Italians, as it had to Englishmen of the Restoration period, because it appeared to be a metaphysically "neutral" position.[102] One of the first Italian scientific societies to be formed in the wake of the Cimento's closing was the Accademia della Traccia (Academy of Traces), which was founded in 1665 at Bologna by Geminiano Montanari, a Bologna mathematics professor.[103] The society's name, which also means tracks or footprints, reflected an overtly Baconian conception of science as a *venatio*. Montanari reported that the company proceeded along Bacon's path "by tracking down the true understanding of nature along the . . . road of experience."[104] The Royal Society also provided the model for an academy founded at Venice by Giovanni Ambrosio Sarotti about 1681. Sarotti, whose family's business contacts brought him to England in the 1670s, was elected a fellow of the Royal Society in 1679. When he returned to Venice, he took with him an air pump and other instruments of the type being used by the Royal Society. Sarotti's academy carried out a series of pneumatical experiments, largely replicating those performed in the Royal Society.[105]

Publicity and Its Problems

Despite its successes in promoting Baconian ideals abroad, the Royal Society continued to face problems implementing them at home. In the first place, esoteric arts such as alchemy stubbornly resisted open disclosure. During the Interregnum, the distiller and translator John French anticipated a general opening up of alchemical secrets to the public. "I rejoyce as at the break of the day after a long tedious night," he wrote, "to see how this solary Art of Alchymie begins for to shine forth out of the clouds of reproach which it hath a long time undeservedly layen under."[106] Such fervent hopes were not realized. The society's efforts to encourage alchemists to reveal their secrets were continually frustrated. Even when the *adepti* did publish their secrets, they did so in language that was scarcely decipherable. Boyle noted that anyone versed in the writings of the alchemists "cannot but discern by their obscure, ambiguous, and almost ænigmatical way of expressing what they pretend to teach, that they have no mind to be understood at all, but by the sons of the art."[107] Indeed, often the virtuosi were themselves obliged to maintain secrecy when dealing with alchemists. In a communication to the *Philosophical Transactions* in 1674, Boyle published the formula for the preparation of "Helmontian laudanum," reporting that he had learned the recipe some years earlier but had been unable to publish it because it was "a great Secret communicated to me by an expert Chymist."[108] Boyle held on to the recipe for fifteen years until he gained permission from Helmont's son to publish it. Secrecy was the price to be paid for maintaining trusting relations with the *adepti*, who provided Boyle with many experiments. Yet Boyle's attitudes toward disseminating alchemical information was extremely ambivalent. Recent research suggests that Boyle was strongly influenced by the traditional view of alchemy as privileged knowledge, and by moral injunctions against revealing alchemical secrets to the vulgar. Boyle's moral scruples surfaced in a paper on the preparation of a mercury and gold amalgam, in which he cited the "political inconveniences which might ensue if it should . . . fall into ill hands" as an excuse for not revealing more about it. In another instance he concealed the formula for an artificial phosphor because he feared possible "mischievous consequences" of its revelation. Boyle wrote that he had suppressed a number of discoveries in order to prevent them from being "perverted to the prejudice of mankind."[109] Larry Principe has convincingly argued that such expressions, rather than demonstrating Boyle's concern for the "public good," were attempts to demonstrate to the *adepti* his ability to keep a secret and, simultaneously, to advertise the goods he had available to barter for their secrets.[110]

Nor did the Royal Society succeed in penetrating the cloak of secrecy surrounding the trades. The society's project for a great collaborative "History of Trades" never materialized, although much effort was spent on it.[111] From its inception, the Committee for the History of Trades was guided by Bacon's idea that the workshop was a sort of laboratory "which takes off the mask and veil from natural objects, which are commonly concealed and obscured under the variety of shapes and external appearances."[112] Boyle, a member of the original trades committee, argued that "the phænomena afforded us by these arts ought to be looked upon as really belonging to the history of nature in its full and due extent."[113] But at the same time, the society's members were preoccupied with the usefulness of their endeavors and envisioned the organization as one whose design was to fulfill the utilitarian aims articulated by the Hartlib circle. Hooke proclaimed that the ends of the society's inquiries were "above all, to ease and dispatch of the labours of mens hands." Although the virtuosi "do not wholly reject Experiments of meer *light* and *theory*," he wrote, "they principally aim at such, whose Applications will *improve and facilitate* the present way of *Manual Arts*." Condemning "useless" scholastic learning, Thomas Sprat wrote that it was for

"*Mechanicks* and *Artificers*" that "the True *Natural Philosophy* should be principally intended."[114]

During the Royal Society's early years, the history of trades project was one of its main concerns. Evelyn and Petty presented catalogs of trades in 1661. The Committee for the History of Trades was appointed in 1664, and from that point on interest in the project steadily increased.[115] Between 1665 and 1680, the history of trades project more or less constantly preoccupied the society. Discussion of trades took place at almost every meeting. Histories of various arts were read and recorded in the society's register, and several were printed in the *Philosophical Transactions*. The society reviewed histories of saltpeter, gunpowder, alum, varnish, clothing, shipping, dyeing, tar pitch, cider, malt, wine, whale fishing, hat making, paper, and tanning. Sprat listed a full page of completed histories, including those of making bread, ironmaking, refining, lead, ceruse, enameling, and masonry.[116] The quality of the individual histories varied enormously. Many were based upon careful research and firsthand observation, were fairly comprehensive in scope, and were probably fairly accurate. William Petty's history of dyeing stood out as an exemplar of what a history of trades might accomplish.[117] Walter Charleton's *Mysterie of Vintners*, read to the society in 1662, was another example that lived up to the Baconian concept of a history of trades.[118] Charleton viewed the vintner's workshop as a laboratory for the study of basic chemical phenomena such as fermentation. His history of wine making was essentially a treatise on fermentation, which Charleton interpreted in Helmontian terms. But if the work illustrates the potential of the history of trades, it also underscores its limitations. For Charleton's "explanation" of fermentation barely advanced beyond a description of the phenomenon. And while he discussed several possible causes for the souring of wine, he admitted that all were little more than guesses.[119] But if his natural philosophy was deficient, it was not because of faulty observations. The *Mysterie of Vintners* concluded with a large collection of remedies currently used by craftsmen for correcting the defects of wine, many of which Charleton probably learned from wine makers. The work as a whole represents the kind of relationship between scholar and craftsman that Bacon must have had in mind when he wrote the *Parasceve*.

Other histories, however, were based upon limited familiarity with the arts. Henshaw's history of saltpeter and gunpowder, for example, was the product mainly of literary sources such as Pliny's *Natural History* and Johann Glauber's *Prosperity of Germany* (1656–1661), a point made with great relish by Henry Stubbe, one of the Royal Society's most virulent critics. Stubbe chose Henshaw's history as an example of the dubious nature of the society's whole enterprise: "The Narration is not only imperfect; but in many parts *false*, so that for ought I can discerne, the *History of Nature* which they propose to themselves, will not merit any more Credit (if so much) then that of Pliny: and these *Experimental Philosophers* instead of *undeceiving* the age as to *inveterate Errors* will *multiply new ones.*"[120]

From the outset, the Committee for the History of Trades faced numerous obstacles. Foremost among them was the reluctance of craftsmen to reveal their secrets. Sprat mused that the Royal Society's aggressive pursuit of craft secrets might "afright private men, from imparting many profitable secrets to them; lest they should thereby become common, and so they be depriv'd of the gain, which else they might be sure of, if they kept them to themselves."[121] His statement underscores one of the main weaknesses of the history of trades project, which necessarily depended upon tradesmen for reliable information. For why would craftsmen want to give up the very secrets that ensured their income? Evelyn wished "with all my heart, that more of our workmen would . . . impart to us what they know of their several trades and manufactures."[122] Despite professing the standard Baconian line that publication of craft secrets would ultimately improve the arts, he postponed publishing his history of engraving "lest by it I should disoblige some, who make those professions their living."[123]

Yet it was not only craftsmen who were reluctant to give up their secrets. The virtuosi themselves had rare secrets that they coveted dearly and were loath to "vulgarize." Evelyn delayed publishing his accounts of etching, painting, and enamel because he feared their publication might "debase much of their esteem by prostituting them to the vulgar."[124] Edmund Wylde, another wealthy dilettante virtuoso and a fellow of the society, claimed to have a technique for softening steel without fire. Despite numerous attempts by the society to have him demonstrate the technique, Wylde, "esteeming it a secret," refused.[125] National security also prevented the virtuosi from openly publishing secrets. Thus Petty's manuscript treatise on shipbuilding was concealed by William Brouncker, president of the society, on the grounds that it was "too great an Arcanum of State to be commonly perused."[126]

Secrecy, of course, negated the whole point of the history of trades. Nevertheless, the Royal Society as a whole was ambivalent over whether to disseminate or withhold certain kinds of knowledge. Although officially the society professed adherence to the principle of public science, it was unable to persuade all of its members to subscribe to the doctrine in practice. In the seventeenth century, medical, alchemical, and technological "secrets" still had value as commodities and as articles of gift exchanges. To publish such secrets would have diminished their value. In 1659, Henry Oldenburg sent Hartlib a recipe for a rare vitriol process "in acknowledgement of ye secret you sent me, wch shall not loose the name of a secret for me." Oldenburg entreated Hartlib "to communicate it to none, but noble Mr Boyle, who, I am sure, upon my desire will impart it to none but MyLady Ranalaugh, wch is a person, yt can keep a secret as well."[127] As this exchange illustrates, secrets were tokens of exclusiveness and privilege, distinguishing those who could "keep a secret" from those who could not. Exchanging secrets within privileged circles gave Oldenburg and Hartlib, both clients of Boyle, tacit membership in elite society.

These attitudes may help explain the Royal Society's waning interest in the history of trades project after about 1680. Historians have suggested that after hopeful beginnings, Restoration science became increasingly elitist in the 1680s and thereafter.[128] To some extent the society's elitist attitudes reflected the changing character of the organization's membership. The predominance of aristocrats, courtiers, and gentlemen in the society marked it as a kind of high-class social club. Despite claims by Sprat and others that the society championed "useful knowledge," it is surprising how few tradesmen actually joined the society. They constituted only about 4 percent of its membership in 1672. After about 1680, the society's membership included growing numbers of scholars, writers, and professional men, who began to play leading roles in the society's deliberations. A new generation of natural philosophers, who were less interested in technological subjects and less ardently Baconian than the original members of the society, tended gradually to take over.[129] Then too, the scope of the history of trades project was simply too vast to be accomplished even by men fired by Baconianism. After all, its promoters were virtuosi, men of a thousand interests who naively thought they could quickly master any craft well enough to be the artisan's instructor. But the history of trades was no task for dilettantes; it required the prolonged, concentrated effort of generations.

Although the decline of the history of trades may have been to some extent a reflection of the growing elitism of Restoration society, the change in the organization's attitude toward mechanical arts was also symptomatic of changing perceptions of its role as a scientific institution. The question the Royal Society faced during the 1670s and 1680s was whether its earlier commitment to a Hartlibian style of Baconianism was feasible during a time of dwindling resources and declining active membership. Would the Royal Society continue to see itself as a "projecting" institution, with a wide variety of members contributing to the great Baconian natural history, as Sprat depicted it, or would it become an exclusive scientific research institute dedicated more to inquiry into theoretical matters? The society's general trend toward a more restricted and exclusive view of its character can be detected in Hooke's change in attitude toward the history of trades. Whereas in the *Micrographia* (1665) he reported that the society's inquiries were intended principally to "*improve and facilitate* the present way of *Manual Arts,*" he took a radically different view in his *General Scheme, or Idea of the Present State of Natural Philosophy*, which dates from the 1680s. In the latter work, Hooke urged confining the history of trades to "such a Description of [the arts] as is only in order to the Use of Philosophical Inquiry, for the Invention of Causes, and for the finding out the ways and means Nature uses, and the Laws by which she is restrain'd in producing divers Effects."[130] These views were consistent with Hooke's changing role as the society's curator of experiments. He began his career as a typical "virtuoso curator," performing a chaotic medley of largely utilitarian experiments on demand. Gradually, however, he shifted his demonstrations from the strictly utilitarian to experiments aimed at deciding theoretical issues—that is, at establishing "the *real*, the *mechanical*, the *experimental* Philosophy."[131] Hooke's case illustrates an important change in the natural philosopher's role: the virtuoso was gradually evolving into a professional scientist.[132]

What can be said about Hooke can be said about the Royal Society as an institution. The 1670s and 1680s constituted a period in the society's history in which fundamental issues regarding the organization's methods and institutional framework were being resolved. Although officially the society professed adherence to the Baconian program, two rival styles of Baconianism emerged within its ranks. One tradition stressed the methodological principles articulated in the *Novum Organum*, where Bacon insisted that experience had to be enlarged by systematic experimentation as opposed to casual observation. The Baconianism of the *Novum Organum* emphasized "experiments of light" designed to reveal the causes of phenomena rather than "experiments of fruit" aimed at improving technology. An alternative Baconian tradition drew its inspiration from Lord Verulam's own example of an experimental history, the *Sylva Sylvarum*, where Bacon accumulated an "undigested heap of particulars" without any particular order, paying scant attention to theory.[133] Such was the style of Baconianism that induced Sprat to characterize the Royal Society's research program as one of "heap[ing] up a mixt Mass of *Experiments*, without digesting them into any perfect model."[134]

The two Baconian styles gave rise to radically different conceptions of the society's aims and how its research should be organized. One approach saw the society as being dedicated primarily to the compilation of histories of nature and the arts. Such histories required diverse talents, not only of experts but also of casual observers. Hence this approach called for a membership policy that would enlist a wide variety of persons into the society's ranks. The other approach insisted upon a more select membership consisting of a dedicated and disciplined corps of quasi-professionals who would be capable of pursuing specific theoretical problems. The two opposing conceptions of the Royal Society gave rise to a series of divergent reform proposals presented to the organization in the 1670s.[135] In a proposal drawn up in 1674, the unidentified fellow "A.B." called for a broad-based and open society. "A.B." subscribed to a traditional Baconian conception of the society as a vehicle for building natural and experimental histories. He believed that experiments should be designed not to resolve theoretical issues, but to produce matters of fact. Hence all kinds of experiments, even trivial ones, were to be preferred over debates about causes: "No Observation or Experiment, if truly made, is therefore to be sleighted, because it may seeme to be but mean. For so far as we have any matter of fact before us, we have really advanced further then we were before. And many things which seeme trivial in them selves, may be a foundation for that which is of greater moment; . . . the meanest observation of what is realy existent in nature, is more valuable than the most illustrious, if ungrounded, phancy." "A.B." held a fairly liberal view of the society's membership. Since the Baconian natural history required diverse talents and many contributors, he envisioned a society open to all. "As for persons," he wrote, "we cannot be too many employ'd, since we have all Nature to turn us in."[136]

Hooke took a completely different view of the society's aims and organization. Although he agreed that natural philosophy should be advanced by cooperative endeavor, he envisioned the Royal Society as a small, highly disciplined band of individuals dedicated to the pursuit of natural knowledge. In a scheme for reforming the society drawn up about 1680, he compared this corps of natural philosophers to the Spanish conquistadors. Like Cortés, who conquered Mexico with a small but disciplined force, "This newfound land [nature] must be conquerd by a Cortesian army well Disciplined and regulated though their number be but small."[137] Each member would take an oath to submit to the society's rules and to "zealously & streanously indeavour to the utmost of his Power and ability . . . to promote the good and interest of the Society." In contrast to the fact-oriented Baconianism propounded by "A.B.," Hooke insisted that the society's business should be "the examination of all Hypotheses and Doctrines that have hitherto been published." Instead of having the curators demonstrate random experiments, he thought the demonstrations should be accompanied by "a compleat and well digested Discourse" upon the theoretical issue being examined.

Although there was considerable diversity among the fellows regarding the society's ultimate objectives, one thing all the proposals agreed upon was the need for secrecy. "A.B." worried about recent attacks upon the society by critics such as Stubbe.[138] It was important, he noted, to win the favorable opinion of the public, which he thought would be best accomplished "not by communicateing, unles what we publish, but by beeing reservd. For to be open, will be to expose every thing, either to be catched up by some, or to be droll'd upon by others, or sleighted by all. Wheras, on the contrary, a secret is wont to be admired, what ever it be. . . . Besides, that it will make Tradesmen & others more free & open, when they understand they may dare to trust their secrets with us."[139] Hooke was positively obsessed with secrecy. His reform proposal would have required all members of the society to take an oath not to disclose anything of the society's deliberations to the public. He wanted to maintain rigorous censorship of the organization's publications, and to keep its archives under lock and key. "Nothing considerable in that kind can bee obtained without secrecy," he concluded, "because els others not qualifyed as abovesaid will share of the benefit."[140]

For many of the virtuosi, the commitment to the Baconian ideology of openness had become a liability. The institutionalization of the new science brought with it a growing consciousness on the part of the virtuosi that science was a special form of knowledge, distinct from the subjects of the university curriculum. Scientists, they believed, had a special role in society.[141] According to Hooke, it was "to improve the knowledge of naturall things, and all useful Arts, . . . not meddling with Divinity, Metaphysics, Moralls, Politics, Grammar, Rhetorick, or Logick."[142] Yet while the establishment of scientific institutions helped to secure that definition of the scientist's role, it weakened the Royal Society's commitment to the Hartlibian ideal of free dissemination of information. The extended and sometimes violent polemics against the Royal Society following the publication of Thomas Sprat's apologetical *History of the Royal Society* caused some fellows to press for a more restrictive policy in matters of membership privileges and the dissemination of information. Hooke advocated limiting the society's membership to those with a demonstrated interest in science and limiting the society's benefits to its fellows alone. "To achieve this," writes Michael Hunter, Hooke "would even have overturned the Society's Baconian commitment to the free dissemination of knowledge."[143] Too much access to scientific knowledge and unrestricted entry into the community of scientists threatened to undermine the success of the Royal Society, Hooke argued.

Hooke's proposals for the reform of the Royal Society came at a time when the institution was caught between mounting criticism from the outside and a decline in its active membership. Without substantial state support such as that enjoyed by the Paris Academy, the Royal Society was poorly equipped to sponsor large-scale research projects. Its voluntary membership policy made it difficult to organize patronage. The Paris Academy, by contrast, became essentially an adjunct of royal authority, entitled to all the prerogatives and privileges that went with being part of official culture in the Old Regime. Yet the academy, which the English virtuosi envied as a model of state-supported science, paid a high price for its elitism. Decked out in all the trappings of the hierarchical society of the Old Regime, it became a target of a revolutionary generation's assault upon rank and privilege. In the French debate over "republican" versus "monarchical" science, Baconianism emerged as the hope of the people. The vogue of natural history in eighteenth-century France can be at least partially explained by the widespread belief that natural history would liberate the intellect from the tyranny of abstruse philosophical systems, in this case Newtonianism. The Royal Academy exercised its prerogative to rule upon the merit of scientific and technological proposals by turning down hundreds of projects put forward by craftsmen, inventors, and scientific amateurs. To those whom the academy rebuffed, the elite body's decisions seemed as arbitrary and authoritarian as the rule of the monarch himself. One dejected scientific amateur accused the organization of having "placed the liberty of opinion in chains." What right did the academy have to pronounce upon projects without a fair hearing, he asked? "Have you the exclusive power to trace out nature's course by placing stumbling blocks against public liberty?"[144] When the Revolution broke out in 1789, the academy came under violent attack, and its activities were immersed in political turmoil for decades.

Somewhat presciently, Thomas Sprat detected the contradictions inherent in the concept of an open scientific society. Might not the Royal Society's openness eventually dilute its ranks with members who had no scientific competence, he asked, and thereby divert it from its main objective? He wondered whether the society, "being so numerous as it is, will not in a short time be diverted from its primitive purpose, seeing there wil be scarce enough men of Philosophical temper always found, to fill it up; and then others will crowd in, who have not the same bent of mind; and so the whole business will insensibly be made, rather a matter of noise and pomp, than of real benefit." If that happened, Sprat mused, if the Royal Society became nothing more than a garrulous social club, it might frighten discoverers and inventors from revealing their secrets to the group, "lest they should thereby become common, and so they be depriv'd of the gain, which else they might be sure of, if they kept them to themselvs."[145]

Sprat did not worry much about the first danger. He did not think England would ever lack the manpower to propel science. The advancement of natural knowledge did not require the services of "perfect philosophers," he observed. It required only honest men, "such as have not their Brains infected by false Images," who would "bring their hands, and their eyes uncorrupted" and serve as witnesses to experiments. Sprat was assured that future ages "will not, for a long time be barren of a Race of Inquisitive minds, when the way is now so plainly trac'd out before them." The second danger, he conceded, was real. Yet he did not see the solution in closing the society's ranks to the less qualified. Consistent with his Baconian convictions, he observed that even if discoverers conspired to keep all their secrets concealed, the Royal Society's method would sooner or later find them out anyway. "How few secrets have there been," he asked, "that have been long conceal'd from the whole World by their Authors?" Sprat had the answer: "There is no question at all, but all, or the greatest part . . . will soon flow into this *public Treasure.*"[146]

CONCLUSION

In Nature's infinite book of secrecy a little I can read.
(William Shakespeare, *Antony and Cleopatra*)

T HE problem of what qualifies as knowledge is intimately connected with the problem of who owns it. That is why the metaphor of the "secrets of nature" has been one of the most prominent and most powerful metaphors in the history of science.[1] To the
extent that science attempts to go beyond the obvious or the "naively"
empirical and to discover a deeper reality than that revealed to the
senses, one might say that finding out the "secrets of nature" is the goal
of all scientific inquiry. The distinction between common sense and a
deeper understanding of nature is one that goes back to the beginnings
of Western thought and has remained a constant feature of natural philosophy ever since. In the *Works and Days*, Hesiod noted that some agricultural techniques were matters of common practice, while others were
more "learned." The latter, more "scientific" rules were known to but a
few, he maintained, implying that he himself possessed esoteric knowledge.[2] Plutarch referred to science as the investigation of the "secrets of
nature," having in mind such inquiries as where the sun sets when it
sinks into the sea, or what becomes of light when its source is extinguished.[3] Long after Halley praised Newton's prowess in "penetrating
. . . into the abstrusest secrets of Nature," important scientific break-
throughs are still being hailed as discoveries of nature's "secrets."[4] Recently the metaphor has been used in connection with the molecular
biological revolution: in decoding the mechanism of genetic replication,
biologists have proclaimed the discovery of the "secret of life."

Contrary to the classical view, metaphors are not merely figurative
substitutions for concrete concepts.[5] According to Max Black, metaphors work interactively with the subjects for which they stand, evoking
associations between different sets of ideas, thus producing new meanings. The implications evoked by the metaphor transform, filter, and organize our view of the metaphor's subject. Conversely, using a word as
a metaphor changes its meaning by virtue of its "connection" to the
subject. "In this 'connection' resides the secret and the mystery of the
metaphor."[6] To speak of the "secrets of nature" is to transfer an entire
system of commonplaces associated with concealment, privacy, exclusiveness, and clandestine activity onto the concept of nature. The metaphor thus organizes our view of nature. It selects and brings forward

certain aspects of nature, while pushing others into the background. Ideas, attitudes, and practices that can be expressed in terms of secrecy-language are rendered prominent, while those that cannot are de-emphasized.

What associations and implications are evoked by the "secrets of nature" metaphor? What does nature look like when seen through its filter? I have suggested that the metaphor has played upon two different senses of the concept of secrecy, one epistemological and the other sociological. One implies that nature is inherently arcane, the other that natural knowledge is privileged. One sense of the metaphor says something about nature itself, while the other makes a statement about the ownership of scientific knowledge and about the practice of esotericism.

The most common meaning of secrecy is that of intentional concealment.[7] By this connotation, to speak of nature's "secrets" is to imply that nature's true character is hidden from public view and common sense. "Nature loves to hide," Heraclitus is supposed to have said.[8] Cicero wrote that physics was a branch of philosophy that concerned "mysteries veiled in concealment by nature herself."[9] To the early Christians, nature's secrets were divinely hidden. According to Lactantius, God made humans last in order to hide the mystery of the creation from them, thus setting the stage for the great drama of the Fall. Whether expressed in secular or religious terms—that is, whether it is God or nature who hides the secret—this conception of nature's secrets makes a fundamental distinction between nature as it appears to the senses and nature as it really is, between nature on the outside and nature on the inside, so to speak. According to this view, the sensible world is like a cloak or a disguise within which "real" nature hides. When coupled with the traditional image of nature as woman, the metaphor took on another dimension: nature is modest; she deceives; she devises various stratagems to fool those who wish to probe her secrets, allowing a glimpse of one aspect of herself but concealing her identity from those (males) who would attempt to know her intimately.[10]

Historically the implication that nature "hides" has taken on a variety of meanings. In the Hermetic tradition, the metaphor described a condition of nature itself. To many Hellenistic thinkers, nature's secrets were so unfathomable that they could be known only by a divine revelation such as the one experienced by Thessalos of Tralles. They were, quite literally, *arcana naturae*. Hence scientific knowledge was a sacrament that had to be protected from those who might abuse or corrupt it. The "initiate" into the occult secrets of natural philosophy was obliged to obey the "law of silence" not to reveal the secrets of nature to any "stranger." This conception of knowledge—the tradition of esotericism, as I have called it—placed the secrets of nature in the realm of sacred and forbidden knowledge. In classical and medieval philosophy, on the other hand, the expression "secrets of nature" referred not to permanent mysteries, but to natural phenomena that were merely difficult to comprehend or to phenomena whose causes were unknown. According to conventional scholastic opinion on occult qualities, the alleged "marvels" produced by magic were "secrets" only in the sense that they resulted from perceptual errors or from unknown or occult causes. In other words, the "secrets of nature" were not intentionally concealed by God or by nature, but were merely examples of the imperfect state of human knowledge. While further investigation might reveal the causes of some of nature's secrets, others were knowable only by experience. Since demonstrative arguments could not be adduced to explain them, the scholastic position put the "secrets of nature" outside the boundaries of *scientia*.

The early modern period added a new implication to the "secrets of nature" metaphor: the idea of a "secret" as a technique or recipe, the sense in which the word was used in the books of secrets. The transference of this implication of a "secret" onto the "secrets of nature" metaphor marked a subtle but revolutionary linguistic shift. According to this sense of the metaphor, the "secrets of nature" were its "inner workings." Thus Francis Bacon spoke of penetrating the secrets of "nature's workshop." Underlying this connotation of the metaphor was the view that nature could be understood in mechanical terms as a set of invisible "techniques" nature employs for producing its various sensible effects. Hence nature's "inner workings" might be replicated as one might replicate a technique by following a recipe. This sense of the "secrets of nature" metaphor set new research goals for early modern science. The new philosophers rejected the scholastic doctrine of occult qualities on the grounds that it was, in Walter Charleton's words, nothing more than a "sanctuary of ignorance." To concede that nature's secrets were unknowable was to lay the intellect in a "deep sleep." For Charleton, the goal of natural philosophy was to "withdrawe that thick Curtain of obscurity, which yet hangs betwixt Nature's Laboratory and Us" so as to get a glimpse of how nature operates.[11] The capacity to artificially reproduce nature's effects was seen as an epistemological guarantee of natural knowledge. As Bacon explained, understanding how works of art are made is akin to taking off nature's veil, "because the method of creating and constructing such miracles of art is in most cases plain, whereas in the miracles of nature it is generally obscure."[12]

In the new philosophies, reproducing natural processes became a kind of touchstone against which claims to natural knowledge would be tested. This conviction accounts in large measure for the influence in Restoration England of Bacon's conception of Solomon's House, where the imitation of nature's works was the goal of science. Robert Boyle argued that the capacity to produce mechanically the "forms and qualities" of nature demonstrated the superiority of corpuscularianism over the scholastic doctrine.[13] Gassendi insisted that just as in the crafts construction was the goal of knowledge, so in natural philosophy: "For although it is not we ourselves who created such things by our industry, we philosophize about them just the way we do about things of which we are the authors. . . . Wherever nature permits, we use anatomy, chemistry, and such aids so that by breaking bodies down as far as possible, dismembering them as it were, we may understand what they were compounded from and how, and whether other things can or could have been compounded in the same manner in the same or different ways."[14] The principle of *verum factum* (maker's knowledge) limits knowledge of the "secrets of nature" to the world of reconstructible phenomena. The ultimate "secrets of nature"—in the sense of causes or essences—are reserved to God alone.

Michael Arbib and Mary Hesse claim that "scientific revolutions are, in fact, metaphoric revolutions."[15] Metaphoric shifts cause us to "see" phenomena differently. When the metaphor of the "secrets of nature" exchanged its meaning as *arcana naturae* for that of mechanical techniques, explanations in terms of occult qualities seemed ludicrously inappropriate. Such accounts quickly fell out of fashion in favor of explanations in terms of the mechanical motion of microscopic particles. The changed concept of nature's "secrets" also set new goals for science. In the new philosophies, it was no longer sufficient to take nature's exterior appearances for granted. "Getting to the bottom of things" meant looking deeper into nature and attempting to examine the motions of minuscule particles. Unlike the occult qualities of scholastic natural philosophy, which were completely hidden from the senses, the mechanical corpuscles of the new philosophy were merely invisible to the naked eye but were in theory *visualizable*. One simply had to use special tools, such as the microscope, to help the senses get a closer look at nature. The mechanical conception of nature's "secrets" dissolved the classical distinction between nature and art. If the "secret" of nature was a matter of how nature was *constructed*, then when the building blocks of nature could not be seen directly, they could be understood by analogy to human artifacts. Nature's "curious" workmanship could be understood in the same terms as the finely wrought "curiosities" of intricate craftsmanship. Thus paper, Bacon observed, "imitates and almost rivals the skin or membrane of an animal, the leaf of a vegetable, and the like pieces of Nature's workmanship."[16]

The other implication of the "secrets of nature" metaphor carried an essentially social meaning. It had to do with the ownership of scientific knowledge. According to this idea of secrecy, knowledge is divided into two parts: an *exoteric* or public teaching and an *esoteric* or secret doctrine.[17] The practice of esotericism goes back at least to the mystery religions of antiquity, whose elaborate initiation rites separated the select few who knew the cult's secrets from the many who did not, thus preserving knowledge as sacred. Yet esotericism was not restricted to religious traditions. It is also encountered in philosophy. Aristotle differentiated his "exoteric discourses" on subjects from those that treated the same matters "on philosophical lines" (*Eudemian Ethics* 1217b20–25). The former works contained Aristotle's "popular philosophical discussions," while the latter included his systematic works addressed to the philosophically minded.[18] We also encountered the exoteric/esoteric distinction in the Islamic exegetical tradition, which distinguished between the apparent meanings of the Koran (*zāhir*) and the hidden truth concealed within it (*bātin*). Corresponding to scriptural batinism, the Islamic "secret sciences" formulated a kind of metaphysical batinism that distinguished between nature's outward appearances and the occult powers lying within. The two forms of esotericism gave rise to a rich corpus of scholarly works on the occult sciences, which both fascinated and repelled medieval and Renaissance intellectuals. To many scholars, magic seemed to hold the promise of giving access to "secrets of nature" that could be exploited for human gain. Esotericism was central to the practice of magic; it was an indispensable element of the magus's identity and an instrument for his self-fashioning. Whether in the court or in the piazza, playing the role of the possessor of secrets was an essential part of the magical act.

The new philosophers loudly rejected the tradition of esotericism and upheld the virtues of open disclosure of scientific knowledge. Boyle denounced the "secretists" who closeted themselves in their private laboratories and asserted that scientists had a moral obligation to disclose their discoveries to the public. In the Royal Society, collective witnessing of experimental facts was advanced as the guarantee of objectivity. Candid and complete reporting of experiments, including failed experiments, insured the credibility of experimental practices and became the foundation of the "consensual" natural philosophy espoused by the Royal Society. Yet experimental knowledge in the Royal Society was never completely open. Although theoretically public, the society's experimental spaces were defined by fixed boundaries that separated "matters of fact" from the extravagant, boastful experiments of the charlatans and professors of secrets below, and from the flamboyant, dazzling demonstrations of courtly science above. For all their differences, the "experiments" performed by the *ciarlatani* in the piazza and the virtuosi in the courts had one thing in common: they were intended to amaze and entertain onlookers. The production of experimental "matters of fact," on the other hand, took place in a neutral middle space and was undertaken by a corporation of gentlemen governed by the code of "civility." In such spaces, "dazzling" experiments were out of place. Matters of fact were *civilized*: they did not offend or insult anyone by their pretentiousness or social hubris.[19]

The ideology crafted in the Royal Society of science as public knowledge has become an integral part of the scientific ethos. Secrecy is universally rejected in modern science. Free and open communication of research is regarded as a sine qua non of scientific progress and an indispensable component of the scientific method.[20] Yet the tension between the ideal of openness and the practical need for secrecy has been a constant feature of modern science. From the early years of the Royal Society down to the present, secrecy has been used to ensure priorities, to guard against plagiarism, to protect competitive positions in the marketplace, and in wartime to keep information from the enemy. Boyle, the most ardent early proponent of openness, advocated secrecy in science under certain conditions and practiced it himself.[21] Modern scientists have come under even greater pressure than were those in Boyle's day to keep results of certain research secret.[22] For one thing, the stakes are higher. Scientists now play for huge rewards in research funds, prizes, recognition, and status. Intense competition and the desire for priority has created an atmosphere in which the exercise of secrecy is in certain situations a practical necessity. Government and business also put pressure on scientists to keep the results of research secret. Corporate funding of scientific research is often contingent upon the researcher's acceptance of restrictions on the dissemination of knowledge, particularly in cases of investigations that may lead to commercial applications. Recently, theoretical advances in the field of cryptography, which threatened to expose military secrets, brought mathematicians and computer scientists into open conflict with the federal government. Such tensions raise fundamental questions about academic freedom and the control of information in a democratic society.[23]

No longer, in modern science, do the "secrets of nature" retain the aura of forbidden knowledge. No longer are nature's secrets arcana. But they are no less esoteric and privileged. If anything, the secrets of nature are more the monopoly of an autonomous corporation of specialists than ever before. The cultural function of secrecy is to articulate a boundary, to circumscribe an interior that is off-limits to outsiders, to mark off a sphere of autonomous power.[24] The separation between insider and outsider is inherent in secrecy. Among other things, secrecy guards a discipline from contamination by outsiders. Thus in the Hermetic tradition initiates were sworn to the law of silence so the secrets of

Hermeticism would not be profaned, while alchemists wrote in ciphers to prevent their secrets from falling into the hands of competitors. In the modern setting, the social function of esotericism has been increasingly performed by the construction of disciplinary boundaries. It can now be seen that the "code of science," including its professed ideal of public knowledge, was an integral part of a set of mechanisms by which the new community of natural philosophers carved out for itself an identity, separating experimental natural knowledge from the disputatiousness of Scholasticism, from the anarchy of popular culture, from the "secretism" of alchemy, and from the sensationalism of courtly science. Institutionalization may have replaced esotericism in science, but sociologically its goals are the same: it is a mechanism for protecting the discipline from external criticism and from pollution by outsiders. The modern scientific community, governed by the principle of peer review, has become essentially closed to input from the outside, even when its research is perceived by outsiders as threatening. The paradox is that "a form of knowledge which is the most open in principle has become the most closed in practice."[25]

Meanwhile, what became of the books of secrets? The sequel to their history after the seventeenth century is a fascinating story in its own right, but it cannot be told here. Many of the sixteenth-century books of secrets continued to be published down through the eighteenth century: the last edition of Alessio's *Secreti* was published in 1780. Recipes from his and other books of secrets continued to appear in domestic handbooks and technical encyclopedias of the nineteenth century. During the Enlightenment, however, the descendants of the professors of secrets were increasingly marginalized as the professional scientific community shored up its borders against charlatans, "crackpots," and intruders. In pre-Revolutionary France, the purveyors of "secret remedies" (*remèdes secrets*) came under the jurisdiction of the Société Royale de Médicine, which was charged with licensing patent medicines. The Société Royale treated the purveyors of "secrets" as contemptuously as the Paris Academy shunned the various circle-squarers and perpetual motion–makers that came before it with their dubious inventions.[26] But the society's attempt to control secret remedies, despite being based upon the "enlightened" principle of stamping out charlatanism and bringing rational medicine to the people, conflicted with two fundamental principles of Enlightenment thought: the right of property and free trade. For was not a *remède secret* a form of private property, one moreover whose commercial value depended upon its being a secret? For the professional elite, however, secrecy was unacceptable. Enlightened medicine could not be occult; in the interests of science and public health, it had to be a visible, controllable realm. The defenders of secret remedies fell back on an argument used by empirics since Nicholas of Poland. Nature was infinitely rich and mysterious, they maintained. It yielded up its secrets to anyone who was willing to be guided by the light of experience. To suppress the contributions of unlettered empirics was to deprive humanity of a precious resource. They insisted that liberty in science demanded equity for the untutored genius. Besides, attempts to regulate secret remedies went against the most hallowed traditions of popular culture. Despite the government's best efforts, the trade in *remèdes secrets* flourished.

Nevertheless, the attack upon popular healers, many of them women, intensified. The tradition of "cookbook medicine" temporarily flourished with the expansion of literacy and the growth of the popular press in the eighteenth and nineteenth centuries. But eventually the physicians appropriated the market and began writing "authorized" domestic medical guides to replace those of the "lady doctors." William Buchan, an Edinburgh M.D., wrote his famous *Domestic Medicine* (1769) in order to root out "Ignorance, Superstition, and Quackery," and to inspire "utter confidence in the physician." Buchan was alarmed by the growth of illicit medicine. "No two characters can be more different than that of the honest physician and the quack," he insisted, "yet they have generally been very much confounded." The problem was that few people were able to distinguish between the empiric who administers a "secret medicine" and the physician who "writes a prescription in mystical characters and an unknown tongue." Buchan shared the Enlightenment's conviction that education was the key to eliminating medical errors among the people. "The most effectual way to destroy quackery in any art or science," he argued, "is to diffuse the knowledge of it among mankind."[27] The academies took up the cause, purging all traces of the books of secrets from the official pharmacopoeias. "Dr Stevens's Water," which was recommended in domestic remedy books throughout the eighteenth century, was still to be found in the London *Pharmacopoeia* of 1724. In the 1740s, when a committee of the Royal College of Physicians set about reforming the *Pharmacopoeia*, the popular cordial, along with other remnants of cookbook medicine, was dropped from the list of official remedies.[28]

In nineteenth-century America, the culture of "how to" thrived. On the thinly settled frontier, where people lived far from professional doctors and manufacturing centers, the ghost of Alessio Piemontese was resurrected in the person of "Doctor" A. W. Chase (1817–1885) of Ann Arbor, Michigan. The American Alessio was an unschooled peddler who began his remarkable rags-to-riches saga by selling household drugs and groceries among the growing settlements of Ohio and Michigan. Chase traveled for a time with a circus, selling "gas beer," which he learned to make by watching a grocer. He collected recipes from housewives, physicians, farmers, and tavern keepers. In Finley, Ohio, he paid eight dollars for a recipe for Good Samaritan Liniment to a man who was selling it in two-ounce vials for twenty-five cents apiece. He obtained the recipe for his Toad Ointment from "an Old Physician who thought more of it than of any other prescription in his possession." The ointment, which was used for sprains, strains, and lame back, brings to mind Alessio's Oil of a Red Dog: it called for putting live toads in boiling water along with other ingredients and cooking them until soft. "Some persons might think it hard on toads," wrote Chase, "but you couldn't kill them quicker in any other way."[29] Among the wonders of modern science Chase sold was his Magnetic Ointment, whose action, although it contained only lard, raisins, and tobacco, "is really magnetic."[30] In 1856, Chase settled in Ann Arbor and printed the first small collection of his recipes, a pamphlet entitled *Dr. Chase's Recipes; or, Information for Everybody*. Twelve years later, the thirty-eighth edition of the book had grown to contain more than six hundred recipes for everything from home cures to making catsup, vinegar, wine, ink, and glue. The work contained practical recipes for merchants, housewives, blacksmiths, jewelers, gunsmiths, farriers, barbers, bakers, dyers, and saloon keepers. But it was for its medical recipes that the work was most often consulted. Although to the regular physicians he was a quack, Chase was perhaps the most widely known, admired, and consulted "doctor" of his time. His medical practice, which he carried on by mail, extended from Iowa to New York. His recipe book was translated into German, Dutch, and Norwegian, and was consulted in Canada, Australia, New Zealand, and South Africa. In 1868, Chase sold the rights to his book and the Steam Printing House in which he published it and moved to Toledo, where he fell in with unscrupulous business partners, lost his fortune, and died in abject poverty in 1885. But his book continued to fill a need in the households of the new nation. Millions of nineteenth-century Americans depended on Chase's home remedies and practical advice. Sixteen different publishers profited from the book's brisk sales. By 1915, some four million copies had been sold. There were years when *Dr. Chase's Recipes* sold second only to the Bible.[31]

A great cultural chasm separates *Dr. Chase's Recipes* from Alessio Piemontese's *Secrets*. Chase, a peddler, had no other purpose than to provide useful information to common people and to make money doing it. He did not imagine his recipes might unlock the "secrets of nature." An even greater chasm separates Alessio from Thessalos of Tralles. Whereas Alessio advertised his "secrets" as having stood the test of experience, Thessalos offered the proof of divine revelation, experience having failed him. The secret of a recipe is its intellectual poverty.

It requires little thought to understand one, little experience to employ one. Recipes collapse lived experience into a series of mechanical acts that, once parsed, anyone can follow. While a "secret" is someone's private property or the property of a group, a recipe doesn't belong to anyone. Once it is published, someone else appropriates it, uses it, varies it, then passes it on. At each stop it gains something or loses something, is improved upon or degraded, and is changed to fit new needs and circumstances. Recipes are built upon the belief that somewhere at the beginning of the chain there is someone who does not use them.

But are they science? That was the question posed by the new philosophy. The answer is not so simple. Recipes are not explanations in the conventional sense. They do not formulate laws of nature. On the other hand, they are algorithms, or rules for making. Even if I cannot explain why something exists in terms of necessary causal laws, if in this chancy world I can make it every time, or nearly so, can it not be said I know it?

APPENDIX

SECRETI ITALIANI: ITALIAN BOOKLETS OF
SECRETS, CA. 1520–1643

THE APPENDIX contains a chronological list by author and/or short title of 83 secrets-pamphlets in Italian published between about 1520 and 1643. These works are described in chapters 3 and 7. Almost all of them are printed on one signature folded into a quarto (eight pages) containing about twenty to thirty recipes; a few are octavos (sixteen pages), containing thirty to a hundred recipes. Sold in the piazzas of the major cities by *ciarlatani* and popular healers, and by peddlers in the countryside, they represent the books-of-secrets tradition at its commonest level. Many of the pamphlets were licensed for sale in multiple cities, although I give only the first city listed on the title page. The pamphlets are located in several different libraries (see location key following the list); however, the bulk come from two collections, one in the Department of Special Collections at the University of Wisconsin, and the other in the Biblioteca Universitaria, Bologna. In cases of pamphlets actually cited in the text, fuller bibliographical details are given in the bibliography.

AUTHOR AND/OR TITLE	PUBLICATION	LOCATION
Opera nova de ricette et secreti	Venice, ca. 1520	W
Opera nova chiamata Secreti Secretorum	Venice, ca. 1520	W
Opera nova intitolata Dificio di ricette	Venice, 1529	NLM
Opera nova excellentissima: . . . di far vari secreti	Venice, 1529	W
Opera nova piacevole	Venice, 1530	W
Opera nova intitolata Dificio di ricette	Venice, 1530	NLM
Opera nova intitolata Dificio di ricette	Venice, 1534	W
Opera nova intitolata Dificio di ricette	Venice, 1538	F
Opera nova nella quale ritroverai molti bellissimi secreti	n.p., ca. 1540	W
Opera nova nella quale ritroverai molti bellissimi secreti	Florence, ca. 1540	NLM
Opera nova intitolata Dificio di ricette	Venice, 1543	FSL
Opera nova intitolata Dificio di ricette	Venice, 1546	NLM
Opera nova intitolata Dificio di ricette	Venice, 1550	F
Opera nova piacevole	Venice, 1550	F
Colleta de molte nobilissime ricette	n.p., ca. 1550	NLM
Opera nova nella quale ritroverai molti bellissimi secreti	Venice, 1553	NLM
Marinello, Giovanni. *Gli ornamenti delle donne*	Venice, 1562	FSL
Marinello, Giovanni. *Le medicine appartenenti alle infermità delle donne*	Venice, 1563	NLM

AUTHOR AND/OR TITLE	PUBLICATION	LOCATION
Vittori, Benedetto. *Prattica d'esperienza*	Venice, 1570	NLM
Giovanni Battista, Fr. *Breve et utile trattato*	Venice, 1575	NLM
Angelico, Vespasiano. *Massimi secreti de l'angelico*	Parma, ca. 1577	NLM
Fontana, Andrea. *Fontana dove n'esce fuori acqua di secreti*	Genoa, 1579	NLM
Lauro, Vincentio. *Il Tesoro nel quale si contiene molti secreti*	Modena, ca. 1580	UW
Ricci, Francesco. *Giardino di secretti rarissima, utili, & piacevoli*	Bologna, ca. 1580	UW
Galasso, Mario. *Novo recettario . . . intitolata Thesoro de poveri*	Milan, 1584	UW
Mutii, P. M., detto il Zanni bolognese. *Nuovo lucidario de secreti*	Bologna, 1585	B
Galasso, Mario. *Giardino medicinale di nuove dato in luce*	Bologna, 1587	UW
Scarmignani, Jacomo. *Tesoro di medicina*	Venice, ca. 1590	UW
Agrippa, Livio. *Discorso ragionale trattato sopra remedii de veneni materiali*	n.p., ca. 1590	UW
Lauro, Vincentio. *Primo giardino, nel qual si contiene Secreti*	Pavia, 1591	UW
Agrippa, Livio. *Secreti medicinali novamente date in luce*	Verona, 1592	NLM
Mazzetto de varii secreti	Macerata, 1595	NLM
Rosaccio, Gioseppe. *Della nobilta et grandezza dell'huomo*	Milan, 1596	UW
Regola nuova d'imparare molti e belliss. givochi	Bologna, ca. 1600	B
Prognostico calculato l'anno MDC	Ferrara, ca. 1600	B
Trabia, Giacomo. *L'Esperienza vincitrice, epilogata in diversi secreti*	Genoa, ca. 1600	B
Fedele, Domenico, detto il Mantoanino. *Con il Poco farete Assai*	Rome, ca. 1600	B
Nuovo discorso sopra i dodeci segni celesti	Bologna, ca. 1600	B
Buger, Giovanni. *Virtus Occulta Perit. Giardino di varii secreti*	Bologna, ca. 1600	B
Amelli, Claudio, detto il gran Piemontese. *Fioretto belissimo*	Bologna, ca. 1600	B
Francesco detto il Biscottino. *Giardino di varii secreti havuta da diversi signori*	Bologna, ca. 1600	B
Pesarino, Il gran Giocator di mano, *Secreti utilissimi e nuovi*	Verona, ca. 1600	B
Avertimenti a tutti quelli che desiderano regolatamente vivere	Florence, 1602	UW
Franzosino, Carlo. *Secreti bellissimi non piu dati in luce*	Viterbo, 1603	UW

AUTHOR AND/OR TITLE	PUBLICATION	LOCATION
Lauro, Vincenzo. *Scelta di varii secreti . . . esperimentati in molte persone*	Bologna, 1603	UW
Germerio, Gulielmo. *Gioia preciosa . . . Opera à chi brama la sanità utilissima*	Venice, 1604	UW
Fonte, Aniello. *Cento secreti medicinali cavata da autori moderni, . . .*	n.p., 1606	UW
Cesare, Giulio. *Thesoro di varii secreti naturali, havuti di diversi signori*	Bologna, 1608	B
Americano. *Il vero e natural fonte . . . d'acqua viva di mirabili e salutiferi segreti*	Rome, 1608	B
Scatalone, Dottor Gratiano. *Il vero, et pretioso thesoro di sanita*	Milan, ca. 1610	UW
Cortese, Angelo. *Tesoro di sanità*	Milan, ca. 1610	UW
Boldini, Antonio Felice. *Il medico de' poveri, o sia il gran stupore de' medici*	Venice, ca. 1610	B
Agrippa, Livio. *Secreti medicinali novamente dati in luce*	Florence, ca. 1610	UW
Magiunii, Antonio. *Il Canevaro . . . col modo di fare Vini diversi*	n.p., ca. 1610	B
Modo facile da difendersi dal gran freddo	Milan, ca. 1610	UW
Giardino di virtu nel qual si contiene alcuni particolari & maravigliosi Secreti	Milan, 1612	B
Benedetto, detto Il Persiano. *I Maravigliosi, et occulti secreti naturali*	Rome, 1613	UW
Leandro, Lorenzo. *Tesoro di varij Secreti . . . Esperimentati con grandissima fatica*	Venice, 1614	UW
Legati, Domenico. *Giardino di vari secreti havuti da diversi Signori*	Milan, ca. 1615	UW
Breve trattato di fisionomia	n.p., ca. 1615	UW
Thesoro di varii secreti naturali, havuti da diversi Signori	Milan, ca. 1615	UW
Pigozzi, Gio.Battista. *Stupendi et maravigliosi secreti*	Bologna, 1615	B
Biagio, detto il Figadet. *Tesoro di secreti, raccolti da diversi valenti huomini*	Bologna, 1617	B
Nuovo, vago, e dilettevole giardino	Milan, 1617	B
Francolino, Tomaso da, detto l'Ortolano. *Tesoro di segreti naturali*	Rome, 1617	B
Vittiario, Giovanni, detto il Tramontano. *Centuria seconda de' secreti*	Viterbo, 1618	UW
Honofri, Fedele. *Centuria di secreti medicinali, e naturali*	Florence, 1619	UW
Sarecino, Giovanni, Medico. *Recettario di Galeno*	Venice, 1619	B

AUTHOR AND/OR TITLE	PUBLICATION	LOCATION
Grisaldi, Fra Paolo, O.P. *Giudicio di fisionomia*	Treviso, 1620	B
Nicolas, Antonio, Francese. *Secreti non più intesi*	Bologna, 1620	B
Pagliarizzo, Dottor Gratiano. *Secreti nuovi e rari*	Milan, ca. 1620	UW
Aliotto, Domenico. *Giuocchi di carte bellissima, Di regola, e memoria*	Venice, ca. 1620	UW
Discorso di fisionomia	n.p. ca. 1620	UW
Bernardo, Francesco, Francese. *Nuova corona di secreti cavati di diversi autori*	Milan, 1620	B
Il Medicinal tesoro di poveri, raccolti & esperimentati da diversi ecc. medici	Bologna, ca. 1620	UW
Guelfo, Gio.Battista, detto il Lombardo. *Amirabi discorso circa la naturalezza*	Florence, 1620	UW
Battista, Gio. *Stupendi et maravigliosi secreti*	Milan, ca. 1620	UW
Francolino, Tomaso. *Tesoro dei secreti naturali*	Rome, ca. 1620	UW
Torelli, Gioan Battista. *Novo giardino di secreti curiosi, et esperimentati*	Bologna, 1628	UW
Le maravigliose virtu dell'avorio calcinato per antichità	Bologna, 1630	B
Milioni, Pietro. *Vago Fioretto di secretini da praticarsi da persone curiose*	Rome, ca. 1630	UW
Fontana, Andrea. *Acque di secreti n'esce fuori acque di Secreti*	Venice, ca. 1630	UW
Cosson, Giovanni, detto il Bontempo Francese. *Secreti novamente ritrovati*	Milan, ca. 1630	UW
Biasio, detto Il Figadet. *Tesoro dei secreti raccolta da diversi valenti huomini*	Bologna, ca. 1630	UW
Scarioni, Francesco. *Centuria prima di secreti medicinali, politici, e naturali*	Perugia, ca. 1630	UW
Tesoro de secreti raccolti da diversi valenti huomini	Venice, ca. 1630	UW
Scarioni, Franceso. *Centuria di secreti politici, chimichi, e naturali*	Siena, 1633	UW
Monte, Ludovico bolognese, della Chitarriglia. *La ghirlandetta fiorita di secreti*	Naples, 1634	UW
Honofri, Fedele. *Centuria di secreti medicinali, e naturali*	Rome, 1637	UW
Legati, Domenico. *Tesoro di varii secreti*	Milan, ca. 1640	UW
Maiorini, Tomaso, detto Policinella. *Frutti soavi colti nel giardino*	Bologna, 1642	B
Honofri, Fedele. *Centuria di secreti medicinali e naturali*	Bologna, 1643	B

LOCATION KEY

B	University of Bologna
F	Ferguson, *Bibliographical Notes*
FSL	Folger Shakespeare Library
NLM	National Library of Medicine
UW	University of Wisconsin
W	Wellcome Historical Medical Library, London

ABBREVIATIONS

BU	*Biographie universelle, ancienne et moderne.* 52 vols. Paris: Michauld, 1811–1828.
DBI	*Dizionario biografico degli italiani.* Rome, 1960–.
DHI	*Dictionary of the History of Ideas.* Edited by Philip R. Wiener. 5 vols. New York: Scribner's, 1973.
DMA	*Dictionary of the Middle Ages.* Edited by Joseph R. Strayer. 13 vols. New York: Scribner's, 1982–1989.
DNB	*Dictionary of National Biography.* Edited by Leslie Stephen and Sidney Lee. 63 vols. New York: Macmillan, 1885–1901.
DSB	*Dictionary of Scientific Biography.* Edited by Charles Coulston Gillispie. 18 vols. New York: Scribner's, 1970–1980.
EI	*Encyclopedia of Islam.* Edited by C. E. Bosworth, E. van Donzel, B. Lewis. and C. Pellat. New ed. Leiden: Brill, 1960–.
HMES	Thorndike, Lynn. *History of Magic and Experimental Science.* 8 vols. New York: Columbia University Press, 1923–1958.
Migne, *PL*	Migne, J. P., ed. *Patrologia cursus completus, series Latinus.* 221 vols. Paris, 1844–1890.
OED	*The Oxford English Dictionary.* Compact ed. 2 vols. Oxford: Oxford University Press, 1971.
Phil. Trans.	*Philosophical Transactions of the Royal Society of London.* London, 1665–.
STC	*Short-Title Catalogue of Books Printed in England, Scotland, and Ireland and of English Books Printed Abroad.* Edited by A. W. Pollard and G. R. Redgrave. London: Bibliographical Society, 1969.
VL	*Die deutsche Literatur des Mittelalters: Verfasserlexikon.* Edited by Kurt Ruh. 2d ed. Berlin: Walter de Gruyter, 1978–.

NOTES

INTRODUCTION
PRINTING, POPULAR CULTURE, AND THE
SCIENTIFIC REVOLUTION

1. Browne, *Works*, 2:58.
2. Ferguson, *Bibliographical Notes*. Ferguson's bibliography was first presented as a series of papers read to the Archaeological Society of Glasgow between 1882 and 1911 and printed in its *Transactions* (1896–1915).
3. Eisenstein, *The Printing Press as an Agent of Change*, pp. 573, 575.
4. Kuhn, "Mathematical versus Experimental Traditions."
5. Shapin and Schaffer, *Leviathan and the Air-Pump*; Shapin, "Robert Boyle's Literary Technology."
6. Zilsel, "Origin of Gilbert's Scientific Method"; Eamon, "Robert Boyle and the Discovery of Colour Indicators."
7. Zilsel, "Sociological Roots of Science"; Rossi, *Philosophy, Technology, and the Arts*.
8. Bacon, *Novum Organum*, in *Works*, 4:96.
9. See in particular the discussion in Davis, *Society and Culture in Early Modern France*, pp. 7ff.; Burke, *Popular Culture in Early Modern Europe*, pp. 258f.
10. Browne, *Religio Medici*, quoted in Bouwsma, "Secularization of Society in the Seventeenth Century," p. 116.
11. Ibid., p. 114.
12. See, for example, White, *Hocus-Pocus*, where the "hocus-pocus" of magic is spelled out in recipes.
13. Pérez-Ramos, *Francis Bacon's Idea of Science*.
14. I am indebted to Brian Copenhaver for his suggestion of the following taxonomy of secrecies, which I have adapted to my purpose.
15. I use the term "epistemic" in the sense employed by Foucault in *The Order of Things*.
16. I paraphrase Darnton, *The Literary Underground of the Old Regime*, p. 40.

CHAPTER ONE
THE LITERATURE OF SECRETS IN THE MIDDLE AGES

1. Lindberg, "Science and the Early Christian Church."
2. For the idea of a "moral economy of science," see Daston, "Moral Economy of Science."
3. Plato, *Phaedo* 67–69. Cf. Wind, *Pagan Mysteries in the Renaissance*, chap. 1.
4. On the encyclopedia tradition, see Stahl, *Roman Science*, pp. 29–42.
5. Ibid., p. 66.

6. Ibid., p. 71.

7. Festugière, *Révélation d'Hermès Trismégiste*, 1:5. Cf. Nock, "Vision of Mandulis Aion," in *Essays on Religion and the Ancient World*, 1:357–400; Dodds, *The Greeks and the Irrational*, chap. 8. For examples of texts, see Luck, *Arcana mundi*.

8. Clagett, *Greek Science in Antiquity*, p. 149. Cf. Festugière, *Révélation d'Hermès Trismégiste*, 1:1–18; and Dodds, *The Greeks and the Irrational*, pp. 236–55.

9. On the *Hermetica*, see Festugière, *Révélation d'Hermès Trismégiste*; and Fowden, *Egyptian Hermes*.

10. Fowden, *Egyptian Hermes*, pp. 57–73. For the philosophical *Hermetica*, see Scott, *Hermetica*; and a recent translation by Copenhaver, *Hermetica*. Many of the technical *Hermetica* are translated in Betz, *Greek Magical Papyri*. In addition, see Festugière, *Hermétisme et mystique païenne*, pp. 28–88.

11. Festugière, *Révélation d'Hermès Trismégiste*, 1:197.

12. Friedrich, ed., *Thessalos von Tralles*; French translation, Festugière, "L'expérience religieuse du médicin Thessalos." See also idem, *Révélation d'Hermès Trismégiste*, 1:56ff.; Smith, "Temple and Magician"; Segal, "Hellenistic Magic"; Cumont, "Le médicin Thessalos"; Meyer-Steineg, "Thessalos von Tralles." Most scholars accept the letter as historical, but see the critical note by Scott, "Ps.-Thessalos of Tralles"; and Dodds, "Supernormal Phenomena in Classical Antiquity," p. 189. Scott believes the letter was a literary fiction based upon Galen's medical writings.

13. Festugière, "L'expérience religieuse du médicin Thessalos," p. 63.

14. Ibid., p. 67.

15. Ibid., pp. 52–53.

16. See Eliade, *History of Religion Ideas*, vol. 2, chap. 26, and the literature cited therein. For transposition of cult mysteries into literary mysteries, see Festugière, *L'Idéal religieux des grecs*, pp. 116–32.

17. Festugière, *Révélation d'Hermès Trismégiste*, pp. 67–88.

18. See Barb, "Survival of Magic Arts"; Betz, *Greek Magical Papyri*, p. xli.

19. Dodds, "Supernormal Phenomena in Classical Antiquity." Fowdon, *Egyptian Hermes*, discusses a fourth-century Greek manual that describes a rite for obtaining a divine revelation (pp. 82–84). The text is in Betz, *Greek Magical Papyri*, pp. 48–54.

20. Festugière, *Révélation d'Hermès Trismégiste*, 1:309–54.

21. Ibid., p. 67.

22. Eliade, *History of Religious Ideas*, 2:299–300.

23. Yates, *Giordano Bruno and the Hermetic Tradition*.

24. "We are swept along on the puffs (*flatu*) of the clever brains of Greece": *Natural History* 29.11.

25. Edelstein, "The Methodists," in *Ancient Medicine*; Lloyd, *Science, Folklore and Ideology*, pp. 182ff.

26. Pliny, *Natural History* 29.9.

27. On the political and social context of classical science, see Lloyd, *Magic, Reason, and Experience*, chap. 4.

28. Many of these beliefs were systematically expounded around 200 B.C. by one Bolos of Mendes (in Egypt), also called "the Democritean," in a series of treatises on "natural properties" (*Physica*). Bolos's works, now lost, were extremely influential in late antiquity, being cited by a host of Greek and Roman writers on natural history. See Wellmann, *Die Φυσικά des Bolos Demokritos*; Festugière, *Révélation d'Hermès Trismégiste*, 1:197ff.; Bidez and Cumont, *Les mages hellénisés*, pp. 107–27.

29. Burkert, *Lore and Science in Ancient Pythagoreanism*; Nock, *Essays on Religion and the Ancient World*: "Greek Magical Papyri," and "Greeks and Magi."

30. Philostratus, *Life of Apollonius of Tyana* 1.2; Pliny, *Natural History* 30.9.

31. Philostratus, *Life of Apollonius of Tyana* 1.2. In addition, see *HMES*, 1:242–67; Bowie, "Apollonius of Tyana."

32. Delatte, *Textes latins et vieux français relatifs aux Cyranides*. See also Festugière, *Révélation d'Hermès Trismégiste*, 1:201–16; and Fowden, *Egyptian Hermes*, pp. 87–89.

33. Delatte, *Textes latins et vieux français relatifs aux Cyranides*, pp. 103, 137.

34. See, for example, the *Corpus Hermeticum*, ed. Nock and Festugière; Bidez and Cumont, *Les mages hellénisés*.

35. Pliny, *Natural History* 26.18; 28.89; 30.1, 17. Pliny's most extensive treatment of magic is found in bk. 30.

36. Segal, "Hellenistic Magic," pp. 356f.

37. Pliny, *Natural History* 30.9.

38. Lloyd, *Science, Folklore and Ideology*, p. 140; Stannard, "Medicinal Plants and Folk Remedies in Pliny"; idem, "Herbal Medicine and Herbal Magic."

39. Pliny, *Natural History* 24.157–67.

40. Ibid. 24.103.

41. Ibid. 7.64. See also Lloyd, *Science, Medicine and Folklore*, pp. 138–39.

42. See Nauert, "Caius Plinius Secundus," for the textual tradition.

43. On Solinus, see Stahl, *Roman Science*, pp. 136–42.

44. An example is the *Medicina Plinii*, a compilation derived from the medical portions of Pliny's *Natural History*. On Pliny in the early Middle Ages, see Nauert, "Caius Plinius Secundus"; Chibnall, "Pliny's *Natural History* in the Middle Ages"; Beddie, "Ancient Classics in the Medieval Libraries."

45. Charles Nauert ("Humanists, Scientists, and Pliny," p. 74) notes that "medieval manuscripts of Pliny contain many glosses to explain specific words, but a careful search has so far revealed no commentaries from the medieval period."

46. Pliny, *Natural History* 25.1. "Those who possess it refuse to teach it, just as though they would themselves lose what they have imparted to others": 25.16.

47. Laistner, *Thought and Letters in Western Europe*, p. 54; Lindberg, "Science and the Early Christian Church."

48. Riddle, "Theory and Practice in Medieval Medicine."

49. Marcellus Empiricus, *De medicamentis liber*.

50. *HMES*, 1:585; Stannard, "Marcellus of Bordeaux."

51. See d'Alverny, "Survivance de la magie antique"; MacKinney, "An Unpublished Treatise on Medicine and Magic."

52. See MacKinney, *Early Medieval Medicine*. For texts of the *receptaria* and *antidotaria*, see Sigerist, *Zur frühmittelalterlichen Rezeptliteratur*; Jörimann, *Frühmittelalterliche Rezeptarien*. For other examples, see Grattan and Singer, *Anglo-Saxon Magic and Medicine*; Walafrid Strabo, *Hortulus*, in Migne, *PL*, 114.1119–30; Voigts, "Anglo-Saxon Plant Remedies"; Stannard, "Medieval Herbals"; idem, "Benedictus Crispus."

53. Quoted in Riddle, "Pseudo-Dioscorides' *Ex herbis feminis*," p. 56.

54. Judging by the number of surviving manuscripts of the work, it was more serviceable than Dioscorides' own more famous work: the Old Latin Dioscorides is found in only three medieval manuscripts, while *Ex herbis feminis* appears in twenty-nine.

55. Rose, "Über die Medicina Plinii"; *HMES*, 1:595–96.

56. *HMES*, 1:594–615.

57. Stannard, "Greco-Roman Materia Medica," p. 467.

58. Voigts, "Significance of the Name Apuleius."

59. Apuleius was widely, though erroneously, regarded as the Latin translator of the Hermetic dialogue *Asclepius*, which supposedly contained the revelations of Thoth (Hermes Trismegistus) to Asclepius, the grandson of the god of medicine: Augustine, *De civitate Dei* 8.23–26; Voigts, "Significance of the Name Apuleius," pp. 218–19.

60. There is an immense literature on the early medieval craft recipe books. Many of the texts are published in Merrifield, *Original Treatises*. For a summary of the textual tradition, see Smith and Hawthorne, *Mappae clavicula*; Multhauf, *Origins of Chemistry*, pp. 153–60.

61. The most reliable edition is that of Halleux, *Les alchimistes grecs*; English translations by Caley, "Leyden Papyrus X," and idem, "Stockholm Papyrus."

62. Halleux, *Les alchimistes grecs*, pp. 24ff.; Multhauf, *Origins of Chemistry*, pp. 98f.

63. Caley, "Leyden Papyrus X," p. 1156, no. 38.

64. Caley, "Stockholm Papyrus," p. 981, no. 2.

65. Ibid., p. 997, no. 139.

66. Rostovzeff, *Social and Economic History of the Roman Empire*, 2:1126.

67. Halleux, *Les alchimistes grecs*, p. 27.

68. Wellman, *Die Φυσικά des Bolos Demokritos*. This work in turn goes back to a lost treatise on metal tinting entitled *Baphika*: Festugière, *Révélation d'Hermès Trismégiste*, 1:197–220, 227–38. See also Fowden, *Egyptian Hermes*, pp. 87–91; Multhauf, *Origins of Chemistry*, pp. 92–101; Halleux, *Les alchimistes grecs*, pp. 62–75; Pfister, "Teinture et alchimie."

69. Festugière, *Révélation d'Hermès Trismégiste*, 1:231–32.

70. Pfister, "Teinture et alchimie," p. 57.

71. Needham, *Science and Civilization in China*, vol. 5, pt. 2, p. 10.

72. Eliade, *History of Religious Ideas*, 2:301–5; idem, *Forge and Crucible*, chap. 13.

73. Smith and Hawthorne, *Mappae clavicula*, p. 37, no. 69.

74. Wilson, *Philosophers, 'Iōsis,' and Water of Life*, pp. 64ff. On the Messalians, see Runciman, *Medieval Manichee*.

75. For the text and its tradition, see Johnson, *Compositiones variae*, pp. 50–62.

76. Smith and Hawthorne, *Mappae clavicula*; Latin text in Phillipps, "Letter"; cf. Johnson, "Some Continental Manuscripts of the *Mappae clavicula*," p. 84.

77. See, for example, Johnson, *Compositiones variae*, pp. 61f.; Roosen-Runge, *Farbgebung und Technik*. For a somewhat different view, see Carroll, "Antique Metal-Joining Formulas in the *Mappae clavicula*."

78. Smith and Hawthorne, *Mappae clavicula*, p. 15.

79. Laistner, *Thought and Letters*, pp. 234–35.

80. Smith and Hawthorne, *Mappae clavicula*, p. 28; Phillipps, "Letter," p. 189. See also the important study by Halleux and Meyvaert, "Les origines de la *Mappae clavicula*"; their reading of the prologue's text differs from Smith and Hawthorne's (the latter based their translation on the Phillipps edition). According to Halleux and Meyvaert, the prologue refers specifically to "Hermetic books" (*in Hermetis libris*), not merely to books in the compiler's possession (*in hec meis*).

81. On this point, see Wilson, *Philosophers, 'Iōsis,' and Water of Life*, p. 66, who suggests that the work's puzzling title is a corruption of the original Greek, *Baphes kleis* (Key to tinting), latinized initially as *Baphae clavicula* and then altered by scribal error to *Mappae clavicula*. See also Halleux and Meyvaert, "Les origines de la *Mappae clavicula*."

82. Smith and Hawthorne, *Mappae clavicula*, p. 32. "Absconde sanctum, et nulli tradendum secretum, neque alicui dederis propheta": Phillips, "Letter," p. 196.

83. In the later Middle Ages the *Mappa* was classified as one of the many "experimental books" (*libri experimentales*) that circulated in scholastic circles and made up the alchemical corpus of the age. See below, chap. 2, and Kibre, "Albertus Magnus, *De Occultis Naturae*," p. 161.

84. Merrifield, *Original Treatises*, 1:183–257. In addition, see Richards, "New Manuscript of Heraclius."

85. Merrifield, *Original Treatises*, 1:182.

86. Ibid., pp. 183, 185, 187, 189.

87. Ibid., pp. 189f. (cf. p. 195); cf. Pliny, *Natural History* 27.59.

88. Merrifield, *Original Treatises*, 1:189.

89. Demus, *Byzantine Art and the West*, chap. 2.

90. Ibid., p. 60.

CHAPTER TWO
KNOWLEDGE AND POWER

1. Metlitzki, *Matter of Araby*, p. 253.

2. Sabra, "Appropriation of Greek Science."

3. See Stock, "Science, Technology, and Economic Progress," pp. 21f.

4. Allen, "Gerbert," pp. 663–68. In addition, see *HMES*, 2:697–718; 2:322.

5. On the transmission of Islamic science to the Latin West, see Lindberg, "Transmission."

6. For Western attitudes toward Islam, see Metliztki, *Matter of Araby*; and Daniel, *Arabs and Mediaeval Europe*.

7. Haskins, *Studies*, p. 135.

8. Ibid., pp. 67–81. In addition, see d'Alverny, "Translations and Translators," pp. 447ff.; Burnett, "Translating of Works from Arabic into Latin."

9. Bacon, *Opus Majus*, 1:52ff., quoting p. 82.

10. Daniel, *Arabs and Mediaeval Europe*, p. 271.

11. Ullmann, *Natur- und Geheimwissenschaften*; *HMES*, 2:214–35, 279–89, 751–824. On the *Hermetica* in Islam, see Plessner, "Hermes Trismegistus"; Kraus, *Jābir*, vol. 2; Massignon, "Inventaire," in Festugière, *Révélation d'Hermès Trismégiste*, 1:384–400; Kiekhefer, *Magic in the Middle Ages*, chap. 6.

12. Ullmann, *Natur- und Geheimwissenschaften*, passim.

13. This point is emphasized by Nasr, *Science and Civilization in Islam*, passim; and more recently by Sabra, "Appropriation of Greek Science."

14. Ullmann, *Natur- und Geheimwissenschaften*, pp. 2–4.

15. Metlitzki, *Matter of Araby*, p. 89; cf. Ullmann, *Natur- und Geheimwissenschaften*, p. 360 and chap. 6. On the Ismaili, see *EI*, 4:63–64, 198–206; *DMA*, 6:614–18.

16. On batinism in Islamic religion, see *EI*, 1:1098ff. (s.v. "bātiniyya"). On the Brethren of Purity, see *EI*, 2:1071–76 (s.v. "Ikhwān al-Safā"); Tibawi, "Ikhwān as-Safā."

17. According to the Ismaili, history progresses through seven stages, each inaugurated by a prophet who announces a new revelation that abrogates the previous one. Seven Imams follow each prophet, the seventh becoming the prophet who proclaims the new age.

18. Nasr, *Science and Civilization in Islam*, p. 37.

19. Pingree, *Picatrix*; German trans., Ritter and Plessner, *Ziel des Weisen*. See also Pingree, "Sources of the *Ghāyat al-Hākim*"; Hartner, "Notes on *Picatrix*."

20. Kraus, "Studien zur Jābir"; idem, *Jābir*, 1:xlviiiff.; *EI*, 2:358. On the connection of Rhazes and Khalid with the Ismaili, see *EI*, 5:112.

21. Corbin, "Livre de Glorieux."

22. Kraus, *Jābir*, 1:xviiff. Kraus suggests that this method was also employed by Roger Bacon.

23. Haskins, *Studies*, p. 73.

24. *HMES*, 2:224.

25. The work, *Medicinalis experimentatio*, or *De medicinis expertis*, was translated into Latin in the thirteenth century. *HMES*, 2:752–56. The text is in Galen, *Opera*, pp. 101–8.

26. *HMES*, 2:227.

27. *Liber Apollonii de principalibus rerum causis*, in Nau, "Un ancienne traduction latine du Bélinous Arabe"; French trans. of the Arabic by Silvestre de Sacy, "Livre du secret." See also Haskins, *Studies*, pp. 79f.; and Kraus, *Jābir*, 2:270–303, who dates the work to the ninth century.

28. In some versions, the corpse of Hermes Trismegistus holds the Emerald Table. See Steele and Singer, "The Emerald Table" (Latin text of a fifteenth-

century manuscript). The authors identify an additional eight fourteenth- and fifteenth-century Latin manuscripts of the work.

29. Kraus, *Jābir*, 2:302f.

30. Silvestre de Sacy, "Livre du secret," p. 123.

31. William of Auvergne, *De legibus*, chaps. 22–26.

32. Peters, *Magician, Witch, and Law*, chap. 2.

33. *Picatrix*, quoted in Garin, "Considerazioni sulla magia," p. 175.

34. Pingree, *Picatrix*, p. 32. Cf. Ritter and Plessner, *Ziel des Weisens*, 58; Compagni, "Picatrix latinus," pp. 302f.

35. *HMES*, 2:267. The most exhaustive studies of the text are Manzalaoui, "Kitāb Sirr al-Asrār"; and Grignaschi, "L'origine et les métamorphoses du 'Sirr al asrār.'" For a census of the medieval manuscripts, see Schmitt, *Pseudo-Aristoteles Latinus*.

36. Manzalaoui, "Kitāb Sirr al-Asrār." Grignaschi, "L'origine et les métamorphoses du 'Sirr al asrâr,'" maintains there was an original Greek core text. These and other problems are discussed in the various articles in Ryan and Schmitt, *Pseudo-Aristotle the "Secret of Secrets."*

37. Bacon, *Secretum secretorum*, pp. 176–77.

38. Marquet suggested that the *Rasāʾil* were "an attempt to arrange and fix the official doctrine of Ismailism": *EI*, 3:1071–76 (s.v. "Ikhwān al-Safā'"); quotation from p. 1072. Whether or not this claim can be substantiated, the later Ismaili accepted the *Rasāʾil* as authentic Ismaili scriptures and speculated that the letters may have been written by one of the Hidden Imams: Stern, "Authorship of the Ikhwān as-Safā," p. 368. In addition, see Manzalaoui, "Kitāb Sirr al-Asrār," pp. 175ff; Grignaschi, "L'origine et les métamorphoses du 'Sirr al asrār'"; Stern, *Studies in Early Ismāʿīlism*; Tibawi, "Ikhwān as-Safā." Stern, *Studies in Early Ismāʿīlism*, chap. 1, believes the Ikhwān to have been an imaginary, utopian fraternity.

39. On John of Spain (Johannes Hispaniensis), see Thorndike, "John of Seville." On Phillip of Tripoli, see Haskins, *Studies*, pp. 137–40; and Manzalaoui, "Philip of Tripoli." The Latin manuscript tradition is treated extensively in Grignaschi, "Diffusion du 'Secretum secretorum.'"

40. The work's Arabic translator, Yahyā ibn al-Batrīq, claims he found the manuscript in a temple of Asclepius after wandering through Persia in search of wisdom: *Secretum secretorum*, p. 39.

41. Ibid., pp. 41f.

42. Ibid., pp. 178–79.

43. Ibid., p. 41.

44. Ross, "Letters of Alexander."

45. Easton, *Roger Bacon*, p. 78.

46. Ibid., pp. 72ff. The idea that philosophy was originally revealed knowledge, passed down through the prophets and then to the pagans, is explicitly stated in the *Secretum secretorum*, p. 64; cf. Bacon's commentary on the transmission of astronomy, pp. 62f. Cf. Bacon, *Opus Majus*, 1:9–14, 52–65; idem, *Opus Tertium*, pp. 79ff. See also Lindberg, "Science as Handmaiden," p. 530.

47. Bacon, *Secretum secretorum*, p. 1.

48. Ibid., pp. 254–55.

49. Bacon, *Opus Majus*, 1:11–12.

50. See, for example, *Opus Tertium*, p. 80. On the subject generally, see Easton, *Roger Bacon*, p. 82; Manzalaoui, "Three Oxford Thinkers," p. 488; Vasoli, "Il Programma riformatore di Ruggero Bacone."

51. Bacon, *Epistola*, p. 544.

52. According to Hime, "Roger Bacon and Gunpowder," Bacon wrote the directions for the manufacture of gunpowder in cipher in the *Epistola*.

53. Bacon, *Secretum secretorum*, pp. 4–5. See also Murray, *Reason and Society*, pp. 114–16.

54. Bacon, *Epistola*, pp. 523ff.

55. Daniel, *Arabs and Mediaeval Europe*, p. 276; Manzalaoui, "Philip of Tripoli," p. 62.

56. ". . . intellectus est capud regimini": Bacon, *Secretum secretorum*, p. 45.

57. Ibid., pp. 141f.

58. Monfrin, "La Place du *Secret des secrets*," p. 93. In addition, see Grignaschi, "Diffusion du 'Secretum secretorum,'" pp. 27–34. For further discussion of the *Secretum*'s influence on political theory, see Murray, *Reason and Society*, pp. 121–24; and Ryan, "The *Secretum secretorum* and the Muscovite Autocracy." On the vernacular dissemination and influence of the *Secretum*, see Hirth, "Deutschen Bearbeitungen"; Gilbert, "Notes on the Influence of the *Secretum*"; and Manzalaoui, *Secretum secretorum: Nine English Versions*.

59. Bacon, *Opus majus*, 2:583–634.

60. Lindberg, "Science as Handmaiden," p. 533; Hackett, "Meaning of Experimental Science."

61. Bacon, *Opus Majus*, 2:634; cf. Easton, *Roger Bacon*, p. 73.

62. Bacon, *Opus Majus*, 2:633.

63. Bacon, *Epistola*, p. 523 (trans. Davis, *Roger Bacon's Letter*, p. 15). Cf. *Secretum secretorum*, p. 1.

64. See Easton, *Roger Bacon*, pp. 169ff; Lindberg, "Science as Handmaiden."

65. Bacon, *Opus Majus*, 1:65.

66. Ibid., 2:620.

67. Ibid., pp. 624f.

68. *Secretum secretorum*, p. 64; cf. Bacon's gloss on this passage, pp. 62f.

69. Bacon, *Opus Majus*, 2:621.

70. Bacon, *Opus Tertium*, p. 59.

71. On the following, see especially Crombie, *Robert Grosseteste*, pp. 25f., 52–54, 91–96, et passim; Pederson, "Development of Natural Philosophy"; and McMorris, "Science as *Scientia*."

72. Hutchison, "What Happened to Occult Qualities?"; Copenhaver, "Scholastic Philosophy and Renaissance Magic."

73. For this and other "secrets" for hardening iron, see Biringuccio, *Pirotechnia*, pp. 371–73.

74. Steneck, *Science and Creation*, pp. 117ff; cf. Stock, *Implications of Literacy*.

75. Agrimi and Crisciani, "Medici e 'vetulae.'" Cf. Jean Gerson, quoted in *HMES*, 4:126.

76. Bacon, *Epistola*, p. 543: "Ergo quod pluribus, hoc est vulgo, in quantum huiusmodi videtur, oportet quod sit falsum." See also Grundmann, "Litteratus—illiteratus."

77. Thus Bernard de Gordon, while generally condemning the lore of the people, occasionally recommended magical cures himself, but only when they were mentioned by standard authorities: DeMaitre, *Bernard de Gordon*, pp. 157 and 161ff.

78. Albertus Magnus, *Book of Minerals*, trans. Wyckoff, p. 47. In defense of Albertus's empiricism, see Shaw, "Scientific Empiricism in the Middle Ages."

79. Schmitt, "Experience and Experiment." Cf. DeMaitre, *Bernard de Gordon*, pp. 128–34.

80. DeMaitre, *Bernard de Gordon*, pp. 126–27.

81. Ibid., pp. 130f.

82. McVaugh, "Two Montpellier Recipe Collections," p. 178.

83. See, for example, Peter of Spain (Pope John XXI), quoted in *HMES*, 2:509f.; Bacon, *Epistola*, 539; Agrimi and Crisciani, "Medici e 'vetulae,'" p. 149; Wardale, "A Low German-Latin Miscellany," p. 15, no. 64.

84. DeMaitre, *Bernard de Gordon*, pp. 130–31.

85. Ringelbergius, *Opera*, p. 606.

86. Siraisi, *Taddeo Alderotti*, p. 303.

87. McVaugh, "The *Experimenta* of Arnald of Villanova," p. 111.

88. Siraisi, *Taddeo Alderotti*, p. 25.

89. Albertus Magnus, *Book of Minerals*, trans. Wyckoff, p. 133.

90. *HMES*, 2:751–812; Braekman, *Magische experimenten*.

91. Eamon, "Technology as Magic"; Weisheipl, "Classification of the Sciences."

92. Gundisalvo, *De divisione philosophiae*, pp. 4, 20; Ovitt, *Restoration of Perfection*, pp. 122–24.

93. Ottaviano, *Un brano inedito di Guglielmo di Conches*, pp. 35–36. On the place of the mechanical arts in these classification schemes, see Ovitt, "The Status of the Mechanical Arts."

94. Hugh of St. Victor, *Didascalicon*, trans. Taylor, p. 154; cf. John of Salisbury, quoted in *HMES*, 2:158.

95. Augustine, *De civitate Dei* 8.23; 10.9.

96. William of Auvergne, *De legibus* 1.24, p. 69.

97. Lactantius, *Divinae institutiones* 2.8.60–63. On curiosity in the Middle Ages, see especially Blumenberg, *Legitimacy of the Modern Age*, pp. 269–360.

98. Economou, *The Goddess Natura*.

99. Ecclus. 3:22–23; other biblical references include Deut. 29:29, Prov. 25:27, 1 Cor. 8:1, Rom. 12:3.

100. Augustine, *Confessions* 10.30; 10.35; cf. Blumenberg, *Legitimacy of the Modern Age*, pp. 309–23.

101. Augustine, *Confessions*, 10.35, 55, trans. Warner, p. 246 (with minor revisions).

102. Blumenberg, *Legitimacy of the Modern Age*; Steiner, "The Faust Legend."

103. Augustine, *De divinatione daemonum* 3.7.

104. Ginzburg, "High and Low"; Clark, "Inversion, Misrule, and the Meaning of Witchcraft."

105. Augustine, *De civitate Dei*, bk. 9.

106. Ibid. 9.9.

107. Ginzburg, "High and Low."

108. Augustine, *Confessions* 5.3.3, trans. Warner, p. 92.

109. Bernard of Clairvaux, *Steps of Humility and Pride*, pp. 57, 66; John Milton, *Paradise Lost* 12.279; Augustine, *De trinitate* 11.5.8.

110. Allen, "Gerbert"; Molland, "Roger Bacon as a Magician"; Comparetti, *Virgil in the Middle Ages*; Eamon, "Technology as Magic."

111. Augustine, *Confessions* 10.35, 55; cf. Blumenberg, *Legitimacy of the Modern Age*, pp. 107f.

112. Peter Damian, *De divina omnipotentia*, trans. Wippel and Wolter, *Medieval Philosophy*, p. 151.

113. Stiefel, *Intellectual Revolution in Twelfth Century Europe*.

114. On the medieval debate over miracles, see Ward, *Miracles and the Medieval Mind*, chap. 1; Radding, "Superstition to Science."

115. One of the propositions (no. 147) banned by the condemnation of 1277; Grant, *Sourcebook in Medieval Science*, p. 215.

116. Quoted in Taylor, *The Medieval Mind*, 1:416f.

117. "Curiositas est investigatio eorum quae ad rem et ad nos non pertinent. Prudentia autem tantum est de his quae ad rem et ad nos pertinent": quoted in Blumenberg, *Legitimacy of the Modern Age*, p. 290.

118. Aquinas, *Summa theologiae* 2a2ae. 167, 1.1.

119. For a convenient summary of late-medieval thought, see Leff, *Dissolution of the Medieval Outlook*.

120. Gerson, *Contra vanam curiositatem*; summarized in Gilson, *History of Christian Philosophy*, pp. 529ff.

121. Gerson, *Contra vanam curiositatem*, quoted in Blumenberg, *Legitimacy of the Modern Age*, p. 638: "Qualiter et quando mundus inceperit aut si finituris sit, sciri nequit ex quibuscumque experientiis quas philosophia sequitur, quoniam hoc in liberrima conditoris voluntate situm est. Philosophi igiture dum hoc secretum divinae voluntatis penetrare, duce experientia, moliuntur, quidni deficiant? Quoniam sicut divina voluntas huius ratio est, ita solis illis scire concessum est, quibus ipsa voluerit revelare."

122. Bernard of Clairvaux, *Steps of Humility and Pride*, p. 60.

123. Bacon, *Epistola*, p. 526; cf. p. 532.

124. Ibid., p. 532; cf. Bacon, *Opus Majus*, 2:587.

125. Bacon, *Epistola*, p. 523.

126. Ibid., p. 543.

127. In general, see Kiekhefer, *Magic in the Middle Ages*; and *HMES*, vol. 3.

128. Aquinas, *De occultis operibus naturae*, trans. McAllister.

129. Peters, *Magician, Witch, and Law*, pp. 93, 117.

130. *HMES*, 2:157ff.

131. For example, Pingree, *Picatrix*, pp. 21, 155–58, 231–32; Ps.-Albertus Magnus, *The Book of Secrets*, ed. Best and Brightman, pp. 38–41.

132. Peters, *Magician, Witch, and Law*, pp. 112–25; Brown, "Society and the Supernatural"; Jones, "Political Uses of Sorcery"; Kiekhefer, *Magic in the Middle Ages*, chap. 5; Round, "Five Magicians."

133. Kyeser, *Bellifortis*, ed. Quarg, pp. 91b–93, 132b, 93b, 94a. See also Eamon, "Technology as Magic," pp. 186–92; and White, "Kyeser's 'Bellifortis.'"

134. Gille, *Engineers of the Renaissance*, p. 59.

135. Bolgar, *The Classical Heritage and Its Beneficiaries*, p. 178. In addition, see Murray, *Reason and Society*, chap. 9.

136. Kyeser, *Bellifortis*, p. 28a.

137. In the printed editions of the work, the title is usually given as *Liber aggregationis*. I have used the Strasbourg, 1625, edition, *Liber aggregationis seu liber secretorum Alberti Magni*, in *De secretis mulierum libellus*, pp. 127–66. For a partial list of manuscripts, see *HMES*, 2:746–48.

138. Bacon, *Opus Majus*, 2:526.

139. For an analysis of the work's sources, see Thorndike, "Further Considerations of the *Experimenta*."

140. *HMES*, 2:730.

141. Ps.-Albertus Magnus, *Liber aggregationis*, in *De secretis mulierum*, p. 127f; idem, *The Book of Secrets*, p. 3.

142. In the Strasbourg, 1625, edition, *De secretis mulierum* is on pp. 3–127, and *De mirabilibus mundi* is on pp. 166–212.

143. Ps.-Albertus Magnus, *De secretis mulierum*, p. 5; Lemay, *Women's Secrets*, p. 59. Lemay's translation also contains a valuable introduction to the text. On the *Secreta mulierum*, see Thorndike, "Further Considerations of the *Experimenta*"; Ferckel, "Die Secreta mulierum und ihr Verfasser"; Kusche, "Zur 'Secreta mulierum'-Forschung." Also valuable is Green, "Women's Medical Practice." Although the *Secreta mulierum* was quite popular, it is doubtful that it was used as a midwife's manual, as Weisner claims: *Working Women in Renaissance Germany*, p. 66.

144. Ps.-Albertus Magnus, *De secretis mulierum*, pp. 116–17; Lemay, *Women's Secrets*, pp. 139–41; cf. Pinto, "Folk Practice of Gynecology," p. 523.

145. Hansen, *Oresme and the Marvels of Nature*, pp. 54–61.

146. Chenu, *Nature, Man, and Society in the Twelfth Century*, p. 14.

147. Aquinas, *De occultis operibus naturae*, ed. McAllister, p. 22.

148. Hansen, *Oresme and the Marvels of Nature*.

149. Ps.-Albertus Magnus, *Liber aggregationis*, p. 166.

150. Ibid., pp. 182–83.

151. For example, the work turns up in monastic, university, and private library catalogs all over Germany: Lehmann, *Mittelalterliche Bibliotekskataloge Deutschlands und der Schwiez*. Almost every physician's library of substantial size had copies of the *Liber aggregationis*. Interestingly, the work was cataloged under both natural philosophy and medicine. Doubtless, a careful survey of other library catalogs would turn up new information concerning the work's distribution.

152. *HMES*, 2:345, 363; William of Auvergne, *De universo* 3.22, in *Opera omnia*, pp. 1059–1061.

153. See, for example, Braekman, *Magische experimenten*; Wardale, "A Low German-Latin Miscellany."

154. Malinowski, *Magic, Science, and Religion*, pp. 17–92.

155. On the problem of the persistence and rationality of magic, see Evans-Pritchard, *Witchcraft, Oracles, and Magic*, pp. 199–204; Thomas, *Religion and the Decline of Magic*, pp. 641–42; Jarvie and Agassi, "The Problem of the Rationality of Magic."

156. See in particular White, *Medieval Technology and Social Change* and "Cultural Climates and Technological Advance"; Dresbeck, "Techne, Labor, et Natura"; Stock, "Science, Technology, and Economic Progress."

157. Bacon, *Opus Majus*, 2:634.

158. On nominalism, see Courtney, "Nominalism in Late-Medieval Religion"; Leff, *Dissolution of the Medieval Outlook*; and Grant, "Condemnation of 1277."

159. Nicholas of Poland, *Antipocras*, ed. Ganszyniec, *Brata Mikołaja z Polski*; ed. Sudhoff, "Antipocras." See also Eamon and Keil, "Nicholas of Poland"; Szpilczynski, "Considérations sur les conceptions de Nicolas de Poland."

160. Nicholas of Poland, *Antipocras*, ed. Ganszyniec, p. 46.

161. For Nicholas's subsequent Polish career, see Eamon and Keil, "Nicholas of Poland." See also Keil, "Medizinische Bildung und Alternativmedizin."

162. Nicholas of Poland, *Experimenta*, ed. Ganszyniec, p. 136. The text is also edited by Johnsson, "Les 'Experimenta Magistri Nicolai.'"

163. Nicholas of Poland, *Experimenta*, ed. Ganszyniec, p. 144.

164. Nicholas of Poland, *Antipocras*, ed. Ganszyniec, p. 60; Sudhoff, p. 47.

165. Nicholas of Poland, *Antipocras*, ed. Ganszyniec, p. 52; Sudhoff, p. 43.

166. Nicholas of Poland, *Antipocras*, ed. Ganszyniec, p. 62: "Cur hoc sit, non legis in me. Qualiter aut quare virtus fluit ex aliqua re, Cum res est clausa, non est michi cognita causa."

167. Ibid., p. 60: "Cur ab Ypocrate de tali propriete non est instructum medicine carpere fructum? Forsan in hac sorte pauper fuit, aut quia forte precauit vates, multi sunt Ypocratis."

168. Ibid., p. 56: "Plebs amat empirica, quia nulli sunt inimica; sed pudor est medicis, quod ouent magnalia vicis, consona facta fora laxant in laudibus ora."

169. McVaugh, "Quantified Medical Theory and Practice at Fourteenth-Century Montpellier"; idem, "Theriac at Montpellier, 1285–1385"; idem, "An Early Discussion of Medicinal Degrees at Montpellier"; DeMaitre, "Theory and Practice in Medical Education at Montpellier." In general, see Bullough, "Development of the Medical University at Montpellier."

170. Nicholas of Poland, *Antipocras*, ed. Ganszyniec, p. 48: "Hic inducitur magister Albertus ad confirmandum que dicta sunt et dicenda. Verbis Albertus verax et dogmate certus hoc sine figmento docet et probat experimento."

171. For example, ibid., p. 58: "Virtus sublimis, que desuper influit imis, a firmamento condescendens, elemento nubit et unitur simul, in re cum sepelitur et manet occulta"; cf. pp. 50, 48.

172. Nicholas of Poland, *Antipocras*, ed. Ganszyneic, p. 50: "Hunc adiens tanga in frustraque plurima frangas, ex hoc non magnes vim perdet ut Anna vel Agnes."

173. Kibre, *Scholarly Privileges in the Middle Ages*.

174. In general, see Rashdall, *The Universities of Europe in the Middle Ages*; Post, "Parisian Masters"; Murray, *Reason and Society*; Kibre, *Scholarly Privileges in the Middle Ages*; and Grundmann, "Sacerdotium—Regnum—Studium."

175. Murray, *Reason and Society*, pp. 237–44.

176. Quoted in ibid., p. 242.

177. Nicholas of Clairvaux, quoted in Thompson, *Literacy of the Laity*, p. 143.

178. Bacon, *Opus Majus*, 1:11.

179. Economou, *The Goddess Natura*; Raby, "*Nuda Natura* and Twelfth-Century Cosmology."

180. In general, see Unwin, *Gilds and Companies of London*; Smith, *English Gilds*. See also Shelby, "'Secret' of the Medieval Masons." The best recent study of craft secrecy and intellectual property is Long, "Invention and the Origins of Patents."

181. Thus the Coventry weavers' guild specified that no weaver was to "teche ne soffer to be taught nother man ne chylde . . . nether his owne sone ne cosyn nor none other except that he be bounde prentes": quoted in Pythian-Adams, *Desolation of a City*, p. 109.

182. Welch, *History of the Worshipful Company of Pewterers of London*, 2:34, 84.

183. Long, "Invention and the Origin of Patents," p. 873.

184. An example of a craft secret held onto privately by a small number of craftsmen was the technique of rhinoplasty, which was practiced by a few surgeons in southern Italy. The secret was said to have been invented by a Sicilian surgeon named Branca, although he probably learned the technique from some Arab surgeon. Because rhinoplasty was practiced by so few surgeons, and was in high demand among upper-class patients, the technique was a valuable secret. The sixteenth-century surgeon Leonardo Fioravanti went to great pains to learn it from a Calabrian surgeon. See Gnudi and Webster, *Gaspare Tagliacozzi*, pp. 110–15 and below, chap. 4.

185. Zilsel, "Sociological Roots of Knowledge." On the mechanical arts in the Middle Ages, see Ovitt, *Restoration of Perfection*; Sternagel, *Die artes mechanicae im Mittelalter*.

186. Bacon, *Opus Majus*, 1:25; cf. Ovitt, *Restoration of Perfection*, p. 119; Beaujouan, *L'interdépendence entre la science scholastique et les techniques utilitaires*; White, "Medieval Engineering and the Sociology of Knowledge."

187. White, "Cultural Climates and Technological Advance," pp. 196ff. Hugh's significant work in this regard was the *Didascalicon*. See also Ovitt, *Restoration of Perfection*, pp. 117–21.

188. Hugh of St. Victor, *Didascalicon*, p. 75.

189. See, in particular, Ovitt, *Restoration of Perfection*, pp. 94–95; and De Rijk, "Three (Four) Human Evils."

190. Theophilus Presbyter, *De diversis artibus*, ed. and trans. Dodwell; *On Divers Arts*, trans. Hawthorne and Smith.

191. Theophilus, *On Divers Arts*, trans. Hawthorne and Smith, pp. 12, 78–79.

192. That these attitudes toward openness were becoming common in the monastic craft tradition by the twelfth century is suggested by the anonymous Berne manuscript, *De clarea* (ca. 1100), a work on making glair to be used as a medium to mix paints for illustrating manuscripts. Anonymous Bernensis explains that it was "not for vanity, but rather as a friendly response to the requirements of many people, that I write this work, which I now desire to carry out, with God's aid, for the purpose of instructing others": Thompson, "The *De clarea* of the So-Called 'Anonymous Bernensis,' " p. 71.

193. The bibliography of the late-medieval technical treatises is quite large. Much of the literature, especially in the German language, has been summarized by Eis, *Mittelalterliche Fachliteratur*, and Keil and Assion, *Fachprosaforschung*. See also the studies in the two Festschriften dedicated to Gerhard Eis: Keil et al., *Fachliteratur des Mittelalters*, and Keil, *Fachprosa-Studien*. Eis's important publications on the subject are listed on pp. 499–534 and 574–86 of the latter two volumes respectively.

194. *Pelzbuch* is the name given the work by Eis, who edited several German manuscripts of it: Eis, *Gottfrieds Pelzbuch*. The Latin text is as yet unedited. A Middle English translation of the text has been edited by Braekman, *Geoffrey of Franconia's Book of Trees and Wine*. On the author, see Keil, "Gottfried von Franken."

195. Eis, *Gottfrieds Pelzbuch*, pp. 26–27, 122–123. Eis discovered a fifteenth-century German text attributed to "ein grosser meister, genannt Meister Richartt," which appears to be a fragment of Richard's work: Zurich MS 102b, published in *Gottfrieds Pelzbuch*, pp. 187–88.

196. Braekman, *Geoffrey of Franconia's Book of Trees and Wine*, pp. 7–11.

197. Eis, *Gottfrieds Pelzbuch*, pp. 138–39, nos. 55–56.

198. Quoted from the Middle English version edition by Braekman, *Geoffrey of Franconia's Book of Trees and Wine*, p. 104.

199. Reporting on a technique he had learned from Pseudo-Aristotle for making sour pomegranates sweet, Gottfried wrote, "Vnd ist, das Aristotiles spricht von malagranatin alsus, so ist is ouch der worheit glich, das ouch andir suwir boume werdin suze in der selbin wis begat": Eis, *Gottfrieds Pelzbuch*, p. 123.

200. Ibid., pp. 64–116; Braekman, *Geoffrey of Franconia's Book of Trees and Wine*, pp. 51–56. Gottfried's friend, the English monk Nicholas Bollard, gives us another example of a cultural broker. Evidently Bollard also wanted to put the art of planting and grafting on a "scientific" footing. Writing the work for "borell folk" (commoners or the unlearned), Bollard began by citing the pseudo-Aristotelian *Secretum secretorum* as an authority for planting and grafting in accordance with celestial signs. However, acknowledging that "there be but fewe" people who understand such arcane matters, he drops the subject and turns to "other thinges more comoun": Braekman, *Geoffrey of Franconia's Book of Trees and Wine*, p. 30.

201. Cod. MS 3227a, Germanisches Nationalmuseum Nürnberg, ed. Hils, "Von dem herten." See also Darmstaeder, *Berg- Probier- und Kunstbüchlein*; idem, "Eisenhartung mit Pflanzen."

202. E.g., "wenne dy sneide blo ist, zo hat is eyne rechte herte"; Hils, "Von dem herten," p. 72.

203. Ploss, "Wielands Schwert Mimung und die alte Stahlhärtung."

204. Hils, "Von dem herten," p. 68. The use of certain plants, such as *Eisenkraut* ("iron-plant"), *Steinwürz* ("stone-wort"), and *Drachenwürz* ("dragon's-wort"), may be explained by their names.

205. Ibid., p. 73: "Nym eynteil rueberetich, vnd eyn teil merretich, vnd eynteil regenwuerme, engerlinge, vnd eynteil buckes blut, wen der bok czu breunsten get."

206. Pliny, *Natural History* 33.148; 37.55.

207. E.g., a recipe prescribing verbena and earthworms crushed, mixed with water, and distilled, and the distillate mixed with ass's milk: Hils, "Von dem herten," p. 74, no. 13.

208. Hassenstein, *Das Feuerwerkbuch von 1420*, p. 47; see also *VL*, 2:730; Hall, "Writings about Technology."

209. Clagget, "Life and Career of Giovanni Fontana"; Birkenmajer, "Lebensgeschichte und wissenschaftlichen Tätigkeit von Giovanni Fontana."

210. Quoted in Prager, "A Manuscript of Taccola," p. 141.

211. Quoted in Reti, "Francesco di Giorgio Martini's Treatise," pp. 291–92.

212. See Long, "Invention and the Origins of Patents."

213. On the early history of patents, see Prager, "History of Intellectual Property"; idem, "Examination of Inventions"; Frumkin, "Origin of Patents"; idem, "Early History of Patents for Invention"; Pohlmann, "Inventor's Right in Early German Law"; Gleitsmann, "Erfinderprivilegien und technologische Wandel im 16. Jahrhundert." The best recent work is Long, "Invention and the Origin of Patents."

214. Prager, "Brunelleschi's Patent," pp. 109–10.

215. Mandich, "Venetian Patents," p. 176.

216. Prager, "History of Intellectual Property," pp. 715, 750.

217. Patterson, *Copyright in Historical Perspective*.

CHAPTER THREE
ARCANA DISCLOSED

1. Ong, *Orality and Literacy*, pp. 112–15.

2. On these themes, see in particular Stock, *Implications of Literacy*.

3. Grundmann, "Litteratus—Illiteratus"; Stock, *Implications of Literacy*, p. 27.

4. Stock, *Implications of Literacy*, p. 31; cf. Lévi-Strauss, *The Savage Mind*, chap. 1.

5. Eisenstein, *The Printing Press as an Agent of Change*, pp. 521ff. I am much indebted to Eisenstein's pioneering work, not least for my chapter title, which I borrow from p. 272.

6. Febvre and Martin, *The Coming of the Book*, p. 159.

7. See Houston, *Literacy in Early Modern Europe*.

8. Eisenstein, *The Printing Press as an Agent of Change*, pp. 520ff.

9. Chrisman, *Lay Culture, Learned Culture*, passim. See also Cook, *The Decline of the Old Medical Regime*, chap. 1.

10. Keil and Assion, *Fachprosaforschung*.

11. Benzing, *Die Buchdrucker des 16. und 17. Jahrhunderts*; Hirsch, *Printing, Selling, Reading*, chap. 7.

12. Engelsing, *Analphabetentum und Lekture*; idem, *Die Bürger als Leser*.

13. Crofts, "Books, Reform, and the Reformation."

14. Telle, "Wissenschaft und Öffentlichkeit"; idem, *Pharmazie und der gemeine Mann*; Zimmerman, *Das Hausarzneibuch*; Schubert and Sudhoff, "Michael Bapst von Rochlitz"; idem, "Die Schriften des Michael Bapst von Rochlitz."

15. Benzing, "Walther H. Ryff und sein literarisches Werk"; Rákóczi, *Walther Hermann Ryffs Populärwissenschaftliche Tätigkeit*.

16. The plagiarism charge was first hurled at Ryff by Leonhard Fuchs, a Tübingen professor of medicine, and by Fuchs's Zurich colleague Conrad Gesner, who repeated the charges in his celebrated *Bibliotheca universalis*, whence it passed unchallenged into the academic tradition; it has been repeated by historians ever since. See Grenzmann, *Traumbuch Artemidori*, pp. 26–37.

17. On printing privileges in Germany, see Pohlmann, "Neue Materialen zum deutschen Urheberschutz"; idem, "Weitere Archivfunde zum kaiserliche Autorenschutz."

18. On literary piracy and controversies over copyrights, see Pohlmann, "Neue Materialen zum deutschen Urheberschutz."

19. Rákóczi, "Ryffs charakteristische Stilmittel."

20. For Ryff's biography, see Benzing, "Walther H. Ryff und sein literarisches Werk"; Lüdke, "Walther Ryff und seine 'Teütsche Apoteck.'"

21. Ryff, *Ein newer Albertus Magnus*.

22. *Der furnembsten, notwendigsten, der gantzen Architectur angehörigen Mathematischen und Mechanischen künst, eygentlicher bericht, und vast klare verstendliche unterrichtung zu rechtem verstandt der lehr Vitruvii*. Ryff's Vitruvius translation has been published in facsimile, with a commentary by Erik Forssman in Ryff, *Zehen Bücher von der Architectur*. In addition, see Röttinger, *Die Hozschnitte des Walther Rivius*. On artistic styles in Nuremberg, see *Gothic and Renaissance Art in Nuremberg*, in particular the article by Brandl, "Art or Craft?"; Strauss, *Nuremberg in the Sixteenth Century*, chap. 6; Panofsky, *Albrecht Dürer*, pp. 242–84.

23. Ryff, *Die gross Chirurgei*; idem, *Stat und Feldbüch bewerter Wundtartznei*. See also Sigerist, *Hieronymous Brunschwig and His Work*; Stannard, "Hans von Gersdorff and Some Anonymous Strasbourg Apothecaries"; Siraisi, *Medieval and Early Renaissance Medicine*, chap. 6.

24. *Nova scientia* (1537) and *Quesiti et inventioni diverse* (1546), translated in Drake and Drabkin, *Mechanics in Sixteenth-Century Italy*. See also Harig, "Walter Hermann Ryff und Nicolo Tartaglia."

25. On Brunfels, best known for his famous herbal, the *Herbarum vivae eicones* (1530), see Roth, "Otto Brunfels"; and Reeds, "Renaissance Humanism and Botany."

26. Dilg, "'Reformation der Apotecken,'" p. 185. See also idem, "Die Strassburger Apothekenordnung"; Philipp, *Das Medizinal- und Apothekenrecht in Nurnberg*; and Kremers and Urdang, *History of Pharmacy*.

27. Thus Paracelsus, who frequently compared his radical medical views to Luther's revolutionary theology, echoed the concerns of many reformers by calling the pharmacies "foul sculleries, from which comes nothing but foul broths": *Selected Writings*, p. 6.

28. On the academic debate over Greek versus Arabic medicine, see Chrisman, *Lay Culture, Learned Culture*, pp. 172ff; Baader, "Medizinisches Reformdenken und Arabismus."

29. Ryff, *Confectbuch* (1544).

30. On this work, see Cushing, *Bio-Bibliography of Andreas Vesalius*, pp. 10–43. A facsimile of the 1541 edition of the work has been published; Russell, "Walter Hermann Ryff and His Anatomy"; Crummer, "Early Anatomical Fugitive Sheets." On Vesalius's *Tabulae*, see also Singer and Rabin, *Prelude to Modern Science*; O'Malley, *Andreas Vesalius*, pp. 85–91.

31. Quoted in Cushing, *Bio-Bibliography of Andreas Vesalius*, p. 17.

32. Ryff, *Kurtz Handbüchlein und Experiment vieler Artzneien*, preface. (Benzing nos. 41–78.)

33. "... des gemeynen mans, auch fleissiger Haushalter": *Confectbuch* (Frankfurt am Main, 1554), preface.

34. Lutz, *Wer war der gemeine Mann?*; Sabean, *Power in the Blood*, pp. 27–37.

35. Telle, "Wissenschaft un Öffentlichkeit"; Schenda, "Der 'gemeine Mann' und sein medikales Verhalten," and the other studies in Telle, *Pharmazie und der gemeine Mann*.

36. Ryff, *Prakticir Büchlein der Leibartzeney*, preface.

37. Goltz, "Die Paracelsisten und die Sprache"; Telle, "Wissenschaft und Öffentlichkeit."

38. *Spiegel der Arzney* (Strasbourg, 1532), quoted in Telle, "Wissenschaft und Öffentlichkeit," p. 40.

39. Quoted in Telle, "Wissenschaft und Öffentlichkeit," p. 34.

40. Dryander, *Artzenei Speigel*, preface.

41. Olschki, *Geschichte der neusprachlichen wissenschaftlichen Literatur*, vol. 2; Goltz, "Die Paracelsisten und die Sprache"; Telle, "Wissenschaft und Öffentlichkeit."

42. Quoted in Cushing, *Bio-Bibliography of Andreas Vesalius*, p. 18.

43. For an account of an ordinary German family's health concerns, see Ozment, *Magdalena and Balthasar*, pp. 110–35.

44. "Cirugia ist ein wirckung der hendt das dem vundartzet zu gehört und nit dem physicus": Brunschwig, *The Book of Cirurgia*, p. 258.

45. Ryff, *Die gross Chirurgei*, fol. a1v.

46. Burke, "A Question of Acculturation?" In addition, see Chartier, "Culture as Appropriation."

47. Haage, "Germanistische Wortforschung." On Ryff's style, see Rákóczi, "Ryffs charakteristische Stilmittel."

48. See Stock, *Implications of Literacy*, pp. 86f. and 524f.

49. In one of his early publications, Cardano printed a list of all his unpublished works, hoping some patron would provide the means to have them printed. It was Petreius who answered, offering to publish any manuscripts that Cardano had ready for the press. Petreius published several works by Cardano, including the celebrated *Ars Magna* (1544) and *De subtilitate* (1550). Shipman, "Johannes Petreius"; Ore, *Cardano*, p. 11.

50. Erasmus, *Adages*, cited by Lowry, *World of Aldus Manutius*, p. 8.

51. For the backgrounds of early printers, see Hirsch, *Printing, Selling, Reading*, pp. 18–22.

52. For details concerning the business operations of early printers, see ibid., chap. 4.

53. Voet, *The Golden Compasses*; Kingdon, "Patronage, Piety, and Printing"; De Roover, "Business Organization of the Plantin Press."

54. Chrisman, *Lay Culture, Learned Culture*, p. 170.

55. On Strasbourg printing, see ibid. On Cammerlander, see Benzing, *Die Drucke Jacob Cammerlanders*; Ritter, *Histoire de l'imprimerie alsacienne*, pp. 288–94.

56. For Cammerlander's religious activities, see Ritter, *Histoire de l'imprimerie alsacienne*, pp. 288, 292–94.

57. Chrisman, *Lay Culture, Learned Culture*, p. 36; on Gulferich, see Benzing, "Hermann Gülfferich."

58. Benzing, "Christian Egenolff zu Strassburg"; idem, "Christian Egenolff zu Frankfurt am Main"; idem, "Christian Egenolff un seine Verlagsproduktion."

59. For Egenolff's career, see Grotefend, *Christian Egenolff*; Mori, "Christian Egenolff"; Ritter, *Histoire de l'imprimerie alsacienne*, pp. 314–15.

60. Grotefend, *Christian Egenolff*, p. 14; Beatus Murner, Frankfurt's first printer, died in 1512: Benzing, *Die Buchdrucker des 16. und 17. Jahrhunderts*, p. 113.

61. For nine years Egenolff operated as Frankfurt's only printer. During his lifetime only four other printers competed with him. Only after Egenolff's death did Frankfurt become a major publishing city: Benzing, *Die Buchdrucker des 16. und 17. Jahrhunderts*, p. 112ff.

62. Dietz, *Geschichte der Frankfurter Büchermesse*; Thompson, *Frankfort Book Fair*; Brauer, "Frankfurt als Stätte deutschen Buchhandels."

63. Grotefend, *Christian Egenolff*, pp. 22–23.

64. See Melanchthon's letter of 1551 to Egenolff: Grotefend, *Christian Egenolff*, p. 28.

65. Shipman, "Johannes Petreius"; Benzing, *Die Buchdrucker des 16. und 17. Jahrhunderts*, p. 334.

66. Petreius published the astronomical and mathematical works of Regiomontanus, Latin treatises on optics by Witelo (*De perspectiva*, 1535) and John Peckham (*Perspectiva communis*, 1543), the major works of Girolamo Cardano, Valerius Cordus's *Dispensatorium* (1546), the first official German

pharmacopoeia, and Geber's *Summa perfectionis* (1545), the most influential
alchemical treatise of the sixteenth century. Many of the manuscripts for these
important publications came from Regiomontanus's library, which became the
property of the Senate of Nuremberg after Regiomontanus's death. In publishing the Regiomontanus manuscripts, Petreius collaborated with the humanist
Johann Schöner, whom he employed as editor. See Shipman, "Johannes
Petreius."

67. Grotefend, *Christian Egenolff*, pp. 16–18; Wigand, "Der Büchernachdruck im 16. Jahrhundert," pp. 227–31. Egenolff's publication was Eucharius Rosslin's popular herbal, the *Kreutterbuch*. Schott accused Egenolff of
plagiarizing the pioneering herbal by Otto Brunfels, *Herbarum vivae eicones*
(1530, 1532). The woodcuts were made by Hans Weiditz. On this text, see
Sprague, "The Herbal of Otto Brunfels"; Roth, "Otto Brunfels."

68. Quoted in Belkin and Caley, *Eucharius Rösslin the Younger*, p. 6.

69. Quoted in Grotefend, *Christian Egenolff*, p. 17.

70. Muchembled, *Popular Culture and Elite Culture*, pp. 283–84. See also
Mandrou, *La Culture populaire aux XVIIe et XVIIIe siècles.*

71. For example, the Italian miller Menocchio, whose case is documented by
Ginzburg, *Cheese and the Worms*, and the German vintner Hans Keil, discussed
by Sabean, *Power in the Blood*, pp. 61–93.

72. Chartier, "Culture as Appropriation," p. 233; see also idem, *Cultural
Uses of Print.*

73. Chartier, *Cultural Uses of Print*, pp. 218–39; Ong, *Presence of the Word*,
pp. 58–59.

74. Perez-Ramos, *Francis Bacon's Idea of Science*, pp. 48–62.

75. Ong, *Presence of the Word*, pp. 28–29, 234.

76. On this subject, see Rossi, *Philosophy, Technology, and the Arts.*

77. For the bibliography of *Kunstbüchlein*, see Darmstaedter, *Berg- Probier-
und Kunstbüchlein*; Ferguson, "Some Early Treatises on Technological Chemistry"; and Paisey, "Sources of the 'Kunstbüchlein.'"

78. Hildebrand, *Magia naturalis*, bk. 4; Telle, "Die 'Magia naturalis'
Wolfgang Hilderbrands"; Schröder-Lembke, "Die Hausväterliteratur."

79. Kertzenmacher, *Alchimi und Bergwerk*. Kertzenmacher is identified in
the text as "ettwan burger zu Mentz, eyn berhumbter Alchimist," fol. Ai. On
this work, see Benzing, *Die Drucke Jacob Cammerlanders*, no. 53; Duveen,
"Notes on Some Alchemical Books"; Chrisman, *Lay Culture, Learned Culture*,
p. 182.

80. Kertzenmacher, *Alchimi und Bergwerk*, preface.

81. *Rechter Gebrauch*, fol. Giiiᵛ.

82. Crisciani, "La 'Quaestio de alchimia' fra duecente e trecento"; Newman,
"Technology and Alchemical Debate"; Varchi, *Questione sull' alchimia.*

83. Newman, "Technology and Alchemical Debate," pp. 426, 440–41;
Moran, *Alchemical World of the German Court.*

84. *Bibliotheca Palatina*, p. 337.

85. Forbes, *Short History of Distillation*, pp. 90–91, 102–3.

86. Agricola, *De re metallica*, p. 248; for other examples, see *Bergwerk- und
Probierbüchlein*, trans. Sisco and Smith.

87. Biringuccio, *Pirotechnia*, pp. 363–67.

88. Ibid., pp. 336–37.

89. The English Paracelsian John Hester made a similar distinction between alchemy as the art of transmutation and alchemy as the art of making chemical drugs: *Key of Philosophy*, pp. 62–63.

90. An example of such a manual is the workshop manual by the fifteenth-century master painter Henry of Lubeck: *The Strasburg Manuscript*, ed. V. and R. Borradaile.

91. Eisenstein, *The Printing Press as an Agent of Change*, pp. 50–51; Hirsch, *Printing, Selling, Reading*, pp. 48–49; Bühler, *The Fifteenth-Century Book*, pp. 69–70, 87.

92. Lehmann-Haupt, *The Göttingen Model Book*; see also the review by Bober, *Speculum* (1974).

93. Bühler, *The Fifteenth-Century Book*, p. 75. In his 1518 edition of the *Probierbüchlein*, Peter Schöfer wrote, "If somebody, in order to make the mountains stand out more clearly and plainly, should like to have the figures brushed or painted, the veins might be shown yellow, the mist and shimmering smoke-colored, and the water blue. To indicate which, I have used the following lettering . . . ," giving a list of "paint-by-letters" instructions: *Bergwerk- und Probierbüchlein*, trans. Sisco and Smith, p. 48.

94. Boltz, *Illuminierbuch*, ed. Benziger, pp. 31–36. The *Artliche Kunst* was translated into English in 1596 under the title *A Booke of Secrets*.

95. A facsimile and translation of the work have been published by Edelstein, "The Allerley Matkel."

96. Brunello, *The Art of Dyeing*, pp. 179–82; Ploss, "Die Färberei in der germanischen Hauswirtschaft."

97. *Allerley Mackel*, fol. Aiii^v.

98. *Von Stahel und Eysen* (Mainz, 1532), preface. The work has been translated by Williams, "A Sixteenth-Century German Treatise"; and by Smith, *Sources for the History of Steel*, pp. 7–19.

99. Willers, "Armor of Nuremberg," p. 103.

100. The etching recipes are in Smith, *Sources for the History of Steel*, pp. 16–17; and Williams, "Sixteenth-Century German Treatise," pp. 71–73. In addition, see Williams, "The Beginnings of Etching." Biringuccio described etching on iron as a "secret" of which he had a somewhat vague understanding: *Pirotechnia*, p. 372.

101. Biringuccio, *Pirotechnia*, pp. 370, 373.

102. Hils, "Von dem herten"; see above, chap. 2.

103. Friedrichs, "Class Formation in the Early Modern German City"; see also Braudel, *Civilization and Capitalism*, 2:316–21, 329–35.

104. Aubin, "Formen und Verbreitung des Verlagsystem."

105. Friedrichs, "Class Formation in the Early Modern German City," p. 34.

106. Recent research is summarized by Houston, *Literacy in Early Modern Europe*; and by Graff, "On Literacy in the Renaissance"; other important studies include Stone, "The Educational Revolution in England"; Cipolla, *Literacy and Development in the West*; Cressy, *Literacy and the Social Order*; and the studies on "Afabetismo e cultura scritta" in *Quaderni storici* (1978). For Germany, see

Engelsing, *Analphabetentum und Lektüre*; Strauss, *Luther's House of Learning*; idem, "Lutheranism and Literacy."

107. Chrisman, *Lay Culture, Learned Culture*, p. 68.

108. Strauss, *Luther's House of Learning*, pp. 200–201.

109. Ibid., pp. 195–96.

110. Engelsing, *Analphabetentum und Lektüre*, p. 33.

111. The number of sixteenth-century German imprints has recently been estimated at between 130,000 and 150,000 (excluding broadsheets): Bezzel, "Die Erschlieaung von Schrifttum des 16. Jahrhunderts," p. A81.

112. Chartier, *Cultural Uses of Print*, p. 152.

113. The inventory was published by Kelchner and Wülcker, *Mess-Memorial des Buchhändlers Michael Harder*.

114. Hackenberg, "Private Book Ownership."

115. Ibid., pp. 53–59, 98–128. In addition, see Hackenberg, "Books in Artisan Homes." By comparison, more than two-thirds of the medical doctors inventoried owned more than a hundred books, while 47 percent owned more than two hundred.

116. Hackenberg, "Books in Artisan Homes," p. 87.

117. Hackenberg, "Private Book Ownership," pp. 128–29. I am grateful to Professor Hackenberg for supplying me with a copy of Wittenheder's inventory. Another barber-surgeon's inventory is discussed by Sokól, "Die Bibliothek eines Barbiers aus dem Jahre 1550."

118. Hackenberg, "Private Book Ownership," p. 111.

119. Ibid., pp. 240–42. The four included two medical books, an arithmetic book, and a history book.

120. Ibid., libraries of a Nuremberg merchant (pp. 227–28) and a Slovenian justice (pp. 206–7).

121. *Bergwerk- und Probierbüchlein*, trans. Smith and Sisco, p. 159; Huffines, "Sixteenth-Century German Printers and the Standardization of New High German."

122. Grendler, *Schooling in Renaissance Italy*, pp. 42–47.

123. Ibid., p. 78.

124. Ginzburg, *Cheese and the Worms*, p. 31.

125. Giovanni Antonia Tagliente, *Libro maistrevole* (Venice, 1524), quoted in Schutte, "Teaching Adults to Read," p. 8. The same article quotes a handbook published in 1479 that advertised itself as being "very useful for learning to read for those who wish to do so without going to school, such as artisans and women," p. 9. See also Lucchi, "La Santacroce, il Salterio e il Babuino."

126. Grendler, *Roman Inquisition*, pp. 5–6.

127. The booklet was printed by the firm of Giovanantonio et fratelli da Sabio. I used the edition of 1529 at the National Library of Medicine.

128. Other editions were printed by the Venetian houses of M. Pagano, G. da Fontana, and G. A. Valvassore.

129. *Dificio*, fol. 12v.

130. See Appendix.

131. Frencken, *T Bouck vā Wondre*; Driessen, "Un Livre flamand de recettes"; Brunello, *The Art of Dyeing*, p. 178.

132. On these editions, see Paisey, "Sources of the 'Kunstbüchlein.'"

133. Ferguson, "Some Early Treatises on Technological Chemistry" (1888), pp. 137–40.

134. On this work, see below, chap. 4.

135. *Een nyeuw tractaet ghenaemt dat batement van recepten*: see Braekman, *Dat Batement van recepten*.

136. Morin, *Catalogue descriptif de la bibliothèque bleue*, nos. 45–48.

137. Coleman, "The People's Health," p. 73.

138. Ong, "Writing Is a Technology," p. 38.

139. Greimas, "La soupe au pistou."

140. Goody, *Domestication of the Savage Mind*, p. 136.

141. *Kunstbüchlein*, title page.

142. Eisenstein, *The Printing Press as an Agent of Change*, pp. 243–47.

143. Goody, *Domestication of the Savage Mind*, p. 140.

144. Davis, *Society and Culture in Early Modern France*; Eisenstein, *The Printing Press as an Agent of Change*.

CHAPTER FOUR
THE PROFESSORS OF SECRETS AND THEIR BOOKS

1. On this edition, see Schmitt, "Francesco Storella and the *Secretum secretorum*." On Storella, see Antonaci, *Francesco Storella*.

2. Garzoni, *Piazza universale* (1616), fol. 80f. On Garzoni and his work, see Cherchi, *Enciclopedismo e politica della rescriturra*, in particular pp. 41–82.

3. Garzoni, *Piazza universale*, fol. 80v. Garzoni based his description of the "professors of secrets" on a contemporary tract by Girolamo Cardano, *De secretis* (1562), in *Opera*, 2:537–51. On this work, see below, chap. 8.

4. Garzoni, *Piazza universale*, fol. 80v.

5. Bairo, *Secreti medicinali*. On Bairo see *DBI*, 5:291–92; Grammatica, "Pietro of Bairo."

6. Zapata, *Li maravigliosi secreti di medicina e chirurgia*, dedication. See also the notice in *BU*, 52:128–29.

7. Brunello, *The Art of Dyeing*, pp. 183–95; Brunello and Facchetti, "Notizie inedite su Giovanni Rosetti"; Lane, *Venice*, pp. 362–64.

8. Rosetti, *Notandissimi secreti*, p. 28.

9. On Della Porta's scientific work, see Muraro, *Giambattista Della Porta mago e scienzato*.

10. For additional information on the "professors of secrets," see Eamon, "Science and Popular Culture."

11. Cook, *The Decline of the Old Medical Regime*, chap. 1.

12. Zanier, "La medicina paracelsiana in Italia"; idem, "Filosofia chimica e pratiche popolari"; Ferrari, "Alcune vie di difusione di Paracelso."

13. See below, chap. 7.

14. Fioravanti, *Compendio dei secreti rationali* (Venice, 1660) p. 43.

15. Rosetti, *Plictho*, p. 89. In the preface to his *Notandissimi secreti*, Rosetti wrote, "Veramente io che speci fatiche, tempo, et la sustantia propria come

ricerca la'arte plebea non mi satisfo di tenirla ne le carcere sotto quella custodia così repente come gli antichi hanno fino a questa hora fatto."

16. Voet, *The Golden Compasses*, 2:264.

17. Alessio Piemontese, *Secreti*, A 'Lettori; trans. William Warde, *Secretes* (1558), "To the Reader," fol. *.ii. In quoting Alessio, I use Warde's translation throughout.

18. For bibliography and analysis of this work, see Ferguson, "Secrets of Alexis"; and Eamon, "The *Secreti* of Alessio Piemontese."

19. Alessio Piemontese, *Secreti*, pp. 54–57, 154–57, 98–99; *Secretes*, fols. 17r–18v, 80r, 42r.

20. Alessio Piemontese, *Secreti*, A 'Lettori; *Secretes*, fol. *.ii.

21. Alessio Piemontese, *Secreti*, A 'Lettori; *Secretes*, fol. *.ii.

22. Paré, *Apology and Treatise*, pp. 24, 138–39.

23. Palmer, "Pharmacy in the Republic of Venice."

24. Alessio Piemontese, *Secretes*, fols. 22r, 26r, 23.

25. Ibid.: "a thing experimented upon many," fol. 37r; "well tried and experimented," fol. 38r; "This has been found true by many men," fol. 39r. Often Alessio was quite specific about documenting his experiences (e.g., "proved in Venice, the year 1504," fol. 38r), or he gives the names of witnesses, such as "a man called Diego, a Portugall," to whom he administered his famous oil of a red dog, fol. 18v.

26. Alessio Piemontese, *Secreti*, A 'Lettori; *Secretes*, fol. *.iii.

27. On this subject, see Riddle, "Oral Contraceptives."

28. For the significance of perfumes and essences in early modern culture, see Camporesi, *Incorruptible Flesh*, pp. 179–207.

29. Cennini, *Craftsman's Handbook*. p. 36; cf. Alessio Piemontese, *Secretes*, fols. 84f. See also Thompson, *Materials and Methods of Medieval Painting*, pp. 145–51. Ultramarine was made from lapis lazuli, not found in Europe but imported from the Orient (mainly Persia) at great cost: Heyd, *Histoire du commerce du levant*, 2:653–54. As the name ultramarine ("beyond the seas") suggests, the process for making ultramarine azure was invented and for a long time known only in the Orient. Contracts between patrons and artists often specified the use of ultramarine as opposed to cheaper substitutes; prudent clients even stipulated ultramarine of a certain grade: Baxandall, *Painting and Experience*, pp. 3–11.

30. Quoted in Thompson, *Materials and Methods of Medieval Painting*, p. 151.

31. Alessio Piemontese, *Secretes*, fols. 107–110; cf. Multhauf, *Origins of Chemistry*, pp. 336–37.

32. Alessio Piemontese, *Secreti*, p. 64; *Secretes*, fol. 22v. Another indication of Ruscelli's hand in the Alessio text is that in the Venice, 1555, edition of the *Secreti*, the initials "G.R." appear on the opening page of bk. 1.

33. Tasso, *Dialogues*, pp. 193–243.

34. On the *poligrafi*, see Grendler, *Critics of the Italian World*.

35. See Grendler, *Roman Inquisition*.

36. See, for example, Ghilini, *Teatro d'huomini letterati*, pp. 126–27.

37. For details concerning the academy, see Eamon and Paheau, "The Accademia Segreta of Girolamo Ruscelli," with a translation of Ruscelli's preface.

38. Ruscelli, *Secreti nuovi*, fol. 1r.

39. Ibid., fol. 2r–v.

40. Ibid., fol. 5r.

41. Ibid., fol. 3v.

42. Ibid., fols. 3v–4r.

43. Cochrane, "The Renaissance Academies"; Maylender, *Storia delle accademie d'Italia*.

44. The sum, substantial as it was, pales beside the finances of the famous Accademia Venetiana, founded in 1557 by Bedoar. Having more than one hundred members, the group was endowed with 100,000 ducats, had other property that brought in at least 1,000 ducats in rents, and received cash gifts amounting to at least 5,300 ducats. Eventually Bedoar overextended himself financially, bankrupting the society. The Venetian government closed the society down in 1561. See Rose, "The Accademia Venetiana."

45. See below, chap. 6.

46. Giannone, *History of Naples*, 2:555. Among these measures were an edict against dueling and reforms to make the justice system more equitable; see 2:527–32. See also Pepe, *Il Mezzogiorno d'Italia sotti gli Spagnoli*, pp. 10–12.

47. Avalos, along with his wife, Maria of Aragon, was the patron of the poet Aretino and the painter Titian. Sanseverino sponsored many cultural activities in Naples and Salerno. He was the patron of the poet Bernardo Tasso and the philosopher Agostino Nifo. His support of the school of Salerno contributed to that ancient university's revival, after decades of decline, in the sixteenth century. On Avalos, see *DBI*, 4:612–16. On Sanseverino, see Fava, "L'ultimi di baroni."

48. Giannone, *History of Naples*, 2:535.

49. Tasso, *Minturno*, in *Dialogues*, p. 198.

50. Jedin, *History of the Council of Trent*, 1:232–43, 262–69, 283–87, 288–92.

51. Jung, "On the Nature of Evangelism"; Church, *The Italian Reformers*, pp. 215–44; Symonds, *Renaissance in Italy*, 2:331–34, 552–53.

52. Jung, "Vittoria Colonna."

53. The poet Jacopo Bonfadio, a former member of the Valdes circle, was executed in Genoa in 1550 on suspicion of heresy.

54. Giannone, *History of Naples*, 2:552.

55. Among the works publicly burned in Naples that year was the anonymous *Beneficio di Christo*, which was written in Naples by a Benedictine monk named Don Benedetto, a disciple of Valdes, and revised by the poet Marcoantonio Flaminio, a member of the Valdes circle in Naples: Prelowski, "The 'Beneficio di Christo.'"

56. Lea, *Inquisition in the Spanish Dependencies*, pp. 70–77.

57. Giannone, *History of Naples*, 2:556.

58. On the "tumult" of 1547, see ibid., pp. 555–64; Lea, *Inquisition in the Spanish Dependencies*, pp. 70–77; Miccio, "Vita di Don Pietro di Toledo," pp. 53–74; and Amabile, *Il Santo Officio della inquisizione in Napoli*, 1:192–206.

59. Giannone, *History of Naples*, 2:555.

60. Minieri-Riccio, "Cenno storico delle accademie" (1879), p. 173.

61. Giannone, *History of Naples*, 2:578–80; see also Romier, *Origines politiques des guerres de religion*, 1:319–21.

62. "L'intention nostra era stata primieramente di studiare & imparar noi stessi, non essendo studio nè altro essercitio alcuno, che più sia vero della Filosofia naturale, che questo di far diligentissima inquisitione, & come una vera anatomia delle cose & dell'operationi della Natura, in se stessa": Ruscelli, *Secreti nuovi*, fol. 3v.

63. Badaloni, "I fratelli Della Porta," p. 699; idem, "Fermenti di vita intellettuale a Napoli," p. 319.

64. Van Deusen, "Telesio"; idem, "The Place of Telesio in the History of Philosophy"; Cassirer, *Individual and Cosmos*, pp. 145–47. Although Telesio's *De rerum natura* was not published until 1563, his ideas were already in circulation in southern Italy by 1547: Van Deusen, "Telesio," p. 4.

65. Telesio, *De rerum natura*, lib. 1. Cassirer, *Individual and Cosmos*, p. 148; cf. Campanella, *Del senso delle cose.*

66. Ruscelli, *Secreti nuovi*, fol. 3v.

67. Ibid., fol. 5r.

68. Telesio, *De rerum natura*, lib. 4, cap. 2, p. 136. Van Deusen, "Telesio," p. 41.

69. Pagel and Rattansi, "Vesalius and Paracelsus," pp. 310–14; see also Hannaway, *Chemists and the Word*, pp. 23f., 40f.

70. Della Porta, *Natural Magick*, pp. 254ff.; idem, *De distillatione*. In *De aëris transmutationibus*, a treatise on meteorology, Della Porta argued that weather phenomena are fundamentally similar to distillation, so that distillation is a "microcosm" of meteorological phenomena. According to Della Porta, natural magic united "actives to passives" in order to imitate nature, a methodology that was clearly influenced by nature philosophy. See also Badaloni, "I fratelli Della Porta," p. 705.

71. On the conjunction of 1524 and the literature it inspired, see *HMES*, 5:178–233; Zambelli, "Fino del mondo"; and the articles in Zambelli, *"Astrologi hallucinati."*

72. Badaloni, "I fratelli Della Porta," pp. 685–89; *HMES*, 5:220f.

73. Badaloni, "Natura e società in Nicolò Franco."

74. Minieri-Riccio, "Cenno storico delle accademie" (1879), pp. 528–29.

75. Varchi, *Questione sull' alchimia*, pp. xxiv, 25. On Varchi, see Pirotti, *Benedetto Varchi e la cultura del suo tempo.*

76. Kristeller, "The School of Salerno," p. 192.

77. Badaloni, "Natura e società in Nicolò Franco"; idem, "Fermenti di vita intellettuale a Napoli."

78. For example, Luca Gaurico's prognostication for 1547 threatened "sollevazioni e movimenti grandi e straordinari di Popoli, incendij, rovine e accidenti orribili": Zambelli, "Many Ends for the World."

79. On Fioravanti, see below, chap. 5.

80. Paracelsus, *Defensiones un Verantwortungen wegen etlicher verunglimpfung seiner Missgönner*, quoted in Pagel, *Paracelsus*, pp. 56–57.

81. Agrippa von Nettesheim, *Of the Vanity and uncertainty of Artes and Sciences*, fol. 148r.

82. Severinus, *Idea medicinae philosophicae*, quoted in Debus, *English Paracelsians*, p. 20.

83. Joubert, *Popular Errors*, p. 44. Joubert noted that in many respects the two arts are "marvelously similiar": for as the soldier lays siege to a city, so a physician besieges an illness.

84. Ibid., p. 45.

85. Ibid, p. 68; in addition, see Lingo, "Empirics and Charlatans."

86. The English translation (by William Warde) of Alessio Piemontese's *Secreti* cost tenpence in 1585: Johnson, "Notes on English Book-prices," p. 97.

87. Bairo, *Secreti medicinali*. On Bairo, see Grammatica, "Pietro of Bairo."

88. The work was translated into Italian by Giovanni Tatti, alias Francesco Sansovino (1521–1583), a Venetian *poligrafo* whose publications consisted mainly of translations of humanist historical works. On Sansovino, see Grendler, "Sansovino and Italian Popular History."

89. Salviani was the principal physician at the Roman medical college between 1551 and 1568, and a personal physician to Popes Julius III and Paul IV, and to Cardinal Cervini, who later became Pope Marcellus II. Since Zapata would have been at least thirty-one when he became Salviani's pupil, his studies were probably informal and sporadic. On Salviani, see *DSB*, 17:89–90.

90. Falloppio, *Secreti maravigliosi* (Venice, 1577), p. 96. On the *ciarlatani*, see Corsini, *Medici ciarlatani*, and below, chap. 7.

91. Della Porta, *Natural Magick*, p. 160. According to Leonardo Fioravanti, "The art of alchemy was a most ingenious investigation of natural philosophy, and of no small importance. For many pretty inventions have been extracted from it, which have been of great ornament to the world, and of great advantage to artisans. Indeed, this art has led the way to the art of glassmaking, . . . and to many other arts necessary for civilized life": *Specchio universale*, fol. 97v.

92. The full title of the work, *Secrets [of] Minerals, Medicines, Artifices, and Alchemy, and Much on the Art of Perfumery, For Every Great Lady*, gives some indication of its broad scope.

93. Among the many alchemists mentioned in her work was a priest named Benedetto, who lodged at Cortese's residence in Olomouc, in Moravia. The priest, who was headed toward Cracow, fell ill and died, leaving a letter containing his alchemical *practica*, which he had addressed to "the discreet and erudite man Stanislaw, moderator of the college of scholars in Cracow": Cortese, *Secreti*, pp. 29ff.

94. Ibid., pp. 19–21.

95. Rossello, *Della summa de' secreti universali in ogni materia . . . si per huomini & donne, di alto ingegno, come ancora per Medici, & ogni sorte di artefici industriosi, & a ogni person virtuosa accomodare*.

96. Rossello, *La seconda parte de secreti universali in ogni materia*, dedication.

97. The Jesuati were founded in the fourteenth century by Giovanni Colombini (ca. 1300–1367), a wealthy Sienese merchant who gave away all his possessions in order to live as a mendicant. The order he founded dedicated itself to

caring for victims of the plague. The order was disbanded in 1668 by Pope Clement IX, supposedly because of abuses connected with the manufacture and distribution of alcoholic beverages. On the Jesuati, see *Catholic Encyclopedia*, 8:458; Dufner, *Geschichte der Jesuaten*.

98. On Rupescissa, see *HMES*, 3:347–69; Multhauf, "John of Rupescissa."

99. The manuscript, now in the Spencer Research Library at the University of Kansas, is labeled Pryce MS E1. In was composed between 1536 and 1562, with additional material by Antonio da Placencia, also a member of the Jesuati order. The manuscript is lavishly illustrated with figures of distillation apparatus.

100. On guaiacum, see Munger, "Guaiacum, the Holy Wood."

101. Cortese, *Secreti*, dedication.

102. Rossello, *Della summa de' secreti universali*, pt. 1, dedication.

103. Falloppio, *Secreti diversi* (Venice, 1563), dedication and proem. An imitation of Alessio's popular *Secreti*, the work contained the usual assortment of medical prescriptions, alchemical and metallurgical recipes, as well as instructions for making perfume, cosmetics, wine, soap, and odoriferous waters.

104. Andrea Marcolini, in his edition of Falloppio's posthumous *De medicatis aquis* (Venice, 1564), first pointed out that the work was spurious (preface).

105. Falloppio, *Secreti diversi*, fols. 2v–4v.

106. The only direct evidence linking Fioravanti with the *Secreti diversi* is the fact that the work's proofreader, Borgaruccio Borgarucci, also edited several of Fioravanti's books.

CHAPTER FIVE
LEONARDO FIORAVANTI, VENDOR OF SECRETS

1. The best biography of Fioravanti, though still inadequate, is Giordano, *Leonardo Fioravanti Bolognese*. For Fioravanti's birthdate, see Furfaro, *La vita e l'opera di Leonardo Fioravanti*. Nothing certain is known about Fioravanti's family or social background. He may have been related to a Bolognese family of architects named Fioravante whose members included an engineer known as the "celebrated Aristotle." Chief architect to the commune of Bologna, Aristotele Fioravante (1414–1486) served the Sforzas in Milan as an engineer and afterwards went to Moscow, hired by Tsar Ivan III to supervise the construction of several churches in the Kremlin: Beltrami, *Vita di Aristotile da Bologna*.

2. Fioravanti, *Tesoro* (1582), fol. 17v; cf. the Proemio to the same work.

3. The letter was one of a collection of testimonials to Fioravanti by various patients and members of the medical profession: *Tesoro*, fol. 216.

4. Fioravanti, *Tesoro*, fol. 2.

5. Ibid., fol. 18r.

6. Ibid., fol. 25v.

7. Ibid., fol. 26r–v.

8. Ibid., fols. 25v–27r.

9. Zambecarri, "The Experiments of Doctor Joseph Zambeccari," trans. Jarcho. See also Webster, "The Helmontian George Thomson," who suggests that Fioravanti's popular *Tesoro* played an important role in reviving interest in splenectomy: pp. 159–60.

10. Pliny, *Natural History* 11.80. The idiom, *courir comme un dirate*, "to run like a spleenless man," still exists in modern French. Pliny also reported a common belief of the time that removal of the spleen "deprives a man of a power of laughing, and that inability to control one's laughter is caused by enlargement of the spleen." The third-century medical poet Quintus Serenus Sammonicus wrote, "A swollen spleen is bad and makes you laugh . . . With spleen removed the laugh will cease, . . . And stern of face you'll be until your dying day," quoted and translated by Jarcho, in Zambecarri, "The Experiments of Doctor Joseph Zambeccari," p. 325.

11. Webster, "The Helmontian George Thomson," p. 159.

12. Aristotle argued that the spleen was an auxiliary blood-making organ, while the Galenists asserted that the spleen cleansed the body of excess melancholic humors.

13. Wear, "The Spleen in Renaissance Anatomy."

14. "Per dire il vero, non è la meglior cosa per imparare, quanto e l'andar per il mondo; percioche ogni giorni si vede cose nuove, & s'imparano varij & diversi secreti importanti": Fioravanti, *Tesoro*, fols. 26v–27r.

15. Ibid., fol. 27r–v.

16. Ibid., fol. 46r.

17. Ibid., fol. 46v. Fioravanti recorded numerous examples of such quarrels among the nobility in the *Tesoro*: fols. 86r, 92r. See also Gnudi and Webster, *Gaspare Tagliacozzi*, pp. 125–26. According to Symonds, the period between 1530 and 1600 "is distinguished by extraordinary ferocity of temper and by an almost unparalleled facility of bloodshed." Symonds attributes the violent temper of the age in part to the Spanish influence, which may help explain why plastic surgery as a specialty was developed in the south, especially in Calabria: *Renaissance in Italy*, 2:559.

18. Fioravanti, *Tesoro*, fols. 46–47. See also Gnudi and Webster, *Gaspare Tagliacozzi*, p. 117, for a description of the procedure.

19. Fioravanti, *Tesoro*, fol. 60v.

20. Ibid., fol. 61r; see also Fioravanti, *Secreti rationali*, fol. 58r–v.

21. Fioravanti, *Tesoro*, fol. 64r.

22. Fioravanti, *Capricci medicinali* (1582), fols. 80v–81r. In 1557, almost two years after his arrival in Rome, Fioravanti received a five-year license to practice surgery and to "administer decoctions of *legno santo*": Corsini, *Medici ciarlatani*, p. 79.

23. ". . . dubitorno di non perdere il grado & la riputation loro, venendo prelati & huomini grandi, che gli approbavano, & se ne servivano con grande istanza": *Capricci medicinali*, fol. 80v.

24. Fioravanti, *Capricci medicinali*, fol. 81r–v. In addition, see Giordano, *Leonardo Fioravanti Bolognese*, pp. 47–49.

25. Fioravanti, *Del regimento della peste*, fol. 70r.

26. Fioravanti, *Tesoro*, fol. 80r.

27. Ibid., fol. 23v. Fabrizio Garzoni, a professor at the Bologna medical school, earned four hundred scudi per year in 1556, while Giulio Cesare Aranzio, an anatomy teacher at Bologna, earned only one hundred lire per year: Gnudi and Webster, *Gaspare Tagliacozzi*, pp. 36n, 60.

28. ". . . s'alcuno si volesse servire di tai nostri remedij mi trovara in Venetia a san Luca dove sempre saro pronto al servitio di tutto": Fioravanti, *Secreti rationali* (1660), p. 43.

29. ". . . & chi non si vorra affaticare in tal materie, . . . potra pigliare detti remedi in Venetia alla speciaria dall'Orso, dove sempre si trovano fatti": *Capricci medicinali*, fol. 2v. See also *Secreti rationali* (1564), fol. 119r; Giordano, *Leonardo Fioravanti Bolognese*, p. 52.

30. Fioravanti, *Compendio di tutta la cirurgia* (1561). Avanzo published the first edition of the work in 1557.

31. Fioravanti, *Secreti rationali* (1564), preface; Ruscelli, *Precetti della militia moderna*, fol. 58v: "Questo maraviglioso instromento mi fece vedere il sopradetto Eccelente Dottore, l'anno del LXIII in casa mia."

32. Fioravanti mentioned, among others, Borgheruccio Borgherucci, a proofreader working for Avanzo, and Battista di Putei, whom he treated for an abscess of the head.

33. Fioravanti described the event in the *Tesoro*, fol. 84v, and in *Secreti rationali*, fol. 54r–v. See also Gentili, "Leonardo Fioravanti Bolognese," pp. 38–39.

34. Palmer, "Pharmacy in the Republic of Venice."

35. Di Francheschi also sold the patent medicines of other empirics, such as Angelo Forte. Ibid., pp. 111–12.

36. The novella appeared in Fioravanti's *Specchio universale*, bk. 2, chap. 14.

37. Giordano, *Leonardo Fioravanti Bolognese*, p. 10.

38. Fioravanti, *Specchio*, fol. 160v.

39. Giorio, "Una fonte del Garzoni."

40. Fioravanti, *Specchio*, fol. 4r–v.

41. Letter from Giandomenico Zavagliano, 4 September 1565, *Tesoro*, fols. 144r–165r. See also Giordano, *Leonardo Fioravanti Bolognese*, p. 52; Gentili, "Leonardo Fioravanti Bolognese," p. 25.

42. Fioravanti, *Del regimento della peste*, fols. 58–60; Giordano, *Leonardo Fioravanti Bolognese*, p. 52.

43. Palmer, "Pharmacy in the Republic of Venice," p. 113.

44. Gentili, "Leonardo Fioravanti Bolognese," pp. 22–32. Fioravanti gave his version of the dispute in *Tesoro*, fol. 83r. See also Giordano, *Leonardo Fioravanti Bolognese*, p. 54.

45. The documents relating to Fioravanti's application, from the archives of the Bologna College of Medicine, are published in Gentili, "Leonardo Fioravanti Bolognese," pp. 27–31. For a fuller explanation of the college's procedures for granting a degree, see Gnudi and Webster, *Gaspare Tagliacozzi*, pp. 152–53.

46. Gentili, "Leonardo Fioravanti Bolognese," p. 24; Gnudi and Webster, *Gaspare Tagliacozzi*, p. 50. For a Bolognese citizen, the application for a medical degree cost seventy scudi, whereas noncitizens paid only twenty-three scudi. Even Tagliacozzi, who was of a wealthy Bolognese family, applied as a nonnative, although he asked for the same privileges as native applicants.

47. "Quem pro idoneo et sufficienti omnes doctores unanimes amiserunt": Gentili, "Leonardo Fioravanti Bolognese," p. 22. Fioravanti maintained that he

398 NOTES TO CHAPTER 5

had already received the degree in 1548. That assertion, of course, raises the question of why he should have had to repeat the process twenty years later.

48. "Et si non furtim hoc sucipuerit, admirabantur quomodo talem promoverimus ad gradum doctoratus et etiam ad dignitatem equestrem," quoting from the summary of the letter in the Bolognese archives: Gentili, "Leonardo Fioravanti Bolognese," p. 30.

49. Fioravanti, *Tesoro*, fol. b7r.

50. Fioravanti identified several of his disciples, including Michel Murso, a Cypriot from Nicosia, the "Doctor and Knight" Dimitrio de Julius della Cava (surgeon), Propertio Bello of Naples (surgeon), Gioanmartin Romano of Salerno, Tarquino Malipiero of Venice, Antonio Palzzuolo (barber), Giulio da San Giuliano, Battista Cesconi (barber): Rostino, *Compendio*, preface.

51. Dall'Oso, "Due lettere inedite," p. 288. See also Latronico, "Una disavventura milanese di Fioravanti."

52. On the earthquake, see Angeio Solerti, *Ferrara e la corte estense*, pp. CLXI–CLXXIII.

53. Fioravanti, *Del regimento della peste*, chap. 1.

54. Symonds, *Renaissance in Italy*, 2:730–48. See also Solerti, *Ferrara e la corte estense*.

55. Dall'Osso, "Due lettere inedite," pp. 289–90.

56. Fioravanti, *Del regimento della peste*, chap. 1. For similar views by contemporaries, see Preto, *Peste e società a Venezia*, pp. 59–75, 166–86.

57. In the dedication of the *Della fisica* to Philip, Fioravanti wrote, "sin duda niguna el es Catolichissimo y Christianissimo sobre todos los otros Reis del mundo."

58. This is revealed in Fioravanti's defense in the face of legal proceedings taken against him in Madrid by the king's prosecuting attorney Martin Ramón: BM Add MS 28,353, fols. 57–61, undated. For a brief discussion, see David Goodman, *Power and Penury*, p. 248.

59. In the royal library, Fioravanti discovered an old manuscript containing some Castilian verses on the "great art," which he copied and published in the *Della fisica*. On the manuscript, see Sarmiento, *Memorias para la historia de la poesía*, pp. 196–97.

60. Fioravanti, *Della fisica*, dedication.

61. According to Barbieri, *I dottori bolognese*, the date was 4 September 1588. See also Gentili, "Leonardo Fioravanti Bolognese," pp. 34–35.

62. The material in the following section is treated in greater detail in Eamon, "'With the Rules of Life and an Enema.'"

63. BM Add. MS 28,353.

64. See Grendler, *Critics of the Italian World*, chap. 5.

65. Gleason, "Sixteenth-Century Italian Evangelism."

66. Grendler, *Critics of the Italian World*, chap. 6.

67. Ibid., pp. 170–77. See also Curcio, *Utopisti e reformatori sociali*.

68. ". . . ma solamente il mio proprio giudicio, & esperienza, che si suol dire esser madre di tutte le cose," *Capricci medicinali*, Raggionamento a' Lettori (Argument to the reader).

69. Fioravanti, *Della fisica*, fol. a8r.

70. Fioravanti, *Del regimento della peste*, fol. 117v: "Dico adunque che bisogna scodarsi il metodo de gli antici."

71. "Dico, che non è altro, che quattro operatione sole, nelle quale consiste tutta la medicina e son queste, cioè quelli che sono troppo caldi, raffredarli, e quelli che son troppo freddi riscaldarli; quelli che sono troppo secchi, humidirli; & quelli che sono troppo humidi, disseccarli; & in questo si riduce tutta la filosofia della notra medicina": Fioravanti, *Della fisica*, p. 14.

72. Fioravanti, *Del regimento della peste*, fol. 119v.

73. ". . . che non seppero altramente fisica, ne metodo niuno, ma solamente hebbero un gran giudicio": *Fioravanti, Capricci medicinali*, fol. 31r.

74. Ibid., fols. 29v–30v.

75. Fioravanti, *Della fisica*, p. 15.

76. Fioravanti, *Capricci medicinali*, fol. 82r–v.

77. Rostino, *Compendio*, pp. 169–70.

78. "Vediamo adunque come i Fisici si usurpono la medicina, & tennero modo tale, che fecero privare tutti gli altri, & loro si fecero laureare": Fioravanti, *Secreti medicinali*, fol. 26.

79. Ibid.

80. Fioravanti, *Capricci medicinali*, fol. 21r–v.

81. Fioravanti, *Secreti rationali*, fol. 38r; *Specchio universale*, fols. 15v–17v.

82. Fioravanti, *La cirurgia*, fols. 129r-31v; *Specchio universale*, fol. 45r–v.

83. Fioravanti, *Della fisica*, fol. a8v.

84. "Si che io, per me credo più ad una minima esperienza, che a tutte le theoriche del mondo insieme": Fioravanti, *Capricci medicinali*, fol. 32v.

85. Ibid., fols. 21v, 86.

86. Ibid., fols. 87v–88r.

87. In Naples Fioravanti met an eighty-year-old man who told him that he occasionally took white hellebore with a cooked apple, "which he then vomited several times and thus was excellently purged": ibid., fol. 54r–v.

88. Fioravanti, *Tesoro*, fols. 80v–81r. See also the chapter "Hypercatharsis" in Camporesi, *Incorruptible Flesh*, pp. 106–30.

89. Fioravanti, *Tesoro*, fol. 32v.

90. Fioravanti, *Capricci medicinali*, fol. 40r–v.

91. Fioravanti, *Tesoro*, fol. 91v.

92. Bovio's works included *Flagello de' medici rationali* (Venice, 1583), *Melampigo overo confusione de medici sofisti che s'intitolano rationali* (Verona, 1585), and *Fulmine contro de' medici putatii rationali* (Verona, 1592), all in *Opere*. See Ingegno, "Il medico de' disperati e abbandonati."

93. Bovio, *Opere*, pp. 4, 52: "scaccio il male, sostento la natura." See also Ingegno, "Il medico de' disperati e abbandonati"; and Dal Fiume, "Un Medico astrologo a Verona."

94. Camporesi, *Incorruptible Flesh*, p. 123.

95. For the following observations, I have relied upon O'Neil, "Discerning Superstition." In addition, see Burke, "Rituals of Healing"; and Monter and Tedeschi, "Toward a Statistical Profile of the Italian Inquisitions."

96. Menghi, *Compendio dell'arte essorcistica* (Venice, 1576), quoted in O'Neil, "Discerning Superstition," p. 336.

97. Ibid., p. 351.

98. Mercurio, *De gli errori populari*, fol. 139v.

99. Douglas, *Natural Symbols*, pp. 93–112 (p. 93). See also idem, *Purity and Danger*, esp. chap. 7, pp. 114–28.

100. Fioravanti, *Capricci medicinali*, fols. 237v–238; Fioravanti, *Specchio universale*, fols. 140–45.

101. Fioravanti, *Del regimento della peste*, p. 8.

102. Fioravanti, *Capricci medicinali*, fol. 21v.

103. Zanier, "La medicina Paracelsiana in Italia."

104. Fioravanti, *Capricci medicinali*, fol. 32.

105. "Ma la esperienza si manifesta & approba per se sola, & non ha bisogno di altro parangone": Fioravanti, *La cirurgia*, fol. 4r.

106. Fioravanti confirms his drugs were sold by pharmacists in Pesaro and other Italian cities (*Tesoro*, fol. 198r). John Hester, the English distiller, sold Fioravanti's drugs at his shop at Paul's Wharf (see below, chap. 6). Further evidence of the longevity of Fioravanti's compositions is found in a letter of 1686 that the anatomist Marcello Malpighi wrote to Francesco Travagini in Venice. Malpighi wrote asking for counsel in regard to the illness of his patient, Count Rinieri Marescotti. Malpighi wanted to know the composition and employment of the "balsamo del famoso cavaliere Fasciotti," which Pope Innocent XI had recommended to Malpighi. Travagini replied that the "balsamo del Sfacchiotti," as he called it, was the one described by Fioravanti in the *Capricci medicinali*: Adelmann, *Correspondence of Marcello Malpighi*, 3:1117–18, 1130 (letters 556 and 562). Fioravanti's balsam was still being used in the eighteenth century, appearing in pharmacopoeias by Jakob Reinhold Spielman (Strasbourg, 1783) and Giovanni Battista Capello (Venice, 1754).

107. Webster, "The Helmontian George Thompson," pp. 159–60. Jean Riolan (1577–1657) mentioned the Fioravanti incident, along with other ancient and modern reports of splenectomies, in *Anthropographia* (Paris, 1626), vol. 2, chap. 23. Despite these reports, Riolan rejected the "opinion of Erasistratus" that animals or humans could live without the spleen.

108. Webster, "The Helmontian George Thompson," p. 159.

109. Fioravanti, *Del regimento della peste*, fols. 99–114.

110. Fioravanti, *Della fisica*, p. 196.

111. Fioravanti, *Tesoro*, fol. b4v.

CHAPTER SIX
NATURAL MAGIC AND THE SECRETS OF NATURE

1. Cortese, *Secreti*, dedication: "non solamente l'huomo si contenta della investigatione, ma certa intutto & per tutto mettendo in opera, di farsi scimia della natura, anzi che superarla, mentre tenta di fare quello, che alla natura e impossibile, & cio sia vero, si puo cavare de Secreti, che tutto giorno si odono & veggono mettere in essecutione."

2. The classic works are Walker, *Spiritual and Demonic Magic*; and Yates, *Giordano Bruno and the Hermetic Tradition*. In addition, see Copenhaver,

"Natural Magic, Hermetism, and Occultism"; idem, "Astrology and Magic"; Zambelli, "Il problema della magia naturale."

3. Bossy, "The Counter-Reformation and the People."

4. Ashworth, "Catholicism and Early Modern Science," pp. 148–51, touches on this theme.

5. Ginzburg, "High and Low," p. 32.

6. On Renaissance magic, see Walker, *Spiritual and Demonic Magic*; Copenhaver, "Scholastic Philosophy and Renaissance Magic"; idem, "Did Science Have a Renaissance?"; idem, "Astrology and Magic"; and Müller-Jahncke, *Astrologisch-magische Theorie und Praxis*.

7. On Agrippa, see Walker, *Spiritual and Demonic Magic*, pp. 90–96; Nauert, *Agrippa and the Crisis of Renaissance Thought*; Zambelli, "A proposito del *Del vanitate scientiarum*"; idem, "Magic and Radical Reformation in Agrippa."

8. See Walker's comments on Pomponazzi, *Spiritual and Demonic Magic*, pp. 107–111.

9. "Fu diligentissimo osservatore e perscrutatore de secreti naturali": *Éloge* of Giovanni Faber (1625), in Gabrieli, "Giovan Battista Della Porta Linceo," p. 425.

10. For Della Porta's biography, I have relied upon Fiorentino, "Della vita e delle opere di Giovan Battista de la Porta," in *Studi e ritratti*; Corsano, "Per la storia del pensiero del tarda rinascimento: III. G. B. Della Porta"; Clubb, *Giambattista Della Porta, Dramatist*; Reinstra, "Giovanni Battista Della Porta"; and the various articles by Gabrieli.

11. Faber, *éloge*, in Gabrieli, "Giovan Battista Della Porta Linceo," p. 424.

12. Letter of Giovanni Battista Longo, in Gabrieli, "Giovan Battista Della Porta Linceo," p. 429.

13. *Éloge* of G. B. Longo (1635), in Gabrieli, "Giovan Battista Della Porta Linceo," p. 429; Gassendi, *Mirrour of true Nobility*, p. 36.

14. Clubb, *Giambattista Della Porta, Dramatist*.

15. Longo *éloge*, Gabrieli, "Giovan Battista Della Porta Linceo," p. 424.

16. Clubb, *Giambattista Della Porta, Dramatist*, p. 99.

17. Badaloni, "I fratelli Della Porta."

18. Quoted in ibid., p. 685.

19. See above, chap. 3; ibid., pp. 685–91; *DBI*, 1:50–52.

20. Badaloni, "I fratelli Della Porta," pp. 689–91; idem, "Natura e società in Nicolò Franco."

21. Fiorentino, *Studi e ritratti*, p. 263; Amabile, *Fra Tommaso Campanella, la sua congiura*, p. 34; Firpo, *Il supplizio di Tommaso Campanella*, p. 54; Clubb, *Giambattista Della Porta, Dramatist*, p. 29; Della Porta, *Criptologia*, ed. Belloni, pp. 15.

22. Badaloni, "I fratelli Della Porta," pp. 698f.

23. Clubb, *Giambattista Della Porta, Dramatist*, p. 10; Rosen, *The Naming of the Telescope*, p. 15.

24. Reinstra, "Giovanni Battista Della Porta," p. 30.

25. Della Porta, *Gli duoi fratelli rivalli*, ed. Clubb, pp. 12–13.

26. Longo, *éloge*, in Gabrieli, "Giovan Battista Della Porta Linceo," p. 429.

27. *Delle celeste fisonomia* (1623), quoted in Reinstra, "Giovanni Battista Della Porta," p. 43.

28. Gliozzi, "Sulla natura dell' 'Accademia de' Secreti.'"

29. Della Porta, *Natural Magick*, preface.

30. "In patria siquidem sua Neapoli Academiam extruxerat secretorum nuncupatam, in quam nemini fas erat insinuare se, qui admirandum aliquod supra vulgi captum non proferrat arcanum, ex quo certissimi, vel ad salutem corporum, vel ad mechanicarum usum, vel ad rerum commutationem effectus sequerentur": Giovanni Imperiali, *Musaeum historiam et physicum* (Venice, 1640), 122–24, quoted in Gliozzi, "Sulla natura dell' 'Accademia de' Secreti,'" p. 538.

31. Ibid., p. 539.

32. Della Porta, *Della magia naturale* (1677), quoted in ibid., pp. 539–40.

33. Ibid.; cf. Clubb, *Giambattista Della Porta, Dramatist*, p. 13.

34. Gliozzi, "Sulla natura dell' 'Accademia de' Secreti,'" p. 539.

35. Thorndike asserted that Pizzimenti later claimed authorship of the *Magia naturalis*, although he did not supply documentation for the assertion: *HMES*, 6:245–46; 418.

36. *DBI*, 2:568–69; Minieri-Riccio, "Cenno storico delle accademie" (1879), p. 167.

37. Minieri-Riccio, "Cenno storico delle accademie" (1879), p. 520. In a dedication to *De refractione* (1593), Della Porta named Pisano as one of his teachers.

38. Gabrieli, "Giovan Battista Della Porta Linceo," pp. 377, 380: "Giovanni Battista Melfi, anche il destillator della mia Academia." Idem, "Bibliografia Lincea, I" (letter of Della Porta to Ulisse Aldrovandi, August 1590): "Favio Giordano è semplicista rarissimo, . . . era uno della academia di mia casa et pars altera vitae."

39. Sarnelli, *Vita*, quoted in Aquilecchia, "Appunti," pp. 3–4.

40. For a summary of the inquisitorial documents relating to Della Porta's case, see Acquilecchia, "Appunti"; and Lopez, *Inquisizione stampa e censura*, pp. 153–59.

41. The evidence for this is a letter, discovered by Pasquale Lopez in the Neapolitan archives, from the cardinal of Pisa to Mario Carafa, the archbishop of Naples, requesting the assistance of the secular authorities in arresting Della Porta and sending him to the Holy Office in Rome: Lopez, *Inquisizione stampa e censura*, 154, and doc. 7, pp. 275–300. Prior to the discovery of this letter, historians had assumed that Della Porta's troubles with the Inquisition began in 1566, forcing him into exile.

42. Clubb, *Giambattista Della Porta, Dramatist*, p. 16; Fiorentino, *Studi e ritratti*, p. 258.

43. Aquilecchia, "Appunti," pp. 5–6.

44. According to Amabile, imprisonment normally preceded such a reexamination. However, the single official document relating to Della Porta's case is too obscure to support such a conclusion. We cannot be entirely sure that Della Porta was actually brought before the Neapolitan Inquisition, much less incarcerated as Amabile supposed. Nevertheless, he faced serious charges, because he

was now being examined by the Holy Office of both Rome and Naples: ibid., pp. 5–6.

45. ". . . per avere scritto intorno alle maraviglie et i segreti della natura," quoted in ibid., p. 6.

46. Wier, *De praestigiis daemonum*, trans. Shea, pp. 225–26. Della Porta described his experiment with the "witch's salve" in *Magia naturalis*, pp. 101–2.

47. The question of why Della Porta was reexamined by the Inquisition has been one of the most disputed issues in Della Porta scholarship, and the debate has gone full circle. Francesco Fiorentino, who was not aware of the document Luigi Amabile later discovered, first suggested that Della Porta was implicated in the Bodin-Wier dispute. When Amabile discovered the document confirming that Della Porta was examined by the Inquisition in 1580, he rejected Fiorentino's hypothesis on the grounds that the Latin edition of Bodin's work was not published until 1581. However, in the 1580 French edition of *Démonomanie*, Bodin noted that after he had completed the work and submitted it to his publisher, he suspended the printing of the book until he could insert a section refuting the opinions of Wier, which had been published in a French translation in 1579. Thus Fiorentino's speculation was essentially correct, although the evidence he adduced to substantiate it was wrong. See Acquilecchia, "Appunti."

48. According to Christopher Baxter, Wier thought the devil had "turned the Roman Church into a front organization": "Johann Weyer's *De praestigiis daemonum*," p. 55.

49. Walker, *Spiritual and Demonic Magic*, pp. 152–56.

50. Baxter, "Jean Bodin's *De la Démonomanie*"; Anglo, "Melancholia and Witchcraft."

51. In the Italian translation of the *Magia naturalis* (Venice, 1560), Della Porta suppressed the account: Bonomo, *Caccia alle streghe*, p. 395.

52. Aquilecchia, "Appunti," pp. 22–26. The text of the prohibition is printed in Fiorentino, *Studi e ritratti*, p. 265.

53. Monter and Tedeschi, "Toward a Statistical Profile of the Italian Inquisitions"; O'Neil, "Discerning Superstition," pp. 45–51; for Venice, see Martin, *Witchraft and the Inquisition in Venice*, esp. pp. 214–18, 260–64.

54. O'Neil, "Magical Healing"; Martin, *Witchcraft and the Inquisition in Venice*. For the shifting meanings of the term "superstition" in the sixteenth and seventeenth centuries, see Burke, *Popular Culture in Early Modern Europe*, chap. 8.

55. Bossy, "The Counter-Reformation and the People," p. 60. On the "reform of popular culture," see in particular Burke, *Popular Culture in Early Modern Europe*, pp. 207–43. In addition, see Ginzburg, *Cheese and the Worms*, pp. 125–26; idem, *The Night Battles*.

56. Thomas, *Religion and the Decline of Magic*, chap. 2; O'Neil, "Magical Healing."

57. Reinburg, "Popular Prayers." Such magical-spiritual remedies were called *segreti* in sixteenth-century Italy: O'Neil, "Magical Healing," p. 91.

58. Thomas, *Religion and the Decline of Magic*, p. 32.

59. For examples of such ordinary village magic, see Kiekhefer, *Magic in the Middle Ages*, pp. 56–94; Thomas, *Religion and the Decline of Magic*.

60. Peters, *Magician, Witch, and Law*, chap. 6; Cohn, *Europe's Inner Demons*, chaps. 9–10.

61. Thomas, *Religion and the Decline of Magic*, p. 49.

62. Peters, *Magician, Witch, and Law*; Cohn, *Europe's Inner Demons*.

63. Pine, *Pietro Pomponazzi*, chap. 3; Cassirer, *Individual and Cosmos*, pp. 103–9; *HMES*, 5:94–110. Pomponazzi's *De naturalium effectuum causis sive de incantationibus* was written in 1520 but not published until 1556.

64. Villari, *La rivolta antispagnola a Napoli*, pp. 100–102.

65. Letter to Cardinal Luigi d'Este, 14 May 1583: Campori, "Gio. Battista Della Porta ed il Cardinale Luigi d'Este," p. 187.

66. Della Porta, *Natural Magick*, pp. 1–2.

67. The work was written ca. 1604: Della Porta, *Criptologia*, ed. Belloni, p. 12. See also Paparelli, "Della Taumatologia," p. 19. Della Porta dedicated this unfinished work to the emperor Rudolf II. Only the index, a sort of expanded prospectus, was completed. The work is reprinted in Muraro, *Giambattista Della Porta mago e scienzato*, pp. 187–99. In addition, see Muraro, "La Taumatologia." On Della Porta's attempts to have the work published, see Della Porta, *Criptologia*, ed. Belloni, 91–101.

68. Paparelli, "Della Taumatologia," p. 19.

69. Della Porta, *Criptologia*, p. 158.

70. Ibid.

71. Paparelli, "Della Taumatologia," p. 19.

72. Della Porta, *Criptologia*, p. 158.

73. In 1564, a goldsmith testified before the Inquisition in Modena that "it is commonly said that white magnets are to be used in amatory incantations by touching the flesh of one or another persons": O'Neil, "Discerning Superstition," p. 138.

74. Della Porta, *Criptologia*, pp. 192–94. This spell was one of the rituals used by the so-called Pauliani, wandering popular healers and snake-handlers who were noted for their antidotes against venomous bites. See below, chap. 7; Corsini, *Medici ciarlatani*, p. 48; Burke, "Rituals of Healing," p. 213.

75. Della Porta, *Criptologia*, pp. 200–201. Della Porta's authority for the enchanted divining stick was the humanist George Agricola, who was in fact extremely skeptical of the technique. A miner, wrote Agricola, "should not make use of an enchanted twig, because if he is prudent and skilled in the natural signs, he understands that a forked stick is of no use to him, for as I have said before, there are natural indications of the veins which he can see for himself without the help of twigs": Agricola, *De re metallica*, pp. 38–41.

76. Della Porta, *Criptologia*, pp. 174–75.

77. Tobit 6:1–8; Della Porta, *Criptologia*, pp. 167–68. Della Porta thought Tobias's fish was the torpedo, famous in Renaissance natural history for the strength of its occult virtue, so powerful that it could stop ships at sea. See Copenhaver, "A Tale of Two Fishes."

78. Della Porta, *Criptologia*, pp. 200–202.

79. On this aspect of Della Porta's research, see Muraro, *Giambattista Della Porta mago e scienzato*, p. 38.

80. Clark, "Inversion, Misrule, and the Meaning of Witchcraft."

81. See Davis, "The Reasons of Misrule," in *Society and Culture in Early Modern France*, pp. 98–123; Kunzle, "World Upside-Down"; Burke, *Popular Culture in Early Modern Europe*, pp. 185–91.

82. Clark, "Inversion, Misrule, and the Meaning of Witchcraft," p. 125.

83. Paparelli, "Della Taumatologia," p. 19.

84. Della Porta, *Criptologia*, p. 78.

85. Della Porta, *Natural Magick*, p. 26.

86. For a concise, intelligent summary, see Walker, *Spiritual and Demonic Magic*, pp. 75–84.

87. Della Porta, *Natural Magick*, pp. 6–7.

88. Ibid., p. 7; cf. Agrippa, *Occult Philosophy*, pp. 30f.

89. Copenhaver, "Astrology and Magic."

90. Della Porta, *Natural Magick*, p. 10.

91. Ibid., p. 13.

92. Ibid., p. 8; cf. Agrippa, *Occult Philosophy*, p. 24.

93. Della Porta, *Natural Magick*, pp. 8–9.

94. Ibid., pp. 16–17. In addition, see Bianchi, *Signatura rerum*; Findlen, "Empty Signs?" (with bibliography).

95. Della Porta, *Natural Magick*, p. 17.

96. For a detailed study of these works and the theoretical system underlying them, see Caputo, "La struttura del segno fisiognomico." In addition, see Bianchi, *Signatura rerum*, pp. 90–92.

97. Della Porta had in mind such works as the *Chiromantia* (1503) of Bartholomeo Cocles and the *Chiromantia* (1560) of Patrizio Tricassio: Della Porta, *Della chirofisionomia*, preface.

98. Copenhaver, "A Tale of Two Fishes," p. 383. See also Doran, "On Elizabethan 'Credulity.'"

99. Della Porta, *Natural Magick*, p. 2.

100. Ibid., pp. 15–16.

101. Findlen, "Jokes of Nature."

102. Della Porta, *Natural Magick*, pp. 27–28.

103. Ibid., p. 27. Della Porta's authority for this passage was the myth of Deucalion and Pyrrha in Ovid, *Metamorphoses* 1.

104. Findlen, "Jokes of Nature."

105. Della Porta, *Natural Magick*, preface.

106. Eulogy of Della Porta by G. B. Longo, in Gabrieli, "Giovan Battista Della Porta Linceo," p. 425. See also Fulco, "Il 'museo' dei fratelli Della Porta."

107. See Gabrieli, "Bibliografia Lincea. I," pp. 262–66 (letters to Ulisse Aldrovandi). On the importance of gift-exchanges in natural history, see Findlen, "The Economy of Scientific Exchange."

108. Della Porta, *Natural Magick*, p. 19.

109. Ibid., pp. 73–74.

110. Ibid., p. 2.

111. Ibid., p. 48.

112. Ibid., p. 52; cf. the discussion of *vis imaginativa* in Walker, *Spiritual and Demonic Magic*, pp. 76–80.

113. Della Porta, *Natural Magick*, pp. 53–54.

114. Ibid., pp. 80–81.
115. Ibid., p. 141.
116. Ibid., pp. 114–15.
117. Ibid., p. 160.
118. Ibid., p. 306.
119. Ibid., p. 309. Smith, *Sources for the History of Steel*, p. 31.
120. Della Porta, *Natural Magick*, p. 311.
121. Ibid., preface.
122. "Principia audistis, scrutamini, operamini, periclitamini": Della Porta, *Magia naturalis*, p. 218.
123. Giovanni Faber, *éloge* of Della Porta, in Gabrieli, "Giovan Battista Della Porta Linceo," p. 425.
124. So according to Bartolomeo Chioccarelli, *De illustribus scriptoribus qui in civitate Neapolis floruerunt* (Naples, 1780), cited by Clubb, *Giambattista Della Porta, Dramatist*, p. xi.
125. Gabrieli, "Bibliografia Lincea. I," p. 268.
126. For my description of court culture, I follow Martines, *Power and Imagination*, pp. 218–40 (here quoting p. 220). In addition, see Bertelli, Cardini, and Garbano Zorzi, *Courts of the Italian Renaissance*; Vasoli, *La cultura delle corti*; Dickens, *Courts of Europe*. On patronage, see Lytle and Orgel, *Patronage in the Renaissance*; Kent, Simons, and Eade, *Patronage, Art, and Society*.
127. Martines, *Power and Imagination*, p. 233.
128. For examples, see Borsook, "Art and Politics at the Medici Court," pts. 1 and 2; and Forster, "Metaphors of Rule."
129. Tasso, *Malpiglio*, in *Dialogues*, p. 171.
130. See Biagioli, "Galileo's System of Patronage"; idem, "Galileo the Emblem-Maker"; and the essays in Moran, *Patronage and Institutions*. For a more detailed elaboration of the argument advanced in the following paragraphs, see Eamon, "Court, Academy, and Printing House," pp. 30–39.
131. Machiavelli, *The Prince*, p. 121.
132. Brown, "Platonism in Fifteenth-Century Florence," p. 395.
133. For medieval attitudes toward curiosity, see above, chap. 2.
134. Impey and MacGregor, *The Origins of Museums*; Salerno, "Arte, scienze e collezioni nel Manierismo"; Lugli, *Naturalia et mirabilia*.
135. Olmi, "Science—Honour—Metaphor," p. 14. On Renaissance collecting, see Findlen, *Possessing Nature*.
136. Berti, *Il Principe dello studiolo*; Olmi, "Science—Honour—Metaphor"; Salerni, "Arte, scienze e collezioni."
137. Gerolamo Porro, *L'orto de semplici di Padova* (Venice, 1591), quoted in Lugli, *Naturalia et mirabilia*, p. 89. See also Tomasi, "Projects for Botanical Gardens"; Prest, *The Garden of Eden*.
138. Houghton, "The English Virtuoso," p. 58. Though devoted principally to the English tradition, Houghton's work is still the best study of the virtuosi. See also Nicholson, "Virtuoso," in *DHI*, 4:486–90.
139. Similarly, Stefano Guazzo ranked those ennobled by virtue above those who gain nobility by birth, because the mind is nobler than the body. For these and other examples, see Holme, "Italian Renaissance Courtesy-Books."
140. Castiglione, *The Courtier*, p. 41.

141. This idea is brought out by Rebhorn, *Courtly Performances,* pp. 34–35.

142. Castiglione, *The Courtier,* p. 43; *Il Cortegiano* 1.26, pp. 66–67.

143. Stephen Greenblatt observes that the sixteenth-century manuals of court behavior were essentially "handbooks for actors, practical guides for a society whose members were nearly always on stage": *Renaissance Self-Fashioning,* pp. 162–63.

144. Della Porta, *Natural Magick,* preface.

145. Ibid., p. 4.

146. Ibid., p. 254.

147. Marie Boas Hall wrote that Della Porta's interest was "that of the party conjurer who deceives the eye by the quickness of his hand or mind": *Scientific Renaissance,* p. 187. Cf. *HMES,* 4:419–20.

148. Findlen, "Jokes of Nature."

149. If in the late sixteenth-century Naples became a theatrical center, writes Louise Clubb, "it was not because there were many playwrights there and few elsewhere—for they were everywhere legion—but in large part because Giambattista Della Porta was a Neapolitan." Clubb, *Giambattista Della Porta, Dramatist,* p. 140.

150. As reported by Francesco Stelluti, who recounted several of Della Porta's predictions of future events: Gabrieli, "Spigolatura Dellaportiane," pp. 512–13.

151. Della Porta, *Natural Magick,* p. 2.

152. Della Porta's relations with the Estensi court are detailed in Campori, "Gio. Battista Della Porta ed il Cardinale Luigi d'Este," with Della Porta's correspondence to d'Este.

153. Letter to Cardinal Federico Borromeo, in Gabrieli, "Bibliografia Lincea. I," p. 268.

154. Campori, "Gio. Battista Della Porta ed il Cardinale Luigi d'Este," p. 171.

155. Ibid., p. 174.

156. Gabrieli, "Giovan Battista Della Porta Linceo," p. 424. See also Galluzzi, "Motivi paracelsiana nella Toscana"; and idem, "Il mecenatismo mediceo."

157. D'Addio, *Gaspare Scioppio,* p. 57; Hessels, *Abrahami Ortelii Epistulae,* no. 147; Evans, *Rudolf II,* chap. 6.

158. The letter, first published by Sarnelli, is also published in Gabrieli, "Giovan Battista Della Porta Linceo," p. 424. See also Fiorentino, *Studi e ritratti,* pp. 279–80.

159. Campori, "Gio. Battista Della Porta ed il Cardinale Luigi d'Este," p. 187. Della Porta claimed to have gotten the recipe from one Angelo Siciliano, who had learned the secret from a friend, who had learned it from a Spanish doctor, who had learned it from a French monk.

160. Paparelli, "Della Taumatologia," p. 12. See also Muraro, *Giambattista Della Porta mago e scienzato,* pp. 21–35.

161. Gabrieli, "Bibliografia Lincea. I," pp. 268–69.

162. Paparelli, "La 'Taumatologia' di Giovambattista Della Porta."

163. Letter of Giovanni Ecchio (Heckius) to Tommaso Mermann, 17 February 1604, in Gabrieli, *Il carteggio Linceo,* p. 30.

164. There is a vast bibliography on Cesi and the early Lincei, including Carutti, *Breve storia*; Gabrieli, "Federico Cesi Linceo"; Olmi, "Federico Cesi e i Lincei"; idem, "La colonia lincea di Napoli"; Westfall, "Galileo and the Accademia dei Lincei." The academy's correspondence is published in Gabrieli, *Il carteggio Linceo*.

165. Francesco Stelluti, quoted in Gabrieli, "Spigolatura Dellaportiane," p. 507.

166. These ideals were articulated in the *Linceografo*, first drafted in 1604 and published in 1624 under the title *Præscriptiones*. The document is printed in Carutti, *Breve storia*, pp. 219–25.

167. Van Kessel, "Joannes van Heeck," p. 118.

168. Ibid., p. 124. Della Porta specifically referred to the Lincei as a "religion": Gabrieli, *Il carteggio Linceo*, p. 226 (letter to Cesi, 31 May 1612).

169. Olmi, "Federico Cesi e i Lincei," pp. 190–91, calls attention to the similarity between the Lincei's militant devotion to science and the militancy of the Jesuit Order.

170. Gabrieli, *Il carteggio Linceo*, p. 40 (17 June 1604); see also Gabrieli, "Federico Cesi Linceo," p. 354.

171. Cesi, "Il Natural Desiderio del Sapere," p. 68.

172. Della Porta, *De distillatione* (Naples, 1604). Della Porta later dedicated two other books to Cesi, *De aeris transmutationibus* (Rome, 1610) and *Elementorum curvilineorum* (Rome, 1610). These latter two were sponsored by the Lincean Academy.

173. Gabrieli, *Il carteggio Linceo*, pp. 155–56 (Cesi to Stelluti, 17 July 1604).

174. Ibid., p. 225 (Della Porta to Cesi, 31 May 1612).

175. Gabrieli, "Giovan Battista Della Porta Linceo," p. 374 (Della Porta to Cesi, 30 March 1612), p. 378 (Cesi to Stelluti, 5 July 1612); Gabrieli, *Il carteggio Linceo*, no. 112.

176. Gabrieli, *Il carteggio Linceo*, p. 210 (Cesi to Galileo, 17 March 1612).

177. Quoted in Drake, *Discoveries and Opinions*, p. 160.

178. Cesi, "Il Natural Desiderio del Sapere," pp. 53–92, quoting p. 71.

179. Ibid., p. 75.

180. Letter to the Tuscan court requesting a license to print the *Tessoro messicano*, one of the first works sponsored by the society, quoted in Morghen, "The Academy of the Lincei," p. 365.

181. Galileo, *Sidereus Nuncius*, trans. Van Helden, p. vii.

182. Ibid., p. 35.

183. Gabrieli, *Il carteggio Linceo*, pp. 114–15 (Della Porta to Cesi, 28 August 1609).

184. Gabrieli, "Giovan Battista Della Porta Linceo," pp. 370–71.

185. Paparelli, "Della Taumatologia," p. 13.

CHAPTER SEVEN
THE SECRETS OF NATURE IN POPULAR CULTURE

1. Ben Jonson, *Volpone*, act 2, sc. 2.
2. Agrippa, *Discorso ragionale trattato sopra remedii de veneni* (ca. 1590).

3. Garzoni, *Hospital of incurable fooles*, p. 6. "Quando fu mai tanta abbon-
danza di quelli che attendono a secreti nuovi, che anche in Bergamo ne com-
parve uno che si vantò di avere un secreto da convertire il turco, e lo volse ven-
dere ad un medico mio amico per una da quaranta, se lui lo voleva? Cosa da far
che, se il Fioravanti da Bologna l'avesse saputa, si disperasse da se stesso, per non
averla posta ne' suoi capricci medicinali, sotto titolo dell'angelico e divino *Elixir
Fioravanti*": *L'Ospidale de' pazzi universale*, in *Opere*, ed. Cherchi, p. 246.

4. Braudel, *Civilization and Capitalism*.

5. On the culture of the piazza, see Burke, *Popular Culture in Early Modern
Europe*, pp. 111f.

6. Garzoni, *Piazza universale*, pp. 745–46.

7. Moryson, *Fynes Moryson's Itinerary*, pp. 424–25.

8. Coryat, *Crudities*, 1:410–11.

9. Lea, *Italian Popular Comedy*, esp. pp. 17–128. See also Smith, *Commedia
dell'Arte*; Nicoll, *World of Harlequin*.

10. Corsini, *Medici ciarlatani*, p. 57. Although Corsini did not document
the quotation, it is from a seventeenth-century reformist tract by Domenico Ottonelli; see Lea, *Italian Popular Comedy*, p. 311. Garzoni's description of a professional commedia dell'arte troupe leaves the impression that the professional
players were not always superior to the street performers whom they imitated:
"When [the actors] enter a city a drum immediately lets everyone know of their
arrival. The woman dressed like a man walks in front, sword in hand, to make the
announcement, inviting the people to a comedy or tragedy in a palace or at the
Pilgrim's Inn, where the mob, naturally curious and eager for novelty, flock to
the hall, which they enter by paying a gazette. There they find a makeshift stage
and a scene scrawled in charcoal with the worst taste. They hear an introductory
concert of braying asses and buzzing hornets, then a prologue by a charlatan, a
clumsy thing like that of Fra Stoppino, with acts as stupid as bad luck, *intermedii*
a thousand times worse, a Magnifico (Pantalone) not worth a penny, a Zanni
who looks like a goose, a Graziano who shits his words, an inane Bawd, a Lover
who twists his arms over every speech, a Spaniard who can't say anything but '*mi
vida*' and '*mi corazon*,' a Pedant who goes off in Tuscan all the time, a Burattino
whose only gesture is putting his cap on his head, a Signora with a voice worse
than a monster, dead in her speech and asleep in her movements, who is constantly at war with Grace and in complete disagreement with Beauty": *Piazza
universale*, pp. 740–41. Garzoni's description strongly suggests that the *ciarlatani* and the commedia actors were interchangeable.

11. Two collections of these pamphlets, most dated between 1580 and
1640, survive. The University of Wisconsin owns a collection of forty-two primarily medical chapbooks. They are housed in the Rare Books Room under the
shelfmark D245, vols. 1 and 2. A similar collection (with many of the same titles) is in the Biblioteca Universitaria of Bologna (described by Ferrari, "Alcune
vie di difusione"). I am grateful to these libraries for supplying me with micro-
film copies of these pamphlets. I was unable to locate the collection at the British
Library (with many titles identical to those in the previous collections) described
by Ferguson, *Bibliographical Notes*, 2:30–33. For a complete list of titles of
these tracts, see the Appendix. Full bibliographical references for titles mentioned in the text are found in the bibliography.

12. T. Maiorini detto il Policinella, *Frutti soavi colti nel giardino* (Bologna, 1642); Francesco detto il Biscottino, *Giardino di varii secreti* (Milano, Genoa, Lucca, n.d.); P. M. Mutii detto il Zanni bolognese, *Nuovo lucidario de secreti* (Bologna, 1585); Americano, *Il vero e natural fonte, dal quale n'esce fuori un fonte d'acqua viva di mirabili, e salutiferi secreti* (Bologna, 1608); T. da Francolino detto l'Ortolano, *Tesoro di segreti naturali* (Rome, Venice, Milan, Sienna, Bologna, 1617); C. Amelli, detto il gran Piemontese, *Fioretto bellissimo con il quale si potrà pigliarsi trattenimento in qual si voglia honorata conversatione, ove si contengono giochi bellissimi e secreti curiosi* (Bologna, n.d.); D. Fedele detto il Manoainino, *Con il Poco farete Assai* (Rome, Bologna, n.d.); F. Boldini detto il Marchesino d'Este, *Il Medico de' poveri, o sia il gran stupere de' medici, epilogata in diversi secreti naturali* (Venice, n.d.); Lodovico Monte bolognese della Chitarriglia, *La ghirlandetta fiorita di varij secreti belissimi* (1633); Il Pesarino, gran Giocator di mano, *Secreti utilissimi e nuovi* (Verona, n.d.); Fioravanti Cortese, *Giardino et fioretto di secreti* (Venice, n.d.); Biagio, detto Il Figadet, *Tesoro di secreti, raccolti da diversi valenti huomini, che ne hanno fatto isperienza* (Bologna, 1617); Giovanni Cosson, detto Il bontempo Francese, *Secreti novamente ritrovati* (Milan, n.d.)

13. Dottor Gratiano Pagliarizzo, *Secreti nuovi e rari* (Bologna, Milan, n.d.). Cf. Dottor Graziano Scatolone, *Il vero, et pretioso thesoro di sanità*. On Graziano, see Lea, *Italian Popular Comedy*, pp. 25–41; Smith, *Commedia dell'Arte*, pp. 35–39.

14. See Lingo, "Empirics and Charlatans," p. 588.

15. Benedetto, detto il Persiano, *I Maravigliosi, et occulti secreti naturali* (Rome, Venice, Bologna, Milan, 1613); Americano, *Il vero e natural fonte, dal quale n'esce fuori un fonte d'acqua viva di mirabili, e salutiferi secreti* (Rome, Brescia, Bologna, 1608). Despite the pretension of the supposed author's name, the latter treatise was hardly exotic, being a reprinting of a pamphlet by the Brescia surgeon and distiller Andrea Fontana, *Fontana dove n'esce fuori acque di secreti* (Venice, Bologna, Parma, Pavia, Modena, n.d.).

16. Giulielmo Germerio, Tolosano, *Gioia preciosa . . . Opera à chi brama la sanità utilissima & necessaria* (Venice, 1604).

17. D. Fedele detto il Mantoianino, *Con il Poco farete assai* (Rome, n.d.).

18. Giacomo Trabia, *L'Esperienza vincitrice, epilogata in diversi secreti*.

19. On this point see Burke, "Rituals of Healing."

20. Coryat, *Crudities*, 1:411–12. In the seventeenth century, Francesco Redi reported that "to demonstrate the power and value of their antidotes, the charlatans eat scorpions and viper's heads and drink their bile": quoted in Viviani, "Ciarlatanismo medico," p. 105.

21. Garzoni, *Piazza universale*, p. 747. In addition, see Corsini, *Medici ciarlatani*, pp. 44–48. Mario Galasso, one of the writers on medical secrets, appears to have been a *pauliano*. His recipe book, entitled *Thesoro de poveri*, begins with a section headed "The true method you should follow if you want to use St. Paul's grace . . . for the benefit of the human body."

22. Burke, "Rituals of Healing," p. 220.

23. Lea, *Italian Popular Comedy*, p. 311.

24. Smith, *Commedia dell'arte*, p. 61.

25. Quoted in Lea, *Italian Popular Comedy*, p. 314.

26. Corsini, *Medici ciarlatani*, pp. 38–39.

27. Angelo Cortese, *Tesoro di sanità* (Milan, n.d.); Biagio detto il Figadet, *Tesoro di secreti*. Similar recipes are found in other booklets.

28. Bakhtin, *Rabelais and His World*, pp. 88–96, 174, et passim.

29. Saccardino's story is told by Ginzburg, "The Dovecote Has Opened Its Eyes." In addition, see idem, "Libertinismo erudito e libertinismo popolare"; Ginzburg and Ferrari, "La colombara ha aperto gli occhi."

30. Unfortunately, the records of Saccardino's trial are lost. However, the events and the trial are described in detail in Campeggi, *Racconti de gli heretici iconomiasti*.

31. Ibid., pp. 81–88, here quoting p. 88.

32. Ibid., p. 69.

33. Saccardino, *Libro nominato la verità della diverse cose*, p. 10.

34. In the *Libro*, Saccardino wrote about the spontaneous generation of organisms by the sun's "natural heat" (*color naturale*). "The elements generate elemental things and every other material; that is, the water creates the fish, the earth the plants, the air the birds and other flying things, all of which are generated by themselves without father or mother": ibid., p. 22.

35. Ibid., pp. 16–17.

36. Quoted in Ginzburg and Ferrari, "La colombara ha aperto gli occhi," p. 635.

37. Fioravanti, *Specchio universale*, fol. 41r–v. See also Camporesi, "Cultura popolare," pp. 87–88.

38. Campeggi, *Racconti de gli heretici iconomiasti*, pp. 125–26.

39. See, for example, Zapata, *Li maravigliosi secreti di medicina e chirurgia*, pp. 1–2. Pseudo-Falloppio also published numerous recipes supposedly obtained from common people, including one for oil of vitriol he got from a Slav who sold aqua vitae in the piazza of San Marco in Venice (*Secreti diversi*, p. 6). Fioravanti thought the unlettered folk had better science than the physicians and philosophers, while Della Porta proposed studying popular superstitions for traces of ancient wisdom.

40. Andrea Fontana, *Fontana dove n'esce fuori acqua di secreti* (Genoa, 1579). The work was reprinted in the early seventeenth century and was reissued in 1608 under the pseudonym "Americano."

41. Andrea Fontana, *Fontana dove n'esce fuori acqua di secreti* (Genoa, 1579); another ed. (Venice, Bologna, Parma, Pavia, Modena, n.d.): "Et farà cambio di Secreti con Secreti."

42. Vincentio Lauro, *Primo giardino, nel quale si contiene Secreti di grande esperienza, & virtù* (Pavia, 1591); idem, *Scelta di varii secreti* (Milan, 1603); idem, *Il Tesoro nel quale si contiene molti secreti de grandissime virtu* (Modena, n.d.).

43. Lorenzo Leandro, *Tesoro di varij Secreti . . . Esperimentati con grandissima fatica, e spesa di tempo, e di danari* (Venice, Brescia, Verona, Vicenza, Ferrara, 1614). The work contains a fairly conventional assortment of methods of "secret" writing (e.g., invisible ink), parlor tricks (e.g., to cook an egg without fire or to make an egg dance on the table), mechanical inventions, and practical jokes.

44. For example, the recipes turn up in a German translation of Columella (1612), in Thomas Vicary's surgical treatise, *The Englishman's Treasure* (1633), and in most German *Hausväterbucher* of the seventeenth century. Alessio's recipes continue to appear in eighteenth-century technical encyclopedias such as Chambers's *Cyclopedia*.

45. Bayle, *Dictionaire historique et critique*, 1:160.

46. The second and third parts of the *Secreti* appeared in 1558 and 1559 in Milan; pt. 4 was published in Venice in 1568.

47. Alexis of Piedmont, *Secretes* (London, 1595), fols. 126v, 145v, 132r, 124r. This edition contains all four parts.

48. Alexis of Piedmont, *Secretes* (1569), "Mayster Alexis of Piemont unto the Reader."

49. Laurent Joubert included this belief among the numerous "popular errors" relating to medicine. Joubert thought it was the "most erroneous of all" errors, possibly because he perceived it was the most widely shared: *Popular Errors*, pp. 44–48.

50. Warde would later become a professor of medicine at Cambridge and a physician to Queen Elizabeth.

51. Alexis of Piedmont, *Secretes* (1558), dedicatory epistle.

52. London, *A catalogue of the most vendable books in England*, sig. Z3r.

53. Although the *Secretes* appealed mainly to middle-class readers, physicians also consulted it. The work is found, for example, in the library of Sir Thomas Smith, the eminent physician and secretary of state under Queen Elizabeth: Strype, *Life of Sir Thomas Smith*, p. 280.

54. Fioravanti, *Les caprices . . . touchant la medecine* (Paris, 1586, 1598).

55. Fioravanti, *Corona; oder, Kron der Artzney* (Frankfurt, 1604, 1618). The work was first published by Johann Berner, who also published a German edition of Fiorvanti's *Della fisica*.

56. Kocher, "John Hester"; Debus, *English Paracelsians*, pp. 64–67. In addition to Debus's fundamental study of English Paracelsianism, see Webster, "Alchemical and Paracelsian Medicine."

57. Hester, *A hundred and fourteene Experiments*, dedication.

58. Fioravanti, *Short discourse uppon chirurgerie*, fol. A.iii^v. There follows a list of all of Fioravanti's works. Hester did not live to complete this ambitious project. His Fioravanti translations included *A Joyfull Jewell* (1579), a translation of Fioravanti's *Regimento della peste* (begun by Thomas Hill, who turned the manuscript over to Hester to see through the press); *A short discours uppon chirurgerie* (London, 1580), a compendium of extracts from Fioravanti's surgical writings; and the *Compendium of the rationall secrets* (London, 1582), a selected translation of Fioravanti's *Secreti rationali* along with extracts from the *Capricci medicinali* and the *Tesoro*. Hester's only work as author, *The pearle of practise; or, Practicers pearle, for phisicke and chirurgerie* (London, 1594), was directly modeled upon Fioravanti's *Tesoro*, being made up largely of case histories of cures made using his own and Fioravanti's medicaments.

59. Hester, *Key of Philosophy*, "To the Reader."

60. In Baker's translation of Gesner, *Newe Jewell of Health*, "To the Reader," sig. A4r: "John Hester dwelling on Powles wharfe, the which is a paynfull traveyler in those matters."

61. "If any be disposed to have any of these afore-sayd compositions redy made," Hester wrote at the end of the *Short discourse uppon chirurgerie*, "for the most part he may have them at Paules Wharfe, by one John Hester practitioner in the Arte of distillation, at the signe of the Furnaises": *Short discours*, fol. 64r. A broadsheet listing the drugs Hester offered for sale shows that his main stock consisted of Fioravanti's remedies. The broadsheet, published ca. 1588, contains a section headed "Certaine Compositions of Leonardo Phirovanti," which lists virtually all of Fioravanti's major medicaments. Hester also sold Fioravanti's incendiary compositions, his "device to make fresh water out of the sea as the ship sayleth," and his pitch to seal ships. The broadsheet (*STC* 13254) bears the title *These Oiles, Waters, Extractions, or Essences, Saltes, and other Compositions; are at Paules wharfe ready made to be solde by John Hester, practitioner in the arte of Distillation; who will also be ready for a reasonable stipend, to instruct any that are desirous to learne the secrets of the same in few dayes.*

62. Hester, *Key of Philosophy*, p. 65.

63. Hester's Fioravanti translations continued to be read down into the seventeenth century. A compilation of Hester's translations, *Three exact pieces of Leonardo Fioravanti*, was published in 1626, in 1652, and in 1653 with the title *An exact collection of the choicest and most rare experiments and secrets in physick and chirurgerie*. Extracts from Warde's Alessio and Hester's Fioravanti were also included in Thomas Vicary, *Englishman's treasure* (London, 1586).

64. Mascall's *Booke of the art and manner how to plant and graffe all sorts of trees* (London, 1572; *STC* 17574) was a translation of a French treatise on gardening by David Brossard, a Benedictine monk of the Abbey of St. Vincent. Other editions appeared in 1575, 1580, 1582, 1590, 1592, and 1652.

65. Mascall, *The husbandlye ordring of poultrie* (London, 1581; *STC* 17589).

66. Mascall, *The government of cattle* (London, 1587; *STC* 17580). Other editions appeared 1591, 1596, 1599, 1605, 1610, 1620, 1627, 1630, and 1680.

67. Mascall, *A profitable boke declaring dyvers approved remedies, to take out spottes and stains* (London, 1583; *STC* 17590). The work, which went through four editions, also included recipes for dyeing and metalworking, which were taken from the *Kunstbüchlein*.

68. Mascall, *Booke of the art and manner how to plant and graffe all sorts of trees*, dedication.

69. Partridge, *Treasurie*, title page.

70. Ibid., sig. D6r.

71. Partridge, *Widdowes treasure*, sigs. A8r, D6r. Partridge's works were among the first in a long tradition of "cookbook medicine" which became especially prominent in the seventeenth and eighteenth centuries. See Blake, "The Compleat Housewife"; Slack, "Mirrors of Health."

72. See Thomas, *Religion and the Decline of Magic*, pp. 212–52, 266–67.

73. Johnson, *Cornucopiæ, or divers secrets* (1595), sig. A3r.

74. Ibid., sigs. A3v, A4r.

75. Hill, *Naturall and Artificiall conclusions*, preface. The work was originally published in 1567 or 1568, although no known copies of this edition survive. On Hill, see Johnson, "Thomas Hill."

76. *Booke of Pretty Conceits* (London, 1586), sig. A3r–v.

77. Johnson, *Cornucopiæ, or divers secrets* (1594), sig. A4v.

78. *Booke of Pretty Conceits*, sigs. A4r, B1r.

79. Ibid., sig. A3r–v.

80. Lupton, *A thousand notable things of sundrie sorts* (London, 1601; first ed. 1579). By 1631, Lupton's work had gone through no less than eight editions.

81. Ibid., pp. 2, 8, 37, 100, 105.

82. Thomas, *Religion and the Decline of Magic*, pp. 202–4; Kaplan, "Greatrakes the Stroker."

83. Zapata, *Li maravigliosi secreti di medicina e chirurgia*, pp. 1–2.

84. Quoted by Davis, "Proverbial Wisdom and Popular Errors," in *Society and Culture in Early Modern France*, p. 261.

85. Ibid., pp. 227–67.

86. Montaigne, "Of Vain Subtleties," in *Essays*, p. 227.

87. Montaigne, *Essays*, p. 580.

88. Aubrey, *Remaines of Gentilisme and Judaisme*, in *Three Prose Works*, p. 132. The word "not" is unfortunately omitted in this edition; see Hunter, *John Aubrey*, p. 168n.2.

89. Aubrey, *Remaines of Gentilisme and Judaisme*, in *Three Prose Works*, pp. 254–55; Bacon, *Of the Wisdom of the Ancients*, in *Works*, 6:695–99. See also Hunter, *John Aubrey*, pp. 195–98; Davis, "Proverbial Wisdom and Popular Errors," in *Society and Culture in Early Modern France*, pp. 227–67.

90. Joubert, *Popular Errors*, pp. 66, 123.

91. Ibid., pp. 69, 172–73.

92. Mercurio, *De gli errori populari*, p. 1.

93. Ibid., p. 175.

94. Ibid., pp. 265–68.

95. See Burke, *Popular Culture*, pp. 209–13.

96. See the documents relating to Borromeo's actions in Taviani, *La commedia dell'arte*, pp. 5–43.

97. See ibid., pp. 85–89, 287–309.

98. Ibid., pp. 165–79, 192–97.

99. Giovan Domenico Ottonelli, *Della Christiana moderatione del theatro* (1646), selections in ibid., pp. 320–526; see pp. 489–90.

100. Mercurio, for example, lumped the *ciarlatani* in a single category with prostitutes, witches, and Jews.

101. Mercurio, *De gli errori populari*, p. 267.

102. Primrose, *Popular Errours*, p. 18. Primrose's treatise, first published in Latin in 1638, was also translated into French and German.

103. Ibid., pp. 42–48.

104. Bacon, *Advancement of Learning*, in *Works*, 3:364.

105. Browne, *Pseudodoxia Epidemica*, in *Works*, 2:52, 40.

106. On Cardano and Mizauld, see below, chap. 8.

107. Browne, *Pseudodoxia Epidemica*, in *Works*, 2:58–59.

108. Ibid., 2:25–30, quoting p. 28.

109. See Burke, *Popular Culture in Early Modern Europe*, chap. 8.

110. Bacon, *De sapientia veterum*, in *Works*, 6:722. See also Jacob, "Science and the Two Cultures."

111. Galileo, *Letter to the Grand Duchess Christina*, in Drake, *Discoveries and Opinions*, pp. 181–82.
112. Ibid., p. 200. In addition, see Jacob, "Science and the Two Cultures," pp. 239–40.
113. See below, chap. 10.
114. Montaigne, "Of Cannibals," in *Essays*, p. 152.
115. Galileo, *Letter to the Grand Duchess Christina*, in Drake, *Discoveries and Opinions*, p. 196.
116. Browne, *Works*, 2:26.

CHAPTER EIGHT
SCIENCE AS A *VENATIO*

1. Bacon, *De augmentis scientiarum*, in *Works*, 1:622.
2. Rossi, *Philosophy, Technology, and the Arts*, p. 42. In addition, see Cavazza, "Metafore venatorie e paradigmi indiziari."
3. Badaloni, "I fratelli Della Porta," p. 688. See above, chap. 4.
4. Della Porta, *De i miracoli et maravigliosi effetti dalla natura prodotti*, p. 2: "i secreti, li quali al tutto stavano rinchiusi nel grembo della natura." Cf. idem, *Magia naturalis*, p. 6: "arcana Naturae gremio penitus latentia."
5. Francecso Stelluti, quoted in Gabrieli, "Spigolatura Dellaportiane," p. 507; Carutti, *Breve storia*, p. 8.
6. Maylender, *Storia delle accademie d'Italia*, 1:478–79.
7. Castiglione, *The Courtier*, p. 38.
8. Evans, *Rudolf II*, p. 196.
9. Lensi Orlandi, *Cosimo e Francesco d'Medici alchemisti*; Berti, *Il Principe dello studiolo*; Galluzzi, "Il Mecenatismo medicео."
10. Schaefer, "Studiolo of Francesco I," pp. 186–98.
11. Galluzzi, "Motivi Paracelsiana nella Toscana"; Covoni, *Don Antonio de' Medici al Casino di San Marco*. For other examples of "secrets" being exchanged in the Medici court, see Magghi, *Bichierografia*, introduction. I thank Mario Biagioli for this reference.
12. Galluzzi, "Motivi Paracelsiana nella Toscana," pp. 56–59.
13. Drake, *Discoveries and Opinions*, pp. 62–63.
14. On gift exchange as an instrument for the creation of patronage relationships, see Biagioli, "Galileo's System of Patronage," and Findlen, "The Economy of Scientific Exchange" (here quoting Biagioli, p. 20).
15. Thorndike, "Newness and Novelty." See also Rossi, *Philosophy, Technology, and the Arts*, pp. 65–70.
16. For references to the *Ne plus ultra* in English literature, see Jones, *Ancients and Moderns*, chap. 6.
17. Lopez de Gómara, *Histoire générale des Indes* (Paris, 1568), quoted in Atkinson, *Nouveaux horizons de la Renaissance*, p. 257. See also Scammell, "The New Worlds and Europe"; and Ryan, "Assimilating New Worlds."
18. Quoted in Wightman, *Science in a Renaissance Society*, p. 74.
19. Browne, *Pseudodoxia Epidemica*, in *Works*, 2:5.
20. Zambecarri, "Experiments of Doctor Joseph Zambeccari," trans. Jarcho, p. 313.

21. Jacques Cartier, *Bref récit et succincte narration* (Paris, 1545), quoted in Atkinson, *Nouveaux horizons de la Renaissance*, p. 256.

22. Jean de Léry, *Histoire d'un voyage en Bresil* (La Rochelle, 1578), quoted in Atkinson, *Nouveaux horizons de la Renaissance*, pp. 286–87.

23. Le Roy, *Of the Interchangeable Course, or Variety of Things in the Whole World*, fol. 127v.

24. Glanvill, *Vanity of Dogmatizing*, 178. Such "conquistadorial" attitudes toward nature were doubtless nourished by reports of the conquistadors themselves, such as that of Hernán Cortés, who wrote home from Mexico describing his inquiries into "the secrets of these parts": quoted in Elliot, *Old World and New*, p. 30.

25. On Renaissance encyclopedism, see Kenny, *Palace of Secrets*, esp. chap. 1.

26. For Mizauld, see the sketches in *BU*, 29:183–84; Niceron, *Mémoires*, 40:211–13; and *HMES*, 5:299–301; 6:216–17. Mizauld's works are listed in Niceron, 40:202–13. Including translations, Mizauld's works totaled forty-one.

27. Mizauld, *Memorabilium aliquot naturae arcanorum silvula* (Paris, 1555; Frankfurt, 1592; etc.).

28. Lemnius, *Secret Miracles*, preface. On Lemnius's life and work, see Van Hoorn, "Levinus Lemnius"; Roshem, "Lévin Lemne."

29. Lemnius, *Secret Miracles*, p. 2.

30. On this point, see Céard, *La nature et les prodiges*, pp. 145–50.

31. Lemnius, *Secret Miracles*, p. 1; see also p. 3.

32. Ibid., pp. 2–3, 25.

33. Park and Daston, "Unnatural Conceptions." In addition, see Céard, *La nature et les prodiges*.

34. Boaistuau, *Certaine Secret Wonders of Nature*, trans. Fenton. Quotations are from Boaistuau's preface and Fenton's dedicatory epistle. The contrast between Boaistuau's and Fenton's views of monsters may lend support to Peter Dear's observations regarding the distinction between the seventeenth-century French and English views of miracles ("Miracles, Experiments").

35. Paré, *On Monsters and Marvels*, pp. 3–4.

36. On Renaissance encyclopedism and the idea of the "circle of knowledge," see Kenny, *Palace of Secrets*, pp. 15–16.

37. Wecker, *De secretis*. Wecker also translated the *Secreti* of Alessio into Latin (1559) and German (1569).

38. Wecker, *De secretis*, dedication: "Inexaustus est naturae thesaurus, & delitescunt proculdubio plurima, quae temporis successu sagacium hominum conatibus eruentur."

39. For the influence of Ramist logic in Germany, see Ong, *Ramus*, pp. 298–300.

40. English owners of Wecker's *De secretis* included Sir Walter Raleigh (Oakeshott, "Raleigh's Library," p. 308), Robert Burton (Gibson and Needham, "Lists of Burton's Library," p. 246), and Brian Twynne, reader in Greek at Oxford (Ovenell, "Twynne's Library," p. 42).

41. Wecker, *The Secrets of Art & Nature*, trans. R. R[ead], preface.

42. On Cardano, see Ingegno, *Saggio*; Fierz, *Girolamo Cardano*; Margolin, "Rationalisme et irrationalisme"; idem, "Analogie et causalité."

43. Cardano, *Opera*, 2:537–50.

44. Ibid., pp. 537–38.

45. Ibid., pp. 540–41.

46. The work was first printed in partial form in Nuremberg by Johannes Petreius. The complete, twenty-one book edition was published the following year in Paris. For a discussion of the work, see Ingegno, *Saggio*, chap. 6; and Maclean, "Interpretation of Natural Signs."

47. Cardano, *De subtilitate*, trans. Cass, p. 75.

48. Scaliger, *Exotericae exercitationes de subtilitate* (1577). For a discussion of the Cardano-Scaliger dispute, see Maclean, "Interpretation of Natural Signs."

49. In the résumé of his publications drawn up in his autobiography, Cardano counted 131 printed books, 170 manuscripts he burned because he considered them useless, and 111 books in manuscript: Cardano, *Book of My Life*, pp. 220–25.

50. See the section titled "Thoughts on How to Perpetuate My Name" in Cardano's autobiography, *Book of My Life*, pp. 32–35.

51. "At non occultando secretum sit: sed secretum occultari meretur": *De secretis*, in *Opera*, 2:544–45.

52. For Cardano's work on the interpretation of dreams, see Fierz, *Girolamo Cardano*, chap. 6.

53. Ginzburg, "Clues: Roots of an Evidential Paradigm."

54. Ibid., p. 102.

55. "The intellectual operations involved—analyses, comparisons, classifications—were formally identical": ibid., p. 103.

56. Ibid., p. 99.

57. Detienne and Vernant, *Cunning Intelligence*.

58. Quoted in Baxandall, *Limewood Sculptors*, p. 32.

59. Detienne and Vernant, *Cunning Intelligence*, pp. 308–18.

60. Gassendi, *Syntagma philosophicum*, in *Opera omnia*, 1:68–69, 120–21; *Selected Works*, pp. 290–93, 367–68. See also Cavazza, "Metafore venatorie e paradigmi indiziari," pp. 107–11.

61. To illustrate how the mind "deduces the existence of something not perceived by the senses from some other thing which is perceived by the senses," Gassendi gives the example of finding out whether pores exist in the skin: "Despite the fact that they are not perceived by the senses, reason proves that they exist on the grounds that if they did not, no passage would lie open for the sweat that passes from the inside outward and is perceived by the senses" (*Opera omnia*, 1:122; *Selected Works*, p. 372).

62. Bacon, *De Sapientia Veterum*, in *Works*, 6:713. Bacon thought the myth of Pan was "big almost to bursting with the secrets and mysteries of Nature." For discussion, see Rossi, *Francis Bacon*, pp. 98, 155.

63. Bacon, *De augmentis scientiarum* 5.2, in *Works*, 1:633, trans. 4:421. Bacon's discussion of *experientia literata* ("learned experience") occurs at 1:623–33, trans. 4:413–21.

64. Gassendi, *Opera omnia*, 1:126.

65. Murdoch, "Analytical Character of Medieval Learning." See also Rossi, "Aristotelians and 'Moderns.'"

66. Daston, "The Factual Sensibility," p. 465.

67. Bacon, "The Refutation of Philosophies," in Farrington, *Philosophy of Bacon*, p. 130.

68. Daston, "The Factual Sensibility," p. 465.

69. Della Porta, *Natural Magick*, preface. Bacon wrote, "It often comes to pass that mean and small things discover great better than great can discover small" (*De augmentis scientarum*, in *Works*, 4:297).

70. Bacon, *Thoughts and Conclusions*, in Farrington, *Philosophy of Bacon*, p. 92.

71. Della Porta, *Natural Magick*, preface.

72. Bacon, *De sapientia veterum*, in *Works*, 6:725–26; *Advancement of Learning*, in *Works*, 3:333.

73. Drake, *Discoveries and Opinions*, p. 258.

74. Galileo, *Two New Sciences*, p. 14.

75. Bacon, *Novum Organum* 1.98, in *Works*, 4:95.

76. Ibid., p. 18.

77. Ibid., p. 16.

78. Ibid., p. 18. Cf. Bacon, *De augmentis scientarum*, in *Works*, 4:408; idem, *Thoughts and Conclusions*, in Farrington, *Philosophy of Bacon*, p. 92.

79. Bacon, *Novum Organum*, in *Works*, 4:18.

80. Bacon, *De augmentis scientiarum*, in *Works*, 4:413. See also Jardine, *Francis Bacon*, pp. 143–49.

81. Jardine, *Francis Bacon*, p. 144.

82. Bacon, *Works*, 4:413. For Bacon's explanation of these types of experiments, summarized in the following paragraphs, see pp. 413–21.

83. Ibid., p. 420.

84. Bacon, *Novum Organum*, in *Works*, 4:25. Bacon's "inductivism" has been the subject of considerable discussion among philosophers of science. For a summary of recent controversies, see McMullin, "Conceptions of Science," pp. 45–54.

85. Horton, "In Defence of Bacon"; Urbach, "Bacon as Precursor to Popper." On Bacon's use of analogy, see Park, "Bacon's 'Enchanted Glass.'"

86. Bacon, *Novum Organum*, bk. 2, in *Works*, 4:155–246. See also Jardine, *Francis Bacon*, pp. 124–26.

87. Bacon, *Works*, 4:203.

88. Ibid., pp. 97, 81.

89. Ibid., pp. 70–71.

90. Bacon, *Refutation of the Philosophers*, in Farrington, *Philosophy of Bacon*, p. 123; cf. Bacon, *Novum Organum*, in *Works*, 4:71–72. See also Rossi, *Francis Bacon*, pp. 11–35.

91. Bacon, *Novum Organum*, in *Works*, 4:87, 104, 107.

92. Bacon, *De augmentis scientiarum*, in *Works*, 4:421.

93. Bacon, *The New Atlantis*, in *Works*, 3:125–66. The description of Solomon's House is on pp. 156–66.

94. Ibid., p. 156. The Latin is from the translation done by William Rawley.

95. Ibid., pp. 164–65.

96. Bacon, *Novum Organum* 1.50, in *Works*, 4:58. Cf. 2:380; and idem, "Thoughts and Conclusions," in Farrington, *Philosophy of Bacon*, p. 86.

97. Quoted in Carutti, *Breve storia*, p. 8.

98. Raleigh, *History of the World* 1.1.7, in *Works*, 2:16–17.

99. Ibid., pp. 284–85 (1.11.2). See also Rattansi, "Alchemy and Natural Magic."

100. Hutchison, "What Happened to Occult Qualities?"; idem, "Dormative Virtues." In addition, see Copenhaver, "The Occultist Tradition"; and Smith, "Knowing Things Inside Out."

101. Hutchison, "What Happened to Occult Qualities?" p. 235.

102. Ibid., pp. 242–48.

103. Charleton, *Physiologia Epicuro-Gassendo-Charltoniana*, p. 341.

104. Glanvill, *Scepsis scientifica*, p. 133.

105. Charleton, *Physiologia Epicuro-Gassendo-Charltoniana*, p. 342–43.

106. Ibid., p. 5. In the same work, Charleton called upon experimental philosophy to "withdraw that thick Curtain of obscurity, which yet hangs betwixt Nature's Laboratory and Us" (p. 342).

107. Glanvill, *Scepsis scientifica*, pp. 132–33.

108. Webster, *Academiarum Examen* (1653), pp. 67–68, in Debus, *Science and Education*, p. 149.

109. Hooke, *General Scheme*, in *Posthumous Works*, pp. 8, 6. On this work and its context, see Hunter and Wood, "Towards Solomon's House."

110. Hooke, *General Scheme*, p. 7.

111. Hooke, *Micrographia*, preface.

112. Hooke, *General Scheme*, pp. 45–62.

113. Hooke, *Micrographia*, preface. Cf. Glanvill, *Vanity of Dogmatizing*, pp. 5–6.

114. Hooke, *Micrographia*, preface.

115. Power, *Experimental Philosophy*, sigs. b2–c3.

116. Hooke, *Posthumous Works*, pp. 7, 4.

117. Glanvill, *Scepsis scientifica*, preface, sig. c1.

118. Ibid., sig. b2v. Evidently the new philosophy gave a new, Baconian meaning to Fioravanti's concept of the "caprices" of nature.

119. Cf. Boyle, *Works*, 2:163–64.

120. Kuhn, "Mathematical versus Experimental Traditions." For a summary of Kuhn's thesis, see my Introduction.

121. For further discussion of this topic, see Hutchison, "What Happened to Occult Qualities?" and Henry, "Occult Qualities and the Experimental Philosophy."

122. On this aspect of the Royal Society's activities, see Hoppen, "Nature of the Early Royal Society."

123. Henry, "Occult Qualities and the Experimental Philosophy," p. 363.

124. Bacon, *Parasceve*, in *Works*, 4:253.

125. For a description of some of these cabinets, see MacGregor, "Cabinet of Curiosities." In addition, see Daston, "The Factual Sensibility." The Royal Society's interest in such matters is discussed by Hunter, "Royal Society's 'Repository.'"

126. Bacon, *Novum Organum*, in *Works*, 4:168.

CHAPTER NINE
THE VIRTUOSI AND THE SECRETS OF NATURE

1. Houghton, "The English Virtuoso." The subject has been more recently explored in an outstanding article by Shapin, "'A Scholar and a Gentleman.'"
2. On the importance of sensibilities in the history of science, see Daston, "Moral Economy of Science."
3. On "civility" and the "moral conventions" governing dispute, see Shapin and Schaffer, *Leviathan and the Air-Pump*, pp. 72–76.
4. Stone, *Crisis of the Aristocracy*, pp. 748–49.
5. See Simon, *Education and Society in Tudor England*; and Kelso, *Doctrine of the English Gentleman*.
6. Elyot, *Boke Named the Governour*.
7. Stone, *Crisis of the Aristocracy*, chap. 8.
8. Peacham, *Compleat Gentleman* (1622), p. 18.
9. See Shapin, "'A Scholar and a Gentleman,'" pp. 287–92.
10. Fletcher, *Intellectual Development of Milton*, p. 83.
11. Houghton, "The English Virtuosi," pp. 64–65. See also Lievsay, *Elizabethan Image of Italy*.
12. Quoted in Lievsay, *Elizabethan Image of Italy*, p. 14.
13. A hint of the aristocracy's defensiveness can be heard in Peacham's recommendation of the study of heraldry on the grounds that familiarity with family ensigns might enable the gentleman to "discerne and know an intruding upstart, shot up with the last nights Muschrome, from an ancient descended and deserving Gentleman": *Compleat Gentleman*, p. 138.
14. Burton, *Anatomy of Melancholy*, 1:244.
15. Ibid., 2:86–89.
16. Peacham, *Compleat Gentleman* (1653), p. 105.
17. Evelyn, *Diary*, 2:396–97, 470–71. Evelyn visited dozens of such cabinets while in Italy. See, for example, his description (2:314–15) of Francesco Gualdo's collection containing antiquities, paintings, and "natural curiosities" including "the knee Bone of a Gyant." In Naples, Evelyn visited Ferdinando Imperato's museum, "full of incomparable rarities" (2:330–31).
18. Evelyn, *Diary and Correspondence*, p. 575.
19. Shadwell, *The Virtuoso*, pp. 78–79.
20. Astell, *Essay in Defense of the Female Sex*, pp. 96–97.
21. See above, chap. 7.
22. Johnson, *Dainty Conceits*, title page.
23. Johnson, *A new Booke of new Conceits*, title page.
24. On these works, see Hall, *Old Conjuring Books*.
25. Evelyn, *Diary and Correspondence*, p. 574.
26. Burton, *Anatomy of Melancholy*, 2:94–95.
27. On Bate, see Taylor, *Mathematical Practitioners*, p. 210.
28. For a survey of this literature, see Zetterberg, "Mathematical Magick." See also idem, "Mistaking of 'the Mathematics' for Magic."
29. Van Etten, *Mathematical Recreations*. This work has customarily been attributed to the Jesuit mathematician Jean Leurechon. However, Hall (*Old*

Conjuring Books, pp. 83–119), presents convincing arguments that the work was actually written by Van Etten, who was one of Leurechon's pupils at the Jesuit school at Pont-à-Mousson.

30. White, *A Rich Cabinet*, title page.
31. On early thermometers, see Middleton, *History of the Thermometer*, pp. 1–26.
32. Bate, *Mysteries of Nature and Art*, pp. A3, 53–54.
33. Van Etten, *Mathematical Recreations*, title page.
34. See Shapiro, *John Wilkins*, pp. 123–40; Webster, *Great Instauration*, pp. 88–99.
35. Charleton, *Immortality of the Human Soul*, pp. 46–47.
36. Evelyn, *Diary*, 3:110–11. A waywiser was an instrument for measuring the distance traveled by road: *OED*.
37. Manuel, *Portrait of Newton*, 39–40, 47–49; Andrade, "Newton's Early Notebook."
38. Quoted in Manuel, *Portrait of Newton*, p. 41.
39. Shapin and Schaffer, *Leviathan and the Air-Pump*, p. 26.
40. See Rossi, *Philosophy, Technology, and the Arts*, pp. 137–45.
41. Bacon, *De augmentis scientiarum*, in *Works*, 4:294–95.
42. Bacon, *Novum Organum*, in *Works*, 4:99.
43. Bacon, *New Atlantis*, in *Works*, 3:156–64.
44. Zetterberg, "Echoes of Nature in Salomon's House, p. 187.
45. On the idea of "maker's knowledge," see Pérez-Ramos, *Francis Bacon's Idea of Science*.
46. Van Etten, *Mathematical Recreations*, pp. 67–68.
47. Bate, *Mysteries of Nature and Art*, pp. 21, 61–62.
48. Wilkins, *Mathematical Magick*, p. 186.
49. Colie, "Drebbel and De Caus."
50. On Plat, see *DNB*, 15:1293–95; Mullett, "Hugh Platt"; and Debus, "Palissy, Plat, and Agricultural Chemistry."
51. Plat, *A briefe apologie of certain newe inventions*.
52. Plat, *Jewell House*, dedication.
53. Plat, *Jewell House* (1653), p. 33.
54. Fourteen editions of *Delights for ladies* appeared between 1602 and 1656.
55. Plat, *Floraes Paradise*, "To the Reader."
56. *Jewell House* (1653), p. 51. On Plat's chemical interests, see Debus, "Palissy, Plat, and Agricultural Chemistry."
57. *Jewell House*, preface.
58. The notebooks, all in the Sloane collection at the British Library, include: Sloane MSS 2172, 2189, 2197, 2249, 2244, 2272 (miscellaneous secrets); 2247 (recipes for dressing and dyeing silk); 2216 (alchemical secrets, headed with the motto, "In vulcano veritas"); 2246 (alchemical secrets); 2212 (alchemical: "Secreta secretiora of Sir Hugh Plat"); 2209 (medical recipes).
59. Sloane MS 2247, fols. 13v–14r.
60. Sloane MS 2249, passim.
61. The rehabilitation of curiosity is the subject of several important studies

by Lorraine Daston, including "Neugier und Naturwissenschaft" and "Moral Economy of Science." In addition, see Blumenberg, *Legitimacy of the Modern Age*, esp. pp. 361–402; and Ginzburg, "High and Low."

62. See *OED*, s.v. "Curioso."

63. See Daston, "Neugier und Naturwissenschaft," from which I derive this threefold scheme of meanings. My definitions differ only slightly from hers.

64. Evelyn, *Diary*, 2:396–97.

65. Power, *Experimental Philosophy* (quoted in *OED*, s.v. "Curiosity").

66. Evelyn, *Diary*, 2:100, 124, 50.

67. Hooke, *Micrographia*, pp. 163, 91, 8.

68. Daston, "Neugier und Naturwissenschaft."

69. Hooke, *Micrographia*, p. 186.

70. Abraham Cowley, "Ode to the Royal Society," in Sprat, *History of the Royal Society* (dedicatory ode).

71. Henry Oldenburg, "The Publisher to the Reader," in Boyle, *Experiments and Considerations Touching Colours* (1664), in *Works*, 1:667. See also Schiebinger, "Feminine Icons"; and Keller, "Baconian Science."

72. Hooke, *Posthumous Works*, pp. 27, 63.

73. Bacon, *Novum Organum*, in *Works*, 4:170–71.

74. Quoted in Barnouw, "Hobbes's Psychology of Thought," p. 533.

75. Bacon, *Works*, 4:71–72.

76. Boyle, *Experiments and Considerations Touching Colours*, in *Works*, 1:662–788, quoting pp. 663–64.

77. Evelyn, *De Vita Propria* (ca. 1697), in *Diary and Correspondence*, 1:33.

78. Van Etten, *Mathematical Recreations*, "By Way of Advertisement."

79. See Daston, "Neugier und Naturwissenschaft."

CHAPTER TEN
FROM THE SECRETS OF NATURE TO PUBLIC KNOWLEDGE

1. Bacon, *Great Instauration*, in *Works*, 4:20; cf. idem, *Advancement of Learning*, in *Works*, 3:264–66.

2. Bacon, *Great Instauration*, in *Works*, 4:88–89.

3. Bacon, *Advancement of Learning*, in *Works*, 3:266.

4. Bacon, *Valerius Terminus*, in *Works*, 3:220.

5. Bacon, *Cogita et Visa*, in *Works*, 3:610; trans. Farrington, *Philosophy of Bacon*, p. 92.

6. Bacon, *Valerius Terminus*, in *Works*, 3:222.

7. See Webster, *Great Instauration*.

8. Hooke, *Micrographia*, preface, p. b2.

9. See Hill, *The World Turned Upside Down*.

10. Quoted in ibid., p. 179.

11. In general, see Westfall, *Science and Religion*, pp. 106–45; for the Restoration period, see Hunter, *Science and Society*, pp. 162–87.

12. John Norris, *Reflections upon the Conduct of Human Life* (1690), quoted in Hunter, *Science and Society*, p. 175.

13. Boyle, *Considerations*, in *Works*, 2:15, 13, 53, 55, 30. On the dating of

the work, most of which was written between 1649 and 1654, see Westfall, "Boyle Papers," p. 65.

14. Bacon, *Novum Organum*, in *Works*, 4:62–63, 14.
15. Bacon, *Advancement of Learning*, in *Works*, 3:289–90.
16. Bacon, *New Atlantis*, in *Works*, 3:165.
17. Prior, "Bacon's Man of Science."
18. Walzer, *Revolution of the Saints*, chaps. 2, 5; Webster, *Great Instauration*.
19. Bacon, *Advancement of Learning*, in *Works*, 3:285–86.
20. Bacon, *Novum Organum*, in *Works*, 4:110.
21. Bacon, *Cogita et Visa*, in *Works*, 3:612.
22. Trevor-Roper, "Three Foreigners"; Webster, *Great Instauration*, pp. 32–99; idem, "Hartlib and the Great Reformation."
23. J. V. Andrae, *Christianae societatis imago* (Strasbourg, 1619); *Reipublicae Christianopolis descriptio* (Strasbourg, 1619); *Christiani amoris dextra porrecta* (Tübingen, 1620). Andrae's works described a utopian community called Christianopolis, where men of learning formed a Christian brotherhood, collaborating to attain universal enlightenment and serving as advisers to the state. On Andrae, see Montgomery, *Cross and Crucible*. On Rosicrucianism, see also Yates, *Rosicrucian Enlightenment*.
24. These works are published, in whole or in part, in Webster, *Hartlib and the Advancement of Learning*, pp. 139–95.
25. Comenius, *Via Lucis, Vestigata & vestiganda, h[oc]. e[st]. Rationabilis disquisitio, quibus modis intellectualis Animorum Lux, Spaientia, per omnes Omnium Hominum mentes, & gentes, jam tandem sub Mundi vesperam feliciter spargi possit* (London, 1668), trans. Compagnac, *The Way of Light*, 51–52, 11, 24, 108. When finally published, the work was dedicated to the Royal Society.
26. See above, n. 23; Yates, *Rosicrucian Enlightenment*, pp. 163–67.
27. *Macaria* is reprinted in Webster, *Hartlib and the Advancement of Learning*, pp. 79–90, to which all citations here refer. Formerly the work was thought to have been by Hartlib; however, see idem, "Authorship and Significance of *Macaria*." Almost nothing is known about Plattes, save that his patron was the engineer William Englebert.
28. Plattes, *Discovery of Infinite Treasure*, "Epistle Dedicatory." Plattes's other work was the *Discovery of Subterraneall Treasure*.
29. Plattes, *Macaria*, pp. 83, 80,
30. Plattes, *Profitable Intelligencer*, fol. A1v.
31. Plattes, *Discovery of Infinite Treasure*, p. 72.
32. Plattes, *Macaria*, p. 89.
33. For a description of some of these projects, see O'Brien, "Commonwealth Schemes." See also Jones, *Ancients and Moderns*, pp. 148–80.
34. See Webster, *Great Instauration*, pp. 67–77. Hartlib gave a detailed description of the Office of Address in his *Considerations tending to the Happy Accomplishment of Englands Reformation* (1647). The work is published in Webster, *Hartlib and the Advancement of Learning*, pp. 119–39. On Renaudot's Bureau, see Solomon, *Public Welfare, Science, and Propaganda*.
35. Quoted in Solomon, *Public Welfare, Science, and Propaganda*, p. 65.

36. Hartlib, *Considerations . . . Englands Reformation*, in Webster, *Hartlib and the Advancement of Learning*, pp. 128, 131–32.

37. Webster, *Great Instauration*, p. 77.

38. Petty, *Advice of W.P.*, pp. 1–2.

39. Cf. Petty's "Proposal for the History of Arts," Sloane MS 2903, fols. 63–65; cf. *Petty Papers*, 1:205–7.

40. Webster, *Great Instauration*, pp. 363–66. See also O'Brien, "Commonwealth Schemes," pp. 34–36, who according to Webster "wrongly attributes this proposal to [Benjamin] Worsley."

41. O'Brien, "Commonwealth Schemes," p. 36; Webster, *Great Instauration*, pp. 366–67.

42. Evelyn, *Diary*, 1 September 1659, 3:232. Evelyn described the society in detail in a letter to Boyle dated 3 September 1659; in Boyle, *Works*, 6:288–91.

43. Hartlib, *Chymical, Medicinal, and Chyrurgical Addresses* (London, 1655). Boyle's tract was published by Rowbottom, "Earliest Published Writing of Robert Boyle." See also Maddison, "Earliest Published Writing of Boyle."

44. O'Brien, "Hartlib's Influence on Boyle." See also Jacob, *Robert Boyle*, pp. 16–38. Boyle's almost filial attachment to Hartlib is betrayed in a letter to Hartlib in which Boyle, in answer to a query from Hartlib, wrote, "I shall freely discloze to Yow my infant Thoughts, & from Your Perusall of them, expect an Advantage not unlike to that which Rosebuds derive from opening their Virgin leaves unto the Sun; who by his looking on them both blowes & Ripens them": quoted in Maddison, "Earliest Published Writing of Boyle," p. 166.

45. Rowbottom, "Earliest Published Writing of Robert Boyle," pp. 380–81.

46. See Evelyn's letter to Mr. Maddox, *Diary and Correspondence*, p. 574.

47. Royal Society, Classified Papers, III(i), fol. 1. Cf. Hooke's more "Baconian" outline in *Posthumous Works*, pp. 24–26. In addition, see Hunter, *Science and Society*, pp. 99–101.

48. Evelyn to Boyle, 9 August 1659, Boyle, *Works*, 6:287–88; Evelyn, *Diary and Correspondence*, p. 578. Evelyn completed only one of the projected histories, that of engraving.

49. Evelyn, *Diary and Correspondence*, p. 590.

50. Bacon, *Advancement of Learning*, in *Works*, 3:332.

51. Hall, "Science, Technology, and Utopia," p. 37. In the same article Hall makes the remark, somewhat puzzling in light of recent research, that "Utopian idealism in England . . . had nothing to do with science," p. 40.

52. Turnbull, "Hartlib's Influence on the Royal Society," p. 127.

53. Evelyn, *Diary*, 3:163. See also Turnbull, *Samuel Hartlib*, p. 24.

54. See O'Brien, "Hartlib's Influence on Boyle," pp. 268f; Turnbull, *Hartlib, Dury, Comenius*.

55. Syfret, "Origins of the Royal Society," p. 97.

56. Webster, *Great Instauration*, p. 501. See also Clucas, "Samuel Hartlib's Ephemerides."

57. This group was described by John Wallis; see Scriba, "Autobiography of John Wallis," pp. 39–40.

58. See Webster, "New Light on the Invisible College."

59. See, for example, Syfret, "Origins of the Royal Society." For a summary

of the problem, see Purver, *Royal Society*; Webster, "Origins of the Royal Society."

60. Hunter, *Science and Society*, p. 26; Webster, *Great Instauration*.

61. Shapiro, "Latitudinarianism and Science." Despite his powerful argument for the "vital contributions" of Puritans to English science, Webster notes that "an obvious characteristic, and a further strength, of the English scientific movement was the diversity of interests displayed by individual members. . . . From an ideological point of view the scientific community was extremely mixed": *Great Instauration*, p. 496.

62. Sprat, *History of the Royal Society*, pp. 53, 328, 38, 332, 374, 341.

63. Glanvill, "Modern Improvements of Useful Knowledge," in *Essays*, 3:36.

64. Sprat, *History of the Royal Society*, p. 63; cf. p. 431, noting the society's composition of "many eminent men of all Qualities."

65. For the society's social composition, see Hunter, *Royal Society and Its Fellows*.

66. Sprat, *History of the Royal Society*, p. 67.

67. Bacon, *Novum Organum*, in *Works*, 4:96.

68. Oldenburg, *Correspondence*. See also Hall, "Oldenburg and the Art of Scientific Communication"; Shapin, "O Henry."

69. Wellisch, "Conrad Gesner"; Durling, "Conrad Gesner's *Liber amicorum*."

70. On Aldrovandi, see Findlen, "The Economy of Scientific Exchange."

71. Quoted in Lenoble, *Mersenne*, p. 1. See also Brown, *Scientific Organizations*, pp. 31–40.

72. Oldenburg, *Correspondence*, 2:27–28.

73. See Andrade, "Birth and Early Days of the *Philosophical Transactions*"; Hall, "Royal Society's Role in the Diffusion of Information."

74. Quoted in Hunter, *Science and Society*, p. 52.

75. Hevelius to Oldenburg, 25 December 1663, in *Correspondence*, 2:138.

76. Petit to Oldenburg, 7 November 1665, in ibid., 2:595.

77. Oldenburg, *Correspondence*, ed. Hall and Hall, 2:xxi–xxiii. See also Cavazza, "Bologna and the Royal Society," pp. 112–13.

78. Hevelius to Oldenburg, 6 January 1665/66, *Correspondence*, 3:6.

79. Glanvill, *Essays*, 3:36–37.

80. Sprat, *History of the Royal Society*, pp. 99–101.

81. See, in particular, Shapin and Schaffer, *Leviathan and the Air-Pump*; Shapin, "Pump and Circumstance"; idem, "House of Experiment."

82. Boyle, *Skeptical Chymist*, in *Works*, 1:461. See also Shapin, "House of Experiment," pp. 377–78.

83. On experimental spaces, see Shapin, "House of Experiment." On civility and experimental science, see Shapin and Schaffer, *Leviathan and the Air-Pump*, pp. 72–76.

84. On the danger of experiments' becoming spectacles, see Golinski, "A Noble Spectacle."

85. Evelyn, *Diary*, 4 August 1681, 4:251–54, quoting pp. 251–52. The experiment was performed by Frederick Slare, F.R.S., who demonstrated several phosphorus specimens in the 1680s. Since Birch, *History of the Royal Society*,

records no meeting for this date, the demonstration may have been performed at an informal meeting. For the context of this and other phosphorus experiments, see Golinsky, "A Noble Spectacle."

86. Evelyn, *Diary*, 4:253. Cf. 7 May 1645, 2:397–98.

87. Golinski, "A Noble Spectacle," pp. 24–26.

88. On the role of conventions in early experimental science, see Shapin and Schaffer, *Leviathan and the Air-Pump*, pp. 22–24, 78–79, 146–54, 225, 329–31.

89. Ibid., p. 336.

90. Hunter and Wood, "Towards Solomon's House," p. 81, quoting Royal Society, Miscellaneous MS 4.72. Cf. Golinski, "A Noble Spectacle," p. 31. On replication, see Shapin and Schaffer, *Leviathan and the Air-Pump*, pp. 225–82.

91. Sprat, *History of the Royal Society*, p. 99.

92. Shapin, "Pump and Circumstance," pp. 490–97; cf. Shapin and Schaffer, *Leviathan and the Air-Pump*, pp. 60–65.

93. Shapin, "Pump and Circumstance," pp. 491–92.

94. Sprat, *History of the Royal Society*, p. 112.

95. Dear, "*Totius in verba*," p. 153.

96. Boyle, *Skeptical Chymist*, in *Works*, 1:460.

97. Oldenburg to Johann Christian von Boineburg, 10 August 1670, *Correspondence*, 7:109.

98. Oldenburg to Georg Stiernhelm 9 December 1669, *Correspondence*, 6:365. Oldenburg was particularly eager to get specimens and reports from "the frozen north."

99. Oldenburg to Malpighi, 28 December 1667, *Correspondence*, 4:92.

100. Malpighi to Oldenburg, 5 July 1669, *Correspondence*, 6:102.

101. Quoted in Hahn, *Anatomy of a Scientific Institution*, pp. 15–16; cf. 24–25.

102. Cavazza, "Bologna and the Royal Society," p. 116. Cf. idem, "Verso la fondazione dell'*Istituto delle Scienze*"; idem, *Settecento inquieto*, pp. 140–48. On Italian natural philosophers' relations with the English and with the Royal Society, see Pighetti, *Influsso scientifico di Robert Boyle*.

103. Cavazza, "Bologna and the Royal Society," pp. 106–7. See also idem, *Settecento inquieto*, pp. 119–48. The academy was also called the Accademia dei Filosofi.

104. Quoted in Cavazza, "Bologna and the Royal Society," p. 107.

105. Pighetti, *Influsso scientifico di Robert Boyle*, pp. 119–54.

106. French, *Art of Distillation*, "To the Reader," fol. A8v.

107. Boyle, *Skeptical Chymist*, in *Works*, 1:459–60. See also Golinski, "Chemistry in the Scientific Revolution," p. 385.

108. Boyle, "An Account of the two Sorts of the Helmontian Laudanum," *Phil. Trans.* 107 (1674): 147–49.

109. Boyle, "Of the Incalescence of *Quicksilver* with *Gold*," p. 529; idem, "The Aerial Noctiluca," in *Works*, 4:379–404, pp. 384, 388, 396. See also Oster, "Boyle as Aristocrat and Artisan," p. 273.

110. Principe, "Boyle's Alchemical Secrecy," p. 71. See also Dobbs, "Studies in the Natural Philosophy of Sir Kenelm Digby"; idem, "From the Secrecy of Alchemy to the Openness of Chemistry."

111. Houghton, "History of Trades"; Ochs, "History of Trades Programme."

112. Bacon, *Parasceve*, in *Works*, 4:257.

113. Boyle, *Considerations Touching the Usefulness of Experimental Natural Philosophy*, in *Works*, 3:436.

114. Hooke, *Micrographia*, preface, fol. g1; Sprat, *History of the Royal Society*, pp. 117–18.

115. Birch, *History of the Royal Society*, 1:407. Christopher Merrett was appointed chairman of the committee.

116. Ibid., 1:41, 51–52, 55, 65, 83 (reprinted in Sprat, *History of the Royal Society*, 285–306), 99–102, 131–32, 144–52, 169–71, 156–62, 324–27, 342, 401; Sprat, *History of the Royal Society*, 258.

117. Petty, "Apparatus to the History of the Common Practices of Dying," in Sprat, *History of the Royal Society*, pp. 284–306.

118. Charleton, *Mysterie of Vintners*, in *Two Discourses*, pp. 141–230.

119. Ibid., p. 159.

120. Stubbe, *Legends no Histories*, p. 35.

121. Sprat, *History of the Royal Society*, p. 71.

122. Quoted in Houghton, "History of Trades," p. 375.

123. Evelyn to Boyle, 9 May 1657, in Boyle, *Works*, 6:287.

124. Ibid., p. 287.

125. Birch, *History of the Royal Society*, 1:416, 428, 434, 460.

126. Aubrey, *Brief Lives*, p. 305.

127. Oldenburg to Hartlib, 25 June 1659, in Oldenburg, *Correspondence*, 1:270; cf. Golinski, "Chemistry in the Scientific Revolution," p. 379.

128. 'Espinasse, "Decline and Fall of Restoration Science."

129. Hunter, *Royal Society and Its Fellows*.

130. Hooke, *Micrographia*, preface, sig. g1; idem, *General Scheme*, in *Posthumous Works*, p. 26.

131. Hooke, *Micrographia*, preface, sig. a2. On Hooke's curatorship, see Pumfrey, "Ideas above His Station."

132. See also Hunter, "Early Problems in Professionalizing Scientific Research."

133. Bacon, *Sylva Sylvarum*, in *Works*, 2:331–680, quoting from the preface by William Rawley, p. 335.

134. Sprat, *History of the Royal Society*, p. 115. On rival Baconianisms, see Hunter and Wood, "Towards Solomon's House," p. 66.

135. These proposals are the subject of Hunter and Wood, "Towards Solomon's House"; see esp. pp. 70–72.

136. Royal Society, Miscellaneous MSS 4.72, in ibid., pp. 81–82.

137. "Proposalls for the Good of the Royal Society," Royal Society, Classified Papers, xx.50, fols. 92–94, in Hunter and Wood, "Towards Solomon's House," p. 87.

138. In the same year that "A.B." wrote his proposal, William Petty observed that the society "has been censured . . . for spending too much time in matters not directly tending to profit and palpable Advantages (as the weighing of Air and the like)": quoted in Syfret, "Early Critics," p. 45.

139. Hunter and Wood, "Towards Solomon's House," p. 86.

140. Ibid., p. 88; cf. p. 74; Hunter, *Science and Society*, p. 57; Shapin, "Who Was Robert Hooke?" p. 275.

141. Ben-David, *Scientist's Role in Society*; Westman, "The Astronomer's Role."

142. Quoted in Hunter, *Science and Society*, p. 146.

143. Ibid., p. 57.

144. Hahn, *Anatomy of a Scientific Institution*, p. 146 and chaps. 3–5.

145. Sprat, *History of the Royal Society*, p. 71.

146. Ibid., p. 74.

CONCLUSION

1. For a brief account, see Hadot, *Idee der Naturgeheimnisse*.

2. Hesiod, *Op.* 780-81, quoted in Lloyd, *Science, Folklore and Ideology*, pp. 119–20.

3. Plutarch, *De curiositate* 517D. Pliny, *Natural History* 2.77, also speaks of the "secrets of nature."

4. Quoted in Easlea, *Fathering the Unthinkable*, p. 26.

5. By "classical" view I mean that of Aristotle, *Poetics* 1457b and *Rhetoric* 1405a. For a useful discussion of metaphors in science, see Bono, "Science, Discourse, and Literature."

6. Black, *Models and Metaphors*, p. 39.

7. Bagley, "Practice of Esotericism"; Simmel, "Sociology of Secrecy"; Bok, *Secrets*, pp. 5–14.

8. Burnet, *Early Greek Philosophy*, p. 133.

9. ". . . a rebus occultis et ab ipsa natura involutis," *Acad.* 1.4.15; cf. 1.5.19: "omnium quae naturae obscuritate occultantur," *De finibus* 5.19.51.

10. The gender implications of the "secrets of nature" metaphor have been explored by Easlea, *Fathering the Unthinkable*, pp. 25–28 et passim; by Keller, "Secrets of God"; idem, "From Secrets of Life to Secrets of Death."

11. Charleton, *Physiologia Epicuro-Gassendo-Charletoniana*, p. 342.

12. Bacon, *Novum Organum*, in *Works*, 4:172.

13. Boyle, *Origin of Forms and Qualities*, in *Works*, 4:49, 112, et passim.

14. Gassendi, *Syntagma*, in *Opera omnia*, 1:122–23; *Selected Works*, pp. 374–75.

15. Arbib and Hesse, *Construction of Reality*, p. 156.

16. Bacon, *Novum Organum*, in *Works*, 4:172.

17. Bagley, "Practice of Esotericism."

18. Cicero, *De finibus* 5.5.12.

19. Shapin, "House of Experiment." See also the hypothesis advanced by Biagioli, "Scientific Revolution, Social Bricolage."

20. Cournand and Zuckerman, "Code of Science," p. 958.

21. Principe, "Boyle's Alchemical Secrecy." On secrecy and priority in the early Royal Society, see Iliffe, "'In the Warehouse.'"

22. On secrecy in modern science, see Bok, *Secrets*, chap. 11; Cade, "Aspects of Secrecy in Science."

23. For a summary of the dispute, see ibid., pp. 166–67.

24. On this point, see Keller, "Secrets of God," p. 232; Simmel, "Sociology of Secrecy"; and Bok, *Secrets*, chap. 8.

25. Shapin and Schaffer, *Leviathan and the Air-Pump*, p. 343.

26. Ramsey, "Traditional Medicine and Medical Enlightenment"; idem, "Property Rights and the Right to Health"; Hahn, *Anatomy of a Scientific Institution*, chap. 5.

27. Buchan, *Domestic Medicine*, pp. xii–xiv.

28. Blake, "The Compleat Housewife," p. 36.

29. Chase, *Dr. Chase's Recipes*, p. 130.

30. Ibid., p. 129.

31. Information on Chase and his book is taken from Fred and Marjorie Kerwin, "Dr. Chase's Wonderful Book," *Ann Arbor Observer*, October 1980, 37–43. I am grateful to Elizabeth Eisenstein for a copy of this article, the only work I have been able to find on Chase.

BIBLIOGRAPHY

THE BIBLIOGRAPHY contains all items cited in the text. It is supplemented by the Appendix, a short-title list of Italian secrets-pamphlets, ca. 1520–1643. When more than one article in a collection of articles are referenced, citations under the author's name are given in abbreviated form; full publication details for the book are given under the editor's name or (in cases when no editor is given) under the book's title.

MANUSCRIPTS

Lawrence, Kansas, Spencer Research Library
  Pryce MS E1, "Libro de i secretti e ricette" (1536–1562).
London, British Museum
  Additional MS 28,353, fols. 57–61 (undated), testimony of Leonardo Fioravanti, Madrid.
  Sloane MS 2903, fols. 63–65, William Petty, "Proposal for the History of Arts."
  Sloane MS 2172, Hugh Plat, secrets and experiments.
  Sloane MS 2189, Hugh Plat, secrets and experiments.
  Sloane MS 2197, Hugh Plat, secrets and experiments.
  Sloane MS 2209, Hugh Plat, miscellaneous medical recipes.
  Sloane MS 2212, Hugh Plat, "Secreta secretiora of Sir Hugh Plat."
  Sloane MS 2216, Hugh Plat, alchemical secrets with motto, "In vulcano veritas."
  Sloane MS 2244, Hugh Plat, secrets and experiments.
  Sloane MS 2246, Hugh Plat, alchemical secrets.
  Sloane MS 2247, Hugh Plat, recipes for dressing and dyeing silk.
  Sloane MS 2249, Hugh Plat, miscellaneous recipes.
  Sloane MS 2272, Hugh Plat, secrets and experiments.
London, Royal Society of London
  Classified Papers, III(i), fol. 1, John Evelyn, outline of a history of trades.

PRINTED WORKS

Primary Sources

Adelmann, Howard B., ed. *The Correspondence of Marcello Malpighi.* 5 vols. Ithaca and London: Cornell University Press, 1975.
Agricola, Georgius. *De re metallica.* Translated by Herbert Clark Hooker and Lou Henry Hoover. New York: Dover Publications, 1950; orig. publ. 1912.
Agrippa, Livio. *Discorso ragionale trattato sopra remedii de veneni materiali, compositi, simplici, corporali, e pestilentiali.* N.p., n.d.
Agrippa von Nettesheim, Henry Cornelius. *Of the Vanity and uncertainty of Artes and Sciences.* Translated by J. Sanford. London, 1569.
———. *Three books of Occult Philosophy.* Translated by J. F. London, 1651.

Albertus Magnus. *The Book of Minerals.* Translated by Dorothy Wyckoff. Oxford: Clarendon Press, 1967.

———, Ps.-. *Alberti Magni De secretis mulierum libellus, . . . eiusdem De virtutibus herbarum, lapidum, & animalium quorundam libellus.* Strasbourg, 1625.

———, Ps.-. *The Book of Secrets of Albertus Magnus of the Virtues of Herbs, Stones, and Certain Beasts. Also a Book of the Marvels of the World.* Edited by Michael R. Best and Frank H. Brightman. London, Oxford, New York: Oxford University Press, 1973.

Allerley Mackel und Flecken aus Gewant, Sammath, Seyden, Güldinen, stücken &c. zu bringen. Mainz, Erfurt, Zwickau, Nuremberg, 1532.

Amelli, Claudio, detto il gran Piemontese. *Fioretto bellissimo con il quale si potrà pigliarsi trattenimento in qual si voglia honorata conversatione, ove si contengono giochi bellissimi e secreti curiosi.* Bologna, n.d.

Americano. *Il vero e natural fonte, dal quale n'esce fuori un fonte d'acqua viva di mirabili, e salutiferi secreti.* Bologna, 1608.

Aquinas, Thomas. *De occultis operibus naturae* (see McAllister, J. B.).

———. *Summa theologiae.* Edited and translated by Thomas Gilby, O.P., et al. 60 vols. London: Blackfriars, 1964.

Artliche Kunste mancherley weyse Dinten und aller hand Farben zubereyten. Nürnberg, Erfurt, Mainz, 1531; Zwickau, 1532.

Astell, Mary. *An Essay in Defense of the Female Sex.* London, 1696.

Aubrey, John. *Aubrey's Brief Lives.* Edited by Oliver Lawson Dick. Harmondsworth: Penguin Books, 1972; orig. publ. 1949.

———. *Three Prose Works.* Edited by John Buchanan-Brown. Carbondale: Southern Illinois University Press, 1972.

Augustine of Hippo. *The Confessions of St. Augustine.* Translated by Rex Warner. New York: Mentor Books, 1963.

———. *De civitate Dei. PL* 41.13–804.

———. *De confessionum libri XIII. PL* 32.659–868.

———. *De trinitate libri quindecim. PL* 42.819–1098.

———. *The Divination of Demons (De divinatione daemonum).* Translated by Ruth W. Brown. In *The Writings of St. Augustine,* vol. 15 (*The Fathers of the Christian Church,* vol. 27), pp. 414–40. Washington, D.C.: Catholic University of America Press, 1955.

———. *On the Holy Trinity.* Translated by A. W. Hadden. In *A Select Library of the Nicene and Post-Nicene Fathers,* ed. Philip Schoff, 3:1–228. Grand Rapids, Mich.: Eerdmans, 1956.

Bacon, Francis. *The Works of Francis Bacon, Baron of Verulam, Viscount of St. Alban, and Lord Chancellor of England.* Edited by James Spedding, Robert Leslie Ellis, and Douglas Denon Heath. 14 vols. New York: Garrett Press, 1968; orig. publ. 1857–1874.

Bacon, Roger. *Epistola de secretis operibus naturae, et de nullitate magiae.* Edited by J. S. Brewer. In *Opera quaedam hactenus inedita,* 1:523–51. London: Longman, Green, Longman, and Roberts, 1859.

———. *Opus Majus.* Translated by Robert Belle Burke. 2 vols. New York: Russell and Russell, 1962.

———. *Opus Tertium.* Edited by J. S. Brewer. In *Opera quaedam hactenus inedita,* vol. 1. London: Longman, Green, Longman, and Roberts, 1859.

————. *Roger Bacon's Letter Concerning the Marvelous Power of Art and of Nature and Concerning the Nullity of Magic.* Translated by Tenney L. Davis. Easton, Pa.: Chemical Publishing Co., 1923.

————. *Secretum secretorum cum glossis et notulis.* Edited by Robert Steele. In *Opera hactenus inedita*, vol. 5. Oxford: Oxford University Press, 1920.

Bairo, Pietro. *Secreti medicinali.* Venice, 1561.

Barbieri, Giacomo. *I dottori bolognese di teologia, filosofia, medicina, ed d'arti liberali.* Bologna, 1623.

Bate, John. *Mysteries of Nature and Art.* London, 1634.

Battista, Gio. figliuolo del gran Medico del Rè di Francia. *Stupendi et maravigliosi secreti, mai più palesati da nessun'altro che da me.* Milan, n.d.

Bayle, Pierre. *Dictionaire historique et critique.* Vol. 1. Amsterdam, 1690.

Benedetto, detto il Persiano. *I Maravigliosi, et occulti secreti naturali. Tradotti di lingua Persiana nella nostra lingua Italica.* Rome, 1613.

Bergwerk- und Probierbuchlein. Translated by A. G. Sisco and C. C. Smith. New York: American Institute of Mining and Metallurgical Engineers, 1949.

Bernard of Clairvaux. *The Steps of Humility and Pride.* Translated by A. Ambrose Conway. In *The Works of Bernard of Clairvaux*, 2:25–82. Washington, D.C.: Cistercian Publications, 1974.

Betz, Hans Dieter, trans. *The Greek Magical Papyri in Translation.* Chicago: University of Chicago Press, 1985.

Biagio, detto il Figadet. *Tesoro di secreti, raccolti da diversi valenti huomini, che ne hanno fatto isperienza.* Bologna, 1617.

Birch, Thomas. *The History of the Royal Society of London for the Improving of Natural Knowledge, from its First Rise.* 4 vols. New York: Johnson Reprint Co., 1968; orig. publ. London, 1656–1657.

Biringuccio, Vannoccio. *Pirotechnia.* Translated by Cyril Stanley Smith and Martha Teach Gnudi. Cambridge: MIT Press, 1959; orig. publ. 1942.

Boaistuau, Pierre. *Certaine Secret Wonders of Nature.* Translated by Edward Fenton. London, 1569.

Boldini, Antonio Felice, detto il Marchesino d'Este. *Il medico de' poveri, o sia il gran stupore de' medici, epilogato in diversi secreti naturali, che alcuno no ne fa stima, cavati da quelle herbe, piante, radiche, fronde, fiori e frutti, create dalla Provvidenza di Dio per beneficio de' mortali, esperimentati da me ... operatore spagirico, e ricavati anco da virtuosi singolarissimi, che in simil professione si essercitano per la salute del genere humano.* Venice, n.d.

Boltz, Valentin. *Illuminierbuch. Wie man allerlei Farben bereiten, mischen und auftragen soll.* Edited by C. J. Benziger. Munich: Georg D. W. Callwey, 1913.

The Booke of Pretty Conceits. London, 1586.

Bovio, Tommaso Zefiriele. *Opere.* Venice, 1626.

Boyle, Robert. "An Account of the two Sorts of the Helmontian Laudanum." *Phil. Trans.* 107 (1674): 147–49.

————. "Of the Incalescence of Quicksilver with Gold." *Phil. Trans.* 122 (1676): 515–33.

————. *The Works of the Honourable Robert Boyle.* Edited by Thomas Birch. 6 vols. Hildesheim: Georg Olms Verlag, 1966; orig. publ. 1772.

Braekman, W. L., ed. *Dat Batement van recepten. Een secreetboek uit de zestiende eeuw. Scripta*, no. 25. Brussels: UFSAL, 1990.

————, ed. *Geoffrey of Franconia's Book of Trees and Wine. Scripta*, no. 24. Brussels: UFSAL, 1989.

————, ed. *Magische experimenten en toverpraktijken uit een middelnederlands handschrift.* Seminarie voor Volkskunde van de Rijksuniversiteit te Gent, IX. Ghent: Seminarie voor Volkskunde, 1966.

Browne, Sir Thomas. *The Works of Sir Thomas Browne.* Edited by Geoffrey Keynes. 4 vols. Chicago: University of Chicago Press, 1964.

Brunschwig, Hieronymus. *The Book of Cirurgia.* Strasbourg, 1497. Edited with study by Henry E. Sigerist. Milan: R. Lier & Co., 1923.

————. *The Vertuose boke of Distyllacyon of the waters of all maner of herbes.* London, 1527.

Buchan, William. *Domestic Medicine, or, A Treatise of the Prevention and Cure of Disease.* 22d ed. Exeter, 1839.

Buger, Giovanni, detto il Genevenino. *Virtus Occulta Perit. Giardino di varii secreti havuti da diversi Signori dove si contengono varie sorte di giochi, secreti, e burle.* Bologna, n.d.

Burton, Robert. *The Anatomy of Melancholy.* Edited by Holbrook Jackson. 3 vols. bound as 1. New York: Vintage Books, 1977; orig. publ. 1932.

Caley, Earle Radcliffe. "The Leyden Papyrus X: An English Translation with Brief Notes." *Journal of Chemical Education* 3 (1926): 1149–66.

————. "The Stockholm Papyrus: An English Translation with Brief Notes." *Journal of Chemical Education* 4 (1927): 979–1002.

Campanella, Tommaso. *Del senso delle cose e della magia.* Edited by Antonio Bruers. Bari: Gius. Laterza & Figli, 1925.

————. *Magia e grazia.* Edited by Romano Amerio. Edizione Nazionale dei Classici del Pensioro Italiano, ser. 2, vol. 5. Rome: Fratelli Bocca, 1957.

Campeggi, Ridolfo. *Racconti de gli heretici iconomiasti giustiziati in Bologna.* Bologna, 1623.

Campori, Giuseppe, ed. "Gio. Battista Della Porta ed il Cardinale Luigi d'Este." *Atti e Memorie delle RR. Deputazione di Storia Patria per le Provincie Modenese e Parmensi* 6 (1872): 165–90.

Cardano, Girolamo. *The Book of My Life.* Translated by Jean Stoner. New York: E. P. Dutton, 1930.

————. *De subtilitate libri XX.* Lyons, 1554.

————. *The First Book of Jerome Cardan's De subtilitate.* Translated by Myrtle Marguerite Cass. Williamsport, Penn.: Bayard Press, 1933.

————. *Opera omnia.* 10 vols. Lyons, 1663.

Castiglione, Baldesar. *The Book of the Courtier.* Translated by Charles S. Singleton. Garden City, N.Y.: Doubleday & Co., 1959.

————. *Il Cortegiano.* Milan: Istituto Editoriale, 1914.

Cesi, Federico. "Il Natural Desiderio del Sapere." In *Scienziati del seicento,* ed. Maria Luisa Altieri Biagi, pp. 53–92. Milan: Rizzoli, 1969.

Chambers, Ephraim. *Cyclopedia: or, An Universal Dictionary of Arts and Sciences.* London, 1791.

Charleton, Walter. *The Immortality of the Human Soul, Demonstrated by the Light of Nature.* London, 1659.

————. *Physiologia Epicuro-Gassendo-Charltoniana, or A Fabrick of Science Natural, Upon the Hypothesis of Atoms.* New York: Johnson Reprint Corp., 1966; orig. publ. 1654.

————. *Two Discourses. I. Concerning the different wits of men; II. Of the mysterie of vintners.* London, 1669.

Chase, A. W. *Dr. Chase's Recipes; or, Information for Everybody.* Ann Arbor, Mich.: By the Author, 1864.

Cicero, Marcus Tullius. *De finibus bonorum et malorum.* 2d ed. Translated by H. Rackam. Loeb Classical Library. Cambridge: Harvard University Press, 1931.

Colleta de molte nobilissime ricette tratte de diversi eccellentiss. Medici. N.p., n.d.

Comenius, Jan Amos. *The Way of Light.* Translated by E. T. Compagnac. Liverpool: The University Press, 1938.

Compagni, Vittoria Perrone. "Picatrix latinus. Concezioni filosofico-religiose e prassi magia." *Medioevo* 1 (1975): 237–337.

Copenhaver, Brian, ed. and trans. *Hermetica. The Greek "Corpus Hermeticum" and the Latin "Asclepius" in a New English Translation, with Notes and Introduction.* Cambridge: Cambridge University Press, 1992.

Corbin, Henry. "Le Livre de Glorieux de Jâbir ibn Hayyân." *Eranos-Jahrbuch* 18 (1950): 47–114.

Corpus Hermeticum. Edited by A.-J. Festugière and A. D. Nock. 4 vols. Paris: Société d'Éditions "Les Belles Lettres," 1945–1954.

Cortese, Angelo. *Tesoro di sanità, Nel quale si contiene alcuni particolari, e maravigliosi secreti. No più da persona alcuna dati in luce.* Milan, n.d.

Cortese, Fioravanti. *Giardino et fioretto di secreti.* Venice, n.d.

Cortese, Isabella. *I Secreti . . . ne' quali si contengono cose minerali, medicinali, artificiose, & alchimiche, & molte de l'arte profumatoria, appartenenti a ogni gran signora.* Venice, 1574.

Coryat, Thomas. *Coryat's Crudities.* 2 vols. Glasgow: James MacLehose and Sons, 1905.

Cosson, Giovanni, detto il Bontempo Francese. *Secreti novamente ritrovati.* Milan, n.d.

Curcio, Carlo, ed. *Utopisti e reformatori sociali del cinquecento.* Bologna: N. Zanichelli, 1941.

Dall'Oso, Eugenio. "Due lettere inedite di Leonardo Fioravanti." *Rivista di storia delle scienze mediche e naturali* 47 (1956): 283–91.

Debus, Allen, ed. *Science and Education in the Seventeenth Century: The Webster-Ward Debate.* London: MacDonald, 1970.

Delatte, Louis, ed. *Textes latins et vieux français relatifs aux Cyranides.* Bibliothèque de la Faculté de Philosophie de l'Université de Liege, Fasc. 93. Paris: Librairie E. Droz, 1942.

Drake, Stillman, trans. *Discoveries and Opinions of Galileo.* New York: Anchor Books, 1957.

Drake, Stillman, and I. E. Drabkin. *Mechanics in Sixteenth-Century Italy.* Madison: University of Wisconsin Press, 1969.

Dryander, Johannes. *Artzenei Speigel gemeyner Inhalt derselbigen, wes bede einem Leib unnd Wundtarzt, in der Theoric, Practic unnd Chirurgei zusteht.* Frankfurt am Main, 1547.

Durling, Richard. "Conrad Gesner's *Liber amicorum*, 1555–1565." *Gesnerus* 22 (1965): 134–59.

Edelstein, Sidney M. "The Allerley Matkel (1532). Facsimile Text, Translation, and Critical Study of the Earliest Printed Book on Spot Removing and Dyeing." *Technology and Culture* 5 (1964): 297–321.

Eis, Gerhard, ed. *Gottfrieds Pelzbuch. Studien zur Reichweite und Dauer der Wirkung des mittelhochdeutschen Fachschriftums.* Südosteuropäische Arbeiten, 38. Munich: Callwey, 1944.

Elyot, Sir Thomas. *The Boke Named the Governour.* Edited by H.H.S. Croft. 2 vols. London: Kegan Paul, Trench and Co., 1883.

Evelyn, John. *The Diary of John Evelyn.* Edited by E. S. De Beer. 6 vols. Oxford: Clarendon Press, 1955.

———. *The Diary and Correspondence of John Evelyn.* Edited by William Bray. London: Charles Scribner's Sons, 1903.

Falloppio, Gabriele, Ps.-. *Secreti diversi et miracolosi: ne' quali si mostra la via facile di risanare tutte le infirmità del corpo humano; et etiandio s'insegna il modo di fare molte altre cose, che à ciascuno sono veramente necessarie.* Venice, 1563.

Farrington, Benjamin. *The Philosophy of Bacon: An Essay on Its Development from 1603 to 1609.* Chicago: University of Chicago Press, 1964.

Fedele, Domenico, detto il Mantoianino. *Con il Poco farete Assai.* Rome, n.d.; Bologna, n.d.

Fioravanti, Leonardo. *Capricci medicinali.* Venice, 1561; Venice 1617.

———. *La cirurgia.* Venice, 1570.

———. *Compendio dei secreti rationali.* Venice, 1564.

———. *Compendio di tutta la cirurgia* (see Rostino, Pietro).

———. *A Compendium of the rationall secrets, of the Worthie Knight and most excellent Doctour of Phisicke and Chirurgerie, Leonardo Phioravante Bolognese.* Translated by John Hester. London, 1582.

———. *Del compendio de i secreti rationali.* Venice, 1564; Venice 1660.

———. *Della fisica.* Venice, 1582.

———. *Dello specchio di scientia universale.* Venice, 1567.

———. *Del regimento della peste.* Venice, 1565; Venice, 1571.

———. *A Joyfull Jewell. Contayning as well such excellent orders, preservatives and precious practises for the Plague, as also such mervelous Medicins for divers maladies, as hitherto have not been published in the English tung.* Translated by John Hester. London, 1579.

———. *Secreti medicinali.* Venice, 1561.

———. *A short discours of . . . Leonardo Phioravanti Bolognese uppon chirurgerie.* Translated by John Hester. London, 1580.

———. *Il tesoro della vita humana.* Venice, 1570; Venice, 1582.

Firpo, Luigi, ed. *Il supplizio di Tommaso Campanella.* Rome: Salerno, 1985.

Fontana, Andrea. *Fontana dove n'esce fuori acqua di secreti.* Genoa, 1579; Venice, n.d.

Francesco detto il Biscottino. *Giardino di varii secreti, havuti da diversi signori, dove si contengono varie sorti di giuochi, secreti, e burle.* Milan, n.d.

Francolino, Tomaso da, detto l'Ortolano. *Tesoro di segreti naturali.* Rome, 1617.

French, John. *The Art of Distillation*. 2d ed. London, 1653.

Frencken, Herman G. T., ed. *T Bouck vā Wondre 1513*. Roermand: H. Timmermans, 1934.

Friedrich, Hans-Viet, ed. *Thessalos von Tralles*. Meisenheim am Glan: Anton Hain, 1968.

Furfaro, Domenico, ed. *La vita e l'opera di Leonardo Fioravanti*. Bologna, 1963.

Gabrieli, Giuseppe, ed. *Il carteggio Linceo della vecchia accademia di Federico Cesi (1603–1630)*. Atti della Reale Accademia Nazionale dei Lincei. Classe di science morali, storiche e filologiche, ser. 6, vol. 7. Rome: G. Bardi, 1938–1941.

Galasso, Mario, della gratia di S. Paolo. *Novo recettario ilquale, e intitolato Thesoro de poveri*. Milan, 1584.

Galen of Pergamon. *Medicinalis experimentatio*. In *Opera*, pp. 101–8. Venice, 1609.

Galilei, Galileo. *Sidereus Nuncius or The Sidereal Messenger*. Translated by Albert Van Helden. Chicago: University of Chicago Press, 1989.

———. *Two New Sciences*. Translated by Stillman Drake. Madison: University of Wisconsin Press, 1974.

Ganszyniec, Ryszard, ed. *Brata Mikołaja z Polski. Pisma Lekarskie*. Prace Naukowe Uniwersytetu Poznańskiego Sekcja Humanistyczna, Nr. 2. Poznań, 1920.

Garzoni, Tommaso. *The Hospital of incurable fooles*. London, 1600.

———. *Opere*. Edited by Paolo Cherchi. Naples: Fulvio Rossi, 1972.

———. *La Piazza universale di tutte le professioni del mondo*. Venice, 1588.

Gassendi, Pierre. *The Mirrour of true Nobility & Gentility*. Translated by William Rand. London, 1657.

———. *Opera omnia*. 6 vols. Stuttgart: Friedrich Frommann Verlag, 1964; orig. publ. 1658.

———. *The Selected Works of Pierre Gassendi*. Edited and translated by Craig B. Bruch. New York: Johnson Reprint Corp., 1972.

Germerio, Giulielmo, Tolosano. *Gioia preciosa . . . Opera à chi brama la sanità utilissima & necessaria*. Venice, 1604.

Gersdorff, Hans von. *Feldtbuch der Wundtartzney*. Strasbourg, 1517.

Gesner, Conrad. *The Newe Jewell of Health*. Translated by George Baker. London, 1576.

Ghilini, Girolamo. *Teatro d'huomini letterati*. Venice, 1647.

Giannone, Pietro. *The Civil History of the Kingdom of Naples*. Translated by James Ogilvie. 2 vols. London, 1729–1731.

Gilbert, William. *De magnete*. London, 1600.

Glanvill, Joseph. *Essays on Several Important Subjects in Philosophy and Religion*. London, 1676.

———. *Scepsis scientifica*. 1655. New York: Garland Press, 1978.

———. *The Vanity of Dogmatizing: or Confidence in Opinions Manifested in a Discourse of the Shortness and Uncertainty of Our Knowledge*. London, 1661.

Glauber, John Rudolph. *The Works of the Highly Experienced and Famous Chymist, John Rudolph Glauber*. Translated by Christopher Pack. London, 1689.

Grant, Edward. *A Sourcebook in Medieval Science.* Cambridge: Harvard University Press, 1974.

Gundisalvo, Domingo. *De divisione philosophiae.* Edited by Ludwig Baur. Beiträge zur Geschichte der Philosophie des Mittelalters, Band IV, Heft 2–3. Münster: Aschendorffschen Buchhandlung, 1903.

Halleux, Robert, ed. *Les alchimistes grecs.* Vol. 1. Paris: "Les Belles Lettres," 1981.

Hansen, Bert, ed. *Nicole Oresme and the Marvels of Nature: A Study of His "De causis mirabilium" with Critical Edition, Translation, and Commentary.* Toronto: Pontifical Institute of Mediaeval Studies, 1985.

Hassenstein, Wilhelm. *Das Feuerwerkbuch von 1420.* Munich: Verlag der deutschen Technik, 1941.

Hessels, John Henry, ed. *Abrahami Ortelii . . . Epistulae.* Cambridge: Typis Academiae, 1887.

Hester, John. *The first (and second) part of the Key of Philosophy.* London, 1596.

———. *One hundred and foureteene experiments and cures of . . . Theophrastus Paracelsus.* London, 1596.

———. *The pearle of practise; or, Practicers pearle, for phisicke and chirurgerie.* London, 1594.

———. *These Oiles, Waters, Extractions, or Essences, Saltes, and Other Compositions; are at Paules wharfe ready made to be solde.* London, 1585.

Hildebrand, Wolfgang. *Neu augirte, weilverbesserte un vielvermehrte Magia naturalis.* Erfurt, 1664.

Hill, Thomas. *A briefe and pleasaunt treatise, entituled, Natural and Artificiall conclusions.* London, 1581.

Hooke, Robert. *Micrographia: or Some Physiological Descriptions of Minute Bodies Made by Magnifying Glasses.* New York: Dover Publications, 1961; orig. publ. 1665.

———. *The Posthumous Works of Robert Hooke.* Edited by Richard Waller. New York: Johnson Reprint Corp., 1969.

Hugh of St. Victor. *The Didascalicon of Hugh of St. Victor.* Translated by Jerome Taylor. New York: Columbia University Press, 1961.

Johannsen, Otto, ed. *Peder Månssons Schriften über technische Chemie und Hüttenwesen: Eine Quelle zur Geschichte der Technik des Mittelalters.* Berlin: VDI-Verlag, 1941.

Johnson, Thomas. *Cornucopiæ, or divers secrets: Werein is contained the rare secrets in Man, Beasts, Foules, Fishes, Trees, Plantes, Stones and such like.* London, 1594.

———. *Dainty Conceits, with a number of rare and witty inventions, never before printed.* London, 1630.

———. *A new Booke of new Conceits, with a number of novelties annexed thereunto. Whereof some be profitable, some necessary, some strange, none hurtful, and all delectable.* London, 1630.

Johnsson, J.W.S. "Les 'Experimenta duodecim Johannis Paulini.'" *Bulletin de la société française d'histoire de la medicine* 12 (1913): 257–67.

———. "Les 'Experimenta Magistri Nicolai.'" *Bulletin de la société française d'histoire de la medicine* 10 (1911): 269–90.

Jörimann, Julius, ed. *Frühmittelalterliche Rezeptarien.* Zurich and Leipzig: Orell Füssli, 1925.

Joubert, Laurent. *Popular Errors.* Translated by Gregory David de Rocher. Tuscaloosa: University of Alabama Press, 1989.

Kelchner, Ernst. and Richard Wülcker. *Mess-Memorial des Frankfurter Buchhändlers Michael Harder. Fastenmesse 1569.* Frankfurt am Main: Joseph Laer, 1873.

Kertzenmacher, Petrus. *Alchimi und Bergwerk. Wie alle farben wasser, olea, salia, und alumina, damit man alle corpora, spiritus und calces preparirt, sublimirt, und fixirt, gemacht sollen werden.* Strasbourg, 1534.

Kunstbüchlein, gerechten gründlichen gebrauch aller kunstbaren Werckleut. Frankfurt am Main, 1535.

Kyeser, Conrad. *Bellifortis.* Edited by Götz Quarg. 2 vols. Dusseldorf: Verlag des Vereins Deutscher Ingenieure, 1967.

Lactantius, *Divinae institutiones.* In *Opera omnia,* ed. Samuel Brandt. Corpus scriptorum ecclesiasticorum latinorum, 19 (1890).

Lauro, Vincentio. *Primo giardino, nel quale si contiene Secreti di grande esperienza, & virtù, à beneficio di corpi humanani.* Pavia, 1591.

———. *Scelta di varii secreti, li quali sono da V.L. stati esperimentati in molte persone.* Milan, 1603.

———. *Il Tesoro nel quale si contiene molti secreti de grandissime virtu a beneficio de corpi humani.* Modena, n.d.

Leandro, Lorenzo. *Tesoro di varij Secreti . . . Esperimentati con grandissima fatica, e spesa di tempo, e di danari.* Venice, 1614.

Lehmann-Haupt, Hellmut, ed. *The Göttingen Model Book.* Columbia: University of Missouri Press, 1972.

Lemay, Helen Rodnite, trans. *Women's Secrets: A Translation of Pseudo-Albertus Magnus's "De Secretis Mulierum" with Commentaries.* Albany: State University of New York Press, 1992.

Lemnius, Levinus. *Occulta naturae miracula.* Antwerp, 1559.

———. *The Secret Miracles of Nature.* London, 1658.

Le Roy, Louis. *Of the Interchangeable Course, or Variety of Things in the Whole World.* Translated by R. A. London, 1594.

London, William. *A catalogue of the most vendable books in England.* London, 1657.

Luck, Georg, ed. and trans. *Arcana mundi: Magic and the Occult in the Greek and Roman Worlds.* Baltimore: Johns Hopkins University Press, 1985.

Lupton, Thomas. *A thousand notable things of sundrie sorts.* London, 1601; first ed. 1579.

McAllister, Joseph Bernard, trans. *The Letter of Saint Thomas Aquinas De Occultis Operibus Naturae ad Quemdam Militem Ultramontanum.* Washington: Catholic University Press, 1939.

Machiavelli, Niccolò. *The Prince.* Translated by George Bull. Harmondsworth: Penguin Books, 1961.

MacKinney, Loren C. "An Unpublished Treatise on Medicine and Magic from the Age of Charlemagne." *Speculum* 18 (1943): 494–96.

Magghi, Giovanni. *Bichierografia libri quattro.* Edited by Paola Barocchi. Florence: Studio per Edizioni Sulu, 1977.

Maiorini, Tomaso, detto Policinella. *Frutti soavi colti nel giardino della virtu, cioè, trenta secreti bellissimi, con una regola per sapere tutto il tempo dell anno.* Bologna, 1642.

Manzalaoui, M. A., ed. *Secretum secretorum: Nine English Versions.* Vol. 1: Text. Early English Text Society, no. 276. Oxford: Oxford University Press, 1977.

Marbode of Rennes. *De lapidibus (1035–1123).* Edited by John Riddle. *Sudhoffs Archiv,* Beiheft 20. Wiesbaden: Franz Steiner, 1977.

Marcellus Empiricus (Marcellus of Bordeaux). *De medicamentis liber.* Edited by Max Niedermann. 2d ed. Corpus Medicorum Latinorum, no. 5. Berlin: Akademische Verlag, 1968.

Mascall, Leonard. *Booke of the art and manner how to plant and graffe all sorts of trees.* London, 1572.

———. *A profitable boke declaring dyvers approved remedies, to take out spottes and staines.* London, 1583.

Mercurio, Scipione. *De gli errori populari d'Italia.* Verona, 1645.

Merrifield, Mary P., ed. and trans. *Original Treatises Dating from the XII to the XVIII Centuries, on the Arts of Painting.* 2 vols. London: J. Murray, 1849.

Miccio, Scipione. "Vita di Don Pietro di Toledo." In *Narrazioni e documenti sulla storia del Regno di Napoli dall'anno 1522 al 1667,* ed. Francesco Palermo. *Archivio storico italiano* 9 (1846): 1–89.

Mizauld, Antoine. *Memorabilium aliquot naturae arcanorum silvula, rerum variarum sympathias, & antipathias, seu naturales concordias & discordias, libellis duobus complectens.* Frankfurt, 1592.

Montaigne, Michel. *The Complete Essays of Montaigne.* Translated by Donald M. Frame. Stanford: Stanford University Press, 1958.

Monte, Ludovico bolognese, della Chitarriglia. *La ghirlandetta fiorita di varij secreti belissimi.* Naples, 1634.

Moryson, Fynes. *Fynes Moryson's Itinerary.* Edited by Charles Hughes. London: Sheratt & Hughes, 1903.

Mutii, Pietro Maria, detto il Zanni bolognese. *Nuovo lucidario de secreti.* Bologna, 1585.

Nau, F., ed. "Un ancienne traduction latine du Bélinous Arabe (Appolonius de Tyane)." *Revue de l'orient Chrétien* 12 (1907): 99–106.

Oldenburg, Henry. *The Correspondence of Henry Oldenburg.* Edited by A. R. Hall and Marie Boas Hall. 11 vols. Madison: University of Wisconsin Press; London: Mansell, 1965–1977.

Opera nova intitolata Dificio di ricette, nella quale si contengono tre utilissimi Ricettari. Venice, 1529.

Opera nova nella quale ritroverai molti belissimi secreti novamente ritrovati per molti eccellentissimi medici. Venice, 1553.

Ottaviano, Carmelo, ed. *Un brano inedito della "Philosophia" di Guglielmo di Conches.* Naples: Morano, 1935.

Pagliarizzo, Dottor Gratiano. *Secreti nuovi e rari, et approvati per salute delli corpi humani.* Bologna, n.d.; Milan, n.d.

Palissy, Bernard. *The Admirable Discourses of Bernard Palissy.* Translated by Aurèle La Rocque. Urbana: University of Illinois Press, 1957.

Paracelsus. *A hundred and foureteene Experiments and cures of . . . Theophrastus Paracelsus.* Translated by John Hester. London, 1596.

————. *Selected Writings.* Translated by Norbert Guterman. 2d ed. Princeton: Princeton University Press, 1969; orig. publ. 1958.

Paré, Ambroise. *The Apology and Treatise of Ambroise Paré.* Edited by Geoffrey Keynes. New York: Dover Publications, 1968; orig. publ. 1952.

————. *On Monsters and Marvels.* Translated by Janis L. Pallister. Chicago: University of Chicago Press, 1982.

Partridge, John. *The treasurie of commodius Conceites, and hidden Secretes.* London, 1584.

————. *The Widdowes treasure.* London, 1655.

Peacham, John. *The Compleat Gentleman.* London, 1622.

Pesarino, Il gran Giocator di mano. *Secreti utilissimi e nuovi.* Verona, n.d.

Petty, William. *The Advice of W.P. to Mr. Samuel Hartlib for the Advancement of Some Particular Parts of Learning.* London, 1647.

————. *The Petty Papers: Some Unpublished Writings of Sir William Petty.* Edited by the marquis of Lansdowne. New York: Augustus M. Kelly, 1967; orig. publ. 1927.

Phillipps, Thomas. "Letter . . . Communicating a Transcript of a MS Treatise on the Preparation of Pigments, and on Various Processes of the Decorative Arts Practised during the Middle Ages, Written in the Twelfth Century, and Entitled Mappae Clavicula." *Archeologia* 32 (1847): 183–244.

Philostratus. *Life of Apollonius of Tyana.* Translated by F. C. Conyears. 2 vols. Loeb Classical Library. Cambridge: Harvard University Press, 1969.

Piemontese, Alessio. *De' secreti parte prima [-terza] Di nuovo riveduta, corretta, & ampliata in più loughi.* Venice, 1560.

————. *De secretis libri sex.* Translated by J. Wecker. Basil, 1559.

————. *Kunstbüch des wolerfarnen Herren Alexii Pedemontani von mancherley nutzlichen undd bewerten Secreten oder Künsten.* Translated by J. Wecker. N.p., 1573.

————. *The second part of the Secretes of Maister Alexis of Piemont.* Translated by William Warde. London, 1563.

————. *The Secretes of the reverende maister Alexis of Piemount.* Translated by William Warde. London, 1558.

————. *I Secreti del reverendo donno Alessio piemontese.* Venice, 1555.

————. *Les secrets de reverend signeur Alexis Piemontois.* Antwerp, 1557.

————. *The thyrde and last parte of the Secretes of the reverende Maister Alexis of Piemont.* Translated by William Warde. London, 1566.

————. *A very excellent and profitable Booke . . . of the expert and Reverend Mayster Alexis, which he termed the fourth and finall booke of his secretes.* Translated by Richard Androse. London, 1569.

Pingree, David, ed. *Picatrix: The Latin Version of the "Ghāyat Al-Hakīm."* London: Warburg Institute, 1986.

Plat, Hugh. *A briefe apologie of certain newe inventions.* London, 1593.

————. *Delights for ladies, to adorne their persons, tables, closets, and distillatories.* London, 1602.

————. *Floraes Paradise, beautified with sundry sorts of delicate fruites and flowers.* London, 1608.

Plat, Hugh. *The Jewell House of Art and Nature. Conteining divers rare and profitable Inventions, together with sundry new experimentes in the Art of Husbandry, Distillation, and Moulding.* London, 1594, 1653.

Plattes, Gabriel. *A Discovery of Infinite Treasure, hidden since the worlds beginning.* London, 1639.

―――. *A Discovery of Subterraneall Treasure, viz. Of all manner of Mines and Mineralls, from the Gold to the Coale; with plaine Directions and Rules for the finding of them in all Kingdomes and Countries.* London, 1639.

―――. *The profitable Intelligencer, Communicating his Knowledge for the Generall good of the Comon-wealth and all Posterity.* London, 1644.

Pliny [the Elder]. *Natural History.* Translated by H. Rackham, W.H.S. Jones, and D. E. Eichholz. 10 vols. Loeb Classical Library. Cambridge: Harvard University Press, 1938–1962.

Plutarch. *De curiositate.* In *Moralia*, 6:469–523. Translated by W. C. Helmbold. Loeb Classical Library. Cambridge: Harvard University Press, 1962.

Porta, Giambattista Della. *Criptologia.* Edited by Gabriella Belloni. Edizione Nazionale di Classici del Pensiero Italiano, ser. 2, 37. Rome: Centro Internazionale di Studi Umanistici, 1982.

―――. *De aëris transmutationibus libri IV.* Rome, 1608.

―――. *De distillatione libri IX.* Rome, 1608.

―――. *De i miracoli et maravigliosi effetti dalla natura prodotti.* Venice, 1560.

―――. *Della chirofisonomia; overo, di quella parte della humana fisonomia, che si appartiene alla mano libri due.* Translated by Pompeo Sarnelli. Naples, 1677.

―――. "Della taumatologia" (see Paparelli, G.).

―――. *Gli duoi fratelli rivalli.* Edited and translated by Louise George Clubb. Berkeley and Los Angeles: University of California Press, 1980.

―――. *Magiae naturalis libri viginti.* Frankfurt, 1591; orig. publ. 1589.

―――. *Magiae naturalis, sive de miraculis rerum naturalium libri IIII.* Antwerp, 1560; orig. publ. 1558.

―――. *Natural Magick.* New York: Basic Books, 1957; orig. publ. 1658.

Power, Henry. *Experimental Philosophy.* London, 1664.

Prager, Frank D., and Guistina Scaglia, eds. *Mariano Taccola and His Book "De Ingeneis."* Cambridge: MIT Press, 1972.

Prelowski, Ruth, trans. "The 'Beneficio di Christo.'" In *Italian Reformation Studies in Honor of Laelius Socinus*, ed. John A. Tedeschi, pp. 197–214. Florence: Monnier, 1965.

Primrose, James. *Popular Errours. Or the Errours of the People in Physick.* Translated by Robert Wittie. London, 1651.

Raleigh, Sir Walter. *The Works of Sir Walter Ralegh, Kt.* 8 vols. New York: Burt Franklin, 1965; orig. publ. 1829.

Rechter Gebrauch d'Alchimei. Mit vil bissher verborgenen nutzbaren unnd lustigen Künsten, Nit allein den für witzigen Alchimismisten [sic] Sonder allen kunstbaren Werckleutten in und ausserhalb feuer. Auch kunst aller menglichen inn vil wege zugebrauchen. Frankfurt am Main, 1531.

Ringelbergius, Joachim. *Opera.* Lyon, 1531.

Ritter, Hellmut, and Martin Plessner, trans. *"Picatrix." Das Ziel des Weisen von Pseudo-Magrītī*. London: Warburg Institute, 1962.

Rosetti, Giovanventura. *Notandissimi secreti de l'arte profumatoria*. Venice, 1555. Venice: Neri Pozza, 1973.

———. *The Plictho*. 1548. Translated by Sidney M. Edelstein and Hector C. Borghetty. Cambridge: MIT Press, 1969.

Rossello, Timotheo. *Della summa de' secreti universali in ogni materia*. Venice, 1559.

———. *La seconda parte de secreti universali in ogni materia*. Venice, 1574.

Rostino, Pietro. *Compendio di tutta la cirurgia . . . Di nuovo ristampato, & dall'eccellente m. Leonardo Fieravanti . . . ampliato, & aggiontovi un nuovo trattato*. Venice, 1561.

Rowbottom, Margaret E. "The Earliest Published Writing of Robert Boyle." *Annals of Science* 6 (1948/50): 376–87.

Ruscelli, Girolamo. *Precetti della militia moderna*. Venice, 1568.

———. *Secreti nuovi di maravigliosa virtù*. Venice, 1567.

Ryff, Walther Hermann. *Confectbuch unnd Hausz-Apoteck, künstlich zubereiten, einmachen, und gebrauchen, wes in ordenlichen Apotecken und Hauszhaltungen zur Artznei, täglicher Notturfft, und auch zum Lust, dienlich unnd nütz*. Frankfurt am Main, 1571.

———. *Die gross Chirurgei, oder volkommene Wundarznei. Chirurgischen Handtwirckung eigentlicher Bericht, und Inhalt alles so der Wundartznei angehörig*. Frankfurt am Main, 1545.

———. *Kurtz Handbüchlein und Experiment vieler Artzneien durch den gantzen Cörper des Menschens*. Frankfurt am Main, 1560.

———. *Omnium humani corporis partium descriptio*. Facs. of 1541 ed. Basel: Sandoz, 1954.

———. *Prakticir Büchlein der Leibartzeney. Wie man in allen Krankheiten, und Leibs Gebrechen, durch bewert Artznei, heylen und helffen sol*. Frankfurt am Main, 1541.

———. *Reformierte deütsche Apoteck darinnen eigentliche Contrafactur der fürnembsten und gebreüchlichsten Kreütter, sampt jhrer Underscheidung, Art, Natur, Krafft, Vermögen unnd Würckung*. Strasbourg, 1573.

———. *Stat und Feldbüch bewerter Wundtartznei*. Frankfurt am Main, 1551.

———, trans. *Ein newer Albertus Magnus, Darin durch sechs kurtzer Büchlein vil haimligkeyten der Natur beschriben*. Strasbourg, 1545.

———, trans. *Zehen Bücher von der Architectur und künstlichen Bauen*. Facs. Edited by E. Forssman. Hildesheim: Georg Olms, 1973.

Saccardino, Costantino. *Libro nominato la varità di diverse cose, quale minutamente tratta di molte salutifere operationi spagiriche, et chimiche; con alcuni veri discorsi delle cagioni delle lunghe infirmità, & come si devono sanar con brevità, & altri utili ragionamenti, quali scuoprono molti inganni, che per interesse spesso, tanto nella Medicina, quanto nelle materie medicinali, intervengano, con le virtù Elementari; Et altri notabili Filosofichi buono pareai, à beneficio universale*. Bologna, 1621.

Scatolone, Dottor Graziano. *Il vero, et pretioso thesoro di sanità*. Milan, n.d.

Scott, Walter, ed. and trans. *Hermetica: The Ancient Greek and Latin Writings*

Which Contain Religious or Philosophical Teachings Ascribed to Hermes Trismegistus. 4 vols. London: Dawson, 1968.

Shadwell, Thomas. *The Virtuoso.* Edited by M. J. Nicholson and D. S. Rodes. Lincoln: University of Nebraska Press, 1966.

Sigerist, Henry E. *Studien und Texte zur frühmittelalterlichen Rezeptliteratur.* Leipzig: Barth, 1923.

Silverstein, Theodore, ed. "Liber Hermetis mercurii triplici de vi rerum principiis." *Archives d'histoire doctrinale et littéraire du moyen age* 22 (1955): 217–302.

Silvestre de Sacy, A. I. "Livre du secret de la créature par le sage Bélinous." *Notices et extraits des manuscripts de la Bibliothèque Nationale et des autres bibliothèques* 4 (1799): 107–58.

Smith, Cyril Stanley. *Sources for the History of the Science of Steel, 1532–1786.* Cambridge: MIT Press, 1968.

Smith, Cyril Stanley, and J. G. Hawthorne, ed. and trans. *Mappae clavicula: A Little Key to the World of Medieval Techniques.* Transactions of the American Philosophical Society, new ser., vol. 64, pt. 4. Philadelphia: American Philosophical Society, 1974.

Sprat, Thomas. *History of the Royal Society of London.* Edited by J. I. Cope and H. W. Jones. London: Routledge and Kegan Paul, 1959; orig. publ. 1657.

The Strasburg Manuscript: A Medieval Painter's Handbook. Translated by Viola and Rosamund Borradaile. New York: Transatlantic Arts, 1976.

Stubbe, Henry. *Legends no Histories: Or, a Specimen of some Animadversions Upon the History of the Royal Society.* London, 1670.

Sudhoff, Karl. "Antipocras, Streitschrift für mystische Heilkunde in Versen des Magisters Nikolaus von Polen." *Sudhoffs Archiv* 9 (1916): 31–52.

Tasso, Torquato. *Dialogues: A Selection with the Discourse on the Art of the Dialogue.* Translated by Carnes Lord and Dain A. Trafton. Berkeley and Los Angeles: University of California Press, 1982.

Taviani, Ferdinando, ed. *La commedia dell'arte e la società barocca. La fascinazione del teatro.* La Commedia dell'arte: Storia e testi documenti, vol. 1. Rome: Bulzoni, 1970.

Telesio, Bernardino. *De rerum natura iuxta propria principia libri IX.* Facs. of 1586 ed. Hildesheim: Georg Olms Verlag, 1971.

Theophilus Presbyter. *De diversis artibus.* Edited and translated by C. R. Dodwell. London and Edinburgh: Thomas Nelson, 1961.

———. *On Divers Arts.* Translated by John G. Hawthorne and Cyril Stanley Smith. New York: Dover Publications, 1974; orig. publ. 1963.

Thompson, Daniel V. "The *De clarea* of the So-Called 'Anonymous Bernensis.'" *Technical Studies in the Fine Arts* 1 (1932): 8–19, 69–81.

Trabia, Giacomo. *L'Esperienza vincitrice, epilogata in diversi secreti, cavati da quelle erbe, che alcuno non ne fa stima, esperimentati da me.* Genoa, n.d.

Van Etten, Heinrich. *Mathematical Recreations.* London, 1633.

Varchi, Benedetto. *Questione sull' alchimia.* Edited by D. Moreni. Florence: Magheri, 1827.

Vittiario, Giovanni, detto il Tramontano. *Centuria seconda de' secreti materiali, medicinal', e curiosi.* Viterbo, 1618.

Von Stahel vnd Eysen. Wie man dieselbigen künstlich weych vnd hert machen sol, Allen Waffen Schmiden, Gold schmiden, Gürtlern, Sigil vnd Stempffel schneidern, sampt allen andern kunstbaren werckleuten, so mit Stahel vnd Eysen, ire arbeyts ubung treyben, Eynem yeden nach gelegenheyt zu geprauchen, fast nützlich zu wissen. Nuremberg, 1532. Reprint. Erfurt, 1532; Mainz, 1532, 1534; Strasbourg, 1539.

Wardale, Walter. "A Low German-Latin Miscellany of the Fourteenth Century." *Niederdeutsche Mitteilungen* 8 (1952): 5–22.

Webster, Charles, ed. *Samuel Hartlib and the Avancement of Learning.* Cambridge: Cambridge University Press, 1970.

Wecker, Johann Jacob. *De secretis libri XVII.* Basel, 1582.

———. *The Secrets of Art & Nature, being The Summe and Substance of Naturall Philosophy.* Translated by R. Read. London, 1660.

Welborn, Mary Catherine. "The errors of the doctors according to Friar Roger Bacon of the Minor Order." *Isis* 18 (1932–1933): 36–62.

White, John. *Hocus-Pocus, or a Rich Cabinet of Legerdemain Curiosities.* London, n.d.

———. *A Rich Cabinet, With Variety of Inventions, unlock'd and opened, for the recreation of Ingenious Spirits at their vacant hours.* London, 1651.

Wier, Johann. *Witches, Devils, and Doctors in the Renaissance: Johann Weyer, De praestigiis daemonum.* Translated by John Shea. Binghamton, N.Y.: Center for Medieval and Renaissance Studies.

Wilkins, John. *Mathematicall Magick, or The wonders that may be performed by Mechanical Geometry.* London, 1648.

William of Auvergne. *De legibus.* In *Opera omnia*, vol. 1. Paris, 1674. Frankfurt am Main: Minerva, 1963.

Williams, Hermann. "A Sixteenth-Century German Treatise *Von Stahel und Eysen*, Translated with Explanatory Notes." *Technical Studies in the Fine Arts* 4 (1935): 63–92.

Wippel, John F., and Allan B. Wolter, ed. and trans. *Medieval Philosophy: From St. Augustine to Nicholas of Cusa.* New York: Free Press, 1969.

Zambecarri, Giuseppe. "Experiments of Doctor Joseph Zambeccari concerning the Excision of Various Organs from Different Living Animals." Translated by Saul Jarcho. *Bulletin of the History of Medicine* 9 (1941): 311–31.

Zapata, Giovanni Battista. *Li Maravigliosi secreti di medicina e chirurgia, nuovamente ritrovati per guarire ogni sorte d'infermità.* Venice, 1577.

Secondary Sources

d'Addio, Mario. *Il pensiero politico di Gaspare Scioppio e il machiavellismo del seicento.* Milan: A. Giuffré, 1962.

Agrimi, Jole, and Chiara Crisciani. "Medici e 'vetulae' dal duecento al quattrocento: Problemi di un ricerca." In *Cultura popolare e cultura dotta nel seicento*, pp. 144–59.

Allen, Roland. "Gerbert, Pope Silvester II." *English Historical Review* 7 (1892): 625–68.

d'Alverny, Marie-Thérèse. "Survivance de la magie antique." In *Antike und Orient im Mittelalter*, ed. Paul Wilpert. Miscellanea Mediaevalia, Veröffentlichungen des Thomas-Instituts der Universität zu Köln, Band I, pp. 154–78. Berlin, New York: Walter De Gruyter, 1971.

d'Alverny, Marie-Thérèse. "Translations and Translators." In *Renaissance and Renewal in the Twelfth Century*, ed. Robert L. Benson and Giles Constable, pp. 421–62. Cambridge: Harvard University Press, 1982.

Amabile, Luigi. *Fra Tomaso Campanella, la sua congiura, i suoi processi e la sua pazzia*. 3 vols. Naples: Morano, 1882.

———. *Il Santo Officio della inquisizione in Napoli*. 2 vols. Città di Castello: S. Lapi, 1892.

Andrade, E. N. da C. "The Birth and Early Days of the *Philosophical Transactions*." *Notes and Records of the Royal Society of London* 20 (1965): 9–27.

———. "Newton's Early Notebook." *Nature* 135 (1935): 360.

Anglo, Sydney. "Melancholia and Witchcraft: The Debate between Wier, Bodin, and Scot." In *Folie et déraison à la Renaissance*, pp. 209–28. Colloque international de la Fédération Internationale des Institutes et Sociétés pour l'Étude de la Renaissance (November 1973). Brussels: Editions de l'Université de Bruxelles, 1976.

———, ed. *The Damned Art: Essays in the Literature of Witchcraft*. London, Henley, and Boston: Routledge and Kegan Paul, 1977.

Antonaci, Antonio. *Francesco Storella, Filosofo salentino del cinquecento*. Università di Bari, Publicazioni dell'Istituto di Filosofia, no. 9. Galentina: Salentina, 1966.

Aquilecchia, Giovanni. "Appunti su G. B. Della Porta e l'inquisizione." *Studi secenteschi* 9 (1968): 3–31.

Arbib, Michael A., and Mary B. Hesse. *The Construction of Reality*. Cambridge: Cambridge University Press, 1986.

Ashworth, William B., Jr. "Catholicism and Early Modern Science." In *God and Nature: Historical Essays on the Encounter between Christianity and Science*, ed. Lindberg and Numbers, pp. 136–66.

Atkinson, Geoffroy. *Les Nouveaux horizons de la Renaissance Française*. Paris: E. Droz, 1935.

Aubin, Hermann. "Formen und Verbreitung des Verlagsystem in der Altnürnberg Wirtschaft." *Beiträge zur Wirtschaftsgeschichte Nürnbergs* 2 (1967): 620–68.

Baader, Gerhard. "Medizinisches Reformdenken und Arabismus im Deutschland des 16. Jahrhunderts." *Sudhoffs Archiv* 63 (1979): 261–96.

Badaloni, Nicola. "Fermenti di vita intellettuale a Napoli dal 1500 alla metà del 600." In *Storia di Napoli: Cultura e letteratura*, 8:307–52. Naples: Edizioni scientifico Italiane, 1980.

———. "I fratelli Della Porta e la cultura magica e astrologica a Napoli nel '500." *Studi storici* 1 (1960): 677–715.

———. "Natura e società in Nicolò Franco." *Società* 16 (1960): 735–77.

Bagley, Paul J. "On the Practice of Esotericism." *Journal of the History of Ideas* 53 (1992): 231–47.

Bakhtin, Mikhail. *Rabelais and His World*. Translated by Helene Iswolsky. Cambridge: MIT Press, 1968.

Barb, A. A. "The Survival of Magic Arts." In *Conflict between Paganism and*

Christianity in the Fourth Century, ed. A. Momigliano, pp. 100–125. Oxford: Clarendon Press, 1963.

Barnouw, Jeffrey. "Hobbes's Psychology of Thought: Endeavors, Purpose and Curiosity." *History of European Ideas* 10 (1989): 519–45.

Baxandall, Michael A. *The Limewood Sculptors of Renaissance Germany*. New Haven: Yale University Press, 1980.

————. *Painting and Experience in Fifteenth-Century Italy: A Primer in the Social History of Pictorial Style*. London: Oxford University Press, 1974.

Baxter, Christopher. "Jean Bodin's *De la Démonomanie des Sorciers*: The Logic of Persecution." In *The Damned Art*, ed. Anglo, pp. 76–105.

————. "Johann Weyer's *De praestigiis daemonum*: Unsystematic Psychopathology." In *The Damned Art*, ed. Anglo, pp. 53–75.

Beaujouan, Guy. *L'interdépendence entre la science scholastique et les techniques utilitaire (XIIe, XIIIe, et XIVe siécles)*. Les Conférences du Palais de la Découverte, ser. D, no. 46. Paris, 1957.

Beddie, J. S. "The Ancient Classics in the Medieval Libraries." *Speculum* 5 (1930): 3–20.

Belkin, Johanna Schwind, and Earle Radcliffe Caley. "Introduction." In *Eucharius Rösslin the Younger "Of Minerals and Mineral Products": Chapters on Minerals from his "Kreutterbüch,"* pp. 1–50. Berlin: Walter de Gruyter, 1978.

Beltrami, Luca. *Vita di Aristotile da Bologna*. Bologna: Libreria Luigi Beltrami, 1912.

Ben-David, Joseph. *The Scientist's Role in Society: A Comparative Study*. Chicago: University of Chicago Press, 1971.

Benzing, Joseph. *Die Buchdrucker des 16. und 17. Jahrhunderts im deutschen Sprachgebiet*. Wiesbaden: Otto Harassowitz, 1963.

————. "Christian Egenolff und seine Verlagsproduktion." *Bösenblatt für Deutschen Buchhandel*, Frankfurt Ausgabe, 77 (1973): A348–A352.

————. "Christian Egenolff zu Frankfurt am Main von Ende 1530 bis 1555." *Das Antiquariat* 11 (1955): 139–40, 162–64, 201–2, 232–36.

————. "Christian Egenolff zu Strassburg und seine Drucke 1528–1530." *Das Antiquariat* 10 (1954): 88–89, 92.

————. *Die Drucke Jacob Cammerlanders zu Strassburg, 1531–1548*. Vienna: Walter Krieg, 1963.

————. "Hermann Gülfferich zu Frankfurt am Main und sein populärmedizinischer Verlag." *Das Antiquariat* 12 (1956): 129–33, 173–75.

————. "Walther H. Ryff und sein literarisches Werk. Eine Bibliographie." *Philobiblion* 2 (1958): 126–54, 203–26.

Bertelli, Sergio, Franco Cardini, and Elvira Garbero Zorzi, eds. *The Courts of the Italian Renaissance*. New York: Facts on File Publications, 1985.

Berti, Luciano. *Il Principe dello studiolo: Francesco I dei Medici e la fine del Rinascimento fiorentino*. Florence: EDAM, 1967.

Betz, Hans Dieter, ed. and trans. *The Greek Magical Papyri in Translation*. Chicago: University of Chicago Press, 1986.

Bezzel, Irmgard. "Die Erschließung von Schrifttum des 16. Jahrhunderts im 'Verzeichnis der im deutschen Sprachbereich erschienenen Drucke des 16. Jahrhunderts.'" *Aus dem Antiquariat* 3 (1978): A81–A85.

Biagioli, Mario. "Galileo's System of Patronage." *History of Science* 27 (1990): 1–62.

———. "Galileo the Emblem-Maker." *Isis* 81 (1990): 230–58.

———. "Scientific Revolution, Social Bricolage, and Etiquette." In *The Scientific Revolution in National Context*, ed. R. Porter and M. Teich, pp. 11–54. Cambridge: Cambridge University Press, 1992.

Bianchi, Massimo Luigi. *Signatura rerum. Segni, magia e conoscenza da Paracelso a Leibniz*. Lessico Intellettuale Europeo, 43. Rome: Ateneo, 1987.

Bibliotheca Palatina. Katalog zur Ausstellung vom 8. Juli bis 2. November 1986, Heiliggeistkirche Heidelberg. Edited by Elmar Mittlar. 2 vols. Heidelberg: Braus, 1986.

Bidez, Joseph, and Franz Cumont. *Les mages hellénisés. Zoroastre, Ostanès et Hystaspe d'après la tradition grecque*. 2 vols. Paris: Société d'Édition "Les Belles Lettres," 1973.

Birkenmajer, Alexander. "Zur Lebensgeschichte und wissenschaftliche Tätigkeit von Giovanni Fontana." *Isis* 17 (1932): 34–53.

Black, Max. *Models and Metaphors: Studies in Language and Philosophy*. Ithaca: Cornell University Press, 1962.

Blake, John B. "The Compleat Housewife." *Bulletin of the History of Medicine* 49 (1975): 30–42.

Blumenberg, Hans. *The Legitimacy of the Modern Age*. Translated by Robert M. Wallace. Cambridge: MIT Press, 1983.

Bober, Harry. Review of *The Göttingen Model Book*, ed. H. Lehmann-Haupt. *Speculum* 49 (1974): 354–58.

Bok, Sissela. *Secrets: On the Ethics of Concealment and Revelation*. New York: Vintage Books, 1983.

Bolgar, R. R. *The Classical Heritage and Its Beneficiaries*. Cambridge: Cambridge University Press, 1954.

Bono, James J. "Science, Discourse, and Literature. The Role/Rule of Metaphor in Science." In *Literature and Science: Theory and Practice*, ed. S. Peterfreund, pp. 59–89. Boston: Northeastern University Press, 1990.

Bonomo, Giuseppe. *Caccia alle streghe. La credenza nelle streghe al secolo XIII al XIX con particolare riferimento all'Italia*. Rome: Palumbo, 1959.

Borsook, Eve. "Art and Politics at the Medici Court. I: The Funeral of Cosimo I de' Medici." *Mitteilungen des Kunsthistorisches Institutes in Florenz* 12 (1965): 31–54.

———. "Art and Politics at the Medici Court. II: The Baptism of Filippo de'Medici." *Mitteilungen des Kunsthistorisches Institutes in Florenz* 13 (1967): 95–114.

Bossy, John. "The Counter-Reformation and the People of Catholic Europe." *Past and Present* 47 (1970): 51–70.

Bouwsma, William J. "The Secularization of Society in the Seventeenth Century." In *A Usable Past: Essays in European Cultural History*, pp. 112–28. Berkeley and Los Angeles: University of California Press, 1990.

Bowie, Ewan Lyall. "Apollonius of Tyana: Tradition and Reality." In *Aufsteig und Niedergang der römischen Welt*, II (Principat): vol. 16, pt. 2, pp. 1652–99. Berlin: De Gruyter, 1978.

Brandl, Rainer. "Art or Craft? Art and the Artist in Medieval Nuremberg." In *Gothic and Renaissance Art in Nuremberg, 1300–1550*, pp. 51–60.

Braudel, Fernand. *The Wheels of Commerce: Civilization and Capitalism, Fifteenth–Eighteenth Century*. Vol. 2. Translated by Siân Reynolds. New York: Harper & Row, 1982.

Brauer, Adalbert. "Frankfurt als Stätte deutschen Buchhandels im Laufe der Jahrhunderte." *Bösenblatt für den Deutschen Buchhandel* 29 (1973): A340–A347.

Brown, Alison. "Platonism in Fifteenth-Century Florence and Its Contribution to Early Modern Political Thought." *Journal of Modern History* 58 (1986): 383–413.

Brown, Harcourt. *Scientific Organizations in Seventeenth-Century France (1620–1680)*. Baltimore: Williams and Wilkins, 1934.

Brown, Peter. "Society and the Supernatural: A Medieval Change." *Daedalus* 184 (1975): 133–51.

Brunello, Franco. *The Art of Dyeing in the History of Mankind*. Translated by Bernard Hickey. Vicenza: Neri Pozza, 1973.

Brunello, Franco, and Franca Facchetti. "Notizie inedite su Giovanni Rosetti ed il suo 'Plictho' sull'arte della tintura." *Laniera* 77 (1963): 1019–23.

Bühler, Curt F. *The Fifteenth-Century Book: The Scribes, the Printers, the Decorators*. Philadelphia: University of Pennsylvania Press, 1960.

Bullough, Vern. "The Development of the Medical University at Montpellier to the End of the Fourteenth Century." *Bulletin of the History of Medicine* 30 (1956): 508–23.

Burke, Peter. *Popular Culture in Early Modern Europe*. New York: Harper & Row, 1978.

———. "A Question of Acculturation?" In *Scienze, credenze occulte, livelli di cultura*, pp. 197–204.

———. "Rituals of Healing in Early Modern Italy." In *The Historical Anthropology of Early Modern Italy: Essays on Perception and Communication*, pp. 207–20. Cambridge: Cambridge University Press, 1987.

Burkert, Walter. *Lore and Science in Ancient Pythagoreanism*. Cambridge: Harvard University Press, 1972.

Burnet, John. *Early Greek Philosophy*. 4th ed. Cleveland: World Publishing Co., 1957.

Burnett, Charles S. F. "Some Comments on the Translating of Works from Arabic into Latin in the Mid-Twelfth Century." *Miscellania Mediaevalia* 17 (1985): 161–71.

Cade, Joseph A. "Aspects of Secrecy in Science." *Impact of Science on Society* 21 (1971): 181–90.

Camporesi, Piero. "Cultura popolare e cultura d'élite fra Medioevo ed età moderna." In *Storia d'Italia*, Annali 4: Intellettuali e potere, ed. Corrado Viviani, pp. 81–157. Turin: Giulio Einaudi, 1981.

———. *The Incorruptible Flesh: Bodily Mutation and Mortification in Religion and Folklore*. Translated by Tania Croft-Murray. Cambridge: Cambridge University Press, 1988.

Caputo, Cosimo. "La struttura del segno fisiognomico. (G. B. Della Porta e l'universo culturale del Cinquecento)." *Il Protagora* 22 (1982): 63–102.

Carroll, Diane Lee. "Antique Metal-Joining Formulas in the *Mappae clavicula.*" *Transactions of the American Philosophical Society* 125, no. 2 (1981): 91–103.

Carutti, Domenico. *Breve storia della Accademia dei Lincei.* Rome: Salviucci, 1883.

Cary, George. *The Medieval Alexander.* Edited by D.J.A. Ross. Cambridge: Cambridge University Press, 1967.

Cassirer, Ernst. *The Individual and the Cosmos in Renaissance Philosophy.* Translated by Mario Domandi. Philadelphia: University of Pennsylvania Press, 1972.

Castellani, Carlo. "Ippolito Salviani." *DSB*, 17:89–90.

The Catholic Encyclopedia: An International Work of Reference on the Constitution, Doctrine, and History of the Catholic Church. Edited by Charles G. Herbermann et al. 17 vols. New York: Encyclopedia Press, 1913–1922.

Cavazza, Marta. "Bologna and the Royal Society in the Seventeenth Century." *Notes and Records of the Royal Society of London* 35 (1980): 105–23.

———. "Metafore venatorie e paradigmi indiziari nella fondazione della scienza sperimentale." *Annali dell'Istituto di discipline filosofiche dell'Università di Bologna* 1 (1980): 107–33.

———. *Settecento inquieto: Alle origini dell'Istituto delle Scienze di Bologna.* Bologna: Il Mulino, 1990.

———. "Verso la fondazione dell'*Istituto delle Scienze*: filosofia 'libera,' baconismo, religione a Bologna (1660–1714)." In *Sull'Identità del pensiero moderno: Studi e saggi*, pp. 97–146. Florence: La Nuova Italia, 1979.

Céard, Jean. *La nature et les prodiges: L'insolite au XVIe siécle, en France.* Geneva: Droz, 1977.

Cennini, Cennino d'Andrea. *The Craftsman's Handbook: The Italian "Il libro dell'arte."* Translated by Daniel V. Thompson, Jr. New York: Dover Publications, 1960; orig. publ. 1933.

Chartier, Roger. *The Cultural Uses of Print in Early Modern France.* Translated by Lydia G. Cochrane. Princeton: Princeton University Press, 1987.

———. "Culture as Appropriation: Popular Culture Uses in Early Modern France." In *Understanding Popular Culture: Europe from the Middle Ages to the Nineteenth Century*, ed. Steven L. Kaplan, pp. 229–53. Berlin, New York, Amsterdam: Mouton, 1984.

Chenu, M.-D. *Nature, Man, and Society in the Twelfth Century.* Translated by J. R. Taylor and L. K. Little. Chicago: University of Chicago Press, 1968.

Cherchi, Paolo. *Enciclopedismo e politica della rescrittura: Tommaso Garzoni.* Pisa: Pacini, 1980.

Chibnall, Marjorie. "Pliny's *Natural History* in the Middle Ages." In *Empire and Aftermath: Silver Latin II*, ed. T. A. Dorey, pp. 57–78. London: Routledge and Kegan Paul, 1975.

Chrisman, Miriam. *Lay Culture, Learned Culture: Books and Social Change in Strasbourg, 1480–1599.* New Haven: Yale University Press, 1982.

Church, Frederic C. *The Italian Reformers, 1534–1564.* New York: Columbia University Press, 1932.

Cipolla, Carlo. *Literacy and Development in the West.* London: Pelican Books, 1969.

Clagett, Marshall. *Greek Science in Antiquity.* New York: Collier Books, 1955.

————. "The Life and Career of Giovanni Fontana." *Annali dell'Istituto e Museo di Storia della Scienze di Firenze* 1 (1976): 5–28.

Clark, Stuart. "Inversion, Misrule, and the Meaning of Witchcraft." *Past and Present* 87 (1980): 98–127.

Clubb, Louise George. *Giambattista Della Porta, Dramatist.* Princeton: Princeton University Press, 1965.

Clucas, Stephan. "Samuel Hartlib's Ephemerides, 1653–59, and the Pursuit of Scientific and Philosophical Manuscripts: The Religious Ethos of an Intelligencer." *The Seventeenth Century* 6 (1991): 33–55.

Cochrane, Eric. "The Renaissance Academies in the Italian and European Setting." In *The Fairest Flower: The Emergence of Linguistic National Consciousness in Renaissance Europe*, pp. 21–39. Florence: Crusca, 1985.

Cohn, Norman. *Europe's Inner Demons: An Enquiry Inspired by the Great Witch-Hunt.* New York: Basic Books, 1975.

Coleman, William. "The People's Health: Medical Themes in Eighteenth-Century French Popular Literature." *Bulletin of the History of Medicine* 51 (1977): 55–77.

Colie, Rosalie L. "Cornelis Drebbel and Salomon De Caus: Two Jacobean Models for Salomon's House." *Huntington Library Quarterly* 18 (1954/55): 245–60.

Comparetti, Domenico. *Vergil in the Middle Ages.* Translated by E.F.M. Benecke. London: Allen and Unwin, 1966; orig. publ. 1908.

Cook, Harold J. *The Decline of the Old Medical Regime in Stuart London.* Ithaca: Cornell University Press, 1986.

Copenhaver, Brian. "Astrology and Magic." In *Cambridge History of Renaissance Philosophy*, ed. C. B. Schmitt and Q. Skinner, pp. 264–300. Cambridge: Cambridge University Press, 1988.

————. "Did Science Have a Renaissance?" *Isis* 83 (1992): 387–407.

————. "Natural Magic, Hermetism, and Occultism in Early Modern Science." In *Reappraisals of the Scientific Revolution*, ed. Lindberg and Westman, pp. 261–302.

————. "The Occultist Tradition in Seventeenth Century Philosophy." In *Cambridge History of Seventeenth-Century Philosophy*, ed. M. Ayers and D. Garber. Forthcoming.

————. "Scholastic Philosophy and Renaissance Magic in the *De vita* of Marsilio Ficino." *Renaissance Quarterly* 37 (1984): 523–54.

————. "A Tale of Two Fishes: Magical Objects in Natural History from Antiquity through the Scientific Revolution." *Journal of the History of Ideas* 52 (1991): 373–98.

Corsano, Antonio. "Per la storia del pensiero del tarda rinascimento: III. G. B. Della Porta." *Giornale critico della filosofia Italiana*, ser. 3, 38 (1959): 76–97.

Corsini, Andrea. *Medici ciarlatani e ciarlatani medici.* Bologna: N. Zanichelli, 1922.

Cournand, André F., and Harriet Zuckerman. "The Code of Science: Analysis and Some Reflections on Its Future." *Studium Generale* 23 (1970): 941–62.

Courtney, William J. "Nominalism in Late-Medieval Religion." In *The Pursuit of Holiness in Late Medieval and Renaissance Religion*, ed. Charles Trinkaus and Heiko Obermann, pp. 26–59. Leiden: Brill, 1974.

Covoni, P. F. *Don Antonio de' Medici al Casino di San Marco*. Florence: Tipografia Cooperativa, 1892.

Cressy, David. *Literacy and the Social Order: Reading and Writing in Tudor and Stuart England*. Cambridge: Cambridge University Press, 1980.

Crisciani, Chiara. "La 'Quaestio de alchimia' fra duecente e trecento." *Medioevo* 2 (1976): 119–68.

Crofts, Richard. "Books, Reform, and the Reformation." *Archiv für Reformationsgeschichte* 71 (1980): 21–35.

Crombie, Alistair C. *Robert Grosseteste and the Origins of Experimental Science, 1100–1700*. Oxford: Clarendon Press, 1958.

Crummer, Leroy. "Early Anatomical Fugitive Sheets." *Annals of Medical History* 5 (1923): 189–209.

Cultura popolare e cultura dotta nel seicento. Relazioni di P. Rossi et al. Milan: F. Angeli, 1983.

Cumont, Franz. "Écrits hermétiques. II. Le médicin Thessalos et les plantes astrales d'Hermès Trismégiste." *Revue de Philologie* 42 (1918): 85–108.

Cushing, Harvey. *A Bio-Bibliography of Andreas Vesalius*. New York: Schuman's, 1943.

Dal Fiume, Antonio. "Un medico astrologo a Verona nel '500: Tomasso Zefiriele Bovio." *Critica storia* 20 (1983): 32–59.

Daniel, Norman. *The Arabs and Mediaeval Europe*. London: Longman, 1975.

Darmstaedter, Ernst. *Berg- Probier- und Kunstbüchlein*. Münchener Beiträge zur Geschichte und Literatur der Naturwissenschaften und Medizin, Heft 2/3. Munich: Münchner Drucke, 1926.

———. "Eisenhartung mit Pflanzen." *Geschichstblätter für Technik, Industrie und Gewerbe* 11 (1927): 163–69.

Darnton, Robert. *The Literary Underground of the Old Regime*. Cambridge: Harvard University Press, 1982.

Daston, Lorraine. "Baconian Facts, Academic Civility, and the Prehistory of Objectivity." *Annals of Scholarship* 8 (1991): 337–63.

———. "The Factual Sensibility." *Isis* 79 (1988): 452–67.

———. "The Moral Economy of Science." *Osiris*. Forthcoming.

———. "Neugier und Naturwissenschaft in der frühen Neuzeit." In *Macrocosmus im Microcosmus: Die Welt in der Stube*. Forthcoming.

———. "Wunder, Naturgesetze und die wissenschaftliche Revolution des 17. Jahrhunderts." *Jahrbuch der Akademie der Wissenschaften in Göttingen* (1991), 99–122.

Davis, Natalie Zemon. *Society and Culture in Early Modern France*. Stanford: Stanford University Press, 1975.

Dear, Peter. "Jesuit Mathematical Science and the Reconstitution of Experience in the Early Seventeenth Century." *Studies in the History and Philosophy of Science* 18 (1987): 133–75.

———. "Miracles, Experiments, and the Ordinary Course of Nature." *Isis* 81 (1990): 663–83.

———. "*Totius in verba*: Rhetoric and Authority in the Early Royal Society." *Isis* 76 (1985): 145–61.

Debus, Allen G. *The Chemical Philosophy: Paracelsian Science and Medicine in the Sixteenth and Seventeenth Centuries.* 2 vols. New York: Science History Publications, 1977.

―――. *The English Paracelsians.* New York: Franklin Watts, 1965.

―――. "Palissy, Plat, and Agricultural Chemistry in the Sixteenth and Seventeenth Centuries." *Archives internationales d'histoire des sciences* 21 (1968): 67–87.

DeMaitre, Luke. *Doctor Bernard de Gordon: Professor and Practitioner.* Toronto: Pontifical Institute of Medieval Studies, 1980.

―――. "Scholasticism in Compendia of Practical Medicine, 1250–1450." *Manuscripta* 20 (1976): 81–95.

―――. "Theory and Practice in Medical Education at Montpellier." *Journal of the History of Medicine* 30 (1975): 103–23.

Demus, Otto. *Byzantine Art and the West.* New York: New York University Press, 1970.

De Rijk, l. M. "Some Notes on the Twelfth-Century Topic of the Three (Four) Human Evils and of Science, Virtue and Technique as Their Remedies." *Vivarium* 5 (1967): 8–15.

De Roover, Raymond. "The Business Organization of the Plantin Press in the Setting of Sixteenth-Century Antwerp." *De Gulden Passer* 24 (1956): 104–20.

Detienne, Marcel, and Jean-Pierre Vernant. *Cunning Intelligence in Greek Culture and Society.* Translated by Janet Lloyd. Chicago: University of Chicago Press, 1991.

Dickens, A. G., ed. *The Courts of Europe: Politics, Patronage and Royalty, 1400–1800.* London: Thames and Hudson, 1974.

Dietz, Alexander. *Zum Geschichte der Frankfurter Büchermesse, 1462–1792.* Frankfurt am Main: R. Th. Hauser, 1921.

Dilg, Peter. "Die 'Reformation der Apotecken' (1536) des Berner Stadtarztes Otto Brunfels." *Gesnerus* 36 (1979): 181–205.

―――. "Die Strassburger Apothekenordnung (um 1500)." *Pharmazeutische Zeitung* 124 (1979): 1681–87.

Dobbs, Betty Jo Teeter. "From the Secrets of Alchemy to the Openness of Chemistry." In *Solomon's House Revisited: The Organization and Institutionalization of Science,* ed. T. Frängsmyr, pp. 75–94. Canton, Mass.: Science History Publications, 1990.

―――. "Studies in the Natural Philosophy of Sir Kenelm Digby. Part III: Digby's Experimental Alchemy—The Book of *Secrets.*" *Ambix* 21 (1974): 1–28.

Dodds, E. R. *The Greeks and the Irrational.* Berkeley and Los Angeles: University of California Press, 1951.

―――. "Supernormal Phenomena in Classical Antiquity." In *The Ancient Concept of Progress,* pp. 158–210. Oxford: Clarendon Press, 1973.

Doran, Madeleine. "On Elizabethan 'Credulity.' With Some Questions concerning the Use of the Marvelous in Literature." *Journal of the History of Ideas* 1 (1940): 151–76.

Douglas, Mary. *Natural Symbols: Explorations in Cosmology.* New York: Vintage Books, 1973.

Douglas, Mary. *Purity and Danger: An Analysis of the Concept of Pollution and Taboo.* London: Routledge and Kegan Paul, 1966.

Dresbeck, LeRoy. "Techne, Labor, et Natura: Ideas and Active Life in the Medieval Winter." *Studies in Medieval and Renaissance History* 2 (1979): 83–119.

Driessen, L.-A. "Un Livre flamand de recettes de teinture de 1513 et la traduction Anglaise de 1583." *Revue générales des matières colorantes,* February 1934, 68–71.

Dufner, Georg. *Geschichte der Jesuaten.* Uomini e Dottrine, no. 21. Rome: Edizioni Storia e Letteratura, 1975.

Duveen, Denis. "Notes on Some Alchemical Books." *The Library,* 5th ser., 1 (1946): 56–61.

Eamon, William. "Court, Academy, and Printing House." In *Patronage and Institutions,* ed. Moran, pp. 25–50.

———. "New Light on Robert Boyle and the Discovery of Colour Indicators." *Ambix* 27 (1980): 204–9.

———. "Science and Popular Culture in Sixteenth-Century Italy: The 'Professors of Secrets' and Their Books." *Sixteenth Century Journal* 4 (1985): 471–85.

———. "The *Secreti* of Alessio Piemontese, 1555." *Res Publica Litterarum* 2 (1979): 43–55.

———. "Technology as Magic in the Late Middle Ages and the Renaissance." *Janus* 70 (1983): 171–212.

———. "'With the Rules of Life and an Enema': Leonardo Fioravanti's Medical Primitivism." In *Renaissance and Revolution: Humanists, Scholars, Craftsmen, and Natural Philosophers in Early Modern Europe,* ed. J. V. Field and F.A.J.L. James, pp. 29–44. Cambridge: Cambridge University Press, 1993.

Eamon, William, and Gundolf Keil. "'Plebs amat empirica': Nicholas of Poland and His Critique of the Medieval Medical Establishment." *Sudhoffs Archiv* 71 (1987): 180–96.

Eamon, William, and Françoise Paheau. "The Accademia Segreta of Girolamo Ruscelli. A Sixteenth-Century Italian Scientific Society." *Isis* 75 (1984): 327–42.

Easlea, Brian. *Fathering the Unthinkable: Masculinity, Scientists and the Nuclear Arms Race.* London: Pluto Press, 1983.

Easton, Stewart C. *Roger Bacon and His Search for a Universal Science.* New York: Columbia University Press, 1952.

Economou, George D. *The Goddess Natura in Medieval Literature.* Cambridge: Harvard University Press, 1972.

Edelstein, Ludwig. *Ancient Medicine.* Edited by Owsei Temkin and C. Lilian Temkin. Baltimore and London: Johns Hopkins University Press, 1967.

Eis, Gerhard. *Mittelalterliche Fachliteratur.* 2d ed. Stuttgart: J. B. Metzler, 1967.

Eisenstein, Elizabeth. *The Printing Press as an Agent of Change.* 2 vols. Cambridge: Cambridge University Press, 1979.

Eliade, Mircea. *The Forge and the Crucible.* Translated by Stephen Corrin. New York: Harper Torchbooks, 1962.

———. *A History of Religion Ideas.* Translated by Willard R. Trask. 2 vols. Chicago: University of Chicago Press, 1978.

Elliot, J. H. *The Old World and New, 1492–1650.* Cambridge: Cambridge University Press, 1970.

Engelsing, Rolf. *Analphabetentum und Lektüre.* Stuttgart: J. B. Metzler, 1973.

———. *Die Bürger als Leser.* Stuttgart: J. B. Metzler, 1974.

'Espinasse, Margaret. "The Decline and Fall of Restoration Science." *Past and Present* 14 (1958): 71–89.

Evans, R.J.W. *Rudolf II and His World: A Study in Intellectual History 1576–1612.* Oxford: Clarendon Press, 1973.

Evans-Pritchard, E. E. *Witchcraft, Oracles, and Magic among the Azande.* Abridged with intro. by Eva Gilles. Oxford: Clarendon Press, 1976.

Fava, Alessandro. "L'ultimi di baroni: Ferrante Sanseverino." *Rassegna storica Salernitana* 4 (1934): 57–82.

Febvre, Lucien, and Henri-Jean Martin. *The Coming of the Book: The Impact of Printing 1450–1800.* Translated by David Gerard. London: Verso Editions, 1976.

Ferckel, Christoph. "Die Secreta mulierum und ihr Verfasser." *Sudhoffs Archiv* 38 (1954): 267–74.

Ferguson, John K. *Bibliographical Notes on Histories of Inventions and Books of Secrets.* 2 vols. London: Holland Press, 1959.

———. "The Secrets of Alexis." *Proceedings of the Royal Society of Medicine, Section on the History of Medicine* 24 (1931): 225–46.

———. "Some Early Treatises on Technological Chemistry." *Proceedings of the Philosophical Society of Glasgow* 19 (1888): 126–59; 25 (1894): 224–35; 43 (1911): 232–58; 44 (1912): 149–81.

Ferrari, Marco. "Alcune vie di difusione in Italia di idee e di testi di Paracelso." In *Scienze, credenze occulte, livelli di cultura,* pp. 21–29.

Festugière, A.-J. "L'expérience religieuse du médicin Thessalos." *Revue biblique* 48 (1939): 45–77.

———. *Hermétisme et mystique païenne.* Paris: Aubier—Montaigne, 1967.

———. *L'Idéal religieux des grecs.* Paris: Librairie Lecoffre, 1932.

———. *La Révélation d'Hermès Trismégiste.* 4 vols. Paris: Librairie Lecoffre, 1950–1954.

Fierz, Markus. *Girolamo Cardano, 1501–1576: Physician, Natural Philosopher, Mathematician, Astrologer, and Interpreter of Dreams.* Translated by Helga Niman. Boston: Birkhauser, 1982.

Findlen, Paula. "The Economy of Scientific Exchange in Early Modern Italy." In *Patronage and Institutions,* ed. Moran, pp. 5–24.

———. "Empty Signs? Reading the Book of Nature in Renaissance Science." *Studies in the History and Philosophy of Science* 21 (1990): 511–18.

———. "Jokes of Nature and Jokes of Knowledge: The Playfulness of Scientific Discourse in Early Modern Europe." *Renaissance Quarterly* 43 (1990): 292–331.

———. *Possessing Nature: Museums, Collecting, and Scientific Culture in Early Modern Italy.* Berkeley and Los Angeles: University of California Press. Forthcoming.

Fiorentino, Francesco. "Della vita e delle opere di Giovan Battista de la Porta." In *Studi e ritratti della rinascenza*, pp. 235–93. Bari: Gius. Laterza & Figli, 1911.

Fletcher, Harris Francis. *The Intellectual Development of John Milton*. 2 vols. Urbana: University of Illinois Press, 1961.

Forbes, Robert James. *Short History of the Art of Distillation*. Leiden: Brill, 1948.

———. *Studies in Ancient Technology*. 9 vols. Leiden: Brill, 1955–.

Forster, Kurt W. "Metaphors of Rule: Political Ideology and History in the Portraits of Cosimo I de' Medici." *Mitteilungen des Kunsthistorische Institutes von Florenz* 15 (1971): 65–104.

Foucault, Michel. *The Order of Things: An Archaeology of the Human Sciences*. New York: Vintage Books, 1973.

Fowden, Garth. *The Egyptian Hermes: A Historical Approach to the Late Pagan Mind*. Cambridge: Cambridge University Press, 1986.

Friedrichs, Christopher R. "Capitalism, Mobility, and Class Formation in the Early Modern German City." *Past and Present* 69 (1975): 24–49.

Frumkin, Maximilian. "Early History of Patents for Invention." *Transactions of the Newcomen Society* 26 (1947–1949): 47–56.

———. "The Origin of Patents." *Journal of the Patent Office Society* 27 (1945): 143–49.

Fulco, Giorgio. "Per Il 'museo' dei fratelli Della Porta." In *Il Rinascimento meridionale. Raccolta di studi pubblicata in onore di Mario Santori*, pp. 3–73. Naples: Società Editrice Napolitana, 1986.

Gabrieli, Giuseppe. "Bibliografia Lincea. I. Giambattista della Porta—Notizia bibliografica dei suo mss. e libri edizione, ecc. con documenti inediti." *Rendiconti della R. Accademia Nazionale dei Lincei. Classe di scienze morali, storiche e filologiche*, ser. 6, 8 (1932): 206–77.

———. "Federico Cesi Linceo." *Nuova antologia*, ser. 7, 277 (1930): 352–69.

———. "Giovan Battista Della Porta Linceo." *Giornale critico della filosofia Italiana* 8 (1927): 360–96, 423–31.

———. "Spigolatura Dellaportiane." *Rendiconti della R. Accademia Nazionale dei Lincei. Classe di scienze morali, storiche e filologiche*, ser. 6, 11 (1935): 491–517.

Galluzzi, Paolo. "L'Accademia del Cimento: 'Gusti' del Principe, filosofia e ideologia dell'esperimento." *Quaderni storici* 48 (1972): 788–844.

———. "Il mecenatismo mediceo e le scienze." In *Idee, istituzione, scienze ed arti nella Firenze dei Medici*, ed. Cesare Vaoli, pp. 189–215. Florence: Giunti Martello, 1980.

———. "Motivi paracelsiana nella Toscana di Cosimo II e di Don Antonio dei Medici: Alchimia, medicina 'chimica' e riforma del sapere." In *Scienze, credenze occulte, livelli di cultura*, pp. 31–62.

Garin, Eugenio. "Considerazioni sulla magia." In *Medioevo e rinascimento. Studi e ricerche*, pp. 170–91. Bari: Editori Laterza, 1961.

Gentili, Giuseppe A. "Leonardo Fioravanti Bolognese alla luce di ignorati documenti." *Revista di storia delle scienze* 42 (1951): 16–41.

Gibson, S., and F.R.D. Needham. "Lists of Burton's Library." *Oxford Biblio-graphical Society, Proceedings and Papers* 1 (1925): 222–46.

Gilbert, Allan H. "Notes on the Influence of the *Secretum secretorum.*" *Speculum* 3 (1928): 84–98.

Gille, Bertrand. *Engineers of the Renaissance.* Cambridge: MIT Press, 1966.

Gilson, Etienne. *History of Christian Philosophy in the Middle Ages.* New York: Random House, 1955.

Ginzburg, Carlo. *The Cheese and the Worms: The Cosmos of a Sixteenth-Century Miller.* Translated by John and Anne Tedeschi. Harmondsworth: Penguin Books, 1982.

———. "Clues: Roots of an Evidential Paradigm." In *Clues, Myths and the Historical Method*, trans. John and Anne Tedeschi, pp. 96–125. Baltimore: Johns Hopkins University Press, 1989.

———. "The Dovecote Has Opened Its Eyes: Popular Conspiracy in Seventeenth-Century Italy." In *The Inquisition in Early Modern Europe: Studies on Sources and Methods*, ed. Gustav Henningsen and John Tedeschi, pp. 190–98. De Kalb: Northern Illinois University Press, 1986.

———. "High and Low: The Theme of Forbidden Knowledge in the Sixteenth and Seventeenth Centuries." *Past and Present* 73 (1976): 28–41.

———. "Libertinismo erudito e libertinismo popolare nella Bologna del seicento." In *Cultura popolare e cultura dotta nel seicento*, pp. 131–34.

———. *The Night Battles: Witchcraft and Agrarian Cults in the Sixteenth and Seventeenth Centuries.* Translated by John and Anne Tedeschi. Harmondsworth: Penguin Books, 1985.

Ginzburg, Carlo, and Ferrari, Marco. "La colombara ha aperto gli occhi." *Quaderni storici* 38 (1978): 631–39.

Giordano, Davide. *Leonardo Fioravanti Bolognese.* Bologna: Editore Licinio Capelli, 1920.

Giorio, Elvina Vidali. "Una fonte del Garzoni: 'Dello specchio di scientia universale' di Leonardo Fioravante." *Lingua nostra* 30 (1969): 39–43.

Gleason, Elisabeth G. "On the Nature of Sixteenth-Century Italian Evangelism: Scholarship, 1953–1978." *Sixteenth Century Journal* 9 (1978): 3–25.

Gleitsmann, Rolf-Jürgen. "'Wir wissen aber, Gott lob, was wir thun': Erfinderprivilegien und technologische Wandel im 16. Jahrhundert." *Zeitschrift für Unternehmensgeschichte* 30 (1985): 69–95.

Gliozzi, Mario. "Sulla natura dell' 'Accademia de' Secreti' di Giovan Battista Porta." *Archives internationales d'histoire des sciences* 12 (1950): 536–41.

Gnudi, Martha Teach, and Jerome Pierce Webster. *The Life and Times of Gaspare Tagliacozzi, Surgeon of Bologna 1545–1599.* New York: Herbert Reichner, 1950.

Golinski, Jan. "Chemistry in the Scientific Revolution: Problems of Language and Communications." In *Reappraisals of the Scientific Revolution*, ed. Lindberg and Westman, pp. 367–96.

———. "A Noble Spectacle: Phosphorus and the Public Culture of Science in the Early Royal Society." *Isis* 80 (1989): 11–39.

Goltz, Dietlinde. "Die Paracelsisten und die Sprache." *Sudhoffs Archiv* 56 (1972): 337–52.

Goodman, David. *Power and Penury: Government, Technology, and Science in Philip II's Spain*. Cambridge: Cambridge University Press, 1988.

Goody, Jack. *The Domestication of the Savage Mind*. Cambridge: Cambridge University Press, 1977.

Gothic and Renaissance Art in Nuremberg, 1300–1550. New York and Munich: Metropolitan Museum of Art and Prestel Verlag, 1986.

Graff, Harvey J. "On Literacy in the Renaissance: Review and Reflections." *History of Education* 12 (1983): 69–85.

Grammatica, A. "Pietro of Bairo, Principal Physician to the Dukes of Savoy in the Sixteenth Century and Professor of Medicine in the University of Turin." *Panminerva medica* 3 (1961): 432–36.

Grant, Edward. "The Condemnation of 1277, God's Absolute Power, and Physical Thought in the Late Middle Ages." *Viator* 10 (1979): 211–44.

Grattan, J.H.G., and Charles Singer. *Anglo-Saxon Magic and Medicine*. London: Oxford University Press, 1952.

Green, Monica. "Women's Medical Practice and Health Care in Medieval Europe." *Signs: Journal of Women in Culture* 14 (1989): 434–73.

Greenblatt, Stephen. *Renaissance Self-Fashioning: From More to Shakespeare*. Chicago: University of Chicago Press, 1980.

Greimas, Algirdas Julien. "La soupe au pistou ou la construction d'un objet de valeur." In *Du sens. II: Essais sémiotiques*, pp. 157–69. Paris: Éditions du seuil, 1983.

Grendler, Paul F. *Critics of the Italian World, 1530–1560: Anton Francesco Doni, Nicoló Franco, and Ortensio Lando*. Madison: University of Wisconsin Press, 1969.

―――. *The Roman Inquisition and the Venetian Press, 1540–1605*. Princeton: Princeton University Press, 1977.

―――. "Sansovino and Italian Popular History." *Studies in the Renaissance* 16 (1969): 139–80.

―――. *Schooling in Renaissance Italy: Literacy and Learning, 1300–1600*. Baltimore: Johns Hopkins University Press, 1989.

Grenzmann, Ludger. *Traumbuch Artemidori: Zur Tradition der ersten Übersetzung ins Deutsche durch W. H. Ryff*. Saecula Spiritalia, Band 2. Baden-Baden: Verlag Valentin Koerner, 1980.

Grignaschi, M. "La Diffusion du 'Secretum secretorum' (Sirr-al-'Asrar) dans l'Europe occidentale." *Archives d'histoire doctrinale et littéraire du moyen age* 47 (1980): 7–69.

―――. "L'origine et les métamorphoses du 'Sirr-al-asrâr.'" *Archives d'histoire doctrinale et littéraire du moyen age* 43 (1976): 7–112.

Grotefend, H. *Christian Egenolff der erste ständige Buchdrucker zu Frankfurt a.M. und seine Vorlaufer*. Frankfurt am Main: K. Th. Völcker's Verlag, 1881.

Grundmann, Herbert. "Litteratus—illiteratus. Der Wandel einer Bildungsnorm von Altertum zum Mittelalter." *Archiv für Kulturgeschichte* 40 (1958): 1–65.

―――. "Sacerdotium—Regnum—Studium. Zur Wertung der Wissenschaft im 13. Jahrhundert." *Archiv für Kulturgeschichte* 34 (1951): 5–21.

Haage, Bernhard. "Germanistische Wortforschung auf dem Gebiet der altdeutschen Fachliteratur der Artes." In *Fachprosaforschung*, ed. Gundolf Keil and Peter Assion, pp. 124–39. Berlin: Erich Schmidt Verlag, 1974.

Hackenberg, Michael. "Books in Artisan Homes of Sixteenth-Century Germany." *Journal of Library History* 21 (1986): 72–91.

———. "Private Book Ownership in Sixteenth-Century German Language Areas." Ph.D. diss., University of California, Berkeley, 1983.

Hackett, Jeremy. "The Meaning of Experimental Science (Scientia experimentalis) in the Philosophy of Roger Bacon." Ph.D. diss., University of Toronto, 1983.

Hadas, Moses. *Hellenistic Culture: Fusion and Diffusion.* New York: Columbia University Press, 1959.

Hadot, Pierre. *Zur Idee der Naturgeheimnisse. Beim Betrachten des Widmungsblattes in den Humboldtschen "Ideen zu einer Geographie der Pflanzen."* Mainz: Akademie der Wissenschaften und der Literatur, 1982.

Hahn, Roger. *The Anatomy of a Scientific Institution: The Paris Academy of Sciences, 1666–1803.* Berkeley and Los Angeles: University of California Press, 1971.

Hall, A. R. "The Scholar and the Craftsman in the Scientific Revolution." In *Critical Problems in the History of Science*, ed. Marshall Clagett, pp. 3–23. Madison: University of Wisconsin Press, 1969.

———. "Science, Technology, and Utopia in the Seventeenth Century." In *Science and Society 1600–1900*, ed. Peter Mathias, pp. 33–53. Cambridge: Cambridge University Press, 1972.

Hall, Bert S. "Der Meister sol auch kennen schreiben und lesen: Writings about Technology ca. 1400–ca. 1600 A.D. and Their Cultural Implications." In *Early Technologies*, ed. Denise Schmandt-Besserat, pp. 47–58. Malibu: Undena Publications, 1979.

Hall, Marie Boas. "Oldenburg and the Art of Scientific Communication." *British Journal for the History of Science* 2 (1965): 277–90.

———. "The Royal Society's Role in the Diffusion of Information in the Seventeenth Century." *Notes and Records of the Royal Society of London* 29 (1975): 173–92.

———. *The Scientific Renaissance, 1450–1630.* London: Collins, 1962.

Hall, Trevor H. *Old Conjuring Books: A Bibliographical and Historical Study.* London: Duckworth, 1972.

Halleux, Robert, and Paul Meyvaert. "Les origines de la *Mappae clavicula*." *Archives d'histoire doctrinale et littéraire du moyen age* 62 (1987): 7–58.

Hannaway, Owen. *The Chemists and the Word: The Didactic Origins of Chemistry.* Baltimore: Johns Hopkins University Press, 1975.

———. "Laboratory Design and the Aim of Science: Andreas Libavius versus Tycho Brahe." *Isis* 77 (1986): 585–610.

Harig, Gerhard. "Walter Hermann Ryff und Nicolo Tartaglia." In *Physik und Renaissance: Zwei Arbeiten zum Entstehen der Klassischen Naturwissenschaften in Europa*, pp. 13–36. Leipzig: Akademische Verlagsgesellschaft, 1984.

Hartner, Willy. "Notes on *Picatrix*." *Isis* 56 (1965): 438–51.

Haskins, Charles Homer. *Studies in the History of Medieval Science.* Cambridge: Harvard University Press, 1954.

Henry, John. "Occult Qualities and the Experimental Philosophy: Active Principles in Pre-Newtonian Matter Theory." *History of Science* 24 (1986): 335–81.

Heyd, W. *Histoire du commerce du levant au moyen-âge.* 2 vols. Amsterdam: Adolf M. Hakkert, 1967.

Hill, Christopher. *The World Turned Upside Down: Radical Ideas during the English Revolution.* Harmondsworth: Penguin Books, 1975.

Hils, Hans-Peter. "Von dem herten: Reflexionen zu einem mittelalterlichen Eisenhärtungsrezept." *Sudhoffs Archiv* 69 (1985): 62–75.

Hime, H.W.L. "Roger Bacon and Gunpowder." In *Roger Bacon Essays,* ed. A. G. Little, pp. 321–36. New York: Russell and Russell, 1914.

Hirsch, Rudolf. *Printing, Selling, and Reading 1450–1550.* Wiesbaden: Otto Harrassowitz, 1967.

Hirth, Wolfgang. "Zu den deutschen Bearbeitungen der Secreta secretorum des Mittelalters." *Leuvense Bijdragen* 55 (1966): 40–70.

Hoffmann, Julius. *Die 'Hausväterliteratur' und die 'Predigten über den christlichen Hausstand.'* Weinheim and Berlin: Verlag Julius Beltz, 1959.

Holme, James W. "Italian Renaissance Courtesy-Books of the Sixteenth Century." *Modern Language Review* 5 (1910): 145–66.

Hoppen, Theodore K. "The Nature of the Early Royal Society." *British Journal for the History of Science* 9 (1976): 1–24, 243–73.

Horton, Mary. "In Defence of Bacon: A Criticism of the Critics of the Inductive Method." *Studies in History and Philosophy of Science* 4 (1973): 241–78.

Houghton, Walter E., Jr. "The English Virtuoso in the Seventeenth Century." *Journal of the History of Ideas* 3 (1942): 51–73, 190–219.

———. "The History of Trades: Its Relation to Seventeenth Century Thought." In *Roots of Scientific Thought,* ed. Wiener and Noland, pp. 354–81.

Houston, Rab. *Literacy in Early Modern Europe: Culture and Education 1500–1800.* London: Longman, 1988.

Huffines, Marion L. "Sixteenth-Century Printers and the Standardization of New High German." *Journal of English and Germanic Philology* 73 (1974): 60–72.

Hunter, Michael. "The Cabinet Institutionalized: The Royal Society's 'Repository' and Its Background." In *The Origins of Museums: The Cabinet of Curiosities in Sixteenth- and Seventeenth-Century Europe,* ed. Impey and MacGregor, pp. 159–68.

———. "Early Problems in Professionalizing Scientific Research: Nehemiah Grew (1641–1712) and the Royal Society, with an Unpublished Letter to Henry Oldenburg." *Notes and Records of the Royal Society of London* 36 (1982): 189–209.

———. *John Aubrey and the Realm of Learning.* New York: Science History Publications, 1975.

———. *The Royal Society and Its Fellows, 1660–1700: The Morphology of an Early Scientific Society.* Calfont St. Giles: British Society for the History of Science, 1982.

———. *Science and Society in Restoration England.* Cambridge: Cambridge University Press, 1981.

Hunter, Michael, and Paul B. Wood, "Towards Solomon's House: Rival Strategies for Reforming the Early Royal Society." *History of Science* 24 (1986): 49–108.

Hutchison, Keith. "Dormative Virtues, Scholastic Qualities, and the New Philosophies." *Annals of Science* 29 (1991): 245–78.

————. "What Happened to Occult Qualities in the Scientific Revolution?" *Isis* 73 (1982): 233–53.

Iliffe, Rob. " 'In the Warehouse': Privacy, Property and Priority in the Early Royal Society." *History of Science* 30 (1992): 29–68.

Impey, Oliver, and Arthur MacGregor, eds. *The Origins of Museums: The Cabinet of Curiosities in Sixteenth- and Seventeenth-Century Europe.* Oxford: Clarendon Press, 1985.

Ingegno, Alfonso. "Il medico de' disperati e abbandonati: Tommaso Zeffiriele Bovio (1521–1609) tra Paracelso e l'alchimia del seicento." In *Cultura popolare e cultura dotta nel seicento*, pp. 164–74.

————. *Saggio sulla filosofia di Cardano.* Florence: La nuova Italia editrice, 1980.

Jacob, James R. " 'By an Orphean Charm': Science and the Two Cultures in Seventeenth-Century England." In *Politics and Culture in Early Modern Europe: Essays in Honor of H. G. Koenigsberger*, ed. Phyllis Mack and Margaret C. Jacob, pp. 231–50. Cambridge: Cambridge University Press, 1987.

————. *Robert Boyle and the English Revolution: A Study in Social and Intellectual Change.* New York: Burt Franklin, 1977.

Jarcho, Saul. "Giuseppe Zambeccari: A Seventeenth Century Pioneer in Experimental Physiology and Surgery." *Bulletin of the History of Medicine* 9 (1941): 144–76.

Jardine, Lisa. *Francis Bacon: Discovery and the Art of Discourse.* London and New York: Cambridge University Press, 1974.

Jarvie, I. C., and Joseph Agassi. "The Problem of the Rationality of Magic." *British Journal of Sociology* 18 (1967): 55–74.

Jedin, Hubert. *A History of the Council of Trent.* Translated by Dom Ernest Graf. 2 vols. St. Louis: Herder, 1957–1961.

Johnson, Francis R. "Notes on English Retail Book-prices, 1550–1640." *The Library*, 5th ser., 5 (1950): 83–112.

————. "Thomas Hill: An Elizabethan Huxley." *Huntington Library Quarterly* 7 (1944): 329–51.

Johnson, Rozelle Parker. *Compositiones variae From Codex 490, Biblioteca Capitolare, Lucca, Italy: An Introductory Study.* Urbana: University of Illinois Press, 1939.

————. "Some Continental Manuscripts of the *Mappae clavicula.*" *Speculum* 12 (1937): 84–103.

Jones, Richard Foster. *Ancients and Moderns: A Study of the Rise of the Scientific Movement in Seventeenth Century England.* 2d ed. Berkeley and Los Angeles: University of California Press, 1965.

Jones, William R. "Political Uses of Sorcery in Medieval Europe." *The Historian* 34 (1972): 670–87.

Jung, Eva-Marie. "On the Nature of Evangelism in Sixteenth-Century Italy." *Journal of the History of Ideas* 14 (1953): 511–27.

————. "Vittoria Colonna: Between Reformation and Counter-Reformation." *Review of Religion* (Spring 1951): 144–59.

Kaplan, Barbara Beigun. "Greatrakes the Stroker: The Interpretations of His Contemporaries." *Isis* 73 (1982): 178–85.

Keil, Gundolf, "Gottfried von Franken." *VL*, 3:125–36.

———. "Medizinische Bildung und Alternativmedizin." In *Nicht Vielwissen sättigt die Seele. Wissen, Erkennen, Bildung, Ausbildung heute*, ed. Winfried Böhm and Martin Lindauer, pp. 245–71. Stuttgart: Ernst Klett, 1988.

———, ed. *Fachprosa-Studien: Beiträge zur mittelalterlichen Wissenschafts- und Geistesgeschichte*. Berlin: Erich Schmidt Verlag, 1982.

Keil, Gundolf, and Peter Assion, eds. *Fachprosaforschung: Acht Vorträge zur mittelalterlichen Artesliteratur*. Berlin: Erich Schmidt Verlag, 1974.

Keil, Gundolf, et al., eds. *Fachliteratur des Mittelalters. Festschrift für Gerhard Eis*. Stuttgart: J. B. Metzler, 1968.

Keller, Evelyn Fox. "Baconian Science: A Hermaphroditic Birth." *Philosophical Forum* 11 (1980): 299–308.

———. "From Secrets of Life to Secrets of Death." In *Body/Politics: Women and the Discourse of Science*, ed. Mary Jacobus, Evelyn Fox Keller, and Sally Shuttleworth, pp. 175–91. London: Routledge, 1989.

———. "Secrets of God, Nature, and Life." *History of the Human Sciences* 3 (1990): 229–42.

Kelso, Ruth. *The Doctrine of the English Gentleman in the Sixteenth Century*. Illinois Studies in Language and Literature, vol. 14, no. 1. Urbana: University of Illinois Press, 1959.

Kenny, Neil. *The Palace of Secrets: Beroalde de Verville and Renaissance Conceptions of Knowledge*. Oxford: Clarendon Press, 1991.

Kent, F. W., Patricia Simons, and J. C. Eade, eds. *Patronage, Art, and Society in Renaissance Italy*. Oxford: Oxford University Press, 1987.

Kibre, Pearl. "Albertus Magnus, *De Occultis Naturae*." *Osiris* 13 (1958): 157–83.

———. *Scholarly Privileges in the Middle Ages: The Rights, Privileges, and Immunities of Scholars and Universities at Bologna, Padua, Paris, and Oxford*. Cambridge: Harvard University Press, 1962.

Kiekhefer, Richard. *Magic in the Middle Ages*. Cambridge: Cambridge University Press, 1989.

Kingdon, Robert M. "Patronage, Piety, and Printing in Sixteenth-Century Europe." In *Festschrift for Frederick B. Artz*, ed. David H. Pinckney and Theodore Ropp, pp. 19–36. Durham, N.C.: Duke University Press, 1964.

Kocher, Paul H. "John Hester, Paracelsan (fl. 1576–93)." In *Joseph Quincy Adams Memorial Studies*, ed. J. G. McManaway, G. E. Dawson, and E. E. Willoughby, pp. 621–38. Washington: Folger Shakespeare Library, 1948.

Kraus, Paul. *Jābir ibn Hayyān*. 2 vols. Cairo: Institut Français d'Archéologie Orientale, 1942–1943.

———. "Studien zur Jābir." *Isis* 15 (1931): 7–30.

Kremers, Edward, and George Urdang. *History of Pharmacy*. 3d ed. revised by Glenn Sonnedecker. Philadelphia: J. B. Lippincott Co., 1963.

Kristeller, Paul Oskar. "The School of Salerno: Its Development and Its Contribution to the History of Learning." *Bulletin of the History of Medicine* 17 (1945): 138–94.

Kuhn, Thomas S. "Mathematical versus Experimental Traditions in the Development of Physical Science." In *The Essential Tension: Selected Studies in Scientific Tradition and Change*, pp. 31–65. Chicago: University of Chicago Press, 1977.

Kunzle, David. "The World Upside-Down: The Iconography of a European Broadsheet Type." In *The Reversible World: Symbolic Inversion in Art and Society*, ed. Barbara Babcock, pp. 39–94. Ithaca: Cornell University Press, 1978.

Kusche, Brigitte. "Zur 'Secreta mulierum'-Forschung." *Janus* 62 (1975): 103–23.

Laistner, M.L.W. *Thought and Letters in Western Europe, A.D. 500 to 900*. Ithaca: Cornell University Press, 1957; orig. publ. 1931.

Lane, Frederic C. *Venice: A Maritime Republic*. Baltimore: Johns Hopkins University Press, 1973.

Latronico, N. "Una disavventura milanese di Fioravanti." *L'Ospedale Maggiore* 29 (1941): 481–82.

Lea, Henry Charles. *The Inquisition in the Spanish Dependencies*. New York: Macmillan, 1908.

Lea, K. M. *Italian Popular Comedy: A Study in the Commedia dell'Arte, 1560–1620*. 2 vols. New York: Russell and Russell, 1962; orig. publ. 1934.

Leff, Gordon. *The Dissolution of the Medieval Outlook: An Essay on Intellectual and Spiritual Change in the Fourteenth Century*. New York: Harper & Row, 1976.

Lehmann, Paul, ed. *Mittelalterliche Bibliothekskataloge Deutschlands und der Schweiz*. 3 vols. Munich: C. H. Beck, 1918–1939.

Lenoble, Robert. *Mersenne, ou La naissance du mécanisme*. Paris: Vrin, 1943.

Lensi Orlandi, Giulio. *Cosimo e Francesco d'Medici alchemisti*. Florence: Nardini, 1978.

Lévi-Strauss, Claude. *The Savage Mind*. Chicago: University of Chicago Press, 1966.

Lievsay, John L. *The Elizabethan Image of Italy*. Ithaca: Cornell University Press, 1964.

Lindberg, David C. "Science and the Early Christian Church." *Isis* 74 (1983): 509–30.

———. "Science as Handmaiden: Roger Bacon and the Patristic Tradition." *Isis* 78 (1987): 518–36.

———. "The Transmission of Greek and Arabic Learning to the West." In *Science in the Middle Ages*, pp. 52–90.

———, ed. *Science in the Middle Ages*. Chicago: University of Chicago Press, 1978.

Lindberg, David C., and Robert S. Westman. *Reappraisals of the Scientific Revolution*. Cambridge: Cambridge University Press, 1990.

Lindberg, David C., and Ronald L. Numbers, ed. *God and Nature: Historical Essays on the Encounter between Christianity and Science*. Berkeley and Los Angeles: University of California Press, 1986.

Lingo, Alison. "Empirics and Charlatans in Early Modern France: The Genesis of the Classification of the 'Other' in Medical Practice." *Journal of Society History* 19 (1986): 583–603.

Lloyd, G.E.R. *Science, Folklore and Ideology: Studies in the Life Sciences in Ancient Greece.* Cambridge: Cambridge University Press, 1983.

———. *Magic, Reason and Experience: Studies in the Origin and Development of Greek Science.* Cambridge: Cambridge University Press, 1979.

Long, Pamela O. "Invention, Authorship, 'Intellectual Property,' and the Origin of Patents: Notes toward a Conceptual History." *Technology and Culture* 32 (1991): 846–84.

———. "The Openness of Knowledge: An Ideal and Its Context in Sixteenth-Century Writings on Mining and Metallurgy." *Technology and Culture* 32 (1991): 313–55.

Lopez, Pasquale. *Inquisizione stampa e censura nel regno di Napoli tra '500 e '600.* Naples: Edizioni del Delfino, 1974.

Lowry, Martin. *The World of Aldus Manutius: Business and Scholarship in Renaissance Venice.* Ithaca: Cornell University Press, 1979.

Lucchi, Piero. "La Santacroce, il Salterio e il Babuino: Libri per imparare e leggere nel primo secolo della stampa." *Quaderni storici* 38 (1978): 593–630.

Lüdke, Carl. "Walther Ryff und seine 'Teütsche Apoteck.'" *Zur Geschichte der Pharmazie* 14 (1962): 25–28.

Lugli, Adalgisa. *Naturalia et mirabilia: Il collezionismo enciclopedico nelle Wunderkammern d'Europa.* Milan: Mazzotta, 1983.

Lutz, Manfred. *Wer war der gemeine Mann? Der dritte Stand in der Krise des Spätmittelalters.* Munich: R. Oldenburg Verlag, 1979.

Lytle, Guy Fitch, and Stephen Orgel, eds. *Patronage in the Renaissance.* Princeton: Princeton University Press, 1981.

MacGregor, Arthur. "The Cabinet of Curiosities in Seventeenth-Century Britain." In *The Origins of Museums: The Cabinet of Curiosities in Sixteenth- and Seventeenth-Century Europe,* ed. Impey and MacGregor, pp. 147–58.

MacKinney, Loren C. *Early Medieval Medicine.* New York: Arno Press, 1979; orig. publ. 1937.

Maclean, Ian. "The Interpretation of Natural Signs: Cardano's *De subtilitate* versus Scaliger's *Exercitationes.*" In *Occult and Scientific Mentalities in the Renaissance,* ed. Michael Vickers, pp. 231–52. Cambridge: Cambridge University Press, 1984.

McMorris, Michael N. "Science as *Scientia.*" *Physis* 23 (1981): 171–96.

McMullin, Ernan. "Conceptions of Science in the Scientific Revolution." In *Reappraisals of the Scientific Revolution,* ed. Lindberg and Westman, pp. 27–92.

McVaugh, Michael. "An Early Discussion of Medicinal Degrees at Montpellier." *Bulletin of the History of Medicine* 49 (1975): 57–71.

———. "The *Experimenta* of Arnald of Villanova." *Journal of Medieval and Renaissance Studies* 1 (1971): 107–18.

———. "Quantified Medical Theory and Practice at Fourteenth-Century Montpellier." *Bulletin of the History of Medicine* 43 (1969): 397–413.

———. "Theriac at Montpellier, 1285–1385 (with an Edition of the *Questiones de tyriaca* of William of Brescia)." *Sudhoffs Archiv* 56 (1972): 115–44.

———. "Two Montpellier Recipe Collections." *Manuscripta* 20 (1976): 175–80.

Maddison, R.E.W. "The Earliest Published Writing of Boyle." *Annals of Science* 17 (1961): 165–73.

Malinowski, Bronislaw. *Magic, Science, and Religion and Other Essays.* New York: Doubleday Anchor Books, 1954.

Mandich, Giulio. "Venetian Patents (1450–1550)." *Journal of the Patent Office Society* 30 (1948): 166–224.

Mandrou, R. *La Culture populaire aux XVIIe et XVIIIe siècles: La Bibliothèque bleue de Troyes.* Paris: Stock, 1964.

Manuel, Frank E. *A Portrait of Isaac Newton.* Cambridge: Harvard University Press, 1968.

Manzalaoui, Mahmoud. "Philip of Tripoli and His Textual Methods." In *Pseudo-Aristotle the "Secret of Secrets,"* ed. Ryan and Schmitt, pp. 55–72.

——. "The Pseudo-Aristotelian Kitāb Sirr al-Asrār: Facts and Problem." *Oriens* 23/24 (1974): 147–257.

——. "The Pseudo-Aristotelian *Sirr al-Asrār* and Three Oxford Thinkers of the Middle Ages." In *Arabic and Islamic Studies in Honor of Hamilton A. R. Gibbs*, ed. George Makdisi, pp. 480–500. Cambridge: Harvard University Press, 1965.

Margolin, Jean-Claude. "Analogie et causalité chez Jerome Cardan." In *Sciences de la Renaissance*, pp. 67–81. VIIIe Congrès International de Tours. Paris: Vrin, 1973.

——. "Rationalisme et irrationalisme dans la pensée de Jerome Cardan." *Revue de l'Université de Bruxelles* 21 (1968–1969): 87–128.

Martin, Ruth. *Witchcraft and the Inquisition in Venice 1550–1650.* Oxford: Basil Blackwell, 1989.

Martines, Lauro. *Power and Imagination: City-States in Renaissance Italy.* New York: Random House, 1980.

Massignon, Louis. "Inventaire de la littérature hermétique arabe." In Festugière, *Révélation d'Hermès Trismégiste*, 1:384–400.

Maylender, Michele. *Storia delle accademie d'Italia.* 5 vols. Bologna: Capelli, 1926–1930.

Metlitzki, Dorothee. *The Matter of Araby in Medieval England.* New Haven: Yale University Press, 1977.

Meyer-Steineg, Theod. "Thessalos von Tralles." *Archiv für Geschichte der Medizin* 4 (1910): 89–108.

Middleton, W. E. Knowles. *A History of the Thermometer.* Baltimore: Johns Hopkins University Press, 1966.

Minieri-Riccio, Camillo. "Cenno storico delle accademie fiorite nella città di Napoli." *Archivio storico per le province napolitane* 3 (1878): 746–58; 4 (1879): 163–77, 379–94, 516–36; 5 (1880): 131–57, 349–73, 578–612.

——. "Notizie delle accademie istitute nelle provincie napolitane." *Archivio storico per le province napolitane* 2 (1877): 382–90, 581–86, 855–68; 3 (1878): 145–63, 293–314.

Molland, George. "Roger Bacon as a Magician." *Traditio* 30 (1974): 445–60.

Monfrin, J. "La Place du *Secret des secrets* dans la littérature française médiévale." In *Pseudo-Aristotle the "Secret of Secrets,"* ed. Ryan and Schmitt, pp. 73–115.

Monter, E. William, and John Tedeschi. "Toward a Statistical Profile of the Italian Inquisitions, Sixteenth to Eighteenth Centuries." In *The Inquisition in*

Early Modern Europe: Studies on Sources and Methods, ed. Gustav Henningsen and John Tedeschi, pp. 130–57. De Kalb: Northern Illinois University Press, 1986.

Montgomery, John Warwick. *Cross and Crucible: Johann Valentin Andrae (1586–1654) Phoenix of the Theologians.* 2 vols. The Hague: M. Nijhoff, 1973.

Moran, Bruce. *The Alchemical World of the German Court: Occult Philosophy and Chemical Medicine in the Circle of Moritz of Hessen (1572–1632). Sudhoffs Archiv,* Beiheft 29. Stuttgart: Franz Steiner, 1991.

————, ed. *Patronage and Institutions: Science, Technology, and Medicine at the European Court 1500–1750.* Bury St. Edmunds: Boydell Press, 1991.

Morghen, R. "The Academy of the Lincei and Galileo Galilei." *Journal of World History* 7 (1963): 365–80.

Mori, Gustav. "Christian Egenolff, der erste ständige Buchdrucker in Frankfurt a.M." *Archiv für Buchgewerbe* 44 (1907): 301–9.

Morin, Alfred. *Catalogue descriptif de la bibliothèque bleue de Troyes.* Histoire et civilisation du livre, no. 7. Geneva: Droz, 1974.

Muchembled, Robert. *Popular Culture and Elite Culture in France 1400–1750.* Translated by Lydia Cochrane. Baton Rouge: Louisiana State University Press, 1985.

Müller-Jahncke, Wolf-Dieter. *Astrologisch-magische Theorie und Praxis in der Heilkunde der Frühen Neuzeit. Sudhoffs Archiv,* Beiheft 25. Stuttgart: Franz Steiner, 1985.

Mullett, Charles F. "Hugh Plat: Elizabethan Virtuoso." *University of Missouri Studies* 21 (1946): 93–118.

Multhauf, Robert P. "John of Rupescissa and the Origins of Medical Chemistry." *Isis* 45 (1954): 357–67.

————. *The Origins of Chemistry.* New York: Franklin Watts, 1967.

Munger, Robert S. "Guaiacum, the Holy Wood from the New World." *Journal of the History of Medicine* 4 (1949): 196–229.

Muraro, Luisa. *Giambattista Della Porta mago e scienzato.* Milan: Feltrinelli, 1978.

————. "La Taumatologia di Giambattista Della Porta e la sconfitta della cultura magica." In *Cultura popolare e cultura dotta nel seicento,* pp. 182–89.

Murdoch, John. "The Analytical Character of Medieval Learning: Natural Philosophy without Nature." In *Approaches to Nature in the Middle Ages,* ed. L. D. Roberts, pp. 171–213. Medieval and Renaissance Texts and Studies, vol. 16. Binghamton, N.Y.: Center for Medieval and Renaissance Studies, 1982.

Murray, Alexander. *Reason and Society in the Middle Ages.* Oxford: Oxford University Press, 1978.

Nasr, Seyyed Hossein. *Science and Civilization in Islam.* New York: New American Library, 1968.

Nauert, Charles G., Jr. *Agrippa and the Crisis of Renaissance Thought.* Illinois Studies in the Social Sciences, no. 55. Urbana: University of Illinois Press, 1965.

————. "Caius Plinius Secundus." In *Catalogus translationum et commentariorum: Mediaeval and Renaissance Latin Translations and Commentaries,* ed.

F. E. Crane and P. O. Kristeller, vol. 4. Washington: Catholic University of America Press, 1980.

———. "Humanists, Scientists, and Pliny: Changing Approaches to a Classical Author." *American Historical Review* 84 (1979): 72–85.

Needham, Joseph. *Science and Civilization in China*, vol. 5, pt. 2: *History of Scientific Thought*. Cambridge: Cambridge University Press, 1962.

Newman, William. "Technology and Alchemical Debate in the Late Middle Ages." *Isis* 80 (1989): 423–45.

Niceron, R. P. "Antoine Mizauld." In *Memoires pour servir a l'histoire des hommes illustres dans la republique des lettres*, 40:201–13. Paris, 1739.

Nicoll, Allardyce. *The World of Harlequin: A Critical Study of the Commedia dell'Arte*. New York: University Press, 1976.

Nock, Arthur Darby. *Essays on Religion and the Ancient World*. Edited by Zeph Stewart. 2 vols. Cambridge: Harvard University Press, 1972.

Oakeshott, Walter. "Sir Walter Raleigh's Library." *The Library*, 5th ser., 23 (1968): 285–327.

O'Brien, John J. "Commonwealth Schemes for the Advancement of Learning." *British Journal of Educational Studies* 16 (1968): 30–42.

———. "Samuel Hartlib's Influence on Robert Boyle's Scientific Development (Parts I and II)." *Annals of Science* 21 (1965): 1–14, 257–76.

Ochs, Kathleen H. "The Royal Society of London's History of Trades Programme: An Early Episode in Applied Science." *Notes and Records of the Royal Society of London* 39 (1985): 129–58.

Olmi, Giuseppe. "La colonia lincea di Napoli." In *Atti del Convegno Galileo e Napoli*. Naples: Guida, 198?.

———. "'In essercitio universale di contemplatione, e prattica': Federico Cesi e i Lincei." In *Università, accademie e società scientifiche in Italia e in Germania dal Cinquecento al Settecento*, ed. L. Boehm and E. Raimondi, pp. 169–236. Bologna: Il Mulino, 1981.

———. "Science—Honour—Metaphor: Italian Cabinets of Curiosities in the Sixteenth and Seventeenth Centuries." In *The Origins of Museums: The Cabinet of Curiosities in Sixteenth- and Seventeenth-Century Europe*, ed. Impey and MacGregor, pp. 5–16.

Olschki, Leonardo. *Geschichte der neusprachlichen wissenschaftlichen Litteratur*. 3 vols. Vaduz: Kraus Reprint, 1965; orig. publ. 1919–1927.

O'Malley, C. D. *Andreas Vesalius of Brussels 1514–1564*. Berkeley and Los Angeles: University of California Press, 1964.

O'Neil, Mary. "Discerning Superstition: Popular Errors and Orthodox Response in Late Sixteenth Century Italy." Ph.D. diss., Stanford University, 1981.

———. "Magical Healing: Love Magic and the Inquisition in Late Sixteenth-Century Modena." In *Inquisition and Society in Early Modern Europe*, ed. Stephen Haliczer, pp. 88–114. Totowa, N.J.: Barnes & Noble, 1967.

Ong, Walter J. *Orality and Literacy: The Technologizing of the Word*. London: Methuen, 1982.

———. *The Presence of the Word: Some Prolegomena for Cultural and Religious History*. New Haven: Yale University Press, 1967.

Ong, Walter J. *Ramus, Method, and the Decay of Dialogue.* Cambridge: Harvard University Press, 1983.

———. "Writing Is a Technology That Restructures Thought." In *The Written Word,* ed. Gerd Baumann, pp. 23–50. Oxford: Clarendon Press, 1986.

Ore, Oystein. *Cardano: The Gambling Scholar.* Princeton: Princeton University Press, 1953.

Oster, Malcolm. "The Scholar and the Craftsman Revisited: Robert Boyle as Aristocrat and Artisan." *Annals of Science* 49 (1992): 255–76.

Ovenell, R. F. "Brian Twynne's Library." *Oxford Bibliographical Society Publications,* n.s., 4 (1950): 3–42.

Ovitt, George, Jr. *The Restoration of Perfection: Labor and Technology in Medieval Culture.* New Brunswick, N.J.: Rutgers University Press, 1987.

———. "The Status of the Mechanical Arts in Medieval Classifications of Learning." *Viator* 14 (1983): 89–105.

Ozment, Steven. *Magdalena and Balthasar: An Intimate Portrait of Life in Sixteenth-Century Europe Revealed in the Letters of a Nuremberg Husband and Wife.* New York: Simon and Schuster, 1986.

Pagel, Walter. *Paracelsus: An Introduction to Philosophical Medicine in the Era of the Renaissance.* Basel and New York: S. Krager, 1958.

Pagel, Walter, and P. M. Rattansi. "Vesalius and Paracelsus." *Medical History* 8 (1964): 309–28.

Paisey, David L. "Some Sources of the 'Kunstbüchlein' of 1535." *Gutenberg-Jahrbuch* (1980), 113–17.

Palmer, Richard. "Pharmacy in the Republic of Venice in the Sixteenth Century." In *The Medical Renaissance of the Sixteenth Century,* ed. A. Wear, R. K. French, and I. M. Lonie, pp. 100–117. Cambridge: Cambridge University Press, 1985.

Panofsky, Erwin. *The Life and Art of Albrecht Dürer.* 2d ed. Princeton: Princeton University Press, 1955.

Paparelli, Gioacchino. "Giambattista Della Porta: Della Taumatologia." *Rivista storica di scienze mediche e naturali* 47 (1956): 1–47.

———. "La taumatologia di Giovambattista della Porta." *Filologia romanza* 2 (1955): 418–29.

Park, Katharine. "Bacon's 'Enchanted Glass.'" *Isis* 75 (1984): 290–302.

Park, Katharine, and Lorraine Daston, "Unnatural Conceptions: The Study of Monsters in Sixteenth- and Seventeenth-Century France and England." *Past and Present* 92 (1981): 20–54.

Patterson, Lyman Ray. *Copyright in Historical Perspective.* Nashville, Tenn.: Vanderbilt University Press, 1968.

Pederson, Olaf. "The Development of Natural Philosophy, 1250–1350." *Classica et Mediaevalia* 14 (1953): 86–155.

Pepe, Gabriele. *Il Mezzogiorno d'Italia sotti gli Spagnoli. La tradizione storiografica.* Florence: Sansoni, 1952.

Pérez-Ramos, Antonio. *Francis Bacon's Idea of Science and the Maker's Knowledge Tradition.* Oxford: Clarendon Press, 1988.

Peters, Edward. *The Magician, the Witch and the Law.* Philadelphia: University of Pennsylvania Press, 1978.

Pfister, R. "Teinture et alchimie dans l'orient hellénistique." *Seminarium Kondakovianum* 7 (1935): 1–59.

Philipp, Egon. *Das Medizinal- und Apothekenrecht in Nurnberg. Quellen und Studien zur Geschichter der Pharmazie*, Band 3. Frankfurt am Main: Pharmazeutischer Verlag, 1962.

Pighetti, Clelia. *L'Influsso scientifico di Robert Boyle nel tardo '600 italiano.* Milan: Angeli, 1988.

Pine, Martin L. *Pietro Pomponazzi: Radical Philosopher of the Renaissance.* Padua: Antenore, 1986.

Pingree, David. "Some of the Sources of the *Ghāyat al-Hākīm*." *Journal of the Warburg and Courtauld Institutes* 43 (1980): 1–15.

Pinto, Lucille B. "The Folk Practice of Gynecology and Obstetrics in the Middle Ages." *Bulletin of the History of Medicine* 47 (1973): 513–23.

Pirotti, Umberto. *Benedetto Varchi e la cultura del suo tempo.* Florence: Olschki, 1971.

Plessner, M. "Hermes Trismegistus and Arab Science." *Studia Islamica* 2 (1954): 45–59.

Ploss, Emil. "Die Färberei in der germanischen Hauswirtschaft." *Zeitschrift für deutsche Philologie* 75 (1956): 1–22.

———. "Wielands Schwert Mimung und die alte Stahlhärtung." In *Beiträge zur Geschichte der deutschen Sprache und Literatur*, ed. H. De Boor and I. Schröbler, 79:110–28. Tübingen: Max Niemeyer, 1957.

Pohlmann, Hansjörg. "The Inventor's Right in Early German Law." *Journal of the Patent Office Society* 43 (1961): 121–39.

———. "Neue Materialen zum deutschen Urheberschutz um 16. Jahrhundert." *Archiv für Geschichte des Buchwesens* 4 (1963): 89–172.

———. "Weitere Archivfunde zum kaiserliche Autorenschutz im 16. und 17. Jahrhunderts." *Archiv für Geschichte des Buchwesens* 6 (1966): 641–80.

Post, Gaines. "Parisian Masters as a Corporation." *Speculum* 9 (1934): 421–45.

Prager, Frank D. "Brunelleschi's Patent." *Journal of the Patent Office Society* 28 (1948): 109–35.

———. "Examination of Inventions from the Middle Ages to 1836." *Journal of the Patent Office Society* 46 (1964): 268–91.

———. "A History of Intellectual Property from 1545 to 1787." *Journal of the Patent Office Society* 26 (1944): 711–60.

———. "A Manuscript of Taccola, Quoting Brunelleschi, on Problems of Inventor and Builders." *Proceedings of the American Philosophical Society* 112, no. 3 (1968): 131–49.

Prest, John M. *The Garden of Eden: The Botanic Garden and the Recreation of Paradise.* New Haven: Yale University Press, 1981.

Preto, Paolo. *Peste e società a Venezia nel 1576.* Vicenza: Pozza, 1978.

Principe, L. M. "Boyle's Alchemical Secrecy: Codes, Ciphers and Concealments." *Ambix* 39 (1992): 63–74.

Prior, Moody. "Bacon's Man of Science." *Journal of the History of Ideas* 15 (1954): 348–70.

Pumfrey, Stephen. "Ideas above His Station: A Social Study of Hooke's Curatorship of the Experiments." *History of Science* 29 (1991): 1–44.

Purver, Margery. *The Royal Society: Concept and Creation.* Cambridge: MIT Press, 1967.

Pythian-Adams, Charles. *Desolation of a City: Coventry and the Urban Crisis of the Late Middle Ages.* Cambridge: Cambridge University Press, 1979.

Raby, F.J.E. "*Nuda Natura* and Twelfth-Century Cosmology." *Speculum* 43 (1968): 72–77.

Radding, Charles M. "Superstition to Science: Nature, Fortune, and the Passing of the Medieval Ordeal." *American Historical Review* 84 (1979): 945–69.

Rákóczi, Katalin. "Walter Hermann Ryffs charakteristische Stilmittel. Ein Vergleich mit H. Brunschwig, H. Gersdorf und O. Brunfels." *Orvostörténeti Közlemények* (Budapest) 30 (1984): 79–88.

———. *Walther Hermann Ryffs Populärwissenschaftliche Tätigkeit.* Inaugural-Diss. Budapest, 1983.

Ramsey, Matthew. "Property Rights and the Right to Health: The Regulation of Secret Remedies in France, 1789–1815." In *Medical Fringe and Medical Orthodoxy 1750–1850,* ed. W. F. Bynum and Roy Porter, pp. 79–105. London: Croom Helm, 1987.

———. "Traditional Medicine and Medical Enlightenment: The Regulation of Secret Remedies in the Ancien Régime." *Historical Reflections/Reflexions historiques* 9 (1982): 215–32.

Rashdall, Hastings. *The Universities of Europe in the Middle Ages,* ed. F. M. Powicke and A. B. Emden. 3 vols. Oxford: Clarendon Press, 1936.

Rattansi, P. M. "Alchemy and Natural Magic in Raleigh's 'History of the World.'" *Ambix* 13 (1966): 122–38.

Rebhorn, Wayne A. *Courtly Performances: Masking and Festivities in Castiglione's "Book of the Courtier."* Detroit: Wayne State University Press, 1978.

Reeds, Karen. "Renaissance Humanism and Botany." *Annals of Science* 33 (1976): 519–42.

Reinburg, Virginia. "Popular Prayers in Late Medieval and Reformation France." Ph.D. diss., Princeton University, 1985.

Reinstra, Miller Howard. "Giovanni Battista Della Porta and Renaissance Science." Ph.D. diss., University of Michigan, 1963.

Reti, Ladislao. "Francesco di Giorgio Martini's Treatise on Engineering and Its Plagiarists." *Technology and Culture* 4 (1963): 287–98.

Richards, John Chatterton. "A New Manuscript of Heraclius." *Speculum* 15 (1940): 255–71.

Riddle, John M. "Oral Contraceptives and Early-Term Abortifacients during Classical Antiquity and the Middle Ages." *Past and Present* 132 (1991): 3–32.

———. "Pseudo-Dioscorides' *Ex herbis feminis* and Early Medieval Medical Botany." *Journal of the History of Biology* 14 (1981): 43–81.

———. "Theory and Practice in Medieval Medicine." *Viator* 5 (1974): 157–84.

Ritter, François. *Histoire de l'imprimerie alsacienne aux XVe et XVIe siècles.* Strasbourg and Paris: Le Roux, 1955.

Romier, Lucien. *Les origines politiques des guerres de religion.* 2 vols. Paris: Perrin, 1913.

Roosen-Runge, Heinz. *Farbgebung und Technik frühmittelalterlicher Buchmalerei: Studien zu den Traktaten "Mappae Clavicula" und "Heraclius."* 2 vols. Munich: Deutscher Kunstverlag, 1967.

Rose, Paul Lawrence. "The Accademia Venetiana: Science and Culture in Renaissance Venice." *Studi Veneziani* 11 (1969): 191–242.

Rose, Valentin. "Über die Medicina Plinii." *Hermes* 8 (1874): 18–66.

Rosen, Edward. *The Naming of the Telescope.* New York: Henry Schuman, 1947.

Roshem, Julien. "Les idées de Lévin Lemne." *La France médicale* 61 (1914): 21–25, 74–77.

Ross, D.J.A. "Letters of Alexander: A New Partial MS of the Unabbreviated Julius Valerius." *Classica et Medievalia* 13 (1952): 38–58.

Rossi, Paolo. "The Aristotelians and the 'Moderns': Hypothesis and Nature." *Annali dell'Istituto e Museo di Storia della Scienza di Firenze* 7 (1982): 3–28.

————. *Francis Bacon: From Magic to Science.* Translated by S. Rabinovitch. Chicago: University of Chicago Press, 1968.

————. *Philosophy, Technology, and the Arts in the Early Modern Era.* Translated by S. Attansio. New York: Harper & Row, 1970.

Rostovzeff, Michael Ivanovitch. *Social and Economic History of the Roman Empire.* 2d ed. revised by P. M. Fraser. 2 vols. Oxford: Clarendon Press, 1957.

Roth, F.W.E. "Otto Brunfels 1489–1534. Ein deutsche Botaniker." *Botanische Zeitung* 11/12 (1900): 191–232.

Röttinger, Heinrich. *Die Holzschnitte zur Architektur und zum Vitruvius Teutsch des Walther Rivius.* Strasbourg: Heitz, 1914.

Round, Nicholas G. "Five Magicians, or The Uses of Literacy." *Modern Language Review* 64 (1969): 793–805.

Runciman, Steven. *The Medieval Manichee: A Study of the Christian Dualist Heresy.* New York: Viking Press, 1961.

————. *The Sicilian Vespers: A History of the Mediterranean World in the Later Thirteenth Century.* Cambridge: Cambridge University Press, 1958.

Russell, K. F. "Walter Hermann Ryff and His Anatomy." *Australian and New Zealand Journal of Surgery* 21 (1951–1952): 66–69.

Ryan, Michael T. "Assimilating New Worlds in the Sixteenth and Seventeenth Centuries." *Comparative Studies in Society and History* 23 (1981): 519–38.

Ryan, W. F. "The *Secretum secretorum* and the Muscovite Autocracy." In *Pseudo-Aristotle the "Secret of Secrets,"* ed. Ryan and Schmitt, pp. 114–23.

Ryan, W. F., and Charles B. Schmitt, ed. *Pseudo-Aristotle the "Secret of Secrets": Sources and Influences.* London: Warburg Institute, 1982.

Sabean, David Warren. *Power in the Blood: Popular Culture and Village Discourse in Early Modern Germany.* Cambridge: Cambridge University Press, 1984.

Sabra, A. I. "The Appropriation and Subsequent Naturalization of Greek Science in Medieval Islam: A Preliminary Statement." *History of Science* 25 (1987): 223–43.

Salerno, Luigi. "Arte, scienze e collezioni nel Manierismo." In *Scritti di storia dell'arte in onore di Mario Salmi,* 3:193–214. Rome: De Luca, 1963.

Sarmiento, Martin. *Memorias para la historia de la poesía y poetas españoles.* Buenos Aires: Emecé, 1942.

Scammell, G. V. "The New Worlds and Europe in the Sixteenth Century." *The Historical Journal* 12 (1969): 389–412.

Schaefer, Scott Jay. "The Studiolo of Francesco I de'Medici in the Palazzo Vecchio in Florence." Ph.D. diss., Bryn Mawr College, 1976.

Schenda, Rudolph. "Der 'gemeine Mann' und sein medikales Verhalten." In *Pharmazie und der gemeine Mann*, ed. Telle, pp. 9–20.

Schiebinger, Londa. "Feminine Icons: The Face of Early Modern Science." *Critical Inquiry* 14 (1988): 661–91.

Schmitt, Charles B. "Experience and Experiment: A Comparison of Zabarella's View with Galileo's in *De motu*." *Studies in the Renaissance* 16 (1969): 80–138.

———. "Francesco Storella and the *Secretum secretorum*." In *Pseudo-Aristotle the "Secret of Secrets,"* ed. Ryan and Schmitt, pp. 124–31.

———. *Pseudo-Aristoteles Latinus: A Guide to Latin Works Falsely Attributed to Aristotle before 1500*. London: Warburg Institute, 1985.

Schmitt, Wolfram. "Zur Literatur der Geheimwissenschaften im späten Mittelalter." In *Fachprosaforschung*, ed. Keil and Assion, pp. 167–82.

Schröder-Lembke, Gertrud. "Die Hausväterliteratur also agrargeschichtliche Quelle." *Zeitschrift für Agrargeschichte und Agrarsoziologie* 1 (1953): 109–19.

Schubert, Eduard, and Karl Sudhoff. "Michael Bapst von Rochlitz, Pfarrer zu Mohorn, ein populärer medizinischer Schriftsteller des 16. Jahrhunderts." *Neues Archiv für Sächsische Geschichte und Altertumskunde* 11 (1890): 78–116.

———. "Die Schriften des Michael Bapst von Rochlitz (1540–1603). Eine bibliographische Studie." *Centralblatt für Bibliothekswesen* 6 (1889): 537–49.

Schutte, Anne Jacobson. "Teaching Adults to Read in Sixteenth-Century Venice: Giovanni Antonio Tagliente's *Libro Maistrevole*." *Sixteenth Century Journal* 17 (1986): 3–16.

Scienze, credenze occulte, livelli di cultura. Istituto Nazionale di Studi sul Rinascimento. Florence: Leo S. Olschki, 1982.

Scott, Alan. "Ps.-Thessalos of Tralles and Galen's De Methodo Medendi." *Sudhoffs Archiv* 75 (1991): 106–110.

Scriba, Christoph J. "The Autobiography of John Wallis, F.R.S." *Notes and Records of the Royal Society of London* 25 (1970): 17–46.

Segal, Alan F. "Hellenistic Magic: Some Questions of Definition." In *Studies in Gnosticism and Hellenistic Religions*, ed. R. Van Den Broek and M. J. Vermaseren, pp. 349–75. Leiden: Brill, 1981.

Shapin, Steven. "The House of Experiment in Seventeenth-Century England." *Isis* 79 (1988): 373–404.

———. "O Henry." *Isis* 78 (1987): 417–24.

———. "Pump and Circumstance: Robert Boyle's Literary Technology." *Social Studies of Science* 14 (1984): 481–520.

———. "'A Scholar and a Gentleman': The Problematic Identity of the Scientific Practitioner in Early Modern England." *History of Science* 29 (1991): 279–327.

———. "Who Was Robert Hooke?" In *Robert Hooke: New Studies*, ed. Michael Hunter and Simon Schaffer, pp. 253–85. Woodbridge: Boydell Press, 1989.

Shapin, Steven, and Simon Schaffer, *Leviathan and the Air-Pump: Hobbes, Boyle, and the Experimental Life*. Princeton: Princeton University Press, 1985.

Shapiro, Barbara. *John Wilkins 1614–1672: An Intellectual Biography*. Berkeley and Los Angeles: University of California Press, 1969.

———. "Latitudinarianism and Science in Seventeenth Century England." *Past and Present* 40 (1968): 16–41.

———. *Probability and Certainty in Seventeenth-Century England: A Study of the Relationships between Natural Science, Religion, History, Law, and Literature.* Princeton: Princeton University Press, 1983.

Shaw, James Rochester. "Scientific Empiricism in the Middle Ages: Albertus Magnus on Sexual Anatomy and Physiology." *Clio Medica* 10 (1975): 53–64.

Shelby, Lon R. "The 'Secret' of the Medieval Masons." In *On Pre-Modern Technology and Science*, ed. B. S. Hall and D. C. Ornet, pp. 201–19. Malibu: Undena Publications, 1976.

Shipman, Joseph C. "Johannes Petreius, Nuremberg Publisher of Scientific Works, 1524–1550." In *Homage to a Bookman: Essays on Manuscripts, Books and Printing Written for Hans P. Kraus on His Sixtieth Birthday*, ed. H. Lehman-Haupt, pp. 147–62. Berlin: Mann, 1967.

Sigerist, Henry E. *Hieronymus Brunschwig and His Work.* New York: Ben Abramson, 1926.

Simmel, Georg. "The Sociology of Secrecy and of Secret Societies." *American Journal of Sociology* 11 (1906): 441–98.

Simon, Joan. *Education and Society in Tudor England.* Cambridge: Cambridge University Press, 1966.

Singer, Charles, and C. Rabin. *Prelude to Modern Science: Being a Discussion of the History, Sources and Circumstances of the "Tabulae Anatomicae Sex" of Vesalius.* Cambridge: Cambridge University Press, 1946.

Siraisi, Nancy G. *Medieval and Early Renaissance Medicine: An Introduction to Knowledge and Practice.* Chicago: University of Chicago Press, 1990.

———. *Taddeo Alderotti and His Pupils: Two Generations of Italian Medical Learning.* Princeton: Princeton University Press, 1981.

Slack, Paul. "Mirrors of Health and Treasures of Poor Men: The Uses of the Vernacular Medical Literature of Tudor England." In *Health, Medicine and Mortality in the Sixteenth Century*, ed. Webster, pp. 237–74.

Smith, A. Mark. "Knowing Things Inside Out: The Scientific Revolution from a Medieval Perspective." *American Historical Review* 95 (1990): 726–44.

Smith, Jonathan Z. "The Temple and Magician." In *God's Christ and His People: Studies in Honour of Nils Alstrup Dahl*, ed. J. Jervell and W. A. Meeks, pp. 233–47. Oslo: Universitetsforlaget, 1977.

Smith, Joshua Toulmin. *English Gilds.* Early English Text Society, orig. ser., no. 40. London: Oxford University Press, 1963.

Smith, Winifred. *The Commedia dell'Arte: A Study in Italian Popular Comedy.* New York: Columbia University Press, 1912.

Sokól, Stanislaw. "Die Bibliothek eines Barbiers aus dem Jahre 1550." *Centaurus* 7 (1961): 197–206.

Solerti, Angelo. *Ferrara e la corte estense nella seconda metá del secolo decimosesto.* Città di Castello: Lapi, 1900.

Solomon, Howard M. *Public Welfare, Science, and Propaganda in Seventeenth Century France: The Innovation of Théophraste Renaudot.* Princeton: Princeton University Press, 1972.

Sprague, T. A. "The Herbal of Otto Brunfels." *Journal of the Linaean Society of London* 48 (1928): 79–124.

Stahl, William H. *Roman Science: Origins, Development, and Influence to the Later Middle Ages.* Madison: University of Wisconsin Press, 1962.

Stannard, Jerry. "Benedictus Crispus, an Eighth Century Medical Poet." *Journal of the History of Medicine* 21 (1966): 24–46.

———. "Bolos of Mendes." *DSB*, 2:535–38.

———. "Greco-Roman Materia Medica in Medieval Germany." *Bulletin of the History of Medicine* 46 (1972): 455–68.

———. "Hans von Gersdorff and Some Anonymous Strasbourg Apothecaries." *Pharmacy in History* 13 (1971): 55–65.

———. "Herbal Medicine and Herbal Magic in Pliny's Time." In *Pline l'Ancien: Temoin de son temps*, ed. J. Pigeaud and J. Oroz, pp. 95–106. Salamanca: Nantes, 1987.

———. "Marcellus of Bordeaux and the Beginnings of Medieval Materia Medica." *Pharmacy in History* 15 (1973): 47–53.

———. "Medicinal Plants and Folk Remedies in Pliny, *Historia Naturalis.*" *History and Philosophy of the Life Sciences* 4 (1982): 3–23.

———. "Medieval Herbals and Their Development." *Clio Medica* 9 (1974): 23–33.

———. "P. A. Mattioli: Sixteenth Century Commentator on Dioscorides." *Bibliographical Contributions* (University of Kansas Libraries) 1 (1969): 59–81.

Steele, Robert, and Dorothea Waley Singer. "The Emerald Table." *Proceedings of the Royal Society of Medicine, Section on the History of Medicine* 21 (1928): 485–501.

Stiefel, Tina. *The Intellectual Revolution in Twelfth Century Europe.* New York: St. Martin's Press, 1985.

Steiner, Arpad. "The Faust Legend and the Christian Tradition." *Proceedings of the Modern Language Association* 54 (1939): 391–404.

Steneck, Nicholas H. *Science and Creation in the Middle Ages: Henry of Langenstein (d. 1397) on Genesis.* Notre Dame: University of Notre Dame Press, 1976.

Stern, S. M. "The Authorship of the Epistles of the Ikhwān as-Safā." *Islamic Culture* 20 (1946): 367–72.

———. *Studies in Early Ismāʾīlism.* Leiden: Brill, 1983.

Sternagel, Peter. *Die artes mechanicae im Mittelalter. Begriffs- und Bedeutungsgeschichte bis zum Ende des 13. Jahrhunderts.* Kallmünz: Lassleben, 1966.

Stock, Brian. *The Implications of Literacy: Written Language and Models of Interpretation in the Eleventh and Twelfth Centuries.* Princeton: Princeton University Press, 1983.

———. "Science, Technology, and Economic Progress in the Early Middle Ages." In *Science in the Middle Ages*, ed. Lindberg, pp. 1–51.

Stone, Lawrence. *The Crisis of the Aristocracy, 1558–1641.* Abridged ed. Oxford: Oxford University Press, 1967.

———. "The Educational Revolution in England, 1560–1640." *Past and Present* 28 (1964): 41–80.

Strauss, Gerald. "Lutheranism and Literacy: A Reassessment." In *Religion and*

Society in Early Modern Europe, 1500–1800, ed. Kaspar von Greyerz, pp. 109–23. London: George Allen & Unwin, 1984.

———. *Luther's House of Learning: Indoctrination of the Young in the German Reformation*. Baltimore: Johns Hopkins University Press, 1978.

———. *Nuremberg in the Sixteenth Century*. New York: Wiley, 1966.

Strayer, J. F. "The Political Crusades of the Thirteenth Century." In *History of the Crusades*. Vol. 2, *The Later Crusades, 1189–1311*, ed. R. L. Wolff and H. W. Hazard, pp. 343–76. Philadelphia: University of Pennsylvania Press, 1962.

Strype, John. *The Life of Sir Thomas Smith, Kt. D.C.L.*. New York: Burt Franklin Reprints, 1974.

Syfret, R. H. "The Origins of the Royal Society." *Notes and Records of the Royal Society of London* 5 (1947): 75–137.

———. "Some Early Critics of the Royal Society." *Notes and Records of the Royal Society of London* 8 (1951): 20–64.

Symonds, John Addington. *Renaissance in Italy*. 2 vols. New York: The Modern Library, n.d.

Szpilczynski, Stanislaw. "Considérations sur les conceptions pseudo-scientifiques des pratiques de Nicolas de Poland." *Le Scalpel* 114 (1961): 230–33.

Taylor, E.G.R. *Mathematical Practitioners of Tudor and Stuart England*. Cambridge: Cambridge University Press, 1954.

Taylor, Henry Osborne. *The Medieval Mind: A History of the Development of Thought and Emotion in the Middle Ages*. 4th ed. 2 vols. London: Macmillan, 1925.

Telle, Joachim. "Das Arzneibuch Johannes Schöners und seine mittelhochdeutschen Quellen." *Centaurus* 17 (1972): 119–41.

———. "Die 'Magia naturalis' Wolfgang Hilderbrands." *Sudhoffs Archiv* 60 (1976): 105–22.

———. "Wissenschaft und Öffentlichkeit im Spiegel der deutschen Arzneibuchliteratur." *Medizinhistorisches Journal* 14 (1979): 32–52.

———, ed. *Pharmazie und der gemeine Mann. Hausarznei und Apotheke in deutschen Schriften der frühen Neuzeit*. Austellungskataloge der Herzog August Bibliothek, Nr. 36. Braunschweig: Waisenhaus, 1982.

Thomas, Keith. *Religion and the Decline of Magic*. New York: Charles Scribner's Sons, 1971.

Thompson, Daniel V. *The Materials and Methods of Medieval Painting*. New York: Dover Publications, n.d.; orig. publ. 1936.

Thompson, James Westfall. *The Frankfort Book Fair*. New York: Burt Franklin, 1968; orig. publ. 1911.

———. *The Literacy of the Laity in the Middle Ages*. Berkeley and Los Angeles: University of California Press, 1939.

Thorndike, Lynn. "Further Considerations of the *Experimenta, Speculum astronomiae*, and *De secretis mulierum* Ascribed to Albertus Magnus." *Speculum* 30 (1955): 413–33.

———. "John of Seville." *Speculum* 34 (1959): 20–38.

———. "Newness and Novelty in Seventeenth Century Science." In *Roots of Scientific Thought*, ed. Wiener and Noland, pp. 443–57.

Tibawi, A. L. "Ikhwān as-Safā and Their *Rasā'il*: A Critical Review of a Century and a Half of Research." *Islamic Quarterly* 2 (1955): 28–46.

Tomasi, Lucia Tongiorgi. "Projects for Botanical and Other Gardens: A Sixteenth-Century Manual." *Journal of Garden History* 3 (1983): 1–34.

Trevor-Roper, H. R. "Three Foreigners: The Philosophers of the Puritan Revolution." In *The Crisis of the Seventeenth Century: Religion, the Reformation and Social Change*, pp. 237–95. New York: Harper & Row, 1968.

Turnbull, G. H. *Hartlib, Dury, Comenius: Gleanings from Hartlib's Papers*. Liverpool: University Press of Liverpool, 1947.

———. *Samuel Hartlib: A Sketch of His Life and His Relations to J. A. Comenius*. Oxford: Oxford University Press, 1920.

———. "Samuel Hartlib's Influence on the Early History of the Royal Society." *Notes and Records of the Royal Society of London* 10 (1953): 101–30.

Ullmann, Manfred. *Die Natur- und Geheimwissenschaften im Islam*. Handbuch der Orientalistik, erste Abteilung, Ergänzungsband 6, 2. Abschnitt. Leiden: Brill, 1972.

Unwin, George. *Gilds and Companies of London*. 4th ed. New York: Barnes and Noble, 1964.

Urbach, Peter. "Francis Bacon as Precursor to Popper." *British Journal of the Philosophy of Science* 33 (1982): 113–32.

Van Deusen, Neil C. "The Place of Telesio in the History of Philosophy," *The Philosophical Review* 44 (1935): 417–34.

———. "Telesio: The First of the Moderns." Ph.D. diss., Columbia University. New York, 1923.

Van Hoorn, C. M. "Levinus Lemnius en Willem Lemnius, Twee Zestiende-Eeuwse Medici." *Archief* 73 (1971): 37–86, 117–50.

Van Kessel, Elisja M. R. "Joannes van Heeck (1579–?), Co-Founder of the Accademia dei Lincei: A Biobibliographical Sketch." *Medelelingen van het Nederlands Institut te Rome* 38 (1976): 109–34.

Vasoli, Cesare. *La cultura delle corti*. Florence: Portolano, 1980.

———. "Il programma riformatore di Ruggero Bacone." *Rivista di filosofia* 2 (1956): 178–96.

Villari, Rosario. *La rivolta antispagnola a Napoli. Le origini (1585–1647)*. Bari: Laterza, 1967.

Viviani, Ugo. "Ciarlatanismo medico." *Rivista di storia critica delle scienze mediche e naturali* 10 (1919): 103–7.

Voet, Leon. *The Golden Compasses: A History and Evaluation of the Printing and Publishing Activities of the Officina Plantiniana at Antwerp*. 2 vols. Antwerp: Vangendt; New York: Schram, 1969–1972.

Voigts, Linda Ehrsam. "Anglo-Saxon Plant Remedies and the Anglo-Saxons." *Isis* 70 (1979): 250–68.

———. "The Significance of the Name Apuleius to the *Herbarum Apulei*." *Bulletin of the History of Medicine* 52 (1978): 214–27.

Walker, D. P. *Spiritual and Demonic Magic from Ficino to Campanella*. Notre Dame: University of Notre Dame Press, 1975.

Walzer, Michael. *The Revolution of the Saints: A Study in the Origins of Radical Politics*. Cambridge: Harvard University Press, 1965.

Ward, Benedicta. *Miracles and the Medieval Mind: Theory, Record and Event, 1000–1215.* Rev. ed. Philadelphia: University of Pennsylvania Press, 1987.

Wear, Andrew. "The Spleen in Renaissance Anatomy." *Medical History* 21 (1977): 43–60.

Webster, Charles. "Alchemical and Paracelsian Medicine." In *Health, Medicine and Mortality in the Sixteenth Century,* pp. 301–34.

———. "The Authorship and Significance of *Macaria.*" *Past and Present* 56 (1972): 34–48.

———. *The Great Instauration: Science, Medicine and Reform 1626–1660.* New York: Holmes & Meier, 1976.

———. "The Helmontian George Thompson and William Harvey: The Revival and Application of Splenectomy to Physiological Research." *Medical History* 15 (1971): 154–67.

———. "New Light on the Invisible College: The Social Relations of English Science in the Mid-Seventeenth Century." *Transactions of the Royal Historical Society,* ser. 5, 24 (1974): 19–42.

———. "The Origins of the Royal Society." *History of Science* 6 (1967): 106–28.

———. "Samuel Hartlib and the Great Reformation." *Acta Comeniana* 2 (1970): 147–64.

———, ed. *Health, Medicine and Mortality in the Sixteenth Century.* Cambridge: Cambridge University Press, 1979.

Weisheipl, James A. "Classification of the Sciences in Medieval Thought." *Mediaeval Studies* 27 (1965): 54–90.

Weisner, Merry E. *Working Women in Renaissance Germany.* New Brunswick, N.J.: Rutgers University Press, 1986.

Welch, Charles. *History of the Worshipful Company of Pewterers of the City of London.* 2 vols. London: Blades, East & Blades, 1902.

Wellisch, Hans. "Conrad Gesner: A Biobibliography." *Journal of the Society for the Bibliography of Natural History* 7 (1975): 151–247.

Wellmann, Max. *Die Φνσικά des Bolos Demokritos und der Magier Anaxilaos aus Larissa.* Abhandlungen der preussischen Akademie der Wissenschaften, Nr. 7. Berlin, 1928.

Westfall, Richard. "Galileo and the Accademia dei Lincei." In *Novità celesti e crisi del sapere,* ed. P. Galluzzi, pp. 189–200. Atti del Convegno Internationale di Studi Galileiani. Florence: Barbera, 1984.

———. *Science and Religion in Seventeenth-Century England.* Ann Arbor: University of Michigan Press, 1973; orig. publ. 1958.

———. "Unpublished Boyle Papers Relating to Scientific Method." *Annals of Science* 12 (1956): 63–73, 103–17.

Westman, Robert S. "The Astronomer's Role in the Sixteenth Century: A Preliminary Study." *History of Science* 18 (1980): 105–47.

White, Lynn, jr. "Cultural Climates and Technological Advance in the Middle Ages." *Viator* 2 (1971): 171–201.

———. "Kyeser's 'Bellifortis': The First Technological Treatise of the Fifteenth Century." *Technology and Culture* 10 (1969): 436–41.

White, Lynn, jr. "Medieval Engineering and the Sociology of Knowledge." *Pacific Historical Review* 44 (1975): 1–21.

———. *Medieval Technology and Social Change*. London: Oxford University Press, 1962.

Wiener, Philip P., and Aaron Noland, ed. *Roots of Scientific Thought: A Cultural Perspective*. New York: Basic Books, 1957.

Wigand, Paul. "Der Büchernachdruck im 16. Jahrhundert." *Wetzlar'sche Beiträge für Geschichte und Rechtsaltertümer* 1 (1840): 227–41.

Wightman, W.P.D. *Science in a Renaissance Society*. London: Hutchinson, 1972.

Willers, Johann. "Armor of Nuremberg." In *Gothic and Renaissance Art in Nuremberg, 1300–1550*, pp. 101–4.

Williams, Hermann. "The Beginnings of Etching." *Technical Studies in the Fine Arts* 3 (1934): 16–18.

Wilson, C. Anne. *Philosophers, 'Iōsis,' and Water of Life*. Proceedings of the Leeds Philosophical and Literary Society, vol. 19, pt. 5. Leeds: Leeds Philosophical and Literary Society, 1984.

Wilson, Catherine. "Visual Surface and Visual Symbol: The Microscope and the Occult in Early Modern Science." *Journal of the History of Ideas* 49 (1988): 85–108.

Wind, Edgar. *Pagan Mysteries in the Renaissance*. Rev. ed. New York: W. W. Norton, 1968.

Yates, Francis. *Giordano Bruno and the Hermetic Tradition*. Chicago: University of Chicago Press, 1964.

———. *The Rosicrucian Enlightenment*. Frogmore, St. Albans: Paladin, 1975; orig. publ. 1972.

Zambelli, Paola. "A proposito del *De vanitate scientarum et artium* di Cornelio Agrippa." *Rivista critica di storia della filosofia* 15 (1960): 166–80.

———. "Fino del mondo o inizio della propaganda? Astrologia, filosofia della storia e propaganda politico-religiosa nel dibattito sulla congiunzione del 1524." In *Scienze, credenze occulte, livelli di cultura*, pp. 291–368.

———. "Magic and Radical Reformation in Agrippa of Nettescheim." *Journal of the Warburg and Courtauld Institutes* 39 (1976): 69–103.

———. "Many Ends for the World: Luca Guarico Instigator of the Debate in Italy." In *"Astrologi hallucinati,"* ed. Zambelli, pp. 239–63.

———. "Il problema della magia naturale nel Rinascimento." *Rivista critica di storia della filosofia* 28 (1973): 271–96.

———. ed. *"Astrologi hallucinati": Stars and the End of the World in Luther's Time*. Berlin and New York: Walter de Gruyter, 1986.

Zanier, Giancarlo. "Filosofia chimica e pratiche popolari fra '500 e '600." In *Cultura popolare e cultura dotta nel seicento*, pp. 175–82.

———. "La medicina Paracelsiana in Italia: Aspetti di un'accoglienza particolare." *Rivista di storia della filosofia* 4 (1985): 627–53.

Zetterberg, J. Peter. "Echoes of Nature in Salomon's House." *Journal of the History of Ideas* 43 (1982): 179–93.

———. "'Mathematical Magick' in England: 1550–1650." Ph.D. diss., University of Wisconsin, 1976.

———. "The Mistaking of 'the Mathematics' for Magic in Tudor and Stuart England." *Sixteenth Century Journal* 11 (1980): 83–97.

Zilsel, Edgar. "The Origin of William Gilbert's Scientific Method." *Journal of the History of Ideas* 2 (1941): 1–32.

———. "The Sociological Roots of Science." *American Journal of Sociology* 47 (1941/42): 544–62.

Ziman, J. M. *Public Knowledge: An Essay concerning the Social Dimension of Science.* Cambridge: Cambridge University Press, 1968.

Zimmerman, Birgit. *Das Hausarzneibuch. Ein Beitrag zur Untersuchung laien-medizinischer Fachliteratur des 16. Jahrhunderts unter besonderer Berück-sichtigung ihres humanmedizinisches-pharmazeutischen Inhalts.* Inaugural-Diss. University of Marburg, 1975.

INDEX